Photochemistry and Radiation Chemistry

ADVANCES IN CHEMISTRY SERIES **254**

Photochemistry and Radiation Chemistry

Complementary Methods for the Study of Electron Transfer

James F. Wishart, EDITOR
Brookhaven National Laboratory

Daniel G. Nocera, EDITOR
Massachusetts Institute of Technology

American Chemical Society, Washington, DC

Library of Congress Cataloging-in-Publication Data

Photochemistry and radiation chemistry: complementary methods for the study of electron transfer/James F. Wishart, editor, Daniel G. Nocera, editor.

p. cm.—(Advances in chemistry series, ISSN 0065–2393; 254)

Includes bibliographical references and indexes.

ISBN 0–8412–3499–X (alk. paper)

1. Radiation chemistry—Congresses. 2. Pulse radiolysis—Congresses. 3. Charge exchange—Congresses.

I. Wishart, James F., 1958– . II. Nocera, Daniel G., 1957– . III. Series.

QD1.A355 no. 254
[QD625]
540 s—dc21
[541.3′8]

98-14002
CIP

The paper used in this publication meets the minimum requirements of American National Standard for Information Sciences—Permanence of Paper for Printed Library Materials, ANSI Z39.48–1984. ∞

Distributed by Oxford University Press

PRINTED IN THE UNITED STATES OF AMERICA

Advisory Board

Foreword

The Advances in Chemistry Series was founded in 1949 by the American Chemical Society as an outlet for symposia and collections of data in special areas of topical interest that could not be accommodated in the Society's journals. It provides a medium for symposia that would otherwise be fragmented because their papers would be distributed among several journals or not published at all.

Papers are reviewed critically according to ACS editorial standards and receive the careful attention and processing characteristics of ACS publications. Volumes in the Advances in Chemistry Series maintain the integrity of the symposia on which they are based; however, verbatim reproductions of previously published papers are not accepted. Papers may include reports of research as well as reviews, because symposia may embrace both types of presentation.

About the Editors

James F. Wishart is a Chemist in the Chemistry Department of Brookhaven National Laboratory (BNL), Upton, New York. He received his Ph.D. degree in inorganic chemistry from Stanford University under Professor Henry Taube, and his S.B. in chemistry from the Massachusetts Institute of Technology. He served as project manager for the construction of the new pulse radiolysis facility at BNL, which is based on a 10 MeV radio-frequency photocathode electron gun, the first of its kind in the world to be dedicated to pulse radiolysis.

His research interests include electron transfer (particularly over long distances using oligopeptides and proteins as structural motifs), radiation chemistry, inorganic reaction mechanisms, the reactivity of energetic inorganic species produced by radiation, the effects of high pressure on chemical reactivity, reactions in supercritical solvents, and the development of new accelerators and new detection techniques for pulse radiolysis.

He co-organized an international symposium for the American Chemical Society Division of Inorganic Chemistry at the 209th National Meeting in Anaheim, California, in 1995.

Daniel G. Nocera received his early education at Rutgers University where he obtained a B.S. degree in 1979. He subsequently moved to Pasadena, California where he began research on the electron transfer reactions of biological and inorganic systems with Professor Harry Gray at the California Institute of Technology. After earning his Ph.D. degree in 1984, he went to East Lansing, Michigan to accept a faculty appointment at Michigan State University (Assistant Professor 1983–1988; Associate Professor 1988–1990; and Professor 1990–1997). In mid-summer 1997, he joined the faculty of the Massachusetts Institute of Technology as a Professor of Chemistry.

Nocera is a physical–inorganic chemist with research spanning the disciplines of chemistry, engineering, and biochemistry. He studies the basic mechanism of energy conversion in biology and chemistry. His research interests include the discovery of new excited state redox mechanisms of transition metal complexes, magnetic and optical properties of layered materials, the role of protons in mediating electron transfer, and the design of optical sensing schemes that rely on a triggered luminescent response from a supramolecular active site. Recently he developed a new method to measure vorticity in highly three-dimensional turbulent flows. The optical diagnostic technique has found application in solving a variety of fluid physics and engineering problems. Nocera has published more than 100 research papers and he has presented more than 150 invited talks. He has received Presidential Young Investigator, Alfred P. Sloan Fellow, Sigma Xi Junior Research, and Dreyfus Awards.

Contents

Preface

Electron transfer is fundamental to mechanisms of chemical and biological energy conversion. Our understanding of electron transfer in biology, chemistry, and materials has been greatly advanced by studies using radiolytic and/or photochemical initiation. Both techniques encompass the fast and even ultrafast regimes, thereby giving the experimentalist a wide temporal window in which to measure kinetics of the electron transfer event. An important theme to have emerged from such studies over the past two decades is that radiation chemistry can be used to investigate electron transfer reactions in ways that are complementary and synergistic to photochemistry.

Despite the complementary nature of photochemistry and radiation chemistry, crossover between the two fields occurs far less frequently than it should. The primary cause is that for practical reasons of infrastructure, pulse radiolysis equipment in the United States and abroad tends to be located in a small number of sites (for example, national laboratories) such that awareness of the availability and applicability of the technique is limited. Consequently, students at most academic institutions have little exposure to the field of radiation chemistry. This educational deficiency is compounded by the fact that, although specialist meetings in radiation chemistry are held on a regular basis, the technique has seldom been featured at general meetings where a wider audience could be reached. Indeed, radiation chemistry had not been featured at a national American Chemical Society meeting between the mid-1970s and the late 1990s. Despite this lack of attention, radiation chemistry remains highly relevant to current-day issues such as the chemical conversion and storage of solar energy and the management and remediation of radioactive waste repositories. For this reason, we believed that a symposium on radiation chemistry and its relation to electron transfer studies would provide a valuable educational service to the chemical community.

In April 1995, we organized a symposium entitled "Complementarity of Radiation Chemistry and Photochemistry in the Study of Electron Transfer" at the 209th American Chemical Society National Meeting in Anaheim, California. A tutorial session on radiation chemistry was held prior to the symposium, to familiarize the audience with radiation chemistry principles, techniques, and applications. The symposium brought together prominent investigators in several areas of electron transfer, including donor–acceptor systems connected by metalloproteins or synthetic spacers, reactivity coupled to electron transfer processes, and electron transfer in unusual media or conditions such as metal clusters, layered systems, high pressure, and subpicosecond timescales. Throughout the presentations, examples abounded of how radiation and photochemical techniques could be applied together to provide more information and clearer understanding of various electron transfer systems. Since that time, comments from several younger chemists encourage us to believe that the educational purpose of the symposium was indeed realized.

This book is our attempt to extend the educational purpose to scientists of all ages who may benefit from knowing more about pulse radiolysis as another potential technique in their arsenal. The content of the book is primarily about electron transfer, however, the lesson about the whole of the two techniques being greater than their sum extends to other areas of chemical reactivity as well. The first three chapters provide the reader with a basic understanding of the effects of ionizing radiation on matter and how various ionizing radiations are generated. The rest of the book provides examples, from all of the aspects of electron transfer presented in the symposium, which demonstrate the cross-fertilization between radiation chemistry and photochemistry. We hope that this book will enable readers to surmount barriers in their own research in ways that they might not have previously considered.

The editors gratefully acknowledge the ACS Division of Inorganic Chemistry, the Petroleum Research Fund, and Associated Universities, Inc. for providing funds in support of the symposium held at the Spring, 1995 ACS National Meeting. We thank all of the symposium participants and we are especially indebted to the authors for their cooperation in the preparation of this volume.

James F. Wishart
Brookhaven National Laboratory
Upton, NY 11973

Daniel G. Nocera
Massachusetts Institute of Technology
Cambridge, MA 02139

Photochemistry and Radiation Chemistry: A Perspective

James F. Wishart

Department of Chemistry, Brookhaven National Laboratory, Upton, NY 11973–5000

Introduction

One hundred years have passed since the discoveries of X-rays by Roentgen, radioactivity by Becquerel, and the quantized, particle nature of the electron by Thomson. These milestones were crucial steps leading to our understanding of the structure of the atom, and consequently to the breathtaking pace of scientific and technological advances in the twentieth century. The instrument of the discoveries of X-rays and radioactivity was the photographic plate, a piece of photochemical technology that dramatizes the fact that the fields of radiation chemistry and photochemistry have been intimately linked from the very beginning. A chapter in this volume (*1*) demonstrates how radiation chemistry continues to pay its debt of gratitude to photography a century later.

Radiation chemistry is concerned with the interactions of ionizing types of radiation, such as high-energy photons (gamma radiation and X-rays), charged particles, and neutrons, with matter. On the subpicosecond time scale, the important issues are the yields and inhomogeneous spatial distributions of initial ionization events and the resulting highly energetic chemical species. On the picosecond time scale, recombination dominates, but a few reactive radical species persist. On longer time scales, these primary radicals can be made to react with many kinds of solutes to produce secondary radicals for subsequent studies of reactivity. Photochemistry, on the other hand, has traditionally been associated with the interaction of matter with lower energy photons (UV and visible light) and the reactions of molecular excited states, although the distinction is somewhat arbitrary and gets blurred by the use of vacuum UV and multiphoton photoionization, and the use of X-rays to study excited states of hydrocarbons.

During the past century, the fields of radiation chemistry and photochemistry developed in parallel, but each at its own pace. Early radiation sources, such as naturally occurring radioisotopes and X-ray tubes, were replaced in the 1940s and 1950s by particle accelerators (Van de Graaff accelerators and cyclotrons) and reactor-produced radioisotopes. At the same time, powerful flash lamps enabled the study of photoinduced kinetics within the time resolution available to instruments of that period. In the early 1960s, pulsed accelerators brought fast kinetics to the field of radiation chemistry and led to an explosion of activity. The development and widespread use of ever-improving laser technologies from the 1970s onward have made photochemistry one of the most widely represented disciplines in the chemical sciences, cutting across the traditional divisions of physical, organic, inorganic, analytical, and biological chemistry. In contrast, the number of researchers in radiation chemistry has declined during the 1990s.

Although particle accelerators and radioisotopes have important commercial uses, which have spurred their development over the years, the commerce that drives laser development is immense, and laser technology has raced beyond that of accelerators for some time. However, in a situation which parallels the interplay of radiation chemistry with photochemistry, during the last decade of this century high-power, ultrafast laser technology has enabled the development of a new generation of accelerators, which use a laser-pulsed photocathode to inject electrons for acceleration. Completing the circle, these photocathode electron guns have been designed to produce highly collimated, monoenergetic electron beams for injection into free-electron lasers that cover UV or infrared regions not conveniently accessed by excited-state photophysics or nonlinear photonic processes. Picosecond photocathode electron guns for pulse radiolysis are now coming on-line; they are expected to perform as well as or better than linear accelerators in some respects, with reduced operating costs and the availability of picosecond-synchronized laser beams for multibeam experiments (pulse-probe, pulse-flash-probe, etc.).

The first two chapters of this volume describe the interactions of ionizing radiation with matter and the instruments used to generate such forms of radiation. Emphasis has been placed on the significant differences between radiation chemistry and photochemistry. On the other hand, detection methods for flash or laser photolysis and pulse radiolysis have much in common. The most common techniques are variations of time-resolved transient absorption spectroscopy (*2*). Emission spectroscopy is a ubiquitous photochemical technique, and it is used by radiation chemists to study geminate recombination processes (*3*). Time-resolved electron paramagnetic resonance techniques are particularly useful for monitoring the reactivity of radical species (*4*). In the technique of fluorescence-detected magnetic resonance (FDMR), the sensitivity and time resolution of optical emission detection have been combined with the specificity of electron paramagnetic resonance to examine the reactivity of radical cations in radiation- and photoinduced experiments (*5*). Reactions of ionic species in polar solvents can be followed by direct-current (dc) conductivity (*2*). Microwave and dc conductivity is used to measure electron and ion migration in nonpolar media such as hydrocarbons (*6*) and liquid crystals (*7*), as well as in gases (*8*). Time-resolved resonance Raman measurements have been allied with pulse radiolysis at the University of Notre Dame, among other places (*9*).

Photochemistry and radiolysis have played a substantial role in providing the experimental basis of our understanding of electron transfer and chemical reactivity mechanisms, particularly on fast time scales. From an experimentalist's perspective, the interplay between the two techniques is manyfold. For example:

- In radiolysis, the excitation energy is transferred to the bulk, and reactant properties are primarily determined by the solvent and solutes. This allows the initiation chemistry (and dosimetry) to remain invariant and reproducible as the reagent system of interest is varied. One potential drawback of this situation is that the limiting rate of intermediate formation is usually controlled by the concentration limit of the precursor species. In photolysis, the energy is transferred directly to the chromophore, which often directly produces the reactive intermediate. Limitations imposed by second-order formation reactions occur less frequently. Accurate actinometry depends on many factors, however.
- Photolytic and radiolytic methods can be used to generate the same intermediate species, or different ones, in the same electron transfer system. The rate-leveling effect of the highly energetic primary radical reactions induced by pulse radiolysis can sometimes result in different distributions of intermediate species than those generated by photolysis. Also, oxidative or reductive radiolytic chemistries can be used interchangeably to approach intermediates from different directions.
- Radiolysis permits reactions to be initiated in systems that do not contain a chromophore or an excited state that would be amenable to photolytic methods. Since a chromophore is not required, the entire spectral window is available for following kinetics.
- In photochemistry, excited-state decay or back-reaction of the electron–hole pair may limit the efficiency of generating the electron-transfer intermediate of interest. Radiolysis experiments can often be designed to generate oxidizing or reducing equivalents exclusively.
- The distinct spatial distribution of ion pairs generated by both techniques permits different aspects of geminate recombination to be investigated.

The power of photochemistry and radiation chemistry as individual techniques is manifest. When the two are combined, experimental difficulties of a single method can sometimes be evaded, and more results and possibly new insights can be obtained from the chemical system under study. Chapters 3–19 provide a wealth of examples of the strengths and weaknesses of the two techniques and the advantages of using both judiciously.

Acknowledgment

This work was performed at Brookhaven National Laboratory under contract DE-AC02-76CH00016 with the U.S. Department of Energy and was supported by its Division of Chemical Sciences, Office of Basic Energy Sciences.

References

1. Khatouri, J.; Mostafavi, M.; Belloni, J. This volume Chapter 18.
2. Patterson, L. K. In *Radiation Chemistry: Principles and Applications;* Farhataziz; Rodgers, M. A. J., Eds.; VCH Publishers: New York, 1987; pp 65–96.
3. Sauer, M. C., Jr.; Jonah, C. D. *Radiat. Phys. Chem.* **1994,** *44,* 281.
4. Beckert, D.; Fessenden, R. W. *J. Phys. Chem.* **1996,** *100,* 1622.
5. (a) Werst, D. W.; Trifunac, A. D. *J. Phys. Chem.* **1991,** *95,* 3466. (b) Percy, L. T.; Bakker, M. G.; Trifunac, A. D. *J. Phys. Chem.* **1989,** *93,* 4393–4396.
6. (a) Nishikawa, M.; Itoh, K.; Holroyd, R. *J. Phys. Chem.* **1988,** *92,* 5262–5266. (b) Munoz, R. C.; Holroyd, R. A.; Nishikawa, M. *J. Phys. Chem.* **1985,** *89,* 2969–2972. (c) Sauer, M. C., Jr.; Shkrob, I. A.; Yan, J.; Schmidt, K.; Trifunac, A. D. *J. Phys. Chem.* **1996,** *100,* 11325.
7. Warman, J. M.; Schouten, P. G.; Gelinck, G. H.; de Haas, M. P. *Chem. Phys.* **1996,** *212,* 183.
8. Shimamori, H.; Fessenden, R. W. *J. Phys. Chem.* **1981,** *74,* 453.
9. Tripathi, G. N. R.; Su, Y.; Bentley, J.; Fessenden, R. W.; Jiang, P.-Y. *J. Am. Chem. Soc.* **1996,** *118,* 2245–2256.

2

Radiation Chemistry: Principles and Applications

Helen Wilkinson Richter

Department of Chemistry, The University of Akron, Akron, OH 44325–3601

Radiation chemistry can be used to study reactions of free radicals and of metal ions in unusual valency states, including electron-transfer reactions. In some instances, radiation chemistry facilitates experiments that can not be studied by photochemistry, owing to differences in the fundamental physical processes in the two methods. Procedures have been developed to accurately determine radiolysis radical yields, and a variety of physical techniques have been used to monitor reactions. In particular, aqueous radiation chemistry has been extensively developed, and many free radicals can be generated in a controlled manner in aqueous solution. There are extensive literature resources for rate constants and for experimental design for a variety of radicals.

The differences in the fundamental physical phenomena leading to the generation of reactive species in photochemical and radiation chemical methods result in differences in the value of each method in the design of a given experimental study. In some experiments, photochemical generation of the reactive species is advantageous, while in others radiation chemical methods make the experiment easier. In some instances, the experiment is possible only with one or the other of the methods.

A case in point is the first demonstration of the reality of the "inverted region" in the dependence of rate constants for outer-sphere electron-transfer reactions on $-\Delta G°$. Decades earlier, electron-transfer theories descending from Marcus had predicted that rate constants initially should increase as $-\Delta G°$ increased, but would reach a maximum and then decrease. In spite of a number of searches for such systems, they were not found. Using pulse radiolysis methodology, Closs and Miller (*1*) produced solvated electrons in solution in 2-methyltetrahydrofuran or isooctane in the presence of solutes consisting of two aromatic groups, A and B, joined by a rigid chemical spacer. In all cases, B was 4-biphenylyl, while the A groups were aromatic species of varying electron

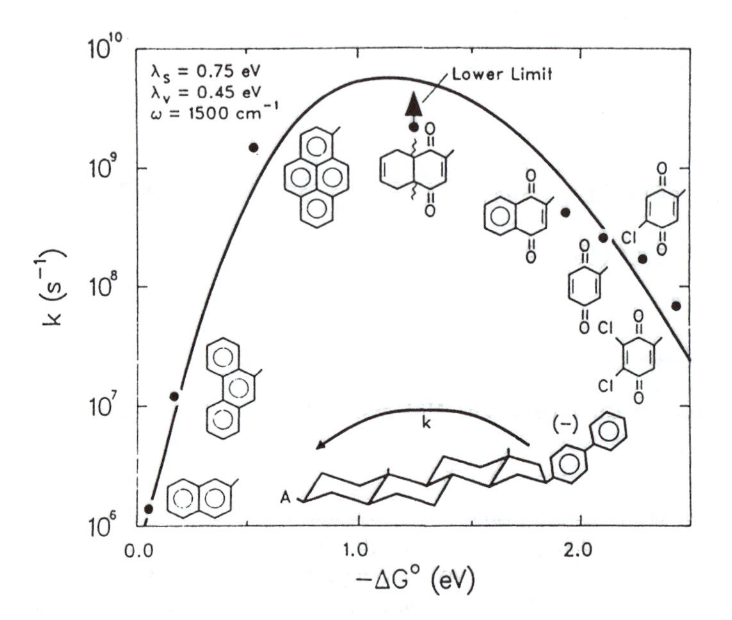

Figure 1. Intramolecular electron-transfer rate constants as a function of free-energy change in 2-methyltetrahydrofuran solution at 296 K. (Reproduced from reference 1. Copyright 1984 American Chemical Society.)

affinity. A and B initially were reduced by e^-_{sol} at diffusion-controlled rates in essentially statistical proportions. Then, intramolecular electron transfer across the spacer proceeded toward equilibrium distributions controlled by the relative electron affinities of A and B. Extraction of rate constants for electron transfer from the reduced B group to A yielded the results shown in Figure 1.

In addition to rate constant measurements and mechanistic determinations for reactions of radical species, pulse radiolysis techniques have been especially successful in measurements of one-electron reduction potentials of redox pairs where one of the partners is unstable so that traditional methods cannot be used. The reduction potentials of a large number of species, in various solvents, have been determined. The techniques and theoretical aspects have been presented by Pedi Neta in the *Journal of Chemical Education* (2).

The focus of this chapter is to present the principles of radiation chemistry, and to present the kinds of experiments in which radiation chemistry provides a good experimental environment, with examples from the literature. To this end, the physical phenomena involved in both photochemical and radiation-chemical free-radical generation are presented, and then the principles and applications of radiation chemistry are more extensively discussed.

Fundamental Physical Interactions in Photochemistry

Photochemistry encompasses the study of chemical changes generated by the absorption of electromagnetic radiation by molecules. Energies of photons vary

Table I. Energies of Electromagnetic Radiation

	Energy Per Photon	*Energy Per Einstein*
Microwave emr (rotational excitation)	0.12 meV	12 J
Infrared emr (vibrational excitation)	0.12 eV	12 kJ
Green light, 530 nm (electronic excitation)	2.34 eV	226 kJ
X-rays, 100 pm	12.4 keV	1.20 GJ
^{60}Co γ-rays	1.2 MeV	116 GJ

NOTE: 1 eV = 1.60219×10^{-19} J; 1 J = 4.184 cal.

from 0.12 meV for a typical microwave region photon to 1.2 MeV for a γ-ray from ^{60}Co (Table I). By tradition, arbitrarily, "photochemistry" is limited to those cases in which the energy of the photons is not large enough to result in ionization of the target molecule. Except in special cases of laser photolysis, a single photon absorbed by a single molecule is responsible for the observed chemistry:

$$A + h\nu \rightarrow A^* \rightarrow \text{products} \quad (1)$$

where h is Planck's constant and ν is frequency. In the absence of complicating equilibria, the absorption of light of wavelength λ_i by a given component of the sample is predicted by the Beer–Lambert law:

$$\text{Absorbance}(\lambda_i) = \log(I_0/I_{tr}) = \epsilon_i cl \quad (2)$$

where I_0 is the incident intensity of light of wavelength λ_i, I_{tr} is the transmitted intensity of this light through the sample, ϵ_i is the molar absorptivity ($dm^3\ mol^{-1}\ cm^{-1}$) of the component at wavelength λ_i, c is the concentration ($mol\ dm^{-3}$) of the component, and l is the thickness of the sample (cm). The distribution of the absorbed photons among the components of the sample is controlled by the nature of the molecular bonds of the components, as reflected in the values of the wavelength-dependent molar absorptivities, ϵ_i, for the individual molecules. As a result of this controlling absorption mode, a minor component in a sample can be responsible for virtually all the energy deposition in the sample. Clearly, this property can be either advantageous or detrimental in the design of a given experiment.

Fundamental Physical Interactions in Radiation Chemistry

In common usage, radiation chemistry encompasses the study of the bombardment of samples with charged particles, which results in ionization and excitation of the sample components, as well as the irradiation of samples with photons whose energies are sufficient to induce ionization of, or ejection of electrons from, components of the irradiated sample. The radiation can be (1) electromagnetic radiation, such as X-rays or gamma rays, or (2) high-energy particles, including electrons (beta-rays), the helium nucleus, He^{2+} (alpha-rays), or heavier nuclei. It is interesting that both photochemists and radiation chemists speak of "irradiating" their samples, and a radiation chemist could in some instances legitimately speak of "photolyzing" a sample, though this is seldom done. In this discussion, *to photolyze* will be used to describe a photolytic treatment of a sample in the sense discussed above, and *to radiolyze* will indicate a radiolytic treatment as defined here.

The physical processes of energy transfer to target species from high-energy photons and from charged particles differ. Likewise, energy absorption by a sample from high-energy and low-energy electromagnetic radiation is governed by different physical processes. When high-energy photons interact with matter, most of the absorbed energy results in ejection of electrons from the atoms of the absorbing material. The interaction depends to a large extent on the atomic makeup of the material and much less on the molecular structure, the controlling factor in irradiation with lower energy photons. Absorption of energy from lower energy electromagnetic radiation, including infrared, visible, UV, and even soft X-rays, depends totally on the molecular structure of the medium. This constitutes the fundamental difference between the effects of ionizing and nonionizing radiation. The photons cannot be partially stopped by the atoms in the medium: the individual photons are entirely absorbed or not absorbed at all, so that energy deposition from high-energy electromagnetic radiation is described by an equation equivalent to the Beer–Lambert law:

$$\text{Absorbance} = \log(I_0/I_{tr}) = \mu l \tag{3}$$

where the proportionality constant μ is called the absorption coefficient of the material, and l is the thickness of the sample. Typical units are centimeters for l and cm^{-1} for μ. Three energy-transfer processes contribute to the energy absorption: the photoelectric effect, the Compton effect, and pair production. The relative contribution of the three processes depends on the energy of the photons.

The most commonly used photons in radiation chemical studies are the gamma rays generated by the disintegration of ^{60}Co nuclei. At the average energy of these photons, 1.2 MeV, Compton absorption predominates. In Compton absorption, photon absorption is followed by ejection of the most loosely bound electron and emission of a photon of lesser energy:

$$A + h\nu_1 \rightarrow (A^+)^* + (e^-)^* + h\nu_2 \tag{4}$$

The contribution of the Compton process to the total absorption coefficient (μ) depends entirely on the number of electrons per gram of sample. As with the photoelectric effect, the observed chemistry is due almost entirely to the action of the ejected energetic electrons as they pass through the sample and dissipate their energy.

When a charged particle such as a high-energy electron penetrates a medium, the predominant mechanism of energy transfer in the energy region of interest in most radiation chemical studies is inelastic interactions with the electrons of the atoms in the sample. These interactions result in ionization or excitation of sample species, as the impinging electron dumps some of its excess energy on its path to thermalization:

$$A + (e^-)_1^* \rightarrow (A^+)^*, A^* + (e^-)_2^* \quad (5)$$

The amount of chemical change produced by a particular high-energy particle depends on the total amount of energy deposited in the sample *and* the rate of the energy deposition. Both factors are determining because a high rate of energy deposition results in high local concentrations of reactive species, which promotes geminate recombination of species. The deposition of energy in radiation-chemical experiments is not so homogeneous as that seen in photochemical experiments. Local inhomogeneities along the particle tracks produce large differences in the yields of reactive species on the time scale of chemical reactions. The "linear energy transfer" (LET), which is the amount of energy transferred to the medium per unit distance traveled, is used to quantify these processes. Typical LET units are keV/μm. The LET is essentially proportional to the density of electrons in the medium. This property has a profound effect on the production of radical and ionic species in solution, and is the source of the difference in "photochemical" and "radiation chemical" experiments.

When samples are irradiated with high-energy photons or charged particles, the average energy deposition (J dm^{-3}) may be no different from that of a photochemical experiment. However, local concentrations can vary enormously. The inhomogeneous distribution of charge pairs and excited species in "tracks" and "spurs" created by the travel of electrons through the sample, electrons produced by high-energy electromagnetic radiation or by the exciting charged particles, produce a very inhomogeneous environment in the immediate aftermath of the irradiation. It makes no difference whether photons or particles are used in the radiolysis: most of the chemistry occurring in a radiolyzed sample results from the deposition of energy by the electrons ejected when a sample species, excited by absorption of a photonionizes, or when a species ionizes as a result of an inelastic collision with a charged particle. These secondary electrons travel through the sample, inducing further ionizations and excitations, until they are thermalized.

There are thus two complications in the irradiation of "concentrated" solutions. First, if a solute is present in high concentration, it can directly absorb a significant portion of the energy deposited and contribute to the radical population. Second, the solute can also affect the primary yield of species by "scavenging" additional primary species in the spurs and tracks, species that would have undergone geminate recombination in dilute solutions. These two effects have been thoroughly documented and must be considered in any experiment. If the solute concentration is at an appropriate level to avoid significant energy absorption, and provided that the radiation has sufficient penetrating power, these inhomogeneities due to spurs and tracks rapidly disappear as the species produced diffuse through the medium. In a properly designed radiation chemical experiment, the effects of tracks and spurs are minimal.

Inhomogeneities of this type do not occur in photolytic methods. However, linear inhomogeneities can arise when the absorbance of the photolyzed sample is so large that a major fraction of incident photons is absorbed, rather than transmitted, and this is a significant consideration in the design of a photochemical experiment. The concentration gradients along the irradiation axis can complicate interpretation of kinetic observations, particularly for radical–radical reactions. Likewise, linear inhomogeneities can arise in radiation chemical experiments when the penetrating power of the particle or electromagnetic radiation is so small that a major proportion of the initial energy is absorbed prior to the exit of the particle from the medium. This effect must also be considered in the design of the irradiation cell and geometry.

Under the usual conditions of a radiation chemical experiment, the disposition of energy among the components of a sample is controlled by the contribution of each component to the density of electrons in the sample. This is true for both high-energy photons and high-energy charged particles. Consequently, in studies of dilute solutions, it is the solvent that absorbs most of the energy. This is the major point of difference in considering whether a radiation chemical or photochemical method is best for a particular application. Even solutes that would absorb virtually all visible or UV light impinging on a sample might absorb virtually none of the ionizing radiation—all the energy from the ionizing radiation could be deposited in the solvent. Consequently, a consideration of optical molar absorptivities of solution components and of solvent may point to a radiation chemical method as the path of choice in some systems.

Product Yields, Dosimetry, and Units

In every free radical experiment, you want to know the concentration of free radicals you have generated. In photochemistry, the yield of a species X is expressed by the quantum yield Φ, defined as the number of species X formed per photon absorbed by the sample:

$$\Phi(\mathrm{X}) = \frac{\text{number of species X formed}}{\text{number of photons absorbed}} \tag{6}$$

The quantum yield, Φ, is wavelength dependent. In radiation chemistry the yield is expressed in terms of the *G* value. Traditionally, in the cgs system, the

G value is expressed as the number of species formed per 100 eV of energy absorbed by the total sample:

$$G(\mathrm{X}) = \frac{\text{number of species X formed}}{\text{energy absorbed in electronvolts} \div 100} \tag{7}$$

In the *Système International d'Unités* (SI), the *G* value of a species X is expressed as the number of micromoles of that species produced per joule of energy absorbed by the sample:

$$G(\mathrm{X}) = \frac{\text{number of micromoles of species X formed}}{\text{energy absorbed in joules}} \tag{8}$$

The magnitudes of the *G* values expressed in these two systems are different since 1 molecule of a species per 100 eV is equivalent to 0.1036 μmol of that species per joule of absorbed energy. SI is used in this article. Spinks and Woods (*3*) present an excellent review of units used for radiation chemical studies in both the cgs system and SI.

The adoption of the term *G value* as the standard for expression of radiation chemical yields can be traced to 1952. On April 8–10 of that year, a meeting was held in the Department of Chemistry at Leeds University. The results, "A General Discussion on Radiation Chemistry", were published in *Discussions of the Faraday Society* in 1952; in the discussion section, Milton Burton (*4*) reported on consultations with several participants on the best method for reporting radiation chemical yields. These participants jointly concluded that the 100-eV yield should be used. This unit was judged to be superior to "ion-pair yield", the *M*/*N* ratio or number of molecules of material changed or of product formed for each ion pair produced. Burton noted that *M*/*N* was particularly difficult to use with condensed phases where the number of ion pairs produced was difficult or impossible to measure. Even though the symbol *G* was "commonly used in radiation chemistry" (*5*) in 1951, there was no standard for reporting yields prior to the Leeds meeting.

Expression of product yields or reactant consumption in terms of energy deposition requires a method of determining the dose received by a sample. The earliest practical chemical dosimeter was the Fricke dosimeter, which utilizes the oxidation of Fe^{2+} to Fe^{3+} in acidic solution. In his work on the Fricke dosimeter, which was published in the 1930s, Fricke reported yields as micromoles of product generated per 10^3 roentgens of dose. By definition, 1 roentgen of X- or gamma-radiation produces in 0.001293 g of air at STP electrons carrying 1 esu of charge. This way of reporting yield is numerically within 5% of the 100-eV yield *G* value. In his 1921 monograph on radiation chemistry, S.C. Lind discussed yields in both gaseous and condensed systems in terms of the *M*/*N* yield. In later work, Lind (*6*) discussed the need for more careful studies of appropriate systems to understand the close relation between the "yield per ion pair", M/N, and the "quantum yield", $M/h\nu$. According to Burton (7), the G value was commonly used during World War II in classified research conducted at the Radiation Chemistry Sections of the Metallurgical Laboratory at the University of Chicago. The selection of G as the symbol for the yield may have come from the interest at the time in relating radiation chemical yields to photochemical yields, as expressed in the work of Lind (5): γ was often used as a symbol for the photochemical quantum yield in the 1940s, and this may have suggested the use of G for the radiation chemical yields.

In photochemical experiments, chemical "actinometers" such as aqueous solutions of ferrioxalate are commonly used to determine the intensity of the light impinging on the sample, that is, the number of photons per second entering the cell. In steady-state radiolysis experiments, for example, with ^{60}Co gamma rays, chemical "dosimeters" are used. In pulse radiolysis experiments, the amount of energy deposited in the sample is determined using a combination of physical and chemical dosimeters. The SI unit for absorbed dose is the gray, where a dose of 1 Gy is the absorption of 1 joule per kilogram of sample. The cgs unit is the rad, where 1 rd is defined as a dose of 100 erg g^{-1}. Since $1\ J = 10^7$ erg, a radiation dose of 1 rd is the same as a dose of 10^{-2} Gy, or 1 krad = 10 Gy.

The physical dosimeter is a current-measuring device which is fixed in place in the electron-beam path on the far side of the sample cell. The device contains a thin foil of Al or Cu, which emits secondary electrons when struck by the high-energy electrons from the accelerator. The current generated is nearly proportional to the total energy deposited in the sample during the pulse. Conversion of the value obtained for the collected current to a value for the dose delivered to the sample is accomplished using a chemical dosimeter. The cell is filled with the dosimetry solution and irradiated: for each pulse, the current and chemical yield are simultaneously measured, yielding a dose calibration curve. Thus, when samples other than the dosimetry solution are placed in the irradiation cell, the delivered dose can be computed from the collected current measurement. The need for a simultaneous dose determination with each pulse via the physical dosimeter arises from the variability of the delivered dose with current instrumentation. In steady-state radiolyses such as with ^{60}Co gamma rays, the dose rate is reproducible, so that a predetermination of the rate with a chemical dosimeter such as the Fricke dosimeter is reliable.

Studies of the Mechanism and Rate Constants of Reactions of Free Radicals and of Ions of Selected Oxidation States

In virtually all radiation chemistry experiments for which the goal is the determination of a mechanism or a reaction rate constant for a free radical or metal ion center, the methods of choice are pulse radiolysis with high-energy electrons, or steady-state radiolysis with high-energy electrons or ^{60}Co gamma rays. In pulse radiolysis experiments, transients have been monitored by optical absorption spectroscopy, conductivity measurements, and electron spin resonance (ESR) spectroscopy. Steady-state electron beams have been coupled with sample flow systems to monitor and characterize transients via ESR spectroscopy. In the following discussion, the focus is primarily on experiments with optical absorbance detection. With ^{60}Co gamma radiolysis, the complete range of analytical methods can be used, as in steady-state photolysis.

The kinetic aspects of the experiment must be analyzed with care to assure that the reaction conditions have been optimized for the study of the desired reaction. It is often the case that not all the primary radicals can be converted to a single radical species, owing to mechanistic and kinetic constraints. Consequently, the requirement for a careful evaluation of the chemical kinetic aspects of the experiment is essential.

Radiation Chemistry of Solvents: Water. The successful design of a radiation chemistry experiment depends upon complete knowledge of the radiation chemistry of the solvent. It is the solvent that will determine the radicals initially present in an irradiated sample, and the fate of all these species needs to assessed. Among the first systems whose radiation chemistry was studied was water, both as liquid and vapor phase, as discussed by Gus Allen in "The Story of the Radiation Chemistry of Water", contained in *Early Developments in Radiation Chemistry* (*8*). Water is the most thoroughly characterized solvent vis-à-vis radiation chemistry. So to illustrate the power of radiation chemical methods in the study of free radical reactions and electron-transfer reactions, I will focus on aqueous systems and hence the radiation chemistry of liquid water. Other solvents can be used when the radiation chemistry of the solvent is carefully considered: as noted previously, Miller et al. (*1*) used pulse radiolysis of solutions in organic solvents for their landmark study showing the Marcus inversion in rate constants.

The radiation chemistry of water has been studied exhaustively. An excellent critical review of the properties, methods of production, and reaction rate constants of hydrated electrons, hydrogen atoms, and hydroxyl radicals ($HO^{\cdot}/O^{\cdot -}$) has been written by Buxton and co-workers (*9*). Reaction rate constants for over 3500 reactions are tabulated in this publication, including reactions with inorganic and organic substrates. Ross and co-workers (*10*) have produced a database containing many pertinent reaction rates. Bombarding liquid water with high-energy photons or charged particles causes excitation and ionization of the water molecules:

$$H_2O \rightsquigarrow H_2O^* \quad (9)$$

$$H_2O \rightsquigarrow H_2O^+ + e^- \quad (10)$$

Within a picosecond, these geminate species are transformed into radical and ionic products. As shown in reactions 11–14, the primary radical species formed

Table II. *G* Values of the Primary Radicals Formed in the Radiolysis of Liquid Water

Radical species	G (*μmol/J*)	G (*number/100 eV*)
Hydrated electron e^-_{aq}	0.27	2.6
Hydroxyl radical $HO^·$	0.28	2.7
Hydrogen atom $H^·$	0.06	0.6

SOURCE: Data from reference 9.

in the radiolysis of liquid water are solvated electrons, hydroxyl radicals, and hydrogen atoms (*see* Table II). H_2O_2, H_2, and H_3O^+ are also formed, with *G* values (μmol J^{-1}) of 0.07, 0.047, and 0.27, respectively.

$$H_2O^* \rightarrow H^· + HO^· \quad (11)$$

$$e^- \rightarrow e^-_{aq} \quad (12)$$

$$H_2O^+ \rightarrow H^+_{aq} + HO^· \quad (13)$$

$$(e^- + H_2O^+) \rightarrow H_2O^* \rightarrow HO^· + H^· \quad (14)$$

It is a point of historical interest to note that no nondissociative excited states of H_2O have been observed experimentally or predicted by calculation. Reaction 14 illustrates the geminate recombination of ion pairs in particle tracks, mentioned previously. To some extent, presolvated electrons and H_2O^+ recombine in the tracks, producing an excited water molecule, which can dissociate to $HO^·$ and $H^·$. At very high solute concentrations, some electrons can be scavenged in their presolvation state by the solute, resulting in a higher yield of scavenged electrons relative to the values from dilute solutions. Yields of products such as $HO^·$ and $H^·$ are correspondingly reduced. Under ordinary conditions of radiolysis, the yields (μmol J^{-1}) of hydrated electron, hydroxyl radical, and hydrogen atom are 0.27, 0.28, and 0.06, respectively (Table II). In strongly alkaline solutions, the hydroxyl radical deprotonates to give its conjugate base, the oxide ion ($O^{·-}$), with $pK_{15} = 11.9$:

$$HO^· \rightleftarrows H^+ + O^{·-} \quad (15)$$

Table III. Rate Constants for Selected Reactions in the Radiolysis of Pure Liquid Water

Reaction	k *(dm³ mol⁻¹ s⁻¹)*
$e^-_{aq} + H_2O \rightarrow H^{\cdot} + HO^-$	1.9×10^1
$e^-_{aq} + HO^{\cdot} \rightarrow HO^-$	3.0×10^{10}
$e^-_{aq} + H^+ \rightarrow H^{\cdot}$	2.3×10^{10a}
$e^-_{aq} + O_2 \rightarrow O_2^{\cdot-}$	1.9×10^{10a}
$H^{\cdot} + H_2O \rightarrow H_2 + HO^{\cdot}$	1×10^1
$H^{\cdot} + HO^- \rightarrow e^-_{aq}$	2.2×10^{7a}
$H^{\cdot} + H^{\cdot} \rightarrow H_2$	$2k = 1.55 \times 10^{10a}$
$H^{\cdot} + O_2 \rightarrow HO_2^{\cdot}$	2.1×10^{10a}
$HO^{\cdot} + HO^{\cdot} \rightarrow H_2O_2$	$2k = 1.1 \times 10^{10a}$
$HO^{\cdot} + HO^- \rightarrow O^{\cdot-} + H_2O$	1.3×10^{10}
$O^{\cdot-} + H_2O \rightarrow HO^{\cdot} + HO^-$	10^8
$O^{\cdot-} + O_2 \rightarrow O_3^-$	3.6×10^{9a}

[a] Recommended value in the source reference.
SOURCE: Data from reference 9.

Rate constants for some reactions occurring in the radiolysis of pure water are collected in Table III.

The optical absorption spectra of these primary radicals are given in Figure 2. Of these, only the hydrated electron has a sufficiently large molar absorptivity in an accessible spectral region, 19,000 dm^3 mol^{-1} cm^{-1} at 720 nm, to make kinetic optical absorption spectroscopy "easy". The other primary radicals can be monitored optically under optimized conditions. In the absence of strongly absorbing species, $HO^{\cdot}$ reactions can be followed by monitoring its absorbance near 250 nm; however, since the molar absorptivity is only 500 dm^3 mol^{-1} cm^{-1}, the optical system has to be well adjusted, and extensive signal averaging is required.

Manipulation of the Primary Radicals. In the radiolysis of aqueous solutions, the initially formed radicals can be manipulated via appropriate chemical additives to produce solutions that largely contain a single radical species. Solutions containing primarily the $HO^{\cdot}$ radical, the hydrated electron, or the hydrogen atom can be produced to study the reactions of individual radicals. As discussed in later sections, solutions of these selected radicals can then be converted to a variety of radical reactants with careful selection of reaction conditions.

The Hydrated Electron. The hydrated electron is the quintessential reducing agent, with a reduction potential of –2.77 V vs. normal hydrogen electrode (NHE) (9). Because of its high molar absorptivity, reactions of e^-_{aq}

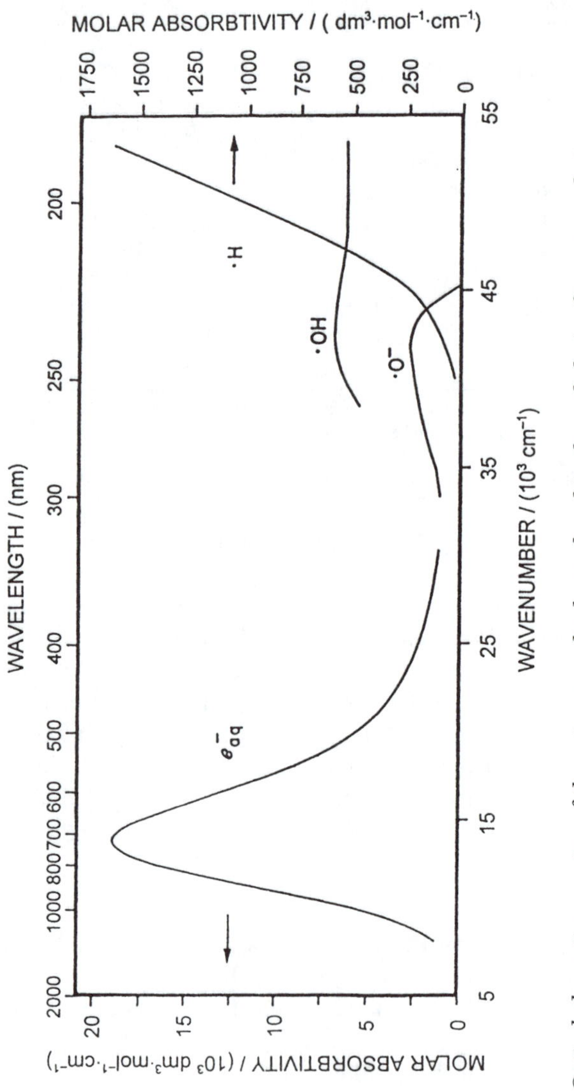

Figure 2. Optical absorption spectra of the primary radicals produced in the radiolysis of aqueous solutions. (Reproduced from reference 9. Copyright 1988 American Chemical Society.)

Table IV. A Selection of Rate Constants for Reduction of Substrates by the Hydrated Electron

Substrate	*Reaction*	*pH*[a]	k *(dm³ mol⁻¹ s⁻¹)*
$Fe(CN)_6^{3-}$	$e^-_{aq} + Fe(CN)_6^{3-} \rightarrow Fe(CN)_6^{4-}$	rv	3.1×10^9
$Ru(bpy)_3^{2+}$[b]	$e^-_{aq} + Ru(bpy)_3^{2+} \rightarrow [Ru^{II}(bpy)_2(bpy^{\cdot-})]^+$	7.0	3.1×10^{10}
$Co(NH_3)_6^{3+}$	$e^-_{aq} + Co(NH_3)_6^{3+} \rightarrow$ products	7	8.8×10^{10}
$IrCl_6^{2-}$	$e^-_{aq} + IrCl_6^{2-} \rightarrow IrCl_6^{3-}$	8	1.2×10^{10}
Ni^{2+}	$e^-_{aq} + Ni^{2+} \rightarrow Ni^+$	Natural	1.4×10^{10}
$Ni(CN)_4^{2-}$	$e^-_{aq} + Ni(CN)_4^{2-} \rightarrow Ni(CN)_4^{3-}$	Natural	3.0×10^9
Cu^+	$e^-_{aq} + Cu^+ \rightarrow Cu$	5.8	2.7×10^{10}
Cu^{2+}	$e^-_{aq} + Cu^{2+} \rightarrow Cu^+$	Natural	3.9×10^{10}
Nitrous oxide	$e^-_{aq} + N_2O \rightarrow N_2 + HO^- + HO^{\cdot}$	rv	8.8×10^9
Oxygen	$e^-_{aq} + O_2 \rightarrow O_2^{\cdot-}$	rv	1.9×10^{10}
Benzene	$e^-_{aq} + C_6H_6 \rightarrow [C_6H_6]^-$	9–11.5	9×10^6
Phenol	$e^-_{aq} + C_6H_6OH \rightarrow$ products	Natural	3.0×10^7

[a] rv indicates that the rate constant is the recommended value for the indicated reaction in the critical review cited.
[b] $Ru(bpy)_3^{2+}$ is the tris(2,2′-bipyridine)ruthenium(II) ion.
SOURCE: Data from references 9 and 10.

are easily followed by monitoring the decay of its absorbance. The hydrated electron quickly reminds one that gratings pass not only the selected light frequency but also integral multiples of that frequency, so that the usually abundant 360-nm light from the light source must be removed with optical filters. Mostly, e^-_{aq} sees the world in black and white: the reaction rate constants are usually either virtually diffusion controlled, or essentially zero. A selection of rate constants for the reactions of the hydrated electron are presented in Table IV. In its reactions with many ionic species, exceptionally large rate constants arise as a result of Scheme I. Coulombic effects, e.g., e^-_{aq} reduces $Ru(bpy)_3^{2+}$ to $[Ru^{II}(bpy)_2(bpy^{\cdot-})]^+$ with a rate constant of 3.1×10^{10} dm³ mol⁻¹ s⁻¹ (reaction 16 in Scheme I) (*11*). The ruthenium remains as Ru^{II}, while one of the 2,2′-bipyridine ligands is reduced to a radical anion: this species has a half-life of ≈3.5 s at pH 10–12.

Typically, e^-_{aq} is generated via radiolysis of N_2-saturated solutions containing 0.1 M of 2-methyl-2-propanol. Any experiment is easier to interpret if you reduce the number of variables. Thus, the alcohol is added to remove the highly reactive $HO^{\cdot}$, which is accomplished by transferring the radical center to a 2-hydroxy-2,2-dimethylethyl radical, as seen in reaction 19.

$$HO^{\cdot} + (CH_3)_3COH \rightarrow H_2O + (CH_2^{\cdot})(CH_3)_2COH \qquad (19)$$

Because of the high alcohol concentrations usually employed, the conversion is complete within a few nanoseconds. The $(CH_2^{\cdot})(CH_3)_2COH$ radical is largely

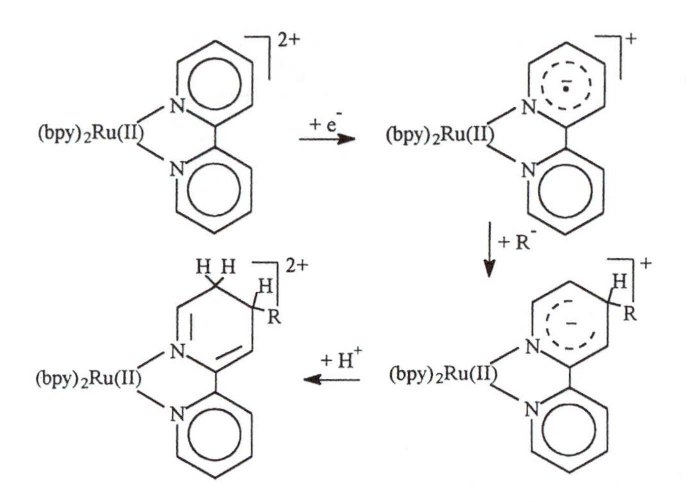

*Scheme I. Transient species arising form reduction of tris(2,2′-bipyridine)ruthenium(*II), as proposed in reference 11.*

unreactive owing to steric hindrance. In addition, its optical absorption spectrum is restricted to the UV below 250 nm. After removal of $HO^{\cdot}$, only e^-_{aq} and $H^{\cdot}$ remain, with $G(e^-_{aq}) = 0.27$ and $G(H^{\cdot}) = 0.06$ (μmol J^{-1}) in neutral solution. The relative concentrations of these two species can be manipulated somewhat by varying solution pH, as discussed below for studies of hydrogen atom reactions.

Caution must be exercised in using 2-methyl-2-propanol to remove $HO^{\cdot}$. The reactions of the 2-hydroxy-2,2-di-methylethyl radical with many species are slow. For example, no reaction was observed in pulse radiolysis experiments with Ni^{2+}, and only a limiting value of $<10^6$ dm^3 mol^{-1} s^{-1} could be assigned. However, like other carbon-centered radicals, $(CH_2^{\cdot})(CH_3)_2COH$ can undergo addition reactions with some metal ions, resulting ultimately in the oxidation or reduction of the metal ion. These adducts typically have significant optical absorbance above 250 nm and into the visible region. Some of these reactions are likewise slow, but others are very fast. For example, $(CH_2^{\cdot})(CH_3)_2COH$ reduces Cu^{2+} to Cu^+ via initial formation of an adduct with a copper–carbon σ bond. This addition reaction is relatively slow, with $k_{20} = 5 \times 10^6$ dm^3 mol^{-1} s^{-1} at pH 6.0 (*12*):

$$(CH_2^{\cdot})(CH_3)_2COH + Cu^{2+} \rightarrow [Cu^{III}CH_2(CH_3)_2COH]^{2+} \qquad (20)$$

The adduct decomposes slowly ($k = 4.5 \times 10^2$ s^{-1}), producing Cu^+ and presumably the alcohol 1-methyl-1,3-propanediol. In contrast, $(CH_2^{\cdot})(CH_3)_2COH$ forms an adduct with Cu^+, forming a metal–carbon bond (reaction 21). The rate constant for this addition reaction is apparently diffusion controlled, with $k_{21} = 4.5 \times 10^9$ dm^3 mol^{-1} s^{-1} (*13*).

$$(CH_2^{\cdot})(CH_3)_2COH + Cu^+ \rightarrow [Cu^{II}CH_2(CH_3)_2COH]^+ \quad (21)$$

The adduct decomposes via β-elimination of hydroxide and release of $(CH_3)_2C = CH_2$ (reaction 22). The reaction is acid catalyzed, with $k_{22} = (5.0 \times 10^4 + 8.6 \times 10^7 [H_3O^+])$ s^{-1} (*13*).

$$[Cu^{II}CH_2(CH_3)_2COH]^+ \rightarrow Cu^{2+} + (CH_3)_2C{=}CCH_2 + HO^- \quad (22)$$

Similarly, $(CH_2^{\cdot})(CH_3)_2COH$ oxidizes Ni^+ to Ni^{2+}, producing 2-methyl-1-propene (*14*):

$$(CH_2^{\cdot})(CH_3)_2COH + Ni^+ \rightarrow (CH_3)_2C{=}CH_2 + HO^- + Ni^{2+} \quad (23)$$

This reaction is likewise very fast, with $k_{23} = 3.0 \times 10^9$ dm^3 mol^{-1} s^{-1}. In this case, an adduct intermediate was not observed, as it was with other aliphatic alcohol carbon-centered radicals.

In the radiolysis of essentially neutral solutions of $[Ru^{II}(bpy)_3]^{2+}$ (where bpy is 2,2′-bipyridine containing $(CH_3)_3COH$, e^-_{aq} reduces $[Ru^{II}(bpy)_3{}^{2+}]$ to $[Ru^{II}(bpy)_2(bpy^{\cdot -})]^+$, as described above. The $(CH_2^{\cdot})(CH_3)_2COH$ radicals generated by $HO^{\cdot}$ from the alcohol rapidly add to the reduced ligand ($k_{17} = (2.5 \pm 0.8) \times 10^9$ dm^3 mol^{-1} s^{-1}), producing a highly absorbing product with $\epsilon_{390} = 11{,}500$ dm^3 mol^{-1} cm^{-1} (Scheme I) (*11*). Protonation of this species to give the final, stable product occurs via general acid catalysis (reaction 18). In this case, using $(CH_3)_3COH$ to "scavenge" $HO^{\cdot}$ simply introduces a new reactant, which may also produce interesting results. In cases where the 2-hydroxy-2,2-di-methylethyl radical causes complications, use of 2-propanol or formate to scavenge $HO^{\cdot}$ and $H^{\cdot}$ can give simpler systems. The radicals produced in the scavenging reactions, with either reagent, are reducing radicals. Use of these materials is discussed later.

The Hydrogen Atom. Hydrogen atom reactions can be studied in isolation by irradiation of N_2-saturated aqueous solutions of 2-methyl-2-propanol adjusted to acid pH. The hydroxyl radicals are deactivated by the alcohol, as in the e^-_{aq} experiments. Since e^-_{aq} is the conjugate base of $H^{\cdot}$, e^-_{aq} can be converted to hydrogen atoms via neutralization with H^+, where $k_{24} = 2.3 \times 10^{10}$ dm^3 mol^{-1} s^{-1} (Table III):

$$H^+_{aq} + e^-_{aq} \rightarrow H^{\cdot} \quad (24)$$

The pK_A of the H atom is 9.6 (*9*). However, conversion of $H^{\cdot}$ to e^-_{aq} requires reaction with HO^-, and this reaction is relatively slow ($k_{25} = 2.2 \times 10^7$ dm^3

mol^{-1} s^{-1}, as shown in Table III), so that highly alkaline solutions are needed to accomplish this.

$$H^{\cdot} + HO^{-} \rightarrow e^{-}_{aq} \quad (25)$$

The reduction potential of $H^{\cdot}$ (−2.1 V versus NHE) (9) is only slightly less negative than that of e^{-}_{aq} (−2.77 V versus NHE) (9), so that it readily reduces many species. Typically, the rate constants are lower for the $H^{\cdot}$ reactions than for the e^{-}_{aq} reactions, especially with positively charged species where Coulombic forces speed the e^{-}_{aq} reactions (Table V). The effects of charge are illustrated by the role reversal in the reaction of these species with $Fe(CN)_6^{3-}$, where the H atom is unhindered by charge repulsion and reacts faster than e^{-}_{aq}. In its reactions with organic compounds, $H^{\cdot}$ can act as a reductant or an oxidant, adding to unsaturated bonds and abstracting H atoms from saturated compounds. The reduction reactions are nearly diffusion controlled, as with benzene and phenol. $H^{\cdot}$ is more effective in reducing these aromatic species than is e^{-}_{aq}, with rate constants more than two orders of magnitude larger than the e^{-}_{aq} constants. In these addition reactions, the behavior of $H^{\cdot}$ is similar to that of $HO^{\cdot}$. Acting as an oxidant, $H^{\cdot}$ abstracts H atoms, transferring the radical center to another atom. The products of reaction with alcohols and a variety of other compounds are essentially the same as the products with $HO^{\cdot}$; however,

Table V. A Selection of Rate Constants for Reactions of the Hydrogen Atom

Substrate	*Reaction*	*pH*[a]	k (dm^3 mol^{-1} s^{-1})
$Fe(CN)_6^{3-}$	$H^{\cdot} + Fe(CN)_6^{3-} \rightarrow Fe(CN)_6^{4-} + H^+$	rv	6.3×10^9
$Ru(bpy)_3^{2+}$	$H^{\cdot} + Ru(bpy)_3^{2+} \rightarrow [Ru(bpy)_3H]^{2+}$	2	9.5×10^9
$Co(NH_3)_6^{3+}$	$H^{\cdot} + Co(NH_3)_6^{3+} \rightarrow Co(NH_3)_6^{2+} + H^+$	1	$<9 \times 10^4$
$IrCl_6^{2-}$	$H^{\cdot} + IrCl_6^{2-} \rightarrow IrCl_6^{3-} + H^+$	Natural	9.2×10^9
Ni^{2+}	$H^{\cdot} + Ni^{2+} \rightarrow$ products	Natural	$<3 \times 10^5$ (no rxn obsd)
$Ni(CN)_4^{2-}$	$H^{\cdot} + Ni(CN)_4^{2-} \rightarrow [Ni(CN)_4H]^{2-}$	Natural	1.8×10^{10}
Cu^+	$H^{\cdot} + Cu^+ \rightarrow CuH^+$	5.6	5.0×10^9
Cu^{2+}	$H^{\cdot} + Cu^{2+} \rightarrow Cu^+ + H^+$	rv	9.1×10^7
Azide ion	$H^{\cdot} + N_3^- \rightarrow HN_3^{\cdot -}$	6.7	1.9×10^9
N_2O	$H^{\cdot} + N_2O \rightarrow N_2 + HO^{\cdot}$	Alkaline	2.1×10^6
Formate	$H^{\cdot} + HCO_2^- \rightarrow H_2 + CO_2^{\cdot -}$	rv	2.1×10^8
Methanol	$H^{\cdot} + CH_3OH \rightarrow H_2 + {}^{\cdot}CH_2OH$	rv	2.6×10^6
2-Propanol	$H^{\cdot} + (CH_3)_2CH(OH) \rightarrow H_2 + (CH_3)_2C^{\cdot}OH$	rv	7.4×10^7
2-Methyl-2-propanol	$H^{\cdot} + (CH_3)_3COH \rightarrow H_2 + (CH_2^{\cdot})(CH_3)_2COH$	2.0	1.7×10^5
Oxygen	$H^{\cdot} + O_2 \rightarrow HO_2^{\cdot}$	rv	1.2×10^{10}
Benzene	$H^{\cdot} + C_6H_6 \rightarrow$ cyclohexadienyl radical	rv	9.1×10^8
Phenol	$H^{\cdot} + C_6H_6OH \rightarrow$ hydroxycyclohexadienyl radical	rv	1.7×10^9

[a] rv indicates that the rate constant is the recommended value for the indicated reaction in the critical review cited.

SOURCE: Data from reference 10.

the rate constants for the H atom actions are typically several orders of magnitude smaller, so that their contributions may be negligible on the time frame of an experiment.

In many instances when the experimental objective is study of reactions of a radical other than $H^{\cdot}$, optimized experimental conditions nonetheless leave the observer with significant numbers of H atoms. For example, studies of e^{-}_{aq} reactions in neutral pH will present the observer with a solution in which 18% of the radicals are $H^{\cdot}$. Thus, the contribution of $H^{\cdot}$ reactions should be addressed.

The Hydroxyl Radical. The hydroxyl radical is a powerful oxidant, which reacts with many substrates at near diffusion-controlled rates (Table VI). The overall reaction often appears to be a simple electron transfer:

$$R^{-} + HO^{\cdot} \rightarrow R^{\cdot} + HO^{-} \qquad (26)$$

However, with some metal ions, the mechanism may involve initial addition of $HO^{\cdot}$ to the metal. For example, the initial product with Cu^{2+} is known to be $[CuOH]^{2+}$. In other instances, the reaction proceeds via initial addition to a ligand, followed by reaction with the metal center. For example, $Ru(bpy)_3{}^{2+}$ reacts via addition of $HO^{\cdot}$ to an aromatic ring. The metal center then acts as an internal reducing agent, yielding HO^{-} and the oxidized complex.

$HO^{\cdot}$ rapidly oxidizes saturated organic compounds via H atom abstraction,

Table VI. A Selection of Rate Constants for Reactions of the Hydroxyl Radical

Substrate	*Reaction*	*pH*[a]	k *(dm^3 mol^{-1} s^{-1})*
$Mo(CN)_8{}^{4-}$	$HO^{\cdot} + Mo(CN)_8{}^{4-} \rightarrow HO^{-} + Mo(CN)_8{}^{3-}$	6.5	5.8×10^9
$Fe(CN)_6{}^{4-}$	$HO^{\cdot} + Fe(CN)_6{}^{4-} \rightarrow HO^{-} + Fe(CN)_6{}^{3-}$	rv	1.05×10^{10}
$Ru(bpy)_3{}^{2+}$	$HO^{\cdot} + Ru(bpy)_3{}^{2+} \rightarrow [Ru(bpy)_3OH]^{2+}$	7.0	6.8×10^9
$IrCl_6{}^{3-}$	$HO^{\cdot} + IrCl_6{}^{3-} \rightarrow HO^{-} + IrCl_6{}^{2-}$	3–4.5	1.3×10^{10}
Ni^{+}	$HO^{\cdot} + Ni^{+} \rightarrow HO^{-} + Ni^{2+}$	Natural	2.0×10^{10}
Cu^{2+}	$HO^{\cdot} + Cu^{2+} \rightarrow [CuOH]^{2+}$	3–7	3.5×10^8
Thiocyanate ion	$HO^{\cdot} + SCN^{-} \rightarrow HO^{-} + SCN^{\cdot}$	rv	1.1×10^{10}
Azide ion	$HO^{\cdot} + N_3{}^{-} \rightarrow HO^{-} + N_3{}^{\cdot}$	7.9–13	1.2×10^{10}
Formate ion	$HO^{\cdot} + HCO_2{}^{-} \rightarrow H_2O + CO_2{}^{\cdot -}$	rv	3.2×10^9
Methanol	$HO^{\cdot} + CH_3OH \rightarrow H_2O + {}^{\cdot}CH_2OH$	rv	9.7×10^8
2-Propanol	$HO^{\cdot} + (CH_3)_2CH(OH) \rightarrow H_2O +$ $(CH_3)_2C^{\cdot}OH$ (85.5%), $({}^{\cdot}CH_2)(CH_3)CHOH$ (13.3%)	rv	1.9×10^9
2-Methyl-2-propanol	$HO^{\cdot} + (CH_3)_3COH \rightarrow H_2O +$ $(CH_2{}^{\cdot})(CH_3)_2COH$	rv	6.0×10^8
Benzene	$HO^{\cdot} + C_6H_6 \rightarrow$ hydroxycyclohexadienyl radical	rv	7.8×10^9
Benzoate	$HO^{\cdot} + C_6H_5CO_2{}^{-} \rightarrow HOC_6H_5CO_2{}^{-}$	rv	5.9×10^9

[a] rv indicates that the rate constant is the recommended value for the indicated reaction in the critical review cited.

SOURCE: Data from reference 9.

as seen in Table VI for several alcohols. With unsaturated organic compounds, $HO^{\cdot}$ typically reacts via addition to a double bond. For example, hydroxycyclohexadienyl radicals are produced when $HO^{\cdot}$ adds to aromatic species:

HO• + (R-benzene) ⟶ (hydroxycyclohexadienyl radical, R, OH) (27)

The ortho, meta, and para adducts can be oxidized to the corresponding ortho, meta, and para phenols by a variety of oxidants. The ipso adduct of aromatic species with appropriate leaving groups can undergo elimination reactions, resulting ultimately in the production of a phenol. For example, the ipso adduct of benzoate eliminates CO_2. Oxidation of the resulting radical anion produces phenol. The ipso adduct of methoxybenzene eliminates CH_3OH, producing the phenol radical, which can be reduced to phenol. The hydroxyl radical adducts of some aromatic compounds undergo acid- or base-catalyzed loss of water, so that the reaction appears to be a simple one-electron oxidation or H-atom abstraction. The ortho, meta, or para adducts of methoxybenzene can undergo an acid-catalyzed loss of water to produce the methoxybenzene radical cation. Hydroxyl radical adducts of hydroxyquinones can undergo acid- or base-catalyzed water loss, producing a semiquinone.

In strongly basic solution, $HO^{\cdot}$ rapidly deprotonates via reaction with hydroxide (reaction 28). With $k_{28} = 1.3 \times 10^{10}$ $dm^3\ mol^{-1}\ s^{-1}$, the reaction half-life is only 5 ns at pH 12

$$HO^{\cdot} + HO^{-} \rightarrow O^{\cdot -} + H_2O \qquad (28)$$

Reactions involving $O^{\cdot -}$ are very different from those of $HO^{\cdot}$. For example, $O^{\cdot -}$ prefers to abstract a hydrogen atom from aromatic compounds instead of adding to the ring as $HO^{\cdot}$ does.

For studies of hydroxyl radical reactions, the hydrated electrons can be converted to hydroxyl radicals by saturating the reaction solution with nitrous oxide. N_2O is substantially soluble in water: with a Henry's law constant (mole fraction scale) at 25 °C of 0.182×10^7, it has a concentration of about 26 mmol dm^{-3} at 25 °C. Solvated electrons are rapidly scavenged by the N_2O (reaction 29), generating $HO^{\cdot}$ with a reaction half-life of just 3.3 ns. Thus, within 16 ns (five half-lives), the conversion is 97% complete.

$$e^-_{aq} + N_2O \rightarrow N_2 + HO^- + HO^{\cdot} \qquad (29)$$

The yield (μmol J^{-1}) of hydroxyl radicals for kinetic studies is thus the initial yield of $HO^{\cdot}$ plus the initial yield of e^-_{aq}, i.e., 0.55. The only other radical present after 16 ns will be the H atom, which represents about 10% of the total radical yield.

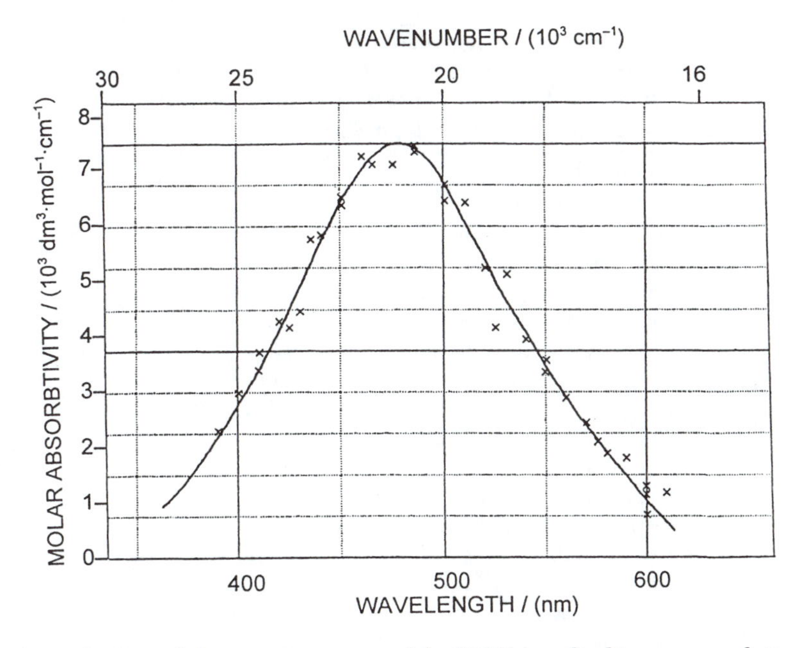

Figure 3. Optical absorption spectrum of the $(SCN)_2^{\cdot-}$ radical in aqueous solution. (Reproduced from reference 17.)

The Thiocyanate Radiolysis Dosimeter. The thiocyanate dosimeter is a reliable, accurate, and convenient means of dose calibration in pulse radiolysis experiments, when coupled with a physical dosimeter of the type described earlier. An aqueous solution of KSCN (10 mmol dm^{-3}) is saturated with N_2O. The e_{aq}^- are quantitatively converted to $HO^{\cdot}$ within about 3 ns, as described above (reaction 29). The hydroxyl radicals then oxidize SCN^-, transferring the radical center to the thiocyanate radical ($SCN^{\cdot}$) (reaction 30). The $SCN^{\cdot}$ radical rapidly couples with a thiocyanate ion, producing $(SCN)_2^{\cdot-}$, a relatively stable radical with a high molar absorptivity (reaction 31):

$$SCN^- + HO^{\cdot} \rightarrow SCN^{\cdot} + HO^- \quad (30)$$

$$SCN^{\cdot} + SCN^- \rightleftarrows (SCN)_2^{\cdot-} \quad (31)$$

The production of the $(SCN)_2^{\cdot-}$ radical has been carefully studied, and the *G* value accurately determined. With its broad absorption band and generous molar absorptivity ($\varepsilon_{472} = 7580$ dm^3 mol^{-1} cm^{-1}), this radiolysis dosimeter system provides an excellent means of calibrating the radiation radical yield (Figure 3). The oxidation of SCN^- by $HO^{\cdot}$ (reaction 30) is virtually diffusion controlled, as is the subsequent complexation of the radical with SCN^-. Since

Table VII. Inorganic Radicals That Can Be Produced in Aqueous Solution, and Whose Reaction Rate Constants have been Measured with a Variety of Substrates

Carbon dioxide radical anion	$CO_2^{\cdot-}$
Carbonate radical	$CO_3^{\cdot-}$
Ozone	O_3
Azide radical	$N_3^{\cdot}$
Amino radical	$NH_2^{\cdot}$
Nitrogen dioxide radical	$NO_2^{\cdot}$
Nitrogen trioxide radical	$NO_3^{\cdot}$
Phosphite radical	$PO_3^{\cdot 2-}$
Phosphate radical	$PO_4^{\cdot 2-}$
Sulfur dioxide radical anion	$SO_2^{\cdot-}$
Sulfite radical	SO_3^{-}
Sulfate radical	$SO_4^{\cdot-}$
Peroxomonosulfate radical	$SO_5^{\cdot-}$
Selenite radical	$SeO_3^{\cdot-}$
Dithiocyanate radical	$(SCN)_2^{\cdot-}$
Dihalogen radical anions	$Cl_2^{\cdot-}$, $Br_2^{\cdot-}$, $I_2^{\cdot-}$
Chlorine dioxide	$ClO_2^{\cdot}$
Bromine dioxide	$BrO_2^{\cdot}$

SOURCE: Data from reference 15.

$K_{31} = 10^5$ dm^3 mol^{-1}, $(SCN)_2^{\cdot-}$ is rapidly formed and significantly stable. Hydrogen-atom reactions in this system are slow and can be ignored.

Designer Radicals: Generation of the Radical of Choice from the Primary Radicals

In aqueous solutions, a large variety of radicals have been generated and their reaction rates and spectra obtained. Often, these radicals provide better radical reagents for generating a particular species for study. Neta and co-workers (*15*) recently tabulated the rate constants for the reactions of 18 inorganic radicals (Table VII). The authors give "cookbook" procedures for generating the radicals from the primary aqueous radicals discussed above. In addition to these inorganic radicals, a variety of carbon-centered radicals can be efficiently generated and used as radical reagents. Neta and co-workers (*16*) recently compiled and evaluated substantial listings for a large number of carbon-centered radicals. In the following pages, three species with high potential as radical reagents are discussed: carbon dioxide radical anion, 1-hydroxy-1-methylethyl radical, and the azide radical. In addition, production of the superoxide radical is described. Hug has collected the optical absorption spectra of numerous inorganic transient species in aqueous solution (*17*).

Table VIII. A Selection of Rate Constants for Reactions of the Carbon Dioxide Radical Anion

Substrate	*Reaction*	*pH*	k *($dm^3\ mol^{-1}\ s^{-1}$)*
$Fe(CN)_6^{3-}$	$CO_2^{\cdot-} + Fe(CN)_6^{3-} \rightarrow CO_2 + Fe(CN)_6^{4-}$	6.0–11.0	7.0×10^8
$Ru(bpy)_3^{2+}$	$CO_2^{\cdot-} + Ru(bpy)_3^{2+} \rightarrow CO_2 + [Ru^{II}(bpy)_2(bpy^{\cdot-})]^+$	3–10	6.0×10^7
$[Ru^{II}(bpy)_2(bpy^{\cdot-})]^+$	$CO_2^{\cdot-} + [Ru^{II}(bpy)_2(bpy^{\cdot-})]^+ \rightarrow Ru^{II}(bpy)_2(bpy{-}CO_2^{2-})$		1.7×10^9
$Ru(NH_3)_6^{3+}$	$CO_2^{\cdot-} + Ru(NH_3)_6^{3+} \rightarrow CO_2 + Ru(NH_3)_6^{2+}$	4.8	2.0×10^9
$Co(NH_3)_6^{3+}$	$CO_2^{\cdot-} + Co(NH_3)_6^{3+} \rightarrow CO_2 + Co(NH_3)_6^{2+}$	6.9	1.1×10^8
$IrCl_6^{2-}$	$CO_2^{\cdot-} + IrCl_6^{2-} \rightarrow CO_2 + IrCl_6^{3-}$	6.0–7.0	1.7×10^9
Ni^+	$CO_2^{\cdot-} + Ni^+ \rightarrow NiCO_2$	5	6.6×10^9
Ni^{2+}	$CO_2^{\cdot-} + Ni^{2+} \rightarrow CO_2 + Ni^+$	Natural (est., no rxn obsd)	100
$Ni(CN)_4^{2-}$	$CO_2^{\cdot-} + Ni(CN)_4^{2-} \rightarrow CO_2 + Ni(CN)_4^{3-}$	Natural	1.2×10^9
Cu^+	$CO_2^{\cdot-} + Cu^+ \rightarrow CuCO_2$	7.3	$>1 \times 10^{10}$
Cu^{2+}	$CO_2^{\cdot-} + Cu^{2+} \rightarrow CO_2 + Cu^+$	7.3	2.0×10^9
N_2O	$CO_2^{\cdot-} + N_2O + H_2O \rightarrow CO_2 + N_2 + HO^- + HO^{\cdot}$	4.4	1.6×10^3
Oxygen	$CO_2^{\cdot-} + O_2 \rightarrow CO_2 + O_2^{\cdot-}$	8	2.0×10^9
Nitrobenzene	$CO_2^{\cdot-} + C_6H_5NO_2 \rightarrow CO_2 + C_6H_5N^{\cdot}O_2^-$	6–7	1.0×10^9

SOURCE: Reference 10.

Carbon Dioxide Radical Anion. $CO_2^{\cdot-}$ is an efficient reducing agent, rapidly reducing a large number of aqueous metal ions and metal complexes without formation of intermediate adducts, as illustrated in Table VIII. Reduction of O_2 by $CO_2^{\cdot-}$ is commonly used to produce superoxide radicals for study of its reactions. $CO_2^{\cdot-}$ also reduces a variety of organic species: it transfers an electron to nitrobenzene, producing a radical anion. The less negative reduction potential of $CO_2^{\cdot-}$(−2.0 V versus NHE) (*16*) relative to that of e^-_{aq} (−2.77 V versus NHE) (*9*) is illustrated by its ability to reduce aromatic compounds only if they contain electron-withdrawing substituents such as $-NO_2$. $CO_2^{\cdot-}$ reduces $Ru(bpy)_3^{2+}$ to $[Ru^{II}(bpy)_2(bpy^{\cdot-})]^+$ (reaction 32), just as e^-_{aq} does. However, the $CO_2^{\cdot-}$ rate constant is smaller by a factor of 500.

$$CO_2^{\cdot-} + Ru(bpy)_3^{2+} \rightarrow CO_2 + [Ru^{II}(bpy)_2(bpy^{\cdot-})]^+ \qquad (32)$$

$$CO_2^{\cdot-} + [Ru^{II}(bpy)_2(bpy^{\cdot-})]^+ \rightarrow Ru^{II}(bpy)_2(bpy{-}CO_2^{2-}) \qquad (33)$$

Like the 2-hydroxy-2,2-di-methylethyl radical, $CO_2^{\cdot-}$ can add to the reduced 2,2′-bipyridyl ligand in the ruthenium complex (reaction 33). In contrast to the electron-transfer reaction, the addition reaction is nearly diffusion controlled. Rates of many of the electron-transfer reactions of $CO_2^{\cdot-}$ are sufficiently below the diffusion-controlled range where correlations between rate constants and reduction potentials can be made, in accordance with Marcus theory, showing that $CO_2^{\cdot-}$ can undergo outer-sphere electron transfers.

$CO_2^{\cdot-}$ is produced by the reduction of CO_2 with e^-_{aq}, or by the abstraction of a hydrogen atom from formate by $H^\cdot$ or $HO^\cdot$ (reactions 34–36).

$$CO_2 + e^-_{aq} \rightarrow CO_2^{\cdot-} \tag{34}$$

$$HCO_2^- + H^\cdot \rightarrow CO_2^{\cdot-} + H_2 \tag{35}$$

$$HCO_2^- + HO^\cdot \rightarrow CO_2^{\cdot-} + H_2O \tag{36}$$

If the objective of an experiment is to convert all the primary radicals to strong one-electron reductants, addition of formate to the solution and saturation with N_2 will accomplish the goal. Both $H^\cdot$ and $HO^\cdot$ will be rapidly converted to $CO_2^{\cdot-}$ via reactions 35 and 36, and e^-_{aq} remains, ready to reduce without complications that could arise from the presence of dissolved CO_2.

In some cases, $CO_2^{\cdot-}$ forms adducts containing metal–carbon bonds, such as with aqueous metal ions like Cu^+ and Ni^+. In the case of Ni^+, the protonated product, $NiCO_2H^+$, is quite stable.

1-Hydroxy-1-methylethyl Radical. The 1-hydroxy-1-methylethyl radical, $(CH_3)_2C^\cdot OH$, and its conjugate base, $(CH_3)_2C^\cdot O^-$, are excellent reducing radicals, which exhibit more selectivity than the hydrated electron (Table IX). The driving force for the reducing power of these radicals is the production of acetone (reactions 39 and 40). These radicals are generated in the radiolysis of N_2O-saturated aqueous solutions containing 2-propanol. e^-_{aq} is converted rapidly to $HO^\cdot$ by reaction with N_2O, as discussed previously. $HO^\cdot$ rapidly oxidizes the alcohol via abstraction of a hydrogen atom from the secondary carbon (reaction 37). Of the hydrogen atoms abstracted by $HO^\cdot$, 85.5% come from the secondary carbon, while the remainder come from one of the methyl groups, producing the $(CH_2^\cdot)(CH_3)CHOH$ radical (*18*). This radical is a weak reducing agent, so that its reactions can generally be ignored. The hydroxylic proton of $(CH_3)_2C^\cdot OH$ is very labile relative to the hydroxylic proton in the parent alcohol: pK_{38} is about 12.03, so that K_{38} is more than five orders of magnitude larger than the corresponding value of the parent alcohol (*19*).

$$HO^\cdot + (CH_3)_2CH(OH) \rightarrow H_2O + (CH_3)_2C^\cdot OH \tag{37}$$

$$(CH_3)_2C^\cdot OH \rightleftarrows (CH_3)_2C^\cdot O^- + H^+ \tag{38}$$

$$S + (CH_3)_2C^\cdot O^- \rightarrow S^{\cdot-} + (CH_3)_2CO \tag{39}$$

$$S + (CH_3)_2C^\cdot OH \rightarrow S^{\cdot-} + (CH_3)_2CO + H^+ \tag{40}$$

Table IX. A Selection of Rate Constants for Reactions of the 1-Hydroxy-1-Methylethyl Radical, $(CH_3)_2C^{\cdot}OH$, and Its Conjugate Base, $(CH_3)_2C^{\cdot}O^-$

Substrate	*Reaction*	*pH*	k *($dm^3\ mol^{-1}\ s^{-1}$)*
$Fe(CN)_6^{3-}$	$(CH_3)_2C^{\cdot}OH + Fe(CN)_6^{3-} \rightarrow$ Products	7.0	4.7×10^9
$Ru(bpy)_3^{2+}$	$(CH_3)_2C^{\cdot}O^- + Ru(bpy)_3^{2+} \rightarrow [Ru^{II}(bpy)_2(bpy^{\cdot -})]^+ + (CH_3)_2CO$	12.7	7.5×10^9
	$(CH_3)_2C^{\cdot}OH + Ru(bpy)_3^{2+} \rightarrow$	7	$<10^6$
$[Ru^{II}(bpy)_2(bpy^{\cdot -})]^+$	$(CH_3)_2C^{\cdot}OH + [Ru^{II}(bpy)_2(bpy^{\cdot -})]^+ \rightarrow [Ru^{II}(bpy)_2(bpy{-}C(CH_3)_2(OH)^-)]^+$		1.4×10^9
$Co(NH_3)_6^{3+}$	$(CH_3)_2C^{\cdot}O^- + Co(NH_3)_6^{3+} \rightarrow (CH_3)_2CO + Co(NH_3)_6^{2+}$	12	5.0×10^9
	$(CH_3)_2C^{\cdot}OH + Co(NH_3)_6^{3+} \rightarrow (CH_3)_2CO + Co(NH_3)_6^{2+} + H^+$	0–3	4.1×10^5
$IrCl_6^{2-}$	$(CH_3)_2C^{\cdot}OH + IrCl_6^{2-} \rightarrow (CH_3)_2CO + IrCl_6^{3-} + H^+$	4–6	4.7×10^9
Ni^+	$(CH_3)_2C^{\cdot}OH + Ni^+ \rightarrow [NiC(OH)(CH_3)_2]^+$	Natural	1.4×10^9
Ni^{2+}	$(CH_3)_2C^{\cdot}OH + Ni^{2+} \rightarrow$ products		$<1 \times 10^6$ No Ni^+ obsd
Cu^+	$(CH_3)_2C^{\cdot}OH + Cu^+ \rightarrow$ products	4.5	5.0×10^9
Cu^{2+}	$(CH_3)_2C^{\cdot}OH + Cu^{2+} \rightarrow (CH_3)_2CO + Cu^+ + H^+$	6	5.0×10^7
Zn^+	$Zn^+ + (CH_3)_2C^{\cdot}OH + H^+ \rightarrow (CH_3)_2CH(OH)$	$+Zn^{2+}$	1.3×10^9
Vitamin K_1[a]	$(CH_3)_2C^{\cdot}OH$ + 2-Me(3-phytyl)NQ $\rightarrow (CH_3)_2CO$ + 2-Me(3-phytyl)$NQ^{\cdot -} + H^+$	7	1.7×10^9

[a] Vitamin K_1 is 2-methyl-3-(3,7,11,15-tetramethyl-2-hexadecenyl)-1,4-naphthalenedione.
SOURCE: Data from reference 16.

Neta and co-workers (*16*) have compiled published rate constants for reactions of a large number of aliphatic carbon-centered radicals, including $CH_3C^{\cdot}(OH)CH_3$ and $(CH_3)_2C^{\cdot}O^-$, with inorganic and organic substrates. As seen from the selection of rate constants in Table IX, the rates of reaction of these radicals can be very high, approaching the diffusion-controlled limit in many cases. With many inorganic and organic substrates, the reaction is a clean outer-sphere electron transfer, as described by reactions 39 and 40. The electron-transfer reactions have been shown in some cases to actually involve an addition–elimination mechanism.

As with formate solutions, all the primary radicals can be converted to strong one-electron reductants in alcohol solutions: radiolysis of N_2-saturated, alkaline solutions of alcohols (R_1R_2CHOH) yields solutions of e^-_{aq} and $R_1R_2C^{\cdot}O^-$. The deprotonated alcohol radical anions have higher rates of electron transfer, and are less likely to undergo addition reactions than the protonated radicals. Hexaamminecobaltate(III) is reduced to hexaamminecobaltate(II) by $(CH_3)_2C^{\cdot}O^-$ or by $CH_3C^{\cdot}(OH)CH_3$, with rate constants of 5.0×10^9 and 4.1×10^5 $dm^3\ mol^{-1}\ s^{-1}$, respectively. The high rate constant for $(CH_3)_2C^{\cdot}O^-$ suggests outer-sphere electron transfer for this complex. This case illustrates a typical aspect of electron-transfer reactions: protonation of the electron-donating atom in a species can dramatically reduce the rate of electron transfer. The radicals display a similar trend in their reactions with $[Ru^{II}(bpy)_3]^{2+}$. Like e^-_{aq}, $(CH_3)_2C^{\cdot}O^-$ acts as a straightforward reducing agent, producing $[Ru^{II}(bpy)_2(bpy^{\cdot-})]^+$ with a near diffusion-controlled rate constant (reaction 41) (*11*).

$$(CH_3)_2C^{\cdot}O^- + Ru(bpy)_3{}^{2+} \rightarrow [Ru^{II}(bpy)_2(bpy^{\cdot-})]^+ + (CH_3)_2CO \quad (41)$$

$$(CH_3)_2C^{\cdot}OH + [Ru^{II}(bpy)_2(bpy^{\cdot-})]^+ \rightarrow [Ru^{II}(bpy)_2(bpy\text{–}C(CH_3)_2(OH)^-)]^+ \quad (42)$$

In contrast, $(CH_3)_2C^{\cdot}OH$ gives no observable reaction with $Ru(bpy)_3{}^{2+}$, so that only a limiting value, $k_{42} < 10^6$ dm^3 mol^{-1} s^{-1}, is obtained. Although $(CH_3)_2C^{\cdot}OH$ does not readily reduce $Ru(bpy)_3{}^{2+}$, it does readily add to the carbanion in $[Ru^{II}(bpy)_2(bpy^{\cdot-})]^+$ (reaction 42), as seen with the 2-hydroxy-2-methylethyl radical and $CO_2{}^{\cdot-}$. Addition of $(CH_3)_2C^{\cdot}O^-$ to the carbanion is not seen because the rapid reduction of $Ru(bpy)_3{}^{2+}$ consumes the radical.

$(CH_3)_2C^{\cdot}OH$ reacts rapidly with Ni^+; however, the reaction occurs not by direct electron transfer but via formation of an adduct with a metal–carbon bond. These examples illustrate the utility of the 1-hydroxy-1-methylethyl radical, and its conjugate base, in studies of electron-transfer reactions in aqueous solutions. They also caution us to always evaluate the possibilities of alternate reaction pathways in our experimental design and interpretation.

Superoxide Radical. Reactions of the superoxide and hydroperoxyl radicals are studied primarily because of the role of the radicals in normal and abnormal biological systems. Bielski and co-workers (*20*) compiled a list of reaction rate constants for $HO_2^{\cdot}$ and $O_2^{\cdot-}$ and other properties of the radicals. Unlike the majority of the radicals discussed here, the self-reactions of $HO_2^{\cdot}$ and $O_2^{\cdot-}$ are slow (Figure 4). In particular, it is probable that no self-reaction occurs with the superoxide radical itself, and that observed decays result from reaction with $HO_2^{\cdot}$ or solution components or contaminants.

For production of the superoxide radical in radiolytic studies, $HO^{\cdot}$, e^-_{aq}, and $H^{\cdot}$ are all utilized via oxygen-saturated formate solutions. The solvated electron and the hydrogen atom produce $O_2^{\cdot-}$ equivalents by direct reactions with O_2 (reactions 43 and 44).

$$e^-_{aq} + O_2 \rightarrow O_2^{\cdot-} \quad (43)$$

$$H + O_2 \rightarrow HO_2 \quad (44)$$

$$HO_2 \rightleftarrows O_2^{\cdot-} + H^+ \quad (45)$$

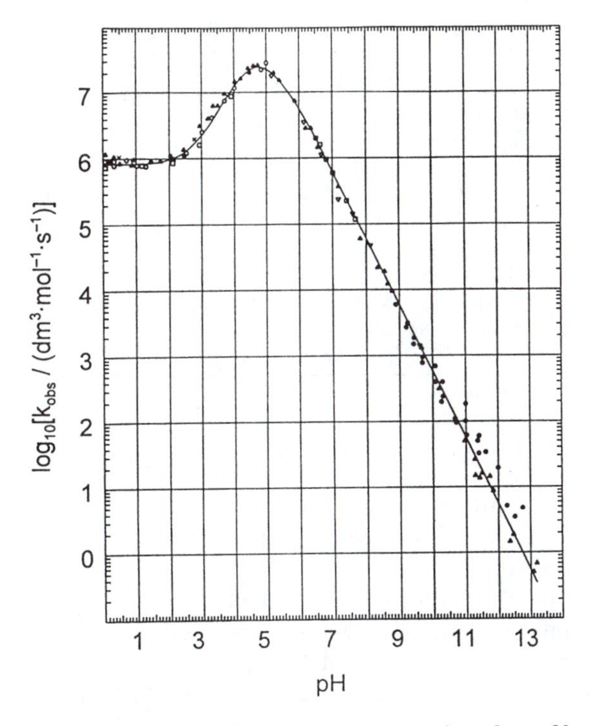

Figure 4. Observed second-order rate constants for the self-reactions of $HO_2^{\cdot}/O_2^{\cdot-}$. (Reproduced from reference 20. Copyright 1985 American Chemical Society.)

The superoxide radical and its conjugate acid, the hydroperoxyl radical, rapidly equilibrate with $pK_{45} = 4.9$. The hydroxyl radical center is transferred to a superoxide radical in a two-step process: $HO^{\cdot}$ abstracts a hydrogen atom from the formate ion, producing the carbon dioxide radical anion, which then transfers an electron to oxygen (reactions 46 and 47).

$$HO^{\cdot} + HCO_2^{-} \rightarrow H_2O + CO_2^{\cdot-} \quad (46)$$

$$CO_2^{\cdot-} + O_2 \rightarrow CO_2 + O_2^{\cdot-} \quad (47)$$

All the reactions leading to $HO_2^{\cdot}$ and $O_2^{\cdot-}$ are rapid, so that all the primary radicals can be converted to these radicals in a matter of a few nanoseconds. Ironically, the superoxide radical reacts with many substrates at such slow rates that they cannot be studied by pulse radiolysis methods, which are adapted for the study of rapid reactions.

Azide Radical. The azide radical ($N_3^{\cdot}$) is a strong one-electron oxidant with a redox potential of +1.3 V versus NHE. Neta and co-workers (*16*) com-

Table X. A Selection of Rate Constants for Reactions of the Azide Radical

Substrate	*Reaction*	*pH*	k *(dm^3 mol^{-1} s^{-1})*
$Fe(CN)_6^{4-}$	$N_3^{\cdot} + Fe(CN)_6^{4-} \rightarrow N_3^- + Fe(CN)_6^{3-}$	9.0	2.5×10^9
$Fe(bpy)_3^{2+}$	$N_3^{\cdot} + Fe(bpy)_3^{2+} \rightarrow N_3^- + Fe(bpy)_3^{3+}$	7.0	2.9×10^9
$Ru(bpy)_3^{2+}$	$N_3^{\cdot} + Ru(bpy)_3^{2+} \rightarrow N_3^- + Fe(bpy)_3^{3+}$	7.0	1.2×10^8
$Mo(CN)_8^{4-}$	$N_3^{\cdot} + Mo(CN)_8^{4-} \rightarrow N_3^- + Mo(CN)_8^{3-}$	9.0	2.2×10^9
$IrCl_6^{3-}$	$N_3^{\cdot} + Ircl_6^{3-} \rightarrow N_3^- + Ircl_6^{2-}$	7.0	4.6×10^8
$IrBr_6^{3-}$	$N_3^{\cdot} + IrBr_6^{3-} \rightarrow N_3^- + IrBr_6^{2-}$	7.0	1.8×10^9
I^-	$N_3^{\cdot} + I^- \rightarrow N_3^- + I^{\cdot}$	Natural	4.5×10^8
Ascorbate ion	$N_3^{\cdot}$ + ascorbate ion $\rightarrow N_3^-$ + ascorbate radical ion + H^+	5.8	3.0×10^9
Phenol	$N_3 + C_6H_5OH \rightarrow N_3^- + C_6H_5O^{\cdot} + H^+$	~6	5×10^7
Phenoxide ion	$N_3^{\cdot} + C_6H_5O^- \rightarrow N_3^- + c_6H_5O^{\cdot}$	11–12	4.3×10^9

SOURCE: Data from reference 10.

piled rate constants for 117 reported reactions of $N_3^{\cdot}$, and they presented an informative discussion of the radical properties. In its reactions with many organic and inorganic substrates, $N_3^{\cdot}$ exhibits rate constants corresponding to reactions that are nearly diffusion controlled (Table X). This radical has the significant advantage relative to $HO^{\cdot}$ of usually reacting by direct electron transfer without formation of intermediate adducts. The azide radical has a weak absorption maximum at 274 nm with ϵ_{277} = 2025 dm^3 mol^{-1} cm^{-1}. Because of the narrow bandwidth of this peak, and its relatively low molar absorptivity, most reported rate constants are based on absorbance growths of products. $N_3^{\cdot}$ is generated by radiolyzing N_2O-saturated solutions of sodium azide. The hydrated electron is converted rapidly to $HO^{\cdot}$ by reaction with the N_2O, as discussed previously. $HO^{\cdot}$ rapidly oxidizes azide ion, transferring the radical center to the azide radical (reaction 48):

$$HO^{\cdot} + N_3^- \rightarrow HO^- + N_3^{\cdot} \tag{48}$$

The rate of oxidation of N_3^- by $HO^{\cdot}$ is nearly diffusion controlled (Table VI), so that with only 10 mmol dm^{-3} of N_3^-, the half-life of reaction 48 is just 6 ns. The overall result is that the yield (μmol J^{-1}) of $N_3^{\cdot}$ is 0.55, with a residual yield of hydrogen atoms of 0.06. The H atoms react only slowly with the basic solution components, and adjustments for their contribution can be made. The rate constant for oxidation of phenol is about two orders of magnitude smaller than that for oxidation of phenoxide, again illustrating the powerful effect of protonation of the electron donating center.

Where to Find Information

A number of compilations and critical evaluations of rate constants for radiolytically and photolytically produced radicals are available. References to these documents are given throughout the text. Most of these have been produced through the Radiation Chemistry Data Center (RCDC) of the Radiation Laboratory at the University of Notre Dame, (NDRL) Notre Dame, IN, under the direction of Alberta B. Ross, and through the National Institute of Standards and Technology (NIST), U.S. Department of Commerce. The RCDC was established in 1965 with support from the U.S. Atomic Energy Commission Division of Technical Information. In the period 1965–1995, the center was jointly supported by the National Bureau of Standards (now NIST) as part of the National Standard Data Reference System. RCDC is now supported solely by the U.S. Department of Energy Office of Basic Energy Sciences. The Center is located within, and is part of, the Radiation Laboratory at the University of Notre Dame. The central historical role played by the scientists at the University of Notre Dame facility, including Dr. Ross, in the development of the field of radiation chemistry is well presented by Kroh (1989) in reference 8.

Reprints of the publications in the *Journal of Physical and Chemical Reference Data* may be purchased from the American Chemical Society. A complete listing of available reprints is given in the back of some issues of the journal, along with ordering instructions.

In addition to hard-copy publications, the RCDC, in association with NIST, has produced the *NDRL-NIST Solution Kinetics Database, Version 2* (*10*). The database may be purchased from NIST, Standard Reference Data, Building 221/Room A320, Gaithersburg, MD 20899. This database runs locally on MS-DOS compatible computers. The current version contains 14,000 rate determinations on 10,800 reactions involving 7800 chemical species. The database can be searched for reactants, products, citation, or author. An updated version, compiling data from the literature through the end of 1994, will be released shortly.

Major portions of the data collections of the RCDC are stored in electronically readable form, including a bibliographic database of over 120,000 references, a chemical registry database, a solution rate constant database, and a triplet–triplet optical absorption database. The data collection is now structured as an Oracle relational database. An implementation of the database query interface based on the World Wide Web (WWW) is currently under development. The WWW site of the RCDC is accessible from the home page of the Notre Dame Radiation Laboratory (Internet address http://www.rad.nd.edu) or directly from the RCDC WWW server at http://allen.rad.nd.edu. The center will answer questions about its offerings and access to them via e-mail at ndrlrcdc.1@nd.edu. The printed compilations of the center are being placed on the WWW, starting with *Rate Constants for the Decay and Reactions of the Lowest Electronically Excited Singlet State of Molecular Oxygen in Solution. An Expanded and Revised Compilation,* by Wilkinson, Helman, and Ross.

Rate constants and other data quoted herein are taken from these materials, unless noted. Original references are given only in a limited number of cases. I have also included a list of several of the most recent books on the general topic of radiation chemistry. Buxton and co-workers (9) present an especially informative discussion of practical aqueous radiation chemistry along with their compilation of more than 3500 reaction rate constants, critically reviewed. Many of the critical compilations present cogent discussions of basic principles relevant to radiation chemistry.

Acknowledgments

This paper is dedicated with admiration to Alberta B. Ross, who for many years was the director of the Radiation Research Data Center at the Radiation Laboratory of the University of Notre Dame. On numerous occasions, I was amazed by the breadth and depth of her knowledge of radiation chemistry and chemistry, and her help has been instrumental in my research and in the work of many others. Her incisive intellect and unending enthusiasm for science have been an inspiration for many researchers. I wish to acknowledge the assistance provided to me by Keith P. Madden, current Director of the RCDC at the Notre Dame Radiation Laboratory, in obtaining information contained in this article. At the University of Akron, Ann D. Bolek, librarian extraordinaire, provided invaluable assistance in retrieval of information and materials.

On behalf of those authors whose work was cited not through direct reference but through reference to data collections, I urge the reader to refer to the data collections and consult the original references. Since this is a tutorial, and not a review article, the work presented was chosen to illustrate concepts and not to provide a critical review.

References

1. Miller, J. R.; Calcaterra, L. T.; Closs, G. L. *J. Am. Chem. Soc.* **1984,** *106,* 3047–3049.
2. Neta, P. *J. Chem. Educ.* **1981,** *58,* 110–113.
3. Spinks, J. W. T.; Woods, R. J. *An Introduction to Radiation Chemistry;* John Wiley & Sons: New York, 1990.
4. Burton, M. *Discuss. Faraday Soc.* **1952,** *12,* 317–318.
5. Burton, M. *J. Chem. Educ.* **1951,** 404–420.
6. Lind, S. C. *J. Phys. Chem.* **1928,** *573–575;* Lind, S. C. *Chem. Rev.* **1930,** 203–213.
7. Burton, M. *J. Phys. Colloid Chem.* **1947,** 611–625.
8. *Early Developments in Radiation Chemistry;* Kroh, J., Ed.; Royal Society of Chemistry: Cambridge, England, 1989.
9. Buxton, G. V.; Greenstock, C. L.; Helman, W. P.; Ross, A. B. *J. Phys. Chem. Ref. Data* **1988,** *17,* 513–886.
10. Ross, A. B.; Mallard, W. G.; Helman, W. P.; Buxton, G. V.; Huie, R. E.; Neta, P. *NDRL–NIST Solution Kinetics Database,* Ver. 2; NIST Standard Reference Data; National Institute of Standards and Technology: Gaithersburg, MD, 1994.
11. Mulazzani, Q. G.; D'Angelantonio, M.; Camaioni, N.; Venturi, M. *J. Chem. Soc. Faraday Trans.* **1991,** *87,* 2179–2185.
12. Freiberg, M.; Meyerstein, D. *J. Chem. Soc. Faraday Trans. 1* **1980,** *76,* 1825–1837.
13. Cohen, H.; Meyerstein, D. *J. Chem. Soc. Faraday Trans. 1* **1988,** *84,* 4157–4160.

14. Kelm, M.; Lilie, J.; Henglein, A.; Janata, E. *J. Phys. Chem.* **1974,** *78,* 882–887.
15. Neta, P.; Huie, R. E.; Ross, A. B. *J. Phys. Chem. Ref. Data* **1988,** *17,* 1027–1284.
16. Neta, P.; Grodkowski, J.; Ross, A. B. *J. Phys. Chem. Ref. Data* **1996,** *25,* 709–1050.
17. Hug, G. L. *Optical Spectra of Nonmetallic Inorganic Transient Species in Aqueous Solution;* NSRDS-NBS 69; National Institute of Standards and Technology: Gaithersburg, MD, 1981.
18. Asmus, K.-D.; Mockel, H.; Henglein, A. *J. Phys. Chem.* **1973,** *77,* 1218–1221.
19. Laroff, G. P.; Fessenden, R. W. *J. Phys. Chem.* **1973,** *77,* 1283–1288.
20. Bielski, B. H. J.; Cabelli, D. E.; Arudi, R. L.; Ross, A. B. *J. Phys. Chem. Ref. Data* **1985,** *14,* 1041–1100.

Bibliography

Asmus, K.-D. *Meth. Enzymol.* **1984,** *105,* 167–178.

Lind, S. C. *The Chemical Effects of Alpha Particles and Electrons;* American Chemical Society Monograph; The Chemical Catalog Company, Inc.: New York, 1921; Second edition, revised 1928.

Matheson, M. S.; Dorfman, L. M. *Pulse Radiolysis;* Massachusetts Institute of Technology: Press. Cambridge, MA, 1969.

Radiation Chemistry: Principles and Applications; Farhataziz, init?; Rodgers, M. A. J., Eds.; VCH Publishers, Inc.: New York, 1987.

3

Accelerators and Other Sources for the Study of Radiation Chemistry

James F. Wishart

Department of Chemistry, Brookhaven National Laboratory, Upton, NY 11973–5000

This chapter is intended as a guide to aid in the design of radiolysis experiments for researchers who are new to the field. It summarizes the features and limitations of the several kinds of particle accelerators and X-ray and gamma radiation sources used in the study of radiation chemistry. The chapter describes (1) the types of ionizing radiation and their interactions with matter, (2) the architecture and use of X-ray and gamma radiation sources, and (3) the electrostatic and radio-frequency methods of particle acceleration. The advantages and disadvantages of four types of accelerators are compared. A description of proton and heavy-ion accelerators completes the chapter.

Radiolysis, the initiation of reactions by ionizing radiation, is a powerful technique for studying chemical systems. In radiolysis, energetic charged particles or photons transfer their energy in varied increments to the molecules of the medium, inducing ionizations and molecular excitations. The rate of energy loss depends on the type of particle, its energy, and the density of the medium. These interactions occur in a confined region around the path of the particles (the "track"), wherein the concentration of molecular fragments is high. Outside the track the solution is unperturbed. Ionization events frequently result in secondary electrons with enough energy to induce further ionizations and excitations within their own tracks or "spurs". At subpicosecond time scales, radiolysis results in an inhomogeneous distribution of ionized and excited solvent molecules, which cross-react to yield solvated electrons, ions, radicals, and molecular products. These primary products can then react with solutes to form transient species of interest.

Photolysis, on the other hand, deposits excitation in specific molecules through light absorption. The spatial pattern of energy deposition mirrors that of the chromophore distribution. In some cases excitation directly generates the species of interest, whereas in others, reaction with a precursor is required. While direct excitation offers advantages in time resolution over schemes involving second-order reactions, in some cases high-background absorption, low quantum yield, or the lack of a suitable photosensitizer system precludes the application of photolytic techniques to study certain chemical systems. Many such systems are amenable to study by radiolysis. Without the need for a strongly absorbing species in solution, a wider range of the spectrum may be available for detection.

This chapter is meant to serve as a guide to the type of accelerators and other radiation sources available for the study of chemical reactions. Radiolysis techniques are in widespread commercial use in applications such as the preparation of materials, sterilization of medical supplies, preservation of foodstuffs, as well as medical therapy. The devices used for these applications will not be covered here, although some of them are also used in chemical kinetics research (*1*).

When designing photolysis experiments, parameters such as wavelength, pulse width, and pulse energy are critical factors. In much the same way, operating parameters of accelerators are critical to the design of radiolysis experiments. The particle type and energy determine the spatial distribution of energy deposition within the sample and the yields of primary radiolysis products. The particle-beam pulse width and pulse structure define the available time resolution. The charge-per-pulse is analogous to laser pulse energy; at lower doses, signal intensity varies with dose, whereas at higher levels, higher-order effects appear (significant bimolecular reactions between radiolysis products, for example).

Types of Ionizing Radiation and Their Means of Interaction with Matter

There are two types of ionizing radiation: energetic photons such as X-rays and gamma rays, and energetic charged particles such as electrons, protons, neutrons, alpha particles, and other atomic nuclei. Low-energy X-rays ($<$100 keV) produce energetic electrons by the photoelectric effect. Compton interactions, which are elastic collisions between photons and loosely bound electrons that produce an energetic electron and another photon of lesser energy, dominate between 0.5 and 30 MeV. Production of electron–positron pairs by X-rays becomes significant above 3 MeV. Energy loss by X-rays to nuclear processes, (γ,n) reactions, become important above 10 MeV and contribute little to radiolysis but do create safety concerns because of nuclear activation to produce radioisotopes. As a result of all these interactions, photon radiolysis reduces to particle radiolysis because most of the energy is dissipated by fast electrons.

In relative terms, the distance between energy deposition interactions (and penetrating power) is very large for X-rays and gamma rays (hence their utility for visualization within opaque objects), and moderate for electrons. For more massive and more highly charged particles, it is short because of their lower velocity (at a given energy) and the strength of their electrostatic interactions with the electrons of the material being penetrated. Heavy ions have a high linear energy transfer (LET) value. As the density of interactions increases, cross reactions between initial radiolysis products become more probable, changing the absolute and relative yields of the ultimate radiolysis products. Although precise examples of this effect are beyond the scope of this chapter, it is important to remember that the different types of radiation result in different distributions of products.

Particle energy is another important criterion. The probability of interaction between a charged particle and the medium it penetrates is a function of energy. Fortunately, the interaction probability in water is relatively constant for an electron in the 1–50-MeV range, resulting in a near-linear relationship between the energy and penetration depth as shown by the dose–depth profiles depicted in Figure 1 for electrons of several energies.

X-ray and Gamma-Ray Sources for Radiolysis

Certain radioactive decay processes can be used to provide a source of gamma radiation for radiolysis. One particularly useful isotope is ^{60}Co, which is produced by thermal neutron bombardment of ^{59}Co. ^{60}Co has a half-life of 5.27 yr and emits gamma radiation with an average energy of 1.2 MeV. Another commonly used gamma source is ^{137}Cs (half-life 30.17 yr, 0.66 MeV).

Radioactive sources must be shielded to protect workers from occupational

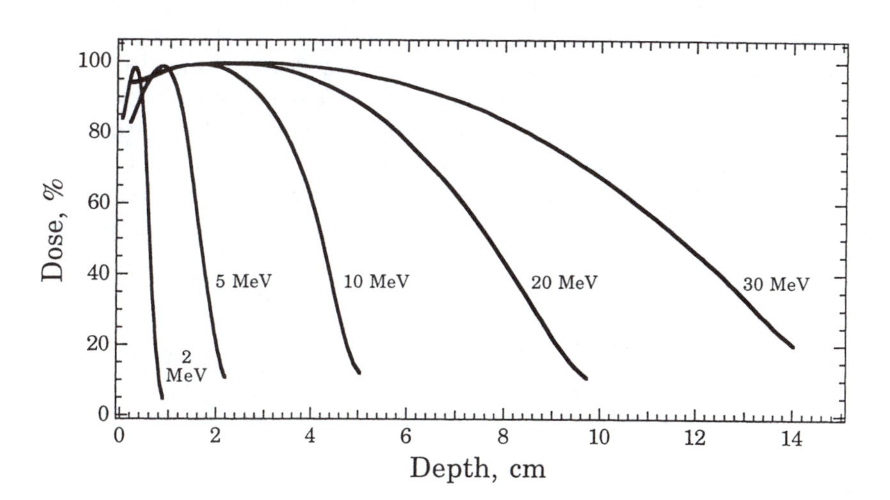

Figure 1. Dose–depth profiles for electrons of several energies in water. (Data from reference 11.)

radiation exposure. Shielding schemes for sources fall into two types, depending on how the source is to be used. The first type, often referred to as the "well" configuration, uses a steel-clad source shaped like a hollow cylinder or a cylindrical array of source rods. The source is completely enclosed by shielding material except for a vertical shaft leading to the center of the source. Samples to be irradiated are lowered into the source for the desired amount of time. The cylindrical source geometry assures a uniform radiation field inside. Sample volume is limited by the design of the source.

The second type of radioactive source structure is the labyrinth. In this design, the source is used to irradiate materials within a room that is separated from occupied areas by a maze of shielding material. The source is stored within a shielding block to allow access to the room for the placement of samples. Once the room is secured, the source is withdrawn from the shield by remote control to irradiate the samples. The access doors and the source control mechanism must be interlocked for the safety of personnel. This type of facility is more elaborate than the well design, but the large volumes of material it can accommodate make it well suited to processing.

Both types of sources are useful for continuous radiolysis or even time-resolved studies on the several-minute time scale or longer, such studies are limited by the time required to move the sample in and out of the source. Competition kinetics is often used to obtain kinetic information about reactions too fast to measure directly by time-resolved methods (2). Continuous gamma radiolysis is also a convenient method of generating radiolysis products for identification and chemical analysis (instead of using pulsed sources, where the average dose rate is much lower).

Particle accelerators may be used to generate pulsed or continuous beams of broadband X-radiation through the process of *bremsstrahlung* (German for "braking radiation"). An electron beam is directed into a thin target of high-atomic-number material such as tungsten, gold, or lead. As the electrons are slowed by their interactions with nuclei in the material, they emit X-rays, mainly in the forward direction with respect to the electron beam. The distribution of X-ray energies is a function of the incident electron energy and the type and thickness of the material under bombardment. Bremsstrahlung generation is an excellent means of generating very short (pico- or nanosecond) pulses of X-rays for time-resolved studies.

X-ray tubes were used by early researchers because they were the most powerful radiation sources at the time. For radiation chemistry, tube voltages as high as 250 kV were used; in comparison, voltages of 150 kV are used for industrial radiography and 60 kV for crystallography. Although X-rays of wavelengths characteristic to the anode material are generated, the bremsstrahlung continuum (which peaks at one-third of the incident electron energy) dominates.

A synchrotron can be used as an intense source of X-radiation over a wide range of energies. Synchrotron radiation is generated when high-energy particles are steered along a circular orbit in a storage ring. X-ray emission from such a source is very broadband; monochromators are often used on synchrotron beamlines. Wigglers (devices with alternating magnetic poles) can be used to increase radiation intensity and enhance a particular energy range. Time-resolved studies are limited by the orbit period of the storage ring and the distribution of the multiple particle bunches simultaneously stored in the ring. Synchrotron X-ray sources excel in their intensity and the ability to tune the energy of the incident radiation. The complexity and physical size of synchrotrons necessitate large facilities. The major synchrotron facilities have a proposal review process for access to beam time and extensive experimental and lodging support for visiting scientists.

Particle Accelerators for Radiolysis

Acceleration of a charged particle requires a force applied by an electric field. The field can be derived from an electrostatic potential, as shown in Figure 2a, or from the sinusoidal electric field component of electromagnetic radiation (typically, but not exclusively, radio frequency (RF) or microwave), as shown in Figure 2b The electrostatic potential gradient is essentially constant in time and space, providing constant acceleration to the particle. The electric field of electromagnetic radiation oscillates in time and space, so the acceleration that charged particle undergoes is time- and location-dependent. The result is that the particles will tend to settle at the point where they receive enough acceleration to keep up with the oscillating field as it moves in time and space. Particles that ride higher on the wave will receive more acceleration and will catch up with ones riding lower. The RF field thus tends to group the electrons into "bunches."

Generally speaking, electrostatic accelerators operate at lower particle energies (1–5 MeV). RF accelerators (e.g., linear accelerators and RF electron guns) can accelerate particles to much higher energies (GeV).

Radiation Safety in Accelerator Facilities (3). During operation, accelerators can generate a large X-ray flux from *bremsstrahlung* produced by the particle beam, from loss during transport as well as from the beam striking the target. The radiation flux depends on many variables, including particle type, energy, current, and the layout of the facility. For accelerators with particle energies below 10 MeV, the radiation hazard is nonexistent once the machine is turned off. (Above that threshold, nuclear activation by the absorption of X-rays can produce radioactive isotopes in common materials.) In order to protect workers, concrete or lead block enclosures may be assembled around the accelerator and beamline, and the entire facility may be located in an underground vault. Typically, experiments are set up while the accelerator is turned off. During operation, a rigorous, redundant interlock system is used

2a

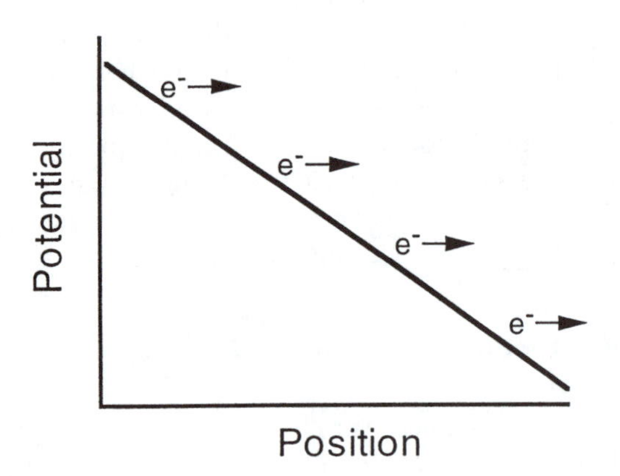

2b

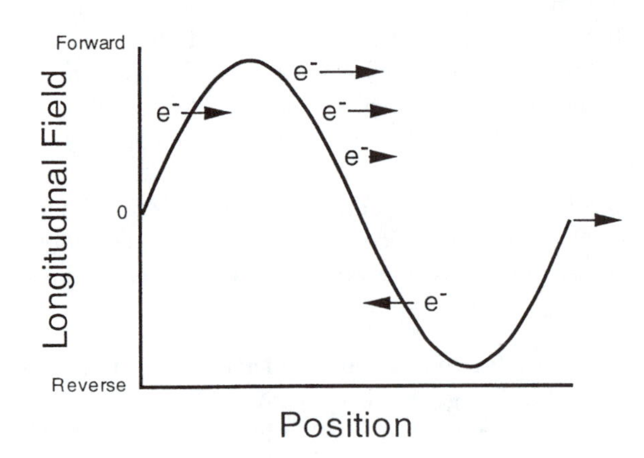

Figure 2. (a) Schematic representation of electrostatic acceleration. (b) Schematic representation of acceleration by an oscillating electric field. Arrow lengths indicate the strength and direction of the acceleration that electrons undergo at different points in the RF phase.

to exclude personnel from the areas in which the accelerator and particle beam are located. Experiment manipulation and data gathering are done by remote control from a nearby room that is protected from the radiation by a given thickness of concrete.

Van de Graaff Generators

Virtually every person who has taken an introductory physics course or visited a science museum has seen a demonstration of a Van de Graaff generator (*4*). Such machines operate on a principle of electrostatics that states that if a charged conductor is brought into internal contact with a hollow conductor, all of the charge on the internal conductor is transferred to the external one, regardless of its potential. A motor-driven pulley at the ground terminal drives a belt of nonconducting material that passes over a second pulley inside a metal sphere or hemisphere (Figure 3). The support structure between the ground

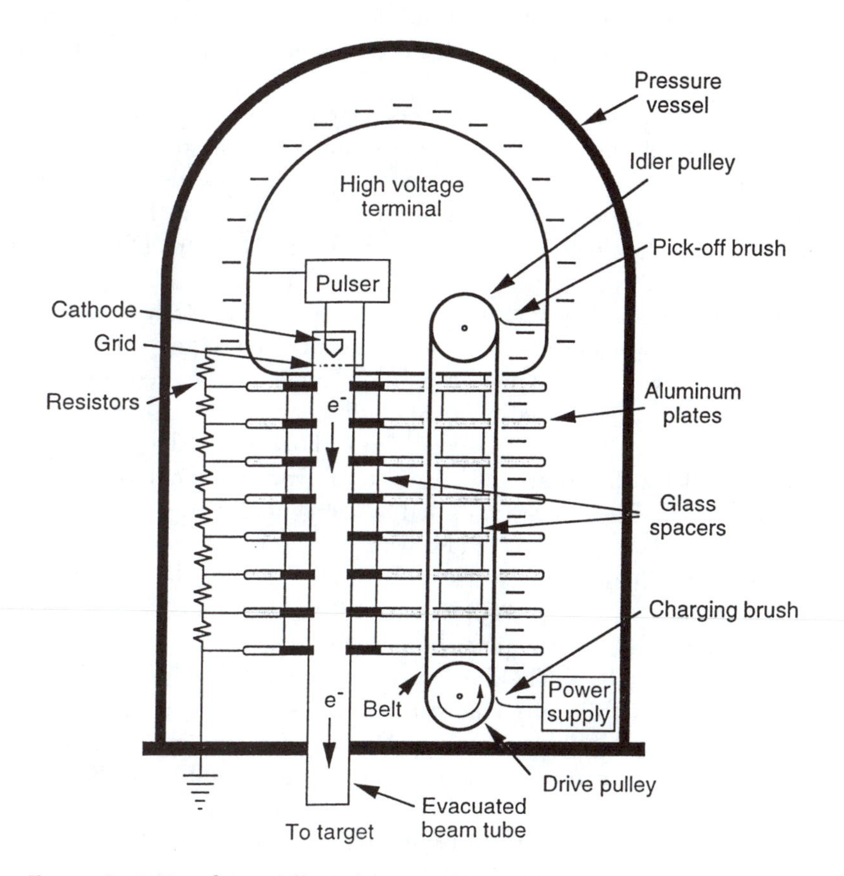

Figure 3. A Van de Graaff accelerator. The actual orientation of the machine may be horizontal or vertical.

terminal and the sphere is insulated. A charge (negative or positive) is applied to the belt at the ground terminal, and the excess charge is removed by a metal brush within the sphere. Charge accumulates on the sphere and a large electrostatic potential develops between the sphere and ground. The generator reaches its maximum potential when the current dissipated to surroundings by corona discharge and insulator leakage matches the charging current. To suppress corona discharge, the accelerator is enclosed in a pressurized tank of insulating gas such as sulfur hexafluoride or a nitrogen–carbon dioxide mixture.

A Van de Graaff accelerator (Figure 3) contains an evacuated beam tube ($\sim 10^{-7}$ Torr), which passes from the high-potential end to ground. The tube itself consists of an alternating stack of metal plates and insulating glass spacers. Each plate is connected to the next through a resistor (typically 1000 MΩ) such that a chain of resistors connects the high-potential end to ground. The resistor chain provides a controlled path for charge dissipation, thereby providing an accurate means of controlling the potential of the machine by variation of the belt charging current. The resistor chain is also a voltage divider: the potential between each consecutive set of plates depends on the value of the resistor between them.

The structure just described generates a large, uniform potential gradient across the length of the beam tube. Electrons or ions emitted from a cathode or ion source placed at the high-potential end of the tube are accelerated to the full potential of the machine (typically 2–5 MeV) when they reach the ground terminal. The electrons or ions are then directed to a target for radiolysis experiments.

Most Van de Graaff generators used for radiolysis are electron machines. Depending on the type of cathode, a Van de Graaff can be equipped to operate in continuous or pulsed emission. A tungsten filament can be used to generate a continuous beam. For pulsed operation, a fine metal grid is placed over an activated metal oxide thermionic cathode. The cathode is biased by a sufficiently positive potential with respect to the grid to override the field gradient and prevent emission. The machine is pulsed by momentarily driving the cathode bias negative to allow it to emit. Pulse widths of 10 ns to 5 μs are common, but the total charge in a pulse will scale with the pulse width. The long drift space at constant potential gradient characteristic of Van de Graaffs precludes currents of more than a few amperes because focusing is not available to counteract space charge effects. The shortest pulse width is limited by the response of the pulsing circuit, although an intricate method for generating a subnanosecond pulse has been developed (5). Very long pulses drain the charge off the sphere, resulting in a decrease in energy toward the end of the pulse. Repetitive pulsing is possible (up to a few kilohertz), but the total beam current must be limited to avoid burning through the beam window at the output of the machine. The width of the revolving belt limits the charging current, and by extension, the beam current.

Advantages. Because the Van de Graaff is an electrostatic device, there is very little electromagnetic emission that could interfere with experimental detection systems. This makes it particularly well suited for magnetic resonance detection systems. The electronic systems are relatively straightforward and the mechanical parts are few, so maintenance is not difficult. System reliability is high, and operating cost is moderate. There are no critical alignments or delicate adjustments needed. Available radiolytic doses range up to 400 krad (a moderately large value, equivalent to 2.4 mM OH radicals in N_2O-saturated water).

Disadvantages. Most machines used for pulse radiolysis operate in the 1–5-MeV energy range. The penetrating power of the electron beam is therefore limited to 0.25 to 2.5 g/cm^2, respectively (equivalent to a 2.5–25-mm path in water). In the lower end of this range, experimental equipment design must accommodate the limited beam-penetration depth. The relatively long limited beam current precludes the study of very fast reactions.

Febetrons

The Febetron is a type of Marx-bank impulse generator. The name Febetron is a trademark of the Field Emission Corporation. It consists of a bank of large capacitors that are charged in parallel. A spark gap is used to generate a high potential by switching the capacitors from parallel to series, whereupon they discharge through a large cathode tube. The Febetron delivers a large charge of electrons over a wide energy distribution. Two models deliver nominal beams of 600 keV or 2 MeV. The limited penetrating power of the electron beams from these devices requires the sample cell to be placed very close to the face of the tube. The very high doses (up to 2000 krad) produced by the Febetron are well suited to the study of radical–radical reactions and gas-phase reactions. Due to the necessary recharging process, however, repetition rates are measured in minutes. The spark gap and electric discharge generate a significant amount of electrical noise for detection systems to pick up. Although the initial machine cost is modest, the cathode tubes are relatively expensive consumables.

Linear Accelerators

Linear accelerators (linacs) use a traveling wave of electromagnetic radiation, typically in the radio or microwave range, to accelerate charged particles. The electron linac itself is made of a highly conducting material such as copper, fabricated into a cylindrical radio-frequency waveguide. Within the waveguide, a series of copper disks with apertures divide the structure into resonant cavities

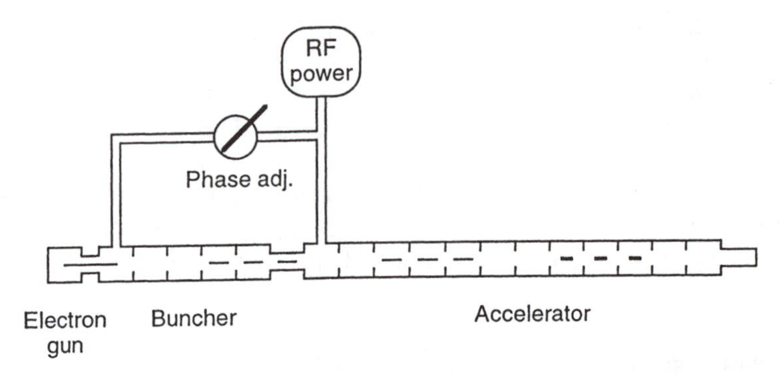

Figure 4. A linear accelerator. This example generates a macropulse consisting of individual bunches. For a single-bunch pulse, a series of subharmonic prebunchers is located between the gun and the buncher to progressively narrow the pulse emitted from the gun, so that it can be inserted into the accelerator in a single RF cycle.

tuned to the frequency of the RF radiation used for acceleration (Figure 4). A several-microsecond pulse of RF power (typically a 5–20 MW peak) is introduced at the "upstream" end of the linac, and it travels down the cavity structure and out into an absorber. Linac sections can be connected sequentially to provide further acceleration. The Stanford Linear Accelerator Center linac, over two miles long, accelerates electrons and positrons to 50 GeV using 240 linac sections (*6*).

Because the electric field of the RF radiation oscillates, particles are only accelerated forward in a limited region of the 2π RF cycle. Particles that enter the linac at the wrong phase will be scattered or propelled backward. To properly inject the particles, they must be collected and pre-accelerated. Linacs use the same types of cathodes and ion sources found in Van de Graaff accelerators. These sources are essentially continuous with respect to the RF frequency. The emitted electrons or ions pass through a series of RF cavities of progressively higher frequency (the prebuncher and buncher), which divide the continuous beam into bunches, accelerate them, and synchronize their injection into the linac at the proper point in the RF cycle. Depending on the design of the bunching system, the entire electron pulse may be grouped into a single RF cycle or spread over several cycles in the form of a macropulse. Such macropulses are highly structured: the pulse width of a single electron bunch is 15 to 30 ps full-width at half-maximum (FWHM), while, the bunches are on the order of a nanosecond apart depending on the RF frequency.

Advantages and Disadvantages. The charge that can be transported in a single bunch is limited by space-charge, mutual repulsion by like-charged particles, but still a very usable charge can be accelerated in one bunch (up to 30 nanocoulombs (nC), depositing 6 krad or 36 μM of OH radicals in N_2O-saturated water).

One major advantage of linacs that operate in a single-bunch mode is the short pulse width, typically 30 ps FWHM, which enables the investigation of very fast reactions. A magnetic pulse compression system installed on the 20-MeV electron linac beamline at Argonne National Laboratory generates 5-ps pulses (7).

As mentioned earlier, linear accelerators are capable of accelerating particles to much higher energies than electrostatic machines. This is beneficial in terms of penetrating power. Electrons in the 8- to 20-MeV range are capable of penetrating ovens, high-pressure cells, or Dewar flasks with energy to spare, resulting in uniform dose distribution and making possible the study of reactions under a wide variety of conditions. Higher particle energies make beam transport easier, allowing the installation of multiple semipermanent experimental stations. One drawback of higher particle energies, however, is the possibility of nuclear activation of materials (to form radioactive isotopes) by bremsstrahlung-induced (γ,n) reactions, in which gamma absorption results in the loss of a neutron. Most of the (γ,n) thresholds of the stable isotopes found in common laboratory materials are above 10 MeV. Activation problems at 20 MeV are still manageable because the cross sections are generally small and the half-lives of the isotopes produced are fairly short.

The high-power pulsed RF system used to generate power for the linac is a potential source of noise in experimental instrumentation. Linac equipment is much more elaborate than the average Van de Graaff or Febetron installation, usually requiring one or more full-time technicians for operation. In contrast, electrostatic accelerators are often operated by experimenters themselves. Generally speaking, operating costs for linacs are consequently higher in equipment and personnel.

RF Photocathode Electron Guns

The RF photocathode electron gun is the newest type of accelerator used for pulse radiolysis. Such devices have been under development since the mid-1980s as electron beam sources for experimental physics facilities and free-electron laser development. They are typically used to produce electron beams in the 4–10-MeV range. The unique quality (low emittance and clean position–momentum relationships) of the electron beams they produce makes extremely sophisticated beam manipulation possible.

Like the linear accelerator, a pulsed high-power RF system provides the acceleration, but the cathode is combined with the resonant cavity accelerating section in one structure (Figure 5). Electron guns are very compact—on the order of 20 to 40 cm long for 5 and 10 MeV guns, respectively. Ten to twenty megawatts of microwave power is used to generate a powerful oscillating electric field within the cavities. The instantaneous field gradient at the cathode

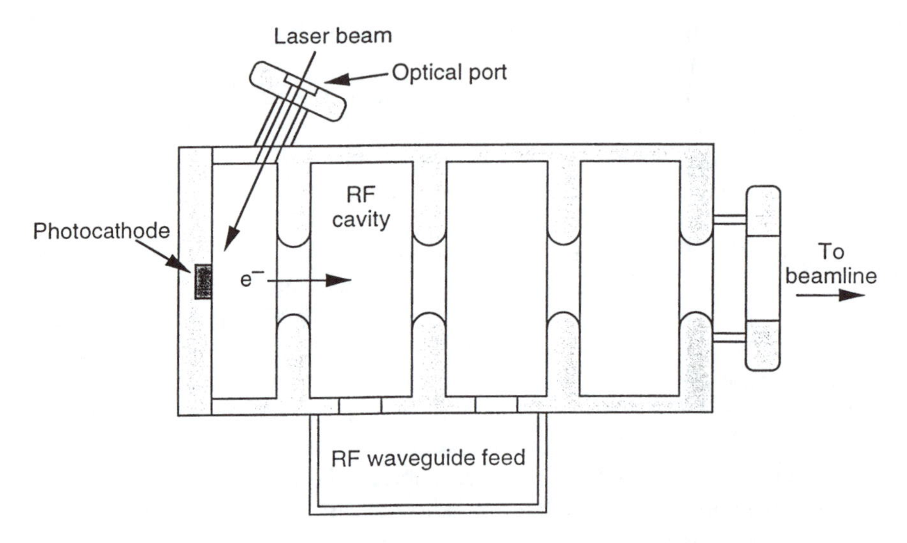

Figure 5. A radio-frequency photocathode electron gun. The overall length is about 30.5 cm (12 in.).

can be as high as 80 to 100 MV/m. To obtain an accelerated beam with optimal characteristics, the launch phase of the electrons from the cathode must be controlled with respect to the RF cycle so that the electrons are accelerated by that tremendous gradient. This is achieved by using a synchronized laser to eject photoelectrons from the cathode at the proper point in time (*8*). Although many types of exotic, highly quantum efficient materials have been tried, simple copper or magnesium metal photocathodes have proven most reliable for everyday use. Modern chirped-pulse-amplified solid state laser systems can provide the power to compensate for their relative inefficiency.

One key feature of the photocathode electron gun is the fact that the temporal and spatial characteristics of the electron pulse are to some extent determined by the incident laser pulse. A 5-ps laser pulse and a well-designed beam transport system will maintain the 5-ps pulse width at the target. Electron gun systems are capable of studying extremely fast reactions.

The availability of reliable femtosecond laser systems makes subpicosecond electron pulses possible (*9*). As a practical matter, however, such pulses would be of limited use for pulse radiolysis because of the so-called "picosecond barrier." Electrons travel through materials at the speed of light in a vacuum, while light itself is slowed by a factor equal to the reciprocal of its refractive index in that medium. The result is that the farther the electron and laser beams travel, the more they get out of synchronization. The light beam samples later and later moments in time relative to the electron-beam passage, resulting in lost temporal resolution. In order to avoid the problem, the interaction region (sample depth) must be limited. For 30-ps pulses, the interaction region is approximately 10 mm. For a 1-ps pulse, the interaction region is only 0.3 mm. As the interaction region decreases, the observable signal is also decreased. Sensitivity improvements and signal averaging can compensate for this decrease at the expense of additional sample consumption.

Advantages. As just discussed, RF photocathode electron guns can deliver very short electron pulses. For physics experiments, the charge-per-pulse is usually less than 1 nC in order to avoid space-charge effects that spoil the characteristics of the beam. In the case of an electron gun designed expressly for pulse radiolysis, such as the one operated by the Chemistry Department of Brookhaven National Laboratory, the charge can be as high as 10–20 nC (4 krad or 24 μM OH radicals in N_2O-saturated water). This is still only a moderate amount of charge; however, because it is delivered in 5–30 ps it corresponds to a high peak current (660–2000 A). The 5–10-MeV beam energy delivered by RF electron guns is within the ideal range, providing sufficient penetrating power without activation problems.

A particular feature of the RF photocathode electron gun is the picosecond-synchronized laser system needed to generate the photoelectrons. Part of the laser's output can be diverted to generate a visible light continuum for time-resolved kinetics or used for combined photolysis–radiolysis experiments. For example, the photochemistry of radiation-induced transients can be studied.

Disadvantages. RF electron guns share many of the disadvantages of linear accelerators, namely, the potential for noise emitted from the pulsed RF system, complexity of design, and the need for technical support staff. In addition, the system requires a reliable, amplified, ultrafast laser system in order to generate electrons, adding another level of complexity. At the time of this writing, the first electron-gun installations built for pulse radiolysis are beginning to come on-line. A track record on performance and reliability has yet to be established.

Beyond Electrons

Although electrons are by far the most common particles used in pulse radiolysis, other ions have significant applications (*10*). Protons and other heavy ions have high-LET values and therefore limited penetration power. The particle tracks are dense with energy deposition interactions, and cross reactions between radical species are highly probable. High-LET radiations provide good tests for theoretical models of energy deposition and reaction schemes of radiolysis products.

Early studies of high-LET radiation were performed using alpha particles from the decay of ^{210}Po (5.3 MeV). Later, ion sources were installed in Van

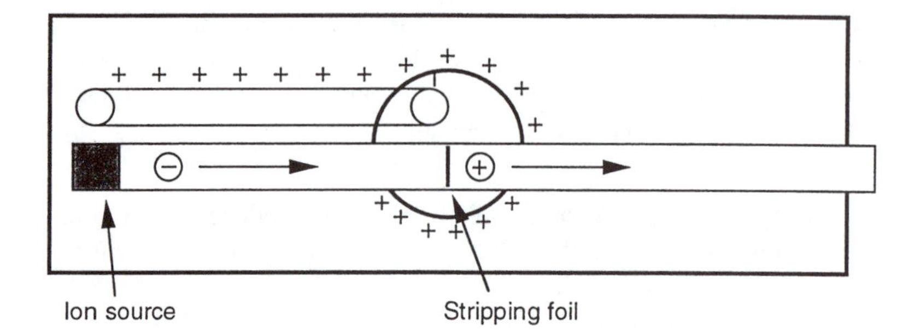

Figure 6. A tandem Van de Graaff accelerator.

de Graaff generators and linear accelerators as those technologies developed. The current supplied by an ion source is usually far less than that supplied by cathodes. To compensate for high LET, the particle energies are higher than typically used for electrons.

Tandem Van de Graaff accelerators (Figure 6) are used to generate a wide variety of ion beams. They often serve as the primary ion source for high-energy particle colliders. The tandem Van de Graaff consists of a beam tube, which extends from the ground terminal, through the high potential terminal, and back to ground. Negative ions, such as hydride (H^-), are emitted by the ion source and accelerated toward the positively charged terminal. When they reach the terminal, they strike a metal foil, which strips away the electrons. The resulting bare proton is accelerated away from the positive terminal toward the other ground terminal. The kinetic energy of the proton leaving the accelerator is twice the potential of the machine.

Cyclotrons are also used as heavy-ion sources. In a cyclotron, ions are confined to circular or spiral orbits by a dipolar magnetic field. The ions are accelerated by an alternating electric field projected by a series of plates above and below the orbital plane of the ions. In the simplest case, there are two sets of half-circular plates called "dees", which alternate in polarity to accelerate the ions.

Conclusions

In designing pulse radiolysis experiments, the fundamental physics and chemistry of radiolysis must be kept in mind. The properties of the accelerators and other radiation sources described in this chapter vary over a wide range. A certain kind of experiment may call for a particular type of accelerator, whereas others may be accomplished by using almost any equipment. Equally important, of course, is the variety of experimental detection systems that are available at accelerator facilities. Some of the available methods include dc or microwave conductivity, single- or multiple-wavelength UV–visible–near-infrared absorption spectroscopy, electron spin resonance, optically detected and other magnetic resonance techniques, and electrochemistry. Certain laboratories specialize in a given method, and certain accelerators are best matched to some techniques and not others, as noted in the preceding discussion.

An attempt to catalog all of the pulse radiolysis laboratories in the United States and abroad and their experimental capabilities is doomed to be incomplete from the start and soon rendered out of date. Moreover, access to the instrumentation depends on the programmatic aim of each institution. Many facilities welcome collaborative work, whereas some are oriented toward a specific program. The introductory chapters of this book may serve as a basis for discussion of experimental design, while the later chapters provide examples of particular applications.

Acknowledgments

I thank Harold Schwarz for eight years of patient tutelage on the subjects of accelerators and radiation chemistry. I also thank Norman Sutin, Carol Creutz, and Robert Marianelli for their unwavering support of the RF electron gun project at Brookhaven National Laboratory. This work was performed at Brookhaven National Laboratory under contract DE–AC02–76CH00016 with the U.S. Department of Energy and was supported by its Division of Chemical Sciences, Office of Basic Energy Sciences.

References

1. Zhao, Z.; Liu, A.; Tong, Z.; Song, Y.; Hu, H.; Sun, W.; Gu, H.; Zhou, R. *Radiat. Phys. Chem.* **1988,** *31,* 319–325.
2. (a) Buxton, G. V.; Greenstock C. L.; Helman, W. P.; Ross, A. B. *J. Phys. Chem. Ref. Data* **1988,** *17,* 513–886, *see* page 519. (b) Espenson, J. H. *Chemical Kinetics and Reaction Mechanisms;* McGraw-Hill: New York, 1981; pp 57–59. (c) Eigen, M.; Johnson, J. S. *Ann. Rev. Phys. Chem.* **1960,** *11,* 307.
3. Swanson, W. P. *Radiological Safety Aspects of the Operation of Electron Linear Accelerators;* IAEA Technical Report 188; International Atomic Energy Agency: Vienna, Austria, 1979.
4. (a) Van de Graaff, R. J. *Phys. Rev.* **1931,** *38,* 1919. (b) Van de Graaff, R. J.; Trump, J. G.; Buechner, W. W. *Rep. Prog. Phys.* **1948,** *11,* 1.
5. Luthjens, L. H.; Vermeulen, M. J. W.; Horn, M. L. *Rev. Sci. Instrum.* **1980,** *51,* 1183.
6. The SLAC linac was upgraded from 20 GeV to 50 GeV in 1989.
7. (a) Mavrogenes, G.; Norem, J.; Simpson, J. *Proceedings of the 1986 Stanford Linear Accelerator Conference;* Stanford Linear Accelerator Center Report SLAC–0303; Stanford Linear Accelerator Center: Stanford CA, pp 429–430. (b) Cox, G. L.; Ficht, D. W.; Jonah, C. D.; Mavrogenes, G. S.; Sauer, M. C., Jr. *Proceedings of the 1989 IEEE Particle Accelerator Conference,* The Institute of Electrical and Electronics Engineers, Inc.: Piscataway, NJ, 1989; pp 912–914.

8. *Lasers for RF Guns: Proceedings;* Srinavasan-Rao, T., Ed.; Brookhaven National Laboratory Formal Publication 52435; Brookhaven National Laboratory: 1994.
9. Trains of ~120 fs FWHM electron pulses have even been generated by an electron gun with a (continuous) thermionic cathode by using an alpha magnet and the excellent phase space control of electron guns to compress the accelerated bunches: *see* Kung, P.; Lihn, H.; Wiedemann, H.; Bocek, D. *Phys Rev. Lett.* **1994,** *73,* 967–970.
10. (a) Schuler, R. H.; LaVerne, J. A. In *New Trends and Developments in Radiation Chemistry;* IAEA–TECDOC–527; International Atomic Energy Agency: Vienna, Austria, 1989; pp 69–88. (b) LaVerne, J. A. *Nucl. Inst. Meth. Phys. Res. B.* **1995,** 302.
11. Boag, J. W. In *Actions Chimiques et Biologiques des Radiations;* Haïssinsky, M., Ed.; Masson et Cie.: Paris, 1963; Vol. 6, p 5.

General References

Pulse Radiolysis; Ebert, M.; Keene, J. P.; Swallow, A. J.; Baxendale, J. H., Eds.; Academic: Orlando, FL, 1965.

Denaro, A. R.; Jayson, G. G. *Fundamentals of Radiation Chemistry;* Butterworth & Co.: London, 1972.

Radiation Chemistry: Principles and Applications; Farhataziz; Rodgers, M. A. J., Eds.; VCH: New York, 1987.

Humphries, S. *Principles of Charged Particle Acceleration:* John Wiley and Sons: New York, 1986.

Lasers for RF Guns: Proceedings; Srinavasan-Rao, T., Ed.; Brookhaven National Laboratory Formal Publication 52435; Brookhaven National Laboratory: 1994.

4

Electron Tunneling in Engineered Proteins

Gary A. Mines, Benjamin E. Ramirez, Harry B. Gray, and Jay R. Winkler

Beckman Institute, California Institute of Technology, Pasadena, CA 91125

Semiclassical theory predicts that the rates of electron transfer (ET) reactions depend on the reaction driving force ($-\Delta G°$), a nuclear reorganization parameter (λ), and the electronic-coupling strength (H_{AB}) between reactants and products at the transition state. ET rates reach their maximum values (k_{ET}°) when the nuclear factor is optimized ($-\Delta G° = \lambda$); these k_{ET}° values are limited only by the strength (H_{AB}^2) of the electronic interaction between the donor (D) and acceptor (A). The dependence of the rates of Ru(His33)cytochrome c *ET reactions on $-\Delta G°$(0.59–1.4 eV) accords closely with semiclassical predictions. The anomalously high rates of highly exergonic ($-\Delta G° \geq 1.4$ eV) ET reactions suggest initial formation of an electronically excited ferroheme in these cases. Coupling-limited Cu^+ to Ru^{3+} and Fe^{2+} to Ru^{3+} ET rates for several Ru-modified proteins are in good agreement with the predictions of a tunneling-pathway model. In azurin, a blue copper protein, the distant D–A pairs are relatively well coupled (k_{ET}° decreases exponentially with Cu–Ru distance; the decay constant is 1.1 Å^{-1}). In contrast to the extended peptides found in azurin and other β-sheet proteins, helical structures have torturous covalent pathways owing to the curvature of the peptide backbone. The decay constants estimated from ET rates for D–A pairs separated by long sections of α helix in myoglobin and the photosynthetic reaction center are between 1.25 and 1.6 Å^{-1}.*

Electron tunneling in proteins occurs in reactions where the electronic interaction between redox sites is relatively weak (*1–5*). Under these circumstances, the transition state for the reaction must be formed many times before there is a successful conversion from reactants to products; the process is electronically nonadiabatic.

$$k_{ET} = \sqrt{\frac{4\pi^3}{h^2\lambda k_B T}}\, H_{AB}^2 \exp\left(-\frac{(\Delta G° + \lambda)^2}{4\lambda k_B T}\right) \qquad (1)$$

where k_{ET} is the electron transfer (ET) rate, h is Planck's constant, k_B is Boltzmann's constant and T is temperature. Semiclassical theory (equation 1) (*6*) predicts that the reaction rate for ET from a donor (D) to an acceptor (A) at fixed separation and orientation depends on the reaction driving force ($-\Delta G°$), a nuclear reorganization parameter (λ), and the electronic-coupling strength (H_{AB}) between reactants and products at the transition state. This theory reduces a complex dynamical problem in multidimensional nuclear-configuration space to a simple expression comprised of just two parameters (λ, H_{AB}). Equation 1 naturally partitions into nuclear (exponential) and electronic (pre-exponential) terms: ET rates reach their maximum values (k°_{ET}) when the nuclear factor is optimized ($-\Delta G° = \lambda$); these k°_{ET} values are limited only by the strength (H^2_{AB}) of the electronic interaction between the donor and acceptor (*7*). When donors and acceptors are separated by long distances (>10 Å), the D–A interaction will be quite small.

The Inverted ET Region

In the region of driving forces greater than λ (the inverted region), ET rates are predicted to decrease with increasing driving force (the inverted effect). Experimental verification of the inverted effect has come from extensive investigations of ET reactions involving both organic (*8–10*) and inorganic (*11–14*) molecules. Some work on biological molecules has been done (*15–20*), including a recent study (*21*) from our laboratory that involved a driving-force range sufficiently wide to probe behavior far in the inverted region. In measurements of the rates of cytochrome c ET reactions whose driving forces varied from 0.54 to 1.89 eV (Table I), inverted behavior was observed; however, at the

Table I. Rate Constants and Driving Forces for Intramolecular ET in RuL_2(X)(His33) Cytochrome c

Complex	*Reaction*	k_{ET} (s^{-1})	$-\Delta G°$ (eV)[a]	
(I) Ru(4,4′,5,5′-$(CH_3)_4$-bpy)$_2$(im)(His)$^{2+}$	$Fe^{2+} \rightarrow Ru^{3+}$	$1.6(2) \times 10^6$	0.54	
(II) Ru(4,4′-$(CH_3)_2$-bpy)$_2$(im)(His)$^{2+}$	$Fe^{2+} \rightarrow Ru^{3+}$	$2.0(2) \times 10^6$	0.70	
(III) Ru(phen)$_2$(im)(His)$^{2+}$	$Fe^{2+} \rightarrow Ru^{3+}$	$3.5(4) \times 10^6$	0.75	
(IV) Ru(phen)$_2$(CN)(His)$^{+}$	$Fe^{2+} \rightarrow Ru^{3+}$	$1.0(1) \times 10^7$	0.78	
(V) Ru(bpy)$_2$(im)(His)$^{2+}$	$Fe^{2+} \rightarrow Ru^{3+}$	$2.6(3) \times 10^6$	0.81	
(VI) Ru(4,4′-(CONH(C_2H_5))$_2$-bpy)$_2$(im)(His)$^{2+}$	$Fe^{2+} \rightarrow Ru^{3+}$	$1.1(1) \times 10^6$	1.00	
(V) Ru(bpy)$_2$(im)(His)$^{2+}$	$^{*}Ru^{2+} \rightarrow Fe^{3+}$	$2.0(5) \times 10^5$	1.3	
(IV) Ru(phen)$_2$(CN)(His)$^{+}$	$^{*}Ru^{2+} \rightarrow Fe^{3+}$	$2.0(5) \times 10^5$	1.4	0.35[b]
(VI) Ru(4,4′-(CONH(C_2H_5))$_2$-bpy)$_2$(im)(His)$^{2+}$	$Ru^{+} \rightarrow Fe^{3+}$	$2.3(2) \times 10^5$	1.44	0.39[b]
(IV) Ru(phen)$_2$(CN)(His)$^{+}$	$Ru^{+} \rightarrow Fe^{3+}$	$4.5(5) \times 10^5$	1.89	0.84[b]

[a] $E°$[cyt c($Fe^{3+/2+}$)] = 0.26 V vs. NHE; $E°$($Ru^{3+/2+}$)[II, V] = 0.96, 1.07 V (pH 7, phosphate); $E°$[$Ru^{3+/2+}L_2$(X)(im)][I, III, IV, VI] = 0.80, 1.01, 1.04, 1.26 V (pH 7, phosphate); E_{00}($^{*}Ru^{2+}$)[V] = 2.1 eV (pH 7, phosphate); E_{00}[$^{*}Ru^{2+}$(phen)$_2$(CN)(im)] = 2.2 eV (pH 7, phosphate); $E°$[$Ru^{2+/+}$4,4′-(CONH$(C_2H_5)_2$-bpy)$_2$(im)$_2$] = −1.18 V (acetonitrile); $E°$[$Ru^{2+/+}$(phen)$_2$(CN)(im)] = −1.63 V (acetonitrile). Errors in $E°$ values are ≤ ±0.03 V.

[b] Assuming formation of the ferroheme ^{3}MLCT excited state.

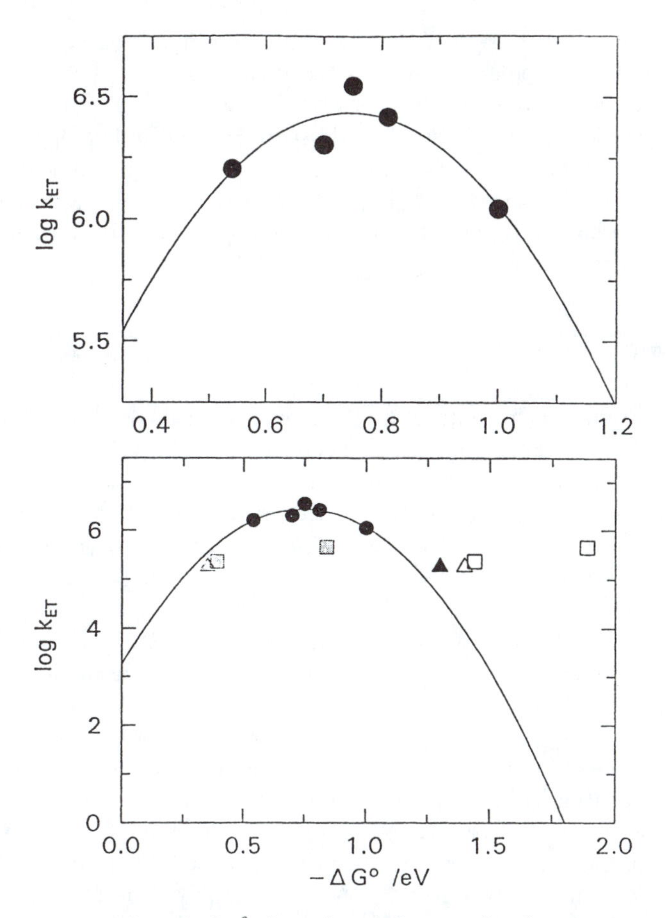

Figure 1. Driving-force ($-\Delta G^\circ$) dependence of intramolecular ET rate constants in Ru(His33)cyt c (Table I). Top: $Fe^{2+} \rightarrow Ru^{3+}$ ET in RuL_2(im)(His33)cyt c (im = imidazole). The curve represents the best fit to equation 1 (H_{AB} = 0.095 cm^{-1}; λ = 0.74 eV). Bottom: Replot of the above $k_{ET}/-\Delta G^\circ$ curve with the addition of $Ru^{+} \rightarrow Fe^{3+}$ (squares) and $^*Ru^{2+} \rightarrow Fe^{3+}$ (triangles) data. The open symbols represent highly exergonic reactions to ground-state products; the gray symbols represent the reaction channel involving formation of the ferroheme metal-to-ligand charge transfer (^{3}MLCT) excited state (~1.05 eV).

highest driving forces, the ET rates are much faster than expected (Figure 1). The leveling of ET rates at high driving forces was attributed to the formation of a ferroheme excited state (~1.05 eV) with a faster rate than the (highly inverted) reaction to give ground-state products.

The phenomenon of rate–energy leveling is common for photoinduced charge separation (22); most examples of inverted behavior involve recombination reactions (23). Invoking the formation of excited-state products is one explanation of rate leveling (*14, 22, 24–26*); photoinduced charge separation generally produces open-shell species (radicals) possessing low-lying excited states, whereas recombination reactions yield closed-shell products (*14*). A key role played by electronic structure is underscored by our finding that a relatively low-lying excited state of a closed-shell product can open a noninverted decay channel deep in the inverted region—the region in which thermal (energy-wasting) recombinations of photogenerated charge-separated states are usually inhibited.

Electronic Coupling

The D–A distance decay of protein ET rate constants depends on the capacity of the polypeptide matrix to mediate electronic couplings. In a seminal paper in 1992, Dutton and co-workers showed (*27*) that an exponential distance-decay constant (1.4 $Å^{-1}$), as originally proposed by Hopfield (*28*), could be used to estimate long-range ET rates in the bacterial photosynthetic reaction center (RC). Although Dutton's rate–distance correlation gives a rough indication of RC coupling strengths (*27, 29*), it seems clear from extensive theoretical and experimental work that the intervening polypeptide structure must be taken into account in attempts to understand distant D–A couplings in other proteins (*1–5, 15, 30–44*).

The medium separating redox sites in a protein are comprised of a complex array of bonded and nonbonded contacts, and ab initio calculation of coupling strengths is a formidable challenge. Beratan, Onuchic, and co-workers developed a generalized superexchange coupling model that accommodates the structural complexity of a protein matrix (*30–34*). In this tunneling-pathway model, the medium between D and A is decomposed into smaller subunits linked by covalent bonds, hydrogen bonds, and through-space jumps. Each link is assigned a coupling decay (ϵ_C, ϵ_H, ϵ_S), and a structure-dependent searching algorithm is used to identify the optimum coupling pathway between the two redox sites. The total coupling of a single pathway is given as a repeated product of the couplings for the individual links (equation 2). A tunneling pathway can be described in terms of an effective covalent tunneling path comprised of n (nonintegral) covalent bonds, with a total length equal to σ_1 (equation 3b).

$$H_{AB} \propto \prod \epsilon_C \prod \epsilon_H \prod \epsilon_S \quad (2)$$

$$H_{AB} \propto (\epsilon_C)n \quad (3a)$$

$$\sigma_1 = n \times 1.4 \text{ Å/bond} \quad (3b)$$

The coupling efficiency for a given tunneling pathway is defined by the ratio of σ_1 to the direct D–A distance, R (*2*). The theoretical minimum value for this ratio is 1, but a more realistic value is 1.2, corresponding to a stretched hydrocarbon bridge, the most efficient σ-tunneling structure. Inefficient pathways will have large values of σ_1/R. For a given structural type, a linear σ_1/R relationship implies that k°_{ET} will be an exponential function of R; the distance–decay constant is determined by the slope of the σ_1/R plot and the value of ϵ_C.

Employing the tunneling-pathway model, Beratan, Betts, and Onuchic (*34*) predicted in 1991 that proteins comprised largely of β-sheet structures would be more effective at mediating long-range couplings than those built from α helices. A β sheet is comprised of extended polypeptide chains interconnected by hydrogen bonds; the individual strands of β sheets define nearly linear coupling pathways spanning 3.4 Å per residue along the peptide backbone. The tunneling length for a β strand exhibits an excellent linear correlation with β-carbon separation (R_β, Figure 2); the best linear fit with zero intercept yields a slope of 1.37 σ_1/R_β (distance–decay constant = 1.0 Å^{-1}) (*2*). Couplings across a β sheet depend upon the ability of hydrogen bonds to mediate the D–A interaction. The standard parameterization of the tunneling-pathway model defines the coupling decay across a hydrogen bond in terms of the heteroatom separation. If the two heteroatoms are separated by twice the 1.4-Å covalent-bond distance, then the hydrogen-bond decay is assigned a value equal to that of

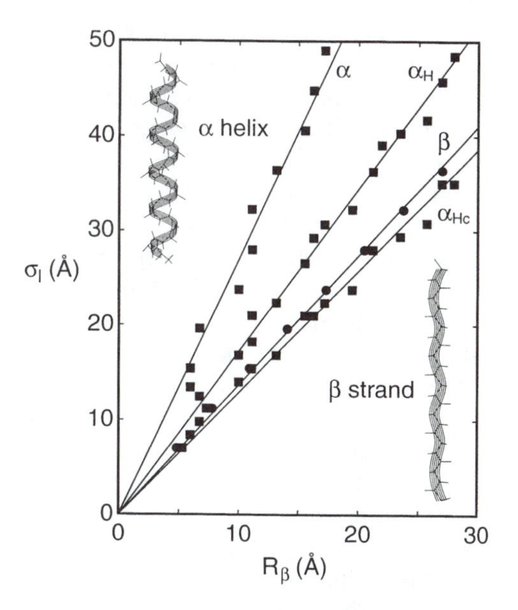

Figure 2. Variation of σ_1 with R_β for β strands (●) and α helices (■) with three different treatments of hydrogen-bond couplings in the helices. (Reproduced with permission from reference 5. Copyright 1996.)

a covalent bond (*32*). Longer heteroatom separations lead to weaker predicted couplings but, as yet, there is no experimental confirmation of this relationship.

In the coiled α-helix structure, a linear distance of just 1.5 Å is spanned per residue. In the absence of mediation by hydrogen bonds, σ_l is a very steep function of R_β, implying that an α helix is a poor conductor of electronic coupling (2.7 σ_l/R_β, distance–decay constant = 1.97 $Å^{-1}$, Figure 2) (*2*). If the hydrogen-bond networks in α helices mediate coupling, then the Beratan–Onuchic parameterization of hydrogen-bond interactions suggests a σ_l/R_β ratio of 1.72 (distance–decay constant = 1.26 $Å^{-1}$, Figure 2) (*3*). Treating hydrogen bonds as covalent bonds further reduces this ratio (1.29 σ_l/R_β, distance–decay constant = 0.94 $Å^{-1}$, Figure 2) (*3*). Hydrogen-bond interactions, then, will determine whether α helices are vastly inferior to or are slightly better than β sheets in mediating long-range ET reactions. It is important to note that the coiled helical structure leads to poorer σ_l/R_β correlations, especially for values of R_β under 10 Å. In this distance region, the tunneling-pathway model predicts little variation in coupling efficiencies for the different secondary structures (Figure 2). The coupling in helical structures could be highly anisotropic. ET along a helix may have a very different distance dependence from ET across helices. In the latter, the coupling efficiency depends on the nature of the interactions between helices. A final point involves the dependence of coupling efficiencies on bond angles. It is well known that β sheets and α helices are described by quite different peptide bond angles (ϕ, ω). Ab initio calculations on saturated hydrocarbons have suggested that different conformations provide different couplings (*45*). Different values of ϵ_c, then, might be necessary to describe couplings in β sheets and α helices.

We have measured the coupling along β strands in Ru-modified derivatives of azurin (*2, 3*). Five azurin mutants have been prepared with His residues at different sites on the strands extending from Met121 (His122, His124, His126) and Cys112 (His109, His107) (Figure 3); $Ru(bpy)_2(im)^{2+}$ (bpy = 2,2′-bipyridine; im = imidazole) has been coordinated to these surface His groups and intraprotein $Cu^+ \rightarrow Ru^{3+}$ ET rates have been measured using photochemical techniques (*2, 3*). The variation of k°_{ET} with direct metal–metal separation (R_M) is well described by an exponential function with a decay constant of 1.1 $Å^{-1}$ (Figure 4). This result is in remarkably good agreement with the slope predicted by the tunneling-pathway model for the coupling decay along a strand of an ideal β sheet. More sophisticated theoretical treatments of the Ru-modified azurins also have succeeded in describing the observed couplings (*46, 47*).

In contrast to the extended peptides found in β sheets, helical structures have torturous covalent pathways owing to the curvature of the peptide backbone. We have studied donor–acceptor pairs separated by α helices in two Ru-modified myoglobins (Mbs), $Ru(bpy)_2(im)$(HisX)-Mb (X = 83, 95) (*3*). The tunneling pathway from His95 to the Mb heme is comprised of a short section of α helix terminating at His93, the heme axial ligand. The coupling for the [$Fe^{2+} \rightarrow Ru^{3+}$(His95)]-Mb ET reaction (*3*) is of the same magnitude as that

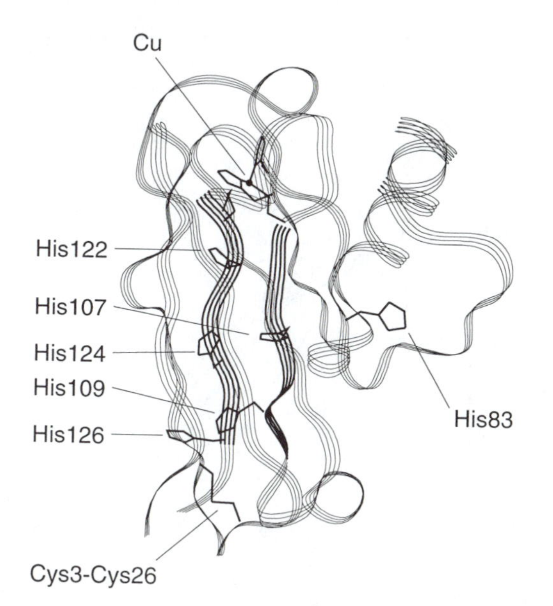

Figure 3. Ribbon structure of Pseudomonas aeruginosa *azurin showing the locations of His residues that have been introduced by site-directed mutagenesis. (Reproduced with permission from reference 5. Copyright 1996.)*

found in Ru-modified azurins with comparable D–A spacings. This result is consistent with the tunneling-pathway model, which predicts very little difference in the coupling efficiencies of α helices and β sheets at small D–A separations (Figure 3). The electronic coupling estimated from the [$Fe^{2+} \rightarrow Ru^{3+}$(His83)]-Mb ET rate, however, is substantially weaker than that found in β-sheet structures at similar separations. Two additional α-helix data points come from work on the bacterial photosynthetic reaction center. The quinones (Q_A, Q_B) and bacteriochlorophyll special pair (BCh_2) of reaction centers are separated by long sections of α helix. Rates of charge-recombination reactions from reduced quinones to the oxidized special pair have been determined (*27*); plots of log k°_{ET} versus R suggest a larger distance–decay constant for α helices (Figure 5). Differences in hydrogen bonding in β sheets and α helices may be responsible for this behavior. Infrared spectra in the amide I (υ_{CO}, CO stretch) region show that hydrogen bonding in α helices (υ_{CO} = 1650–1660 cm^{-1}) is significant (nonhydrogen-bonded peptides, υ_{CO} = 1680–1700 cm^{-1}) but is not as strong as that in β sheets (υ_{CO} ~ 1630 cm^{-1}) (*48, 49*). If spectroscopically derived hydrogen-bond strengths reflect electronic-coupling efficiencies, then long-range couplings at given distances along α helices will be weaker than those at corresponding distances along β strands.

Experimental evidence supports the tunneling-pathway prediction that different protein secondary structures mediate electronic couplings with different

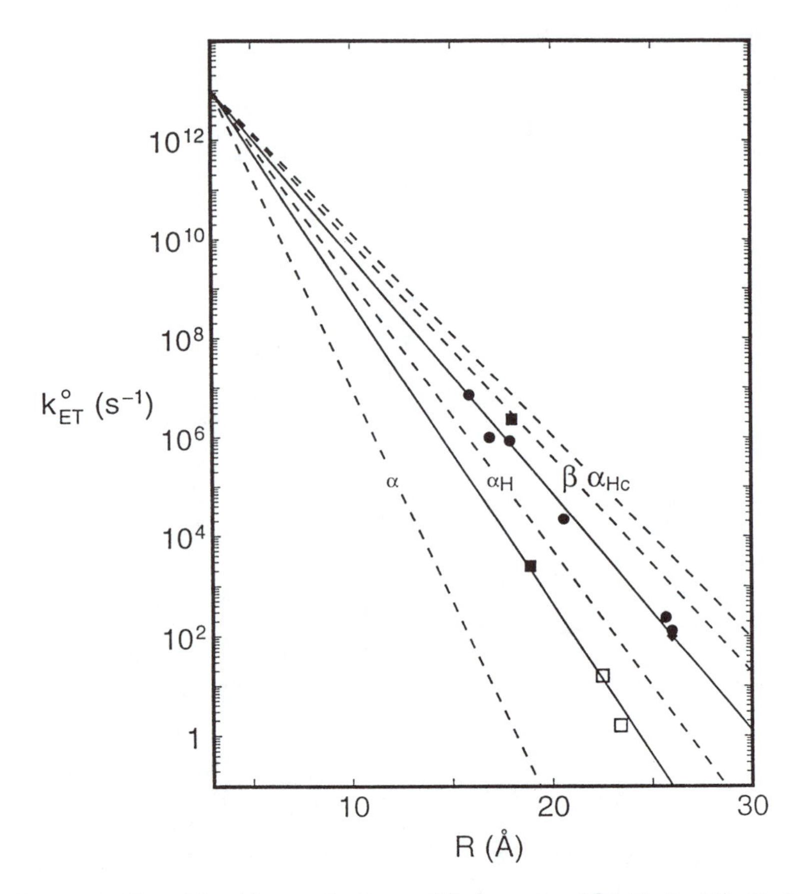

Figure 4. Plot of log k°_{ET} vs. R: Ru-modified azurins (●) (2, 3, 46)*; Cys3-Cys26(S_2^-) → Cu^{2+} ET in azurin (◆) (50, 51); Ru-modified Mb (●) (3); and the RC (□) (27). Dashed lines are distance decays predicted using the tunneling-pathway model for β strands and α helices. Solid lines are the best linear fits with an intercept at 13, and they correspond to distance decays of 1.1 Å^{-1} for azurin and 1.4 Å^{-1} for Mb and the RC. (Reproduced with permission from reference 5. Copyright 1996.)*

efficiencies (*2, 3, 50, 51*). We can define different ET coupling zones in a rate versus distance plot (Figure 5). The β zone, representing efficient mediation of electronic coupling, is bound by coupling-decay constants of 0.9 and 1.15 Å^{-1}. We call this the β zone because the tunneling-pathway model predicts that ET rates in β-sheet proteins will fall in this region. All of the ET rates measured with Ru-modified azurins fall in this zone. The α (or α-helix) zone describes systems with coupling-decay constants between 1.25 and 1.6 Å^{-1}. ET rates from Ru(His83)-modified myoglobin and the two RC Q-BCh_2 pairs lie in this zone. ET rate data are available for a Ru-modified myoglobin (His70)

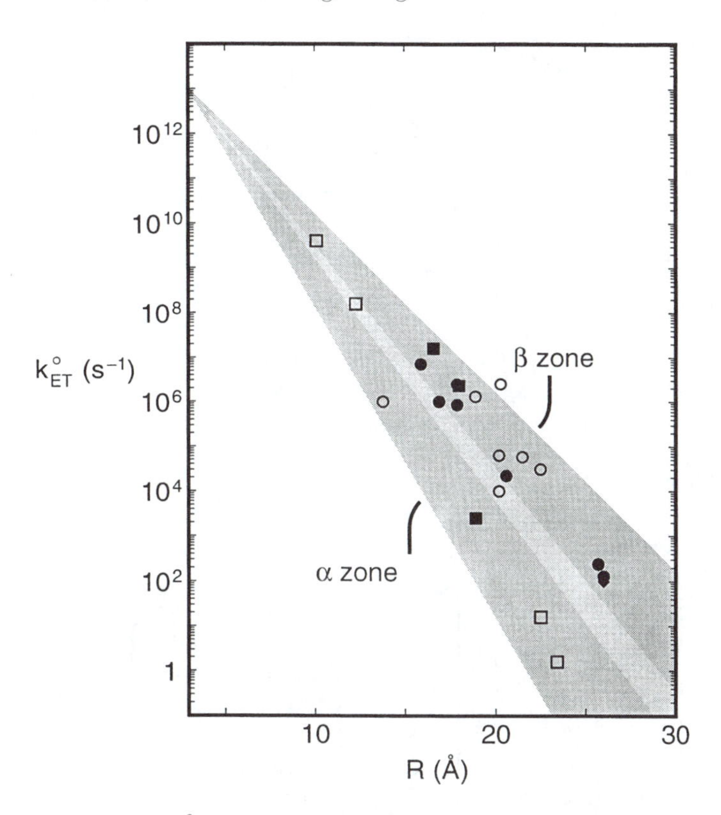

Figure 5. Plot of log k°_{ET} *vs.* R, *illustrating the different ET coupling zones. Zones are bounded by the following distance–decay lines: α zone, 1.25 and 1.6* $Å^{-1}$; *β zone, 1.15 and 0.9* $Å^{-1}$. *The light shaded region is the interface between the α and β zones. For Ru-bpy-modified proteins, metal–metal separation distances are used. Distances between redox sites in the RC are reported as edge–edge separations. Ru-modified azurin data (●) (2, 3, 46); [Ru-label site,* k°_{ET} s^{-1}, *R Å] His122,* 7.1×10^6, *15.9; His124,* 2.2×10^4, *20.6; His126,* 1.3×10^2, *26.0; His109,* 8.5×10^5, *17.9; His107,* 2.4×10^2, *25.7; His83,* 1.0×10^6, *16.9. Ru-modified myoglobin data (■) (3); His83,* 2.5×10^3, *18.9; His95,* 2.3×10^6, *18.0; His70,* 1.6×10^7, *16.6. Ru-modified cytochrome* c *data (○) (2); His39,* 3.3×10^6, *20.3; His33,* 2.7×10^6, *17.9; His66,* 1.3×10^6, *18.9; His72,* 1.0×10^6, *13.8; His58,* 6.3×10^4, *20.2; His62,* 1.0×10^4, *20.2; His54,* 3.1×10^4, *22.5; His54(-Ile52),* 5.8×10^4, *21.5. Cys3-Cys26(*S_2^-*) →* Cu^{2+} *ET in azurin (◆) (50, 51);* 1.0×10^2, *26. RC data (□) (27); [donor to* BCh_2^+, k°_{ET} s^{-1}, *R Å]* Q_A^-, 1.6×10^1, *22.5;* Q_B^-, *1.6, 23.4;* BPh^-, 4.0×10^9, *10.1; cytcochrome* c_{559}, 1.6×10^8, *12.3. (Reproduced with permission from reference 5. Copyright 1996.)*

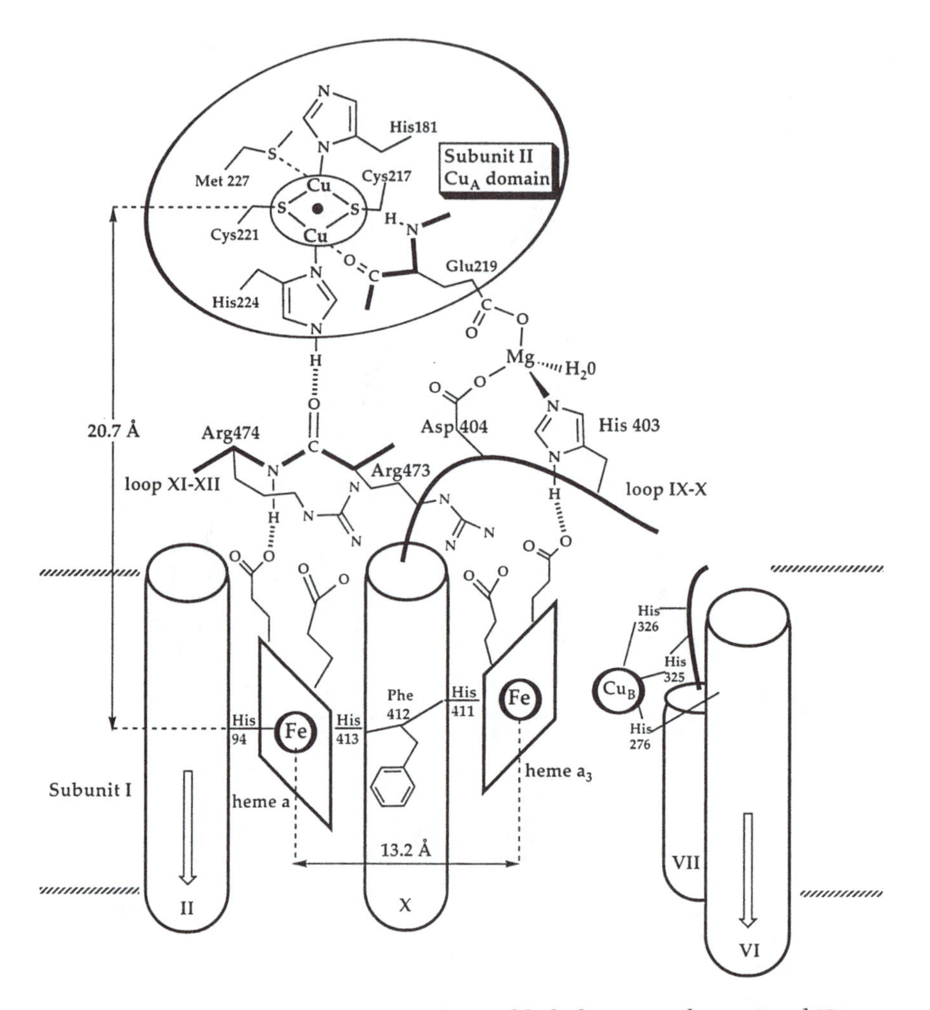

Figure 6. Schematic representation of several links between subunits I and II in cytochrome c *oxidase. The oval loop at the top represents the soluble or exposed domain of subunit II that contains the Cu_A center. The cylinders correspond to transmembrane α helices of subunit I (arrows indicate the direction of the peptide chain). The porphyrin rings of the hemes of cytochromes* a *and* a_3 *are drawn as squares with the propionate groups highlighted. Two loops (loop IX–X and loop XI–XII) connecting helices in subunit I are also shown. Hydrogen bonds to H—N and C═O of a peptide unit in loop XI–XII connect subunits I and II and form a good Cu_A to heme* a *electron transfer pathway from the imidazole of a histidine (His224) to one of the heme propionates. The electron transfer distance from the center of the Cu_A binuclear complex to the Fe in heme* a *is 20.7 Å. Two arginines (Arg473 and Arg474) form salt bridges with propionates of hemes* a_3 *and* a*; and a Mg complex is linked to both Cu_A and heme* a_3 *and could serve as a communicator between subunits I and II. The amino acid numbers refer to the* Paracoccus denitrificans *enzyme (53).*

where the intervening medium is not a simple section of α helix; the His70-Mb ET rate lies in the β-sheet zone (*3*). In the photosynthetic reaction center, two BCh_2^+ hole-filling reactions occur over relatively short distances where the differences between the β-sheet and α-helix zones are less distinct: the observed rates lie between the two zones (*27*).

The bond connections in multisubunit redox enzymes such as cytochrome c oxidase may play a key role in directing and regulating electron flow. Inspection of the structure of the oxidase reveals that ET from Cu_A (subunit II) to cytochrome a (subunit I) occurs over a 20.7-Å distance through a direct coupling pathway consisting of 14 covalent bonds and 2 intersubunit hydrogen bonds (Figure 6) (*52–54*). Based on the relative bond couplings extracted from work on Ru-modified proteins, the 20.7-Å Cu_A/cyt a ET rate falls in the efficient (β) coupling zone of Figure 4 (k°_{ET} between 4×10^4 and 8×10^5 s^{-1}). With these k°_{ET} values, the reorganization energy for Cu_A to cyt a ET must be between 0.15 and 0.5 eV (*54*). Apparently, the combination of a low reorganization energy and an efficient ET pathway allows electrons to flow rapidly with only a small change in free energy from the Cu_A center of subunit II to cytochrome a in subunit I of the oxidase.

Acknowledgments

Our work on electron transfer in proteins is supported by the National Science Foundation, the National Institutes of Health, and the Arnold and Mabel Beckman Foundation.

References

1. Wuttke, D. S.; Bjerrum, M. J.; Winkler, J. R.; Gray, H. B. *Science (Washington, D.C.)* **1992,** *256,* 1007–1009.
2. Langen, R.; Chang, I.-J.; Germanas, J. P.; Richards, J. H.; Winkler, J. R.; Gray, H. B. *Science (Washington, D.C.)* **1995,** *268,* 1733–1735.
3. Langen, R.; Colón, J. L.; Casimiro, D. R.; Karpishin, T. B.; Winkler, J. R.; Gray, H. B. *J. Biol. Inorg. Chem.* **1996,** *1,* 221–225.
4. Winkler, J. R.; Gray, H. B. *Chem. Rev.* **1992,** *92,* 369–379.
5. Gray, H. B.; Winkler, J. R. *Annu. Rev. Biochem.* **1996,** *65,* 537–561.
6. Marcus, R. A.; Sutin, N. *Biochim. Biophys. Acta,* **1985,** *811,* 265–322.
7. Newton, M. D. *J. Phys. Chem.* **1988,** *92,* 3049–3056.
8. Closs, G. L.; Miller, J. R. *Science (Washington, D.C.)* **1988,** *240,* 440–447.
9. Wasielewski, M. R.; Niemczyk, M. P.; Svec, W. A.; Pewitt, E. B. *J. Am. Chem. Soc.* **1985,** *107,* 1080–1082.
10. Gould, I. R.; Ege, D.; Mattes, S. L.; Farid, S. *J. Am. Chem. Soc.* **1987,** *109,* 3794–3796.
11. Fox, L. S.; Kozik, M.; Winkler, J. R.; Gray, H. B. *Science, (Washington, D.C.)* **1990,** *247,* 1069–1071.
12. Chen, P.; Duesing, R.; Tapolsky, G.; Meyer, T. J. *J. Am. Chem. Soc.* **1989,** *111,* 8305–8306.

13. Macqueen, D. B.; Schanze, K. S. *J. Am. Chem. Soc.* **1991,** *113,* 7470–7479.
14. McCleskey, T. M.; Winkler, J. R.; Gray, H. B. *J. Am. Chem. Soc.* **1992,** *114,* 6935–6937.
15. Bjerrum, M. J.; Casimiro, D. R.; Chang, I.-J.; Di Bilio, A. J.; Gray, H. B.; Hill, M. G.; Langen, R.; Mines, G. A.; Skov, L. K.; Winkler, J. R.; Wuttke, D. S. *J. Bioenerg. Biomembr.* **1995,** *27,* 295–302.
16. McLendon, G.; Hake, R. *Chem. Rev.* **1992,** *92,* 481–490.
17. Simmons, J.; McLendon, G.; Qiao, T. *J. Am. Chem. Soc.* **1993,** *115,* 4889–4890.
18. Scott, J. R.; Willie, A.; Mark, M.; Stayton, P. S.; Sligar, S. G.; Durham, B.; Millett, F. *J. Am. Chem. Soc.* **1993,** *115,* 6820–6824.
19. Brooks, H. B.; Davidson, V. L. *J. Am. Chem. Soc.* **1994,** *116,* 11201–11202.
20. Jia, Y. W.; Dimagno, T. J.; Chan, C. K.; Wang, Z. Y.; Du, M.; Hanson, D. K.; Schiffer, M.; Norris, J. R.; Fleming, G. R. *J. Phys. Chem.* **1993,** *97,* 13180–13191.
21. Mines, G. A.; Bjerrum, M. J.; Hill, M. G.; Casimiro, D. R.; Chang, I.-J.; Winkler, J. R.; Gray, H. B. *J. Am. Chem. Soc.* **1996,** *118,* 1961–1965.
22. Rehm, D.; Weller, A. *Isr. J. Chem.* **1970,** *8,* 259–271.
23. Suppan, P. *Top. Curr. Chem.* **1992,** *163,* 95–130.
24. Kikuchi, K.; Katagiri, T.; Niwa, T.; Takahashi, Y.; Suzuki, T.; Ikeda, H.; Miyashi, T. *Chem. Phys. Lett.* **1992,** *193,* 155–160.
25. Kikuchi, K.; Niwa, T.; Takahashi, Y.; Ikeda, H.; Miyashi, T. *J. Phys. Chem.* **1993,** *97,* 5070–5073.
26. Siders, P.; Marcus, R. A. *J. Am. Chem. Soc.* **1981,** *103,* 748–752.
27. Moser, C. C.; Keske, J. M.; Warncke, K.; Farid, R. S.; Dutton, P. L. *Nature (London)* **1992,** *355,* 796–802.
28. Hopfield, J. J. *Proc. Natl. Acad. Sci. U.S.A.* **1974,** *71,* 3640–3644.
29. Farid, R. S.; Moser, C. C.; Dutton, P. L. *Curr. Opin. Struct. Biol.* **1993,** *3,* 225–233.
30. Beratan, D. N.; Onuchic, J. N.; Hopfield, J. J. *J. Chem. Phys.* **1987,** *86,* 4488–4498.
31. Onuchic, J. N.; Beratan, D. N. *J. Chem. Phys.* **1990,** *92,* 722–733.
32. Onuchic, J. N.; Beratan, D. N.; Winkler, J. R.; Gray, H. B. *Annu. Rev. Biophys. Biomol. Struct.* **1992,** *21,* 349–377.
33. Beratan, D. N.; Betts, J. N.; Onuchic, J. N. *J. Phys. Chem.* **1992,** *96,* 2852–2855.
34. Beratan, D. N.; Betts, J. N.; Onuchic, J. N. *Science (Washington, D.C.)* **1991,** *252,* 1285–1288.
35. Skourtis, S. S.; Regan, J. J.; Onuchic, J. N. *J. Phys. Chem.* **1994,** *98,* 3379–3388.
36. Siddarth, P.; Marcus, R. A. *J. Phys. Chem.* **1990,** *94,* 8430–8434.
37. Siddarth, P.; Marcus, R. A. *J. Phys. Chem.* **1990,** *94,* 2985–2989.
38. Siddarth, P.; Marcus, R. A. *J. Phys. Chem.* **1992,** *96,* 3213–3217.
39. Siddarth, P.; Marcus, R. A. *J. Phys. Chem.* **1993,** *97,* 13078–13082.
40. Siddarth, P.; Marcus, R. A. *J. Phys. Chem.* **1993,** *97,* 2400–2405.
41. Gruschus, J. M.; Kuki, A. *Chem. Phys. Lett.* **1992,** *192,* 205–212.
42. Gruschus, J. M.; Kuki, A. *J. Phys. Chem.* **1993,** *97,* 5581–5593.
43. Friesner, R. A. *Structure* **1994,** *2,* 339–343.
44. Evenson, J. W.; Karplus, M. *Science (Washington, D.C.)* **1993,** *262,* 1247–1249.
45. Liang, C.; Newton, M. D. *J. Phys. Chem.* **1992,** *96,* 2855–2866.
46. Regan, J. J.; Di Bilio, A. J.; Langen, R.; Skov, L. K.; Winkler, J. R.; Gray, H. B.; Onuchic, J. N. *Chem. Biol.* **1995,** *2,* 489–496.
47. Gehlen, J. N.; Daizadeh, I.; Stuchebrukhov, A. A.; Marcus, R. A. *Inorg. Chim. Acta* **1996,** *243,* 271–282.
48. Schellman, J. A.; Schellman, C. In *The Proteins,* 2nd ed.; Neurath, H., Ed.; Academic: Orlando, FL, 1962; Vol. 2.
49. Susi, H. *Meth. Enzymol.* **1972,** *26,* 455–472.
50. Farver, O.; Skov, L. K.; Vandekamp, M.; Canters, G. W.; Pecht, I. *Eur. J. Biochem.* **1992,** *210,* 399–403.

51. Farver, O.; Pecht, I. *Biophys. Chem.* **1994,** *50,* 203–216.
52. Tsukihara, T.; Aoyama, H.; Yamashita, E.; Tomizaki, T.; Yamaguchi, H.; Shinzawa-Itoh, K.; Nakashima, R.; Yaono, R.; Yoshikawa, S. *Science (Washington, D. C.)* **1995,** *269,* 1071–1074.
53. Iwata, S.; Ostermeier, C.; Ludwig, B.; Michel, H. *Nature (London)* **1995,** *376,* 660–669.
54. Ramirez, B. E.; Malmström, B. G.; Winkler, J. R.; Gray, H. B. *Proc. Natl. Acad. Sci. U.S.A.* **1995,** *92,* 11949–11951.

5

Pulse Radiolysis: A Tool for Investigating Long-Range Electron Transfer in Proteins

I. Pecht[1] and O. Farver[2]

[1] **Department of Immunology, Weizmann Institute of Science, Rehovot 76100, Israel**
[2] **Department of Chemistry, Royal Danish School of Pharmacy, Copenhagen, Denmark**

One of the fundamental processes in biological energy-conversion systems is that of electron transfer. It takes place among and within proteins over considerable distances between active sites containing transition metal ions or organic cofactors and is generally characterized by relatively weak electronic interactions among them. Considerable research efforts have and are being invested in resolving the factors that control the rates of long-range electron transfer in proteins. Different experimental methods have been employed in these studies, ranging from fast mixing (stopped flow, rapid freeze EPR) and T-jump chemical relaxation to flash photolysis and pulse radiolysis. The latter two methods employ introduction of very short electromagnetic radiation pulses, absorbed by solutes in the former and by solvent in the latter. These in turn produce the electron donors or acceptors initiating the reaction of interest. The application of pulse radiolysis to studies of electron transfer within proteins is briefly reviewed to indicate its advantages. Results of its application to two types of copper-containing electron mediating proteins are presented and discussed.

Pulse radiolysis is a method first introduced in the 1960s, and it has found a broad range of important applications in chemistry and biochemistry, extending far beyond the scope of free radicals and radiation chemistry to which it was first applied (*1*). One application that is of considerable significance is the investigation of electron transfer processes in proteins. The method is based on the excitation and ionization of solvent molecules by short pulses of high-energy electrons. Thus, although pulse radiolysis is the electron analog of the flash photolysis method, photoexcitation of specific solutes rather than of the bulk solvent molecules distinguishes these two methods and provides the former with some concrete advantages.

Introducing high-energy electrons (e.g., 5–10 MeV) into dilute aqueous solutions of a given solute causes primary changes only in the solvent. Thus, water molecules undergo conversion into OH radicals and hydrated electrons, and to a lesser extent, H atoms, H_2, and H_2O_2 are also produced. (The yields are usually presented as *G* values, that is, the number of entities produced per 100 eV of absorbed energy: e^-_{aq} = 2.9, OH = 2.8, H = 0.55, H_2O_2 = 0.75, and H_2 = 0.45.) The hydrated electrons and OH radicals are exceptionally reactive and present thermodynamic extremes of reducing and oxidizing potentials, respectively. Hence, they provide the possibility of initiating a wide range of electron transfer processes.

As will be detailed later in this chapter, these highly reactive agents, though having their own applications, are usually converted to less aggressive agents via protocols devised by radiation chemists. This is illustrated by one useful procedure, that of converting the e^-_{aq} (with a reduction potential of –2.8 V) to a considerably milder reductant, the CO_2^- radical ($E°$ = –1.8 V). First the former is converted, in N_2O-saturated solutions, into an additional equivalent of OH radicals by the following reaction:

$$e^-_{aq} + N_2O \rightarrow N_2 + OH + OH^-$$

The two equivalents of OH radicals then react with formate ions, also present in the solution, to produce two equivalents of the CO_2^- radical:

$$HCO_2^- + OH \rightarrow H_2O + CO_2^-$$

By analogy, other reducing and oxidizing agents can be produced and employed (*2*). An additional, technically important advantage of pulse radiolysis over flash photolysis is noteworthy, namely, the whole range of optical spectrum is usually available for monitoring the induced reactions by the former method. This is the case because the reactive species is derived from the solvent rather than from the excitation of a given solute, as is the case in flash photolysis. Taken together, the wide range of chemical reactivity of the produced reagents, combined with a time resolution that usually extends from nanoseconds to minutes and the convenience of spectrophotometric monitoring of the reactions, has made pulse radiolysis the method of choice for investigation of a wide range of chemical processes.

The potential of pulse radiolysis for studying biological redox processes, particularly of macromolecules, was recognized rather early. It was initially employed for investigating radiation-induced damage and, later on, as an effective tool for resolving electron transfer processes to and within proteins. Cytochrome c, a well-characterized electron-mediating protein, was the first to be examined for its reactivity with hydrated electrons by this method. Two groups pioneered this study (*3, 4*) and found the reduction of the Fe^{3+} site to be a diffusion-controlled process (5×10^{10} $M^{-1}s^{-1}$). These two and several other groups (e.g., 5–7) then extended the investigation of this protein. Soon thereafter, other electron-mediating proteins with distinct redox centers like copper or nonheme iron (*8, 9*) were also examined for their reactivity with e^-_{aq}, yielding similar, rather high, diffusion-limited rate constants. These results illustrated the applicability of the method, yet at the same time they indicated the drawbacks of the hydrated electron as a reductant with excessive driving force and therefore limited specificity, which leads to additional side reactions. Hence, new studies increasingly employed milder reducing or oxidizing agents than the e^-_{aq} or OH radicals, respectively (cf. ref. 10).

Two main interests were guiding studies that applied pulse radiolysis to redox proteins: the pursuit of reaction mechanisms of these proteins, and the more general problem of resolving the parameters that determine specific rates of electron transfer within proteins (*10–12*). Obviously, these two motives overlap and complement each other, as we shall see. The fast progress attained in the last two decades in resolving three-dimensional structures of an increasing number of redox-active proteins provided the insights essential for a meaningful analysis and interpretation of the kinetic results.

Azurin as a Model System for Intramolecular Electron Transfer

Azurins are "blue" single-copper proteins that mediate electrons in the energy conversion systems of several bacterial strains (*13, 14*). Although azurins isolated from distinct bacteria are highly homologous in their sequences, differences do exist, conferring upon these proteins variation in reactivity and redox potentials (*13*). All azurins sequenced to date contain a disulfide bridge (Cys3–Cys26) at one end of their β-sandwich structure, 2.6 nm from the copper binding site present at the opposite end of the barrel-shaped protein (Figure 1) (*13, 14*). Using pulse-radiolytically produced CO_2^- radicals, this disulfide is reduced to the $RSSR^-$ radical, and this transient species was found to decay by an intramolecular electron transfer process to the copper(II) center (*15, 16*).

This process has been investigated in greater detail in both wild type (wt) and single-site mutated azurins (*15–19*). The former were isolated from different bacteria, exhibited a range of differences in their properties (*15, 16*), yet were less amenable to analysis than the latter. Hence, changes in specific parameters of the protein seen in single-site mutants of *Pseudomonas aeruginosa* (Pae) azurin constitute a main topic of these studies. For example, mutants that have redox potentials close to that of the wild type (wt) (E° = 304 mV) but differ in the residues proximal to the copper center were investigated. In one of these, the Met64 residue of the wild type protein is substituted by Glu

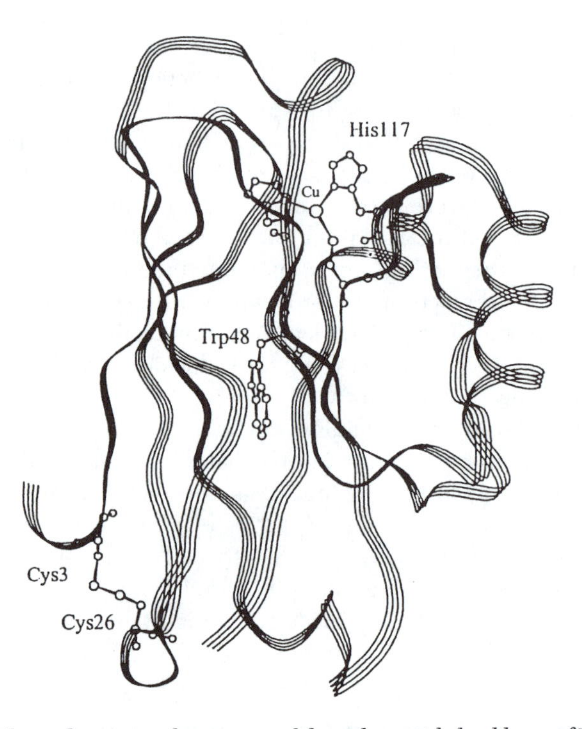

Figure 1. Three-dimensional structure of the polypeptide backbone of Pseudomonas aeruginosa *azurin, with some amino acid residues of particular interest included. Coordinates were obtained from reference 21.*

([M64E]azurin) with a Cu(II)–Cu(I) redox potential of 278 mV (*18*). In two other mutants, Ser has been introduced instead of either Ile7 ([I7S]azurin) or Phe110 ([F110S]azurin) with redox potentials at pH 7.0 of 301 and 314 mV, respectively (*17, 19*).

Pulse-radiolytic reductions are routinely performed in N_2O-saturated aqueous solutions containing 10 mM phosphate and 100 mM formate ions. Under these conditions, the primary products of water decomposition by the pulse of accelerated electrons are practically all converted into CO_2^- radical ions (*10*) that react rapidly with the two redox-active sites present in azurin: the Cu(II) ion and the disulfide bridge (*8, 15*). This is illustrated by absorption changes observed solutions of oxidized, single-site mutant (F100S) of Pae azurin following the pulse-radiolytically produced CO_2^- radicals (Figure 2). When the reaction is monitored at the main absorption band of the Cu(II) ($\varepsilon 625$ = 5700 M^{-1} cm^{-1}), an initial, relatively fast phase of decrease in absorption is noted. This decay was found to be a second-order process corresponding to a direct, diffusion-controlled bimolecular reduction of the Cu(II) center by CO_2^- radicals. However, as is illustrated by Figure 2a, a slower phase of Cu(II) reduction

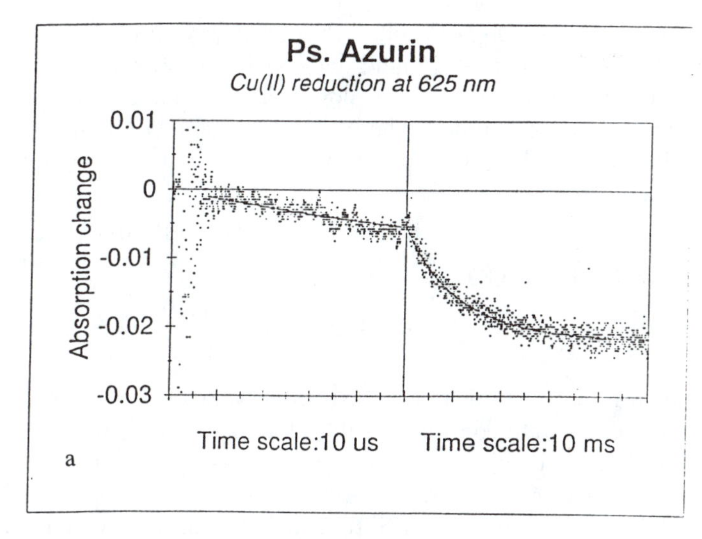

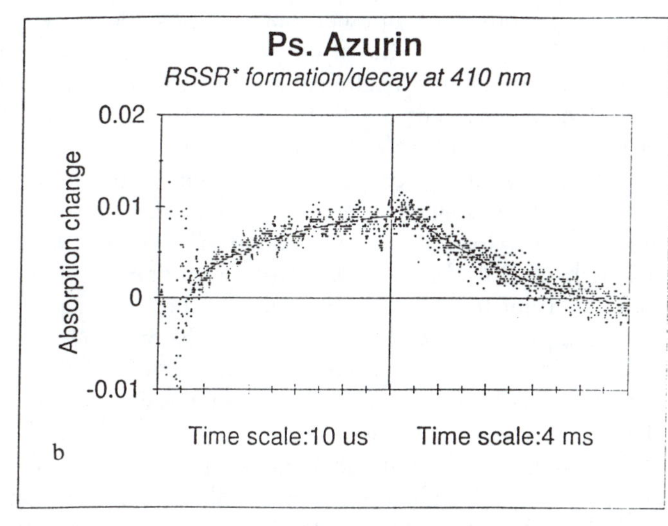

Figure 2. Time-resolved absorption changes observed at 625 nm (a), and 410 nm (b) in Pseudomonas aeruginosa *azurin upon reduction by CO_2^- radicals.* T = *285 K, pH = 4.0, and pulse width = 0.5 μs of 5 MeV electrons.*

is also observed. At 410 nm (Figure 2b) a fast increase in absorption is observed which is assigned to the formation of $RSSR^-$ radical ion, known to have a strong absorption band centered at this wavelength ($\varepsilon 410 = 10{,}000\ M^{-1}\ cm^{-1}$) (*1, 2, 10*). Because the formation process of this radical is bimolecular, with a rate constant similar to that of the fast Cu(II) reduction phase, these are interpreted as being due to the parallel reduction by CO_2^- radicals of the Cu(II) site, and of the partially exposed single disulfide bridge (Cys3–Cys26) (*15, 16*). The transient 410-nm absorption was found to decay in a slower time domain with a rate constant identical to that of the slow phase observed for the reduction of the 625-nm band. This unimolecular process is assigned to an intramolecular electron transfer from the disulfide radical anion to the Cu(II) center.

$$\text{RSSR–Az–Cu(II)} + CO_2^{-} \begin{cases} \nearrow \text{RSSR}^{-}\text{–Az–Cu(II)} \rightarrow \text{RSSR–Az–Cu(I)} + CO_2 \\ \searrow \text{RSSR–Az–Cu(I)} \end{cases}$$

Although in most azurins only a first-order, slow electron transfer is observed for some, the rate constants of this slow reaction phase monitored at both 410 and 625 nm do show a slight dependence on the protein concentrations. This observation is most probably due to a mechanism whereby electron transfer (ET) takes place between $RSSR^{-}$ and Cu(II) in two parallel reactions, one an inter- and the other an intramolecular process. In order to resolve between these two ET processes, the observed rate constants, determined at a given temperature, were plotted as a function of the [Cu(II)]azurin concentrations, and from the straight lines obtained with nonzero intercepts, the rate constants for the unimolecular process were calculated.

Kinetic and thermodynamic parameters determined for the internal ET process in different wt and single-site mutated Pae azurins are summarized in Table I. For those azurins for which no three-dimensional structures have so far been determined, hypothetical structures were calculated. In line with results of direct structure determination of single-site mutants, these calculations suggest that the mutations have a rather limited influence on the protein's overall structure, except at the immediate loci of the substitution (*20–23*). The algorithm developed for identifying long-range ET (LRET) pathways (*24, 25*) has been applied to all the azurins where this internal process was studied, and the results of the calculations suggest that similar hypothetical routes may be operative in all. Two main pathways were identified and are illustrated in Figure 3. One pathway (Figure 3, left) proceeds through the polypeptide backbone from Cys3 to Asn10, then through a hydrogen bond from the carbonyl of Asn10 to NH of the His46, the imidazole of which is a copper ligand. Another pathway (Figure 3, right) proceeds directly from Cys3 via a hydrogen bond to Thr30 and further from Val31 to Trp48, by a 0.40-nm through-space jump. Val49 and Phe111 are connected by another hydrogen bond, and then by a backbone connection to the Cys112 copper ligand.

Two features of the intramolecular LRET in azurins deserve attention. The first is that the copper site is characterized by an exceptionally low Frank–Condon barrier for electron transfer (*14, 20*). The second is that the Cys3–Cys26 disulfide bridge plays, most probably, only a structural role in this protein (*20*). Hence, the LRET induced from $RSSR^{-}$ to the Cu(II) center is

Table I. Kinetic and Thermodynamic Data for the Intramolecular Reduction of Cu(II) by $RSSR^-$ at 298 K and pH 7

Azurin	k_{298} (s^{-1})	E′ (*mV*)	$-\Delta G°$ ($kJ\ mol^{-1}$)	ΔH ($kJ\ mol^{-1}$)	ΔS ($J\ K^{-1}\ mol^{-1}$)
Wild type					
Pseudomonas aeruginosa[a]	44 ± 7	304	68.9	47.5 ± 4.0	−56.5 ± 7.0
P. fluor[b]	22 ± 3	347	73.0	36.3 ± 1.2	−97.7 ± 5.0
Alc. spp.[a]	28 ± 1.5	260	64.6	16.7 ± 1.5	−171 ± 18
Alc. faec.[b]	11 ± 2	266	65.2	54.5 ± 1.4	−43.9 ± 9.5
Mutant (Pseudomonas aeruginosa)					
D23A[c]	15 ± 3	311	69.6	47.8 ± 1.4	−61.4 ± 6.3
F110S[d]	38 ± 10	314	69.9	55.5 ± 5.0	−28.7 ± 4.5
F114A[e]	72 ± 14	358	74.1	52.1 ± 1.3	−36.1 ± 8.2
H35Q[f]	53 ± 11	268	65.4	37.3 ± 1.3	−86.5 ± 5.8
I7S[d]	42 ± 8	301	68.6	56.6 ± 4.1	−21.5 ± 4.2
M44K[f]	134 ± 12	370	75.3	47.2 ± 0.7	−46.4 ± 4.4
M64E[d]	55 ± 8	278	66.4	46.3 ± 6.2	−56.2 ± 7.2
M121L[e]	38 ± 7	412	79.3	45.2 ± 1.3	−61.5 ± 7.2
W48L[e]	40 ± 4	323	70.7	48.3 ± 0.9	−51.5 ± 5.7
W48M[e]	33 ± 5	312	69.7	48.4 ± 1.3	−50.9 ± 7.4

[a] Reference 15.
[b] Reference 17.
[c] Reference 44.
[d] Reference 43.
[e] Reference 19.
[f] Reference 18.

apparently not part of azurin's physiological function, and the medium, separating the electron donor and acceptor, did not undergo evolutionary selection for optimal performance of this process. Azurins therefore provide an interesting model system where structure–function relationships for LRET within a protein matrix can be examined in considerable detail (*16, 26*). The distance between electron donor (D) and acceptor (A) is expected to be the same for all azurins studied so far. In contrast, the driving force for the intramolecular LRET can be varied. In some azurin mutants recently examined (e.g., M64E-, 17S- and F110S), these values are rather close (Table I), the microenvironments of both D and A are also unaltered, and the reorganization energies are expected to be the same. This provides a case for probing the role that the medium separating D and A may have on the LRET rate constants.

The semiclassical Marcus equation can be applied to electron transfer between spatially fixed and oriented redox centers (*26*). Values obtained by employing this theoretical framework are in good agreement for both wild type and single-site mutants of Pae azurin (*15*), thus supporting the applicability of this analysis of the LRET process. The mechanism of intramolecular electron transfer through matrices of biological macromolecules, mainly proteins, has attracted considerable current interest (*25*). Moreover, the question of whether

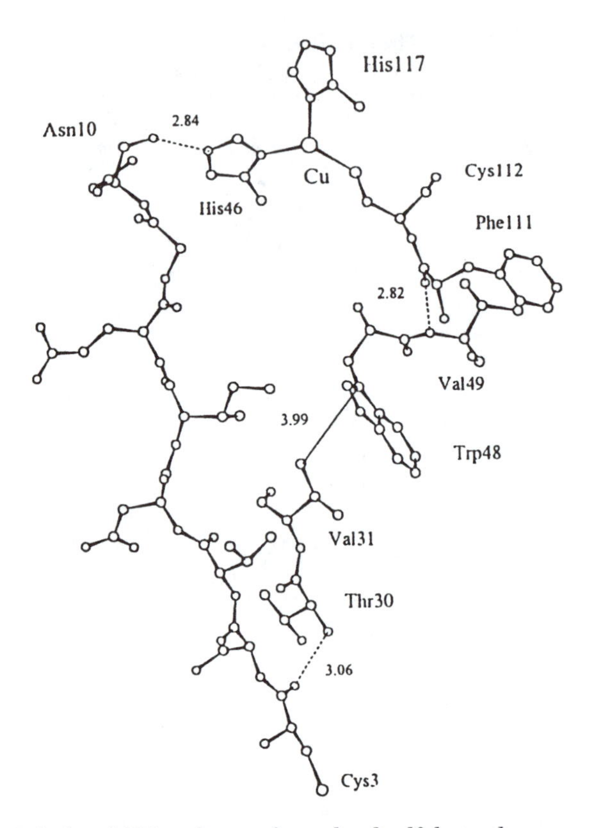

Figure 3. Calculated ET pathways from the disulfide to the copper center in Pseudomonas aeruginosa *azurin. Calculation was performed by using the methodology of Beratan et al. (24). Hydrogen bonds are shown by broken lines and the through-space jump by a thin line. Some distances (in angstroms) are also indicated in the figure.*

electron transfer proceeds via through-space or through-bond pathways is still an issue of major discussion (*26–29*). From the refined crystallographic structure of the wt azurin (*20, 21*), we calculate for the shortest edge-to-edge distance, $r - r_0 = 2.46$ nm, between the copper ligating thiolate of Cys112 and the sulfur of Cys26. Using this value leads to an electronic coupling decay constant $\beta = 10 \pm 0.5$ nm^{-1}. Moser et al. calculated, from their analysis of a large number of biological electron transfer systems based on a through-space model, a β value of 14 nm^{-1} (*28*). This difference seems too large to be accounted for by experimental uncertainties. Furthermore, results of our studies of LRET in different wt azurins (*15, 17*) yield a calculated maximal (i.e., for $\Delta G° = -\lambda$) LRET rate constant of 300 s^{-1}, whereas when using the correlation line of Moser et al. (*28*) with a β of 14 nm^{-1} this rate constant should be two orders of magnitude smaller (*17, 28, 29*). This discrepancy strongly suggests that it is the through-bond LRET model which applies to this examined process in azurin, where the donor and acceptor are probably coupled by superexchange via the bridging orbitals (*16–19*). The magnitude of the ionization energy of the tunneling electron is crucial for determining the dominant LRET pathways. Its relative value compared with the energy of the bridging orbitals determines the through-bond decay and the preference for electron or hole tunneling. The charge-transfer bands of the two redox centers in azurin ($RSSR^-$ and Cu(II)) have energies of 2–3 eV, which are low compared with the HOMO–LUMO gap in saturated organic compound (>7 eV). We therefore expect hole transfer via the occupied bonding orbitals to be the dominant mechanism.

Considerable experimental evidence has been accumulated for electron transfer over long distances through saturated bonds whose electronic interactions decrease exponentially with the distance. In the highly interconnected β-sheet structure of azurin, the redox centers will couple strongly with the protein. The LRET pathway calculations for azurin indicate that the two most likely routes mentioned in the preceding discussion are (1) only through the peptide and hydrogen bonds (pathway 1, Figure 3, left) and/or (2) pathway 2 (Figure 3, right), which also includes a through-space jump involving Trp48. The calculations show pathway 1 to be equally probable as pathway 2. Analysis of the LRET process in a range of different azurins is thus in very good agreement with a through-bond tunneling model. Furthermore, they illustrate the usefulness of applying pulse radiolysis to measurements of electron transfer processes in single-site mutated proteins. They enabled resolution of the role of driving force and reorganization energies in determining the ET rates. Moreover, these results clearly demonstrate the applicability of the general Marcus theory (*26*) for assessing these parameters, and that the electron transfer mechanism can be rationalized in terms of a through-bond model.

Oxidases as Models for Optimized LRET

Studies of intramolecular ET in oxidases provide interesting examples of how pulse radiolysis is employed to obtain insights into both (1) these enzymes' respective mechanisms of action and (2) electron transfer along protein polypeptide matrices that were most probably selected by evolution (*9, 10, 30–32*). Thus, early attempts to study the electron uptake mechanism by the blue oxidase, ceruloplasmin, showed that a diffusion-controlled decay process of the e^-_{aq} in solutions of this protein is paralleled by the formation of transient optical absorptions due to electron adducts of protein residues, primarily of cystine disulfide bonds (*30*). The monomolecular decay of the latter absorption was found to have the same rate constant as that at which the type 1 Cu(II) absorption band was reduced. These results were interpreted as being the combined result of the high reactivity of the e^-_{aq} and the relatively inaccessible type 1 Cu(II) site, yielding an indirect, intramolecular electron transfer pathway from surface-exposed residues (*30*).

A more recent and elegant application of pulse radiolysis to the study of the mechanism of action of a complex, multicentered redox enzyme is that of xanthine oxidase (*31, 32*). This enzyme has been extensively investigated by different kinetic methods, including stopped flow and flash photolysis. Use of pulse-radiolytically produced radicals (*N*-methylnicotinamide or 5-deazalumiflavin) enabled direct monitoring of details of the internal electron flow, starting from the enzyme's molybdenum center, which is the preferred site of electron uptake, to the iron–sulfur center, and finally ending at the flavin site, which is where dioxygen is reduced (*32*). Intramolecular ET rates of up to 8.5 $\times$ 10^3 s^{-1} were observed, and it would be of great interest to have these rates examined in terms of the three-dimensional structure of this enzyme when it becomes available.

The high-resolution structure of the blue copper enzyme ascorbate oxidase (AO) prompted several laboratories to examine the internal electron flow from the substrate oxidation site [type 1 Cu(II)] to that of dioxygen reduction. Again, both flash photolysis and pulse radiolysis were employed, providing a comparison of their respective features.

Blue copper oxidases catalyze the specific one-electron oxidation of substrates (e.g., L-ascorbate) by O_2, which is reduced to water (*33*). Their minimal catalytic unit consists of four copper ions bound to distinct sites in the protein, designated type 1, 2, and 3, respectively (T1, 2, 3). Early pulse radiolysis studies of the blue oxidases laccase and ceruloplasmin have mainly focused on the kinetics of T1[Cu(II)] reduction and contributed to the notion that the T1 site in these enzymes acts as the electron uptake site from substrates (*8, 9, 30*). The ground for more meaningful studies of both action mechanism and LRET in this important group of enzymes was set by the high-resolution determination of the three-dimensional structure of ascorbate oxidase. This provided detailed insights into structure and spatial relationships among the enzyme's four different copper coordinating sites (*34–36*).

AO was shown to exist as a dimer of identical 70-kDa subunits, each containing a catalytic unit of one T1, one T2, and one T3 copper site. The latter two were found to be proximal, forming a trinuclear center. The reduction potentials of the T1 and T3 Cu(II)–Cu(I) couples are practically identical—350 mV at pH 7.0 (*33*)—whereas the T2 potential is considerably lower (<300 mV). The catalytic cycle of this enzyme is suggested to proceed by a sequential mechanism, where single electrons are transferred from the reducing substrates to the T1[Cu(II)]. ET from the latter then takes place to the trinuclear copper center, which serves as the dioxygen binding and reducing site. Intramolecular ET from T1[Cu(I)] to T3[Cu(II)] therefore seems to be the essential step required for the four electrons (and four protons) necessary for O_2 reduction to two water molecules.

Three groups have independently studied the intramolecular ET from T1[Cu(I)] to T3[Cu(II)] in AO by photochemical and pulse radiolysis methods and have reported similar rate constants for this process (*37–41*). Photochemically produced lumiflavin semiquinone was shown to reduce T1[Cu(II)] in a fast second-order process that was followed by a partial reoxidation of the T1[Cu(I)] site in a first-order reaction with a rate constant of 160 s^{-1} (*37*). The flavin absorption in the near-UV region prevented monitoring changes at 330 nm, where the T3[Cu(II)] absorbs. Still, the authors interpreted the latter process as being due to intramolecular ET between the T1 and T3 sites. This was further supported by experiments using AO where type 2 copper was removed. Pulse radiolysis study of AO enabled the independent direct monitoring of the process as observed at both 610-nm (T1) and 330-nm (T3) bands (*38*). The CO_2^- radicals were used as primary electron donors, and they reduced the type 1 Cu(II) site in a bimolecular diffusion-controlled process (Figure 4). The ensuing processes of T1[Cu(I)] reoxidation and T3[Cu(II)] reduction occurred with identical rate constants that were concentration-independent, confirming their assignment as intramolecular ET from the T1 to T3 sites. However, two to three distinct phases were observed in this intramolecular ET process, having rate constants of 200 s^{-1} for the fastest phase and 2 s^{-1} for the slowest (Figure 4). Similar ET rates were later determined in another pulse radiolysis study performed using different organic radicals as reductants, yet monitoring the 610-nm chromophore only (*39*).

An underlying assumption in our studies of AO (*38, 40*) is that the structure of the blue oxidases has most probably been optimized by evolution for the internal LRET involved in its catalytic cycle, and it therefore represents a complementary system to that of azurins for examination of intraprotein ET rates. The following important questions then arise: Is there a control of the intramolecular ET rate in AO during its multielectron reduction and oxidation? Does the internal ET rate depend on the number of reduction equivalents taken up by the molecule? How does the rate of electron transfer relate to the conformational changes that were resolved by the structural studies (*33–36*)? Does the presence of substrates (organic reductants or dioxygen) affect (e.g., by allostery) the internal ET rates? These questions gain more significance when it is noted that steady-state kinetic measurements of AO activity yield turnover numbers of 12,000–15,000 s^{-1} (*37*), which are considerably faster than the observed values just mentioned for intramolecular T1 to T3 ET rates.

The notion that AO molecules reduced to different degrees may exhibit distinct reactivities can be ruled out because experiments starting with the fully oxidized enzyme and ending with less than 5% oxidized T1 sites yielded essentially identical intramolecular ET rates (*38*). Also, no difference in these internal ET rate constants was observed with enzymes "activated" or "pulsed" by turning over 1 mM ascorbate in the presence of 0.25 mM O_2 prior to the determination of the intramolecular ET rates. In contrast to these experiments, which were all performed under anaerobic conditions, a more recent pulse-radiolytic study employed AO solutions containing small and controlled concentrations of O_2 (15–65 μm), and conspicuous differences in the kinetic behavior were observed. A new and faster intramolecular ET phase was discovered which

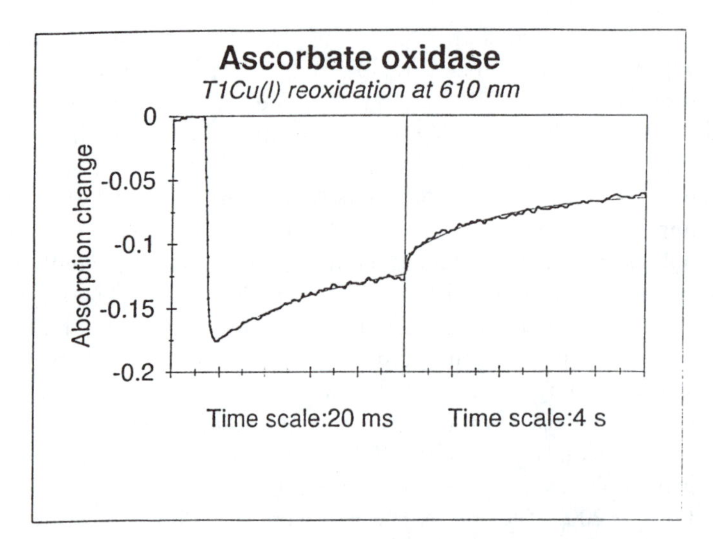

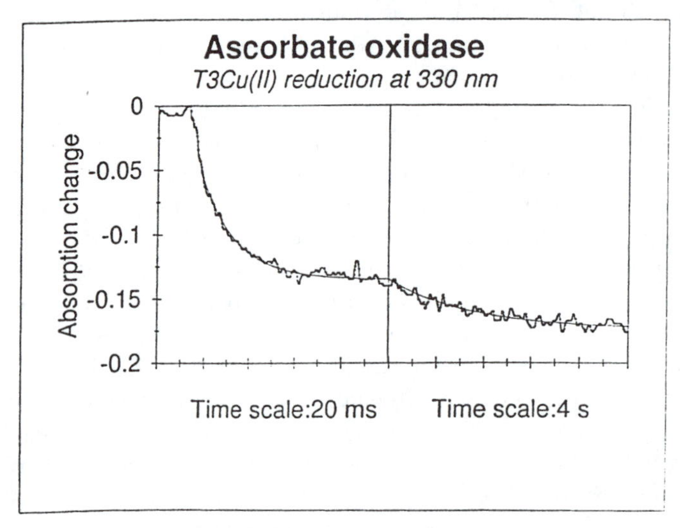

Figure 4. Time-resolved absorption changes monitored at 610 nm (A), and 330 nm (B) in ascorbate oxidase upon reduction by CO_2^- radicals. T = *285 K, pH = 5.5, and pulse width = 0.5 μs of 5 MeV electrons.*

depended on the presence of dioxygen (*40*). The rate constant of this phase, monitored at both 610 nm (T1) and 330 nm (T3) was about 1100 s^{-1} (293 K, pH 5.8) and was maintained at this value as long as O_2 remained in the solution. Large spectral changes took place at the 330-nm range, most probably because of interaction between the trinuclear site and dioxygen following the intramolecular ET from T1[Cu(I)] to T3[Cu(II)]. Oxygen reduction intermediates coordinated to the trinuclear center would probably increase the ET driving force significantly, and this may result in an enhanced rate of the intramolecular process. Our calculations show that a 100-mV increase in the reduction potential of the T3 center would lead to the observed increase in ET rate. Similar effects have been observed in the catalytic reaction of cytochrome oxidase with O_2 (*42*).

The highest rate constant observed for intramolecular ET in AO is 1100 s^{-1}, which still is considerably smaller than the turnover number of about 14,000 s^{-1}. Thus, interaction between dioxygen and the trinuclear site is not sufficient to ensure maximal enzymatic activity. Under optimal conditions, the concentration of reducing substrate (e.g., ascorbate) is sufficiently high to maintain a steady state of fully reduced copper sites. Thus, an antithetical approach was very recently taken by Tollin and co-workers studying the reoxidation of fully reduced AO by a laser-generated triplet state of 5-deazariboflavin (*41*). Subsequent to the assumed one-electron oxidation of the reduced trinuclear cluster, a rapid, biphasic intramolecular ET occurs from T1[Cu(I)] (and presumably) to the oxidized trinuclear center. The faster of the two observed rate constants (9500 and 1400 s^{-1}, respectively) is comparable to the turnover number determined for AO under steady-state conditions and renders it likely that this is the rate-limiting step in catalysis.

One of the T1–copper ligands in the three-dimensional model of AO is the Cys507 thiolate, while imidazoles of the two neighboring His506 and His508 coordinate to the T3 copper ions. This led to the proposal that the shortest ET pathway from T1[Cu(I)] to T3[Cu(II)] takes place via Cys507 and either His506 or His508 (*34*) (cf. Figure 5). Both pathways consist of nine covalent bonds yielding a total distance of 1.34 nm. Performing the pathway calculations as developed by Beratan et al. (*24*) confirms this notion and suggests a further alternative path via a hydrogen bond between the carbonyl oxygen of Cys507 and His506. A final noteworthy point is the similarity between a structural feature observed in AO and in plastocyanin: in both cases, the cysteine thiolate ligand of the T1–copper is utilized in ET to or from this metal ion.

Conclusions

Pulse radiolysis has been employed successfully to resolve mechanisms of action of redox proteins and of electron transfer within their polypeptide matrix. The limitations on the use of this method, set by the requirement for expensive electron accelerators, are more than compensated for by experimental advantages, as illustrated by the results described in this chapter. Future applications to the study of engineered proteins and other model systems would certainly extend our understanding of both of these aspects of redox processes in biological macromolecules.

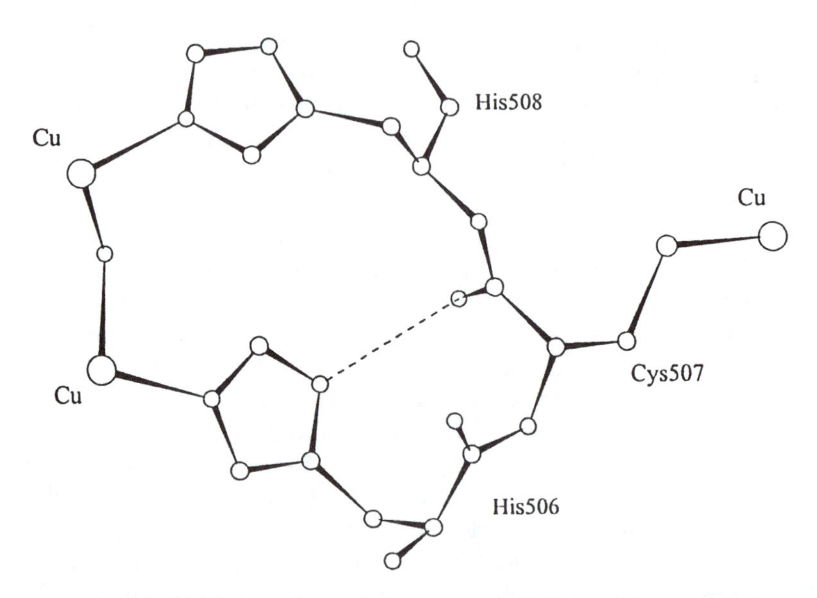

Figure 5. Calculated ET pathways between T1 Cu (right) and the T3 copper pair (left). The hydrogen bond is indicated as a broken line. The calculations were based on the three-dimensional structure determined by Messerschmidt et al. (34).

Acknowledgments

Support of our work by a grant from the German–Israeli Foundation (I–0320–211.05/93) is gratefully acknowledged. Part of the work was supported by the Danish Natural Science Research Foundation (The Bioinorganic Program). The coordinates of the azurin structures were kindly provided by Robert Huber and Herbert Nar.

References

1. *Fast Processes in Radiation Chemistry and Biology;* Adams, G. E.; Fielden, E. M.; Michael, B. D., Eds.; Wiley: New York, 1975.
2. Buxton, G. V.; Sellers, R. M. *J. Chem. Soc. Faraday Trans.* **1973,** *69,* 555–559.
3. Land, E. J.; Swallow, A. J. *Arch. Biochem. Biophys.* **1971,** *145,* 365–372.
4. Faraggi, M.; Pecht, I. *FEBS Lett.* **1971,** *13,* 221–224.
5. Wilting, J.; Braams, R.; Nauta, H.; van Buuren, K. J. H. *Biochim. Biophys. Acta* **1972,** *283,* 543–547.
6. Lichtin, N. N.; Shafferman, A.; Stein, G. *Science (Washington, D.C.)* **1973,** *179,* 680–683.
7. Pecht, I.; Faraggi, M. *Proc. Natl. Acad. Sci. U.S.A.* **1972,** *69,* 902–906.
8. Faraggi, M.; Pecht, I. *Biochem. Biophys. Res. Comm.* **1971,** *45,* 842–848.
9. Pecht, I.; Faraggi, M. *Nature (London)* **1971,** *233,* 116–118.
10. Klapper, M. H.; Faraggi, M. *Quart. Rev. Biophys.* **1979,** *12,* 465–519.

11. Pecht, I.; Goldberg, M. In *Fast Processes in Radiation Chemistry and Biology;* Michael, B. D., Ed.; Wiley: New York, 1975; pp 274–284.
12. Faraggi, M.; Klapper, M. H. In *Excess Electrons in Dielectric Media;* Ferradini, C.; Jay-Gerin, J.-P., Eds; CRC: Boca Raton, FL, 1990; pp 397–423.
13. Adman, E. T. *Adv. Protein Chem.* **1991,** *42,* 195–197.
14. Clarke, M. J. et al., Eds. *Struct. Bonding* **1991,** *75,* 1.
15. Farver, O.; Pecht, I. *Proc. Natl. Acad. Sci. U.S.A.* **1989,** *86,* 6968.
16. Farver, O.; Pecht, I. *Biophys. Chem.* **1994,** *50,* 203.
17. Farver, O.; Pecht, I. *J. Am. Chem. Soc.* **1992,** *114,* 5764.
18. Farver, O.; Skov, L. K.; Van de Kamp, M.; Canters, G. W.; Pecht, I. *Eur. J. Biochem.* **1992,** *210,* 399.
19. Farver, O.; Skov, L. K.; Pascher, T.; Karlsson, B. G.; Nordling, M.; Lundberg, L. G.; Vänngård, T.; Pecht, I. *Biochemistry* **1993,** *32,* 7317.
20. Baker, E. N. *J. Mol. Biol.* **1988,** *203,* 1071.
21. Nar, H.; Messerschmidt, A.; Huber, R.; Van de Kamp, M.; Canters, G. W. *J. Mol. Biol.* **1991,** *218,* 427.
22. Van Pouderoyen, G.; Mazumdar, D. M.; Hunt, N. I.; Hill, H. A. O.; Canters, G. W. *Eur. J. Biochem.* **1994,** *222,* 583.
23. Gilardi, G.; Mei, G.; Rosato, N.; Canters, G. W.; Finazzi-Agro, A. *Biochemistry* **1994,** *33,* 1425.
24. Beratan, D. N.; Betts, J. N.; Onuchic, J. N. *Science (Washington, D. C.)* **1991,** *252,* 1285.
25. Onuchic, J. N.; Beratan, D. N.; Winkler, J. R.; Gray, H. B. *Annu. Rev. Biophys. Biomol. Struct.* **1992,** *21,* 349.
26. Marcus, R. A.; Sutin, N. *Biochim. Biophys. Acta* **1985,** *811,* 265.
27. Jortner, J.; Bixon, M. *Mol. Cryst. Liq. Cryst.* **1993,** *234,* 29.
28. Moser, C. C.; Keske, J. M.; Warncke, K.; Farid, R.S.; Dutton, P. L. *Nature (London)* **1992,** *355,* 802.
29. Farid, R. S.; Moser, C. C.; Dutton, P. L. *Curr. Opinion Struc. Biol.* **1993,** *3,* 225.
30. Faraggi, M.; Pecht, I. *J. Biol. Chem.* **1973,** *248,* 3146–3149.
31. Anderson, R. F.; Hille, R.; Massey, V. *J. Biol. Chem.* **1986,** *261,* 15870–15876.
32. Hille, R.; Anderson, R. F. *J. Biol. Chem.* **1991,** *266,* 5608–5615.
33. Kroneck, P. M. H.; Arnstrong, F. A.; Merkle, H.; Marchesini, A. *Archeological Chemistry IV;* Allen, R. O., Ed.; Advances in Chemistry 220; American Chemical Society: Washington, DC, 1989; pp 223–248.
34. Messerschmidt, A.; Rossi, A.; Ladenstein, R.; Huber, R.; Bolognesi, M.; Gatti, G.; Marchesini, A.; Petruzelli, R.; Finazzi-Agro, A. *J. Mol. Biol.* **1989,** *206,* 513–529.
35. Messerschmidt, A.; Ladenstein, R.; Huber, R.; Bolognesi, M.; Avigliano, L.; Petruzelli, R.; Rossi, A.; Finazzi-Agro, A. *J. Mol. Biol.* **1992,** *224,* 179–205.
36. Messerschmidt, A.; Luecke, H.; Huber, R. *J. Mol. Biol.* **1993,** *230,* 997–1014.
37. Meyer, T. E.; Marchesini, A.; Cusanovich, M. A.; Tollin, G. *Biochemistry* **1991,** *30,* 4619–4623.
38. Farver, O.; Pecht, I. *Proc. Natl. Acad. Sci. U.S.A.* **1992,** *89,* 8283–8287.
39. Kyritsis, P.; Messerschmidt, A.; Huber, R.; Salmon, G. A.; Sykes, A. G. *J. Chem. Soc. Dalton Trans.* **1993,** 731–735.
40. Farver, O.; Wherland, S.; Pecht, I. *J. Biol. Chem.* **1994,** *269,* 22933–22936.
41. Tollin, G.; Meyer, T. E.; Cusanovich, M. A.; Curir, P.; Marchesini, A. *Biochim. Biophys. Acta* **1993,** *1183,* 309–314.
42. Brunori, M.; Antonini, G.; Malatesta, F.; Sarti, P.; Wilson, M. T. *FEBS Lett.* **1992,** *314,* 191–195.
43. Farver, O.; Skov, L. K.; Gilardi, G.; van Poudenoyen, G.; Canters, G. W.; Wherland, S.; Pecht, I. *Chem. Phys.* **1996,** *204,* 271–277.
44. Farver, O.; Bunander, N.; Skov, L. K.; Pecht, I. *Inorg. Chem. Acta* **1996,** *243,* 127–133.

6

Study of Oxyferryl Heme Reactivity Using Both Radiation and Photochemical Techniques

A. M. English[1], T. Fox[1], G. Tsaprailis[1], C. W. Fenwick[1], J. F. Wishart[2], J. T. Hazzard[3], and G. Tollin[3]

[1] **Department of Chemistry and Biochemistry, Concordia University, Montreal, Quebec, Canada H3G 1M8**
[2] **Department of Chemistry, Brookhaven National Laboratory, Upton, NY 11973**
[3] **Department of Biochemistry, University of Arizona, Tucson, AZ 85721**

Flash photolysis and pulse radiolysis were used to generate reductants in situ *to study the electron transfer (ET) reactivity of the* $Fe^{IV}{=}O$ *heme centers in myoglobin and cytochrome c* peroxidase. Reduction of a_5Ru^{III} *groups covalently bound to surface histidines allowed* intramolecular $Ru^{II} \rightarrow Fe^{IV}{=}O$ *ET rates to be measured. Protonation of the oxene ligand was found to be largely rate determining in myoglobin, consistent with the lack of proton donors in its heme pocket. The large distance (21–23 Å) between surface histidines and the heme in wild-type cytochrome c peroxidase prevented the determination of the rate-limiting step(s) involved in* $Fe^{IV}{=}O$ *reduction in this peroxidase, and strategies for attachment of an artificial redox center closer to its heme are outlined. From the work performed to date, pulse radiolysis appears to be a more versatile technique than flash photolysis for the study of* $Fe^{IV}{=}O$ *heme reactivity in proteins.*

Oxyferryl heme centers ($Fe^{IV}{=}O$) are now believed to be reactive intermediates in all heme enzymes that undergo redox catalysis. $Fe^{IV}{=}O$ species have been observed or predicted in heme peroxidases (*1*), catalases (*2*), oxygenases (*3*), and cytochrome *c* oxidase (*4*). Myoglobin (Mb), which normally functions as a reversible O_2-binding protein (*5*) and does not undergo redox catalysis, will, however, react with H_2O_2 to generate an $Fe^{IV}{=}O$ center (*6, 7*).

The detailed mechanisms of formation and decay of the transient $Fe^{IV}{=}O$ intermediates in heme oxygenases and oxidases, such as cytochrome P_{450} and cytochrome *c* oxidase, respectively, are complex and the subjects of much controversy (*3, 4*). The mechanism of formation of the stable $Fe^{IV}{=}O$ centers in small heme peroxidases such as cytochrome *c* peroxidase (CCP) and horseradish peroxidase (HRP) is much better understood (*8, 9*). The steps involved in peroxidase catalysis can be summarized by the following scheme:

$$Fe^{III},P + H_2O_2 \rightarrow Fe^{IV}{=}O,P^{\cdot +} + H_2O \quad (1)$$

$$Fe^{IV}{=}O,P^{\cdot +} + 1e^- \rightarrow Fe^{IV}{=}O,P \quad (2)$$

$$Fe^{IV}{=}O,P + 1e^- + 2H^+ \rightarrow Fe^{III},P + H_2O \quad (3)$$

The resting form (Fe^{III},P) reacts rapidly ($>10^7\ M^{-1}\ s^{-1}$) with H_2O_2 to generate the two-electron oxidized intermediate termed compound I ($Fe^{IV}{=}O,P^{\cdot +}$), where $P^{\cdot +}$ is a cation radical that is located either on the porphyrin or protein (*1*). Compound I is generally reduced back to the resting form in one-electron reduction steps via compound II ($Fe^{IV}{=}O$,P). The electron donors can be a large variety of species, including the macromolecule ferrocytochrome *c*, which is the physiological reducing substrate for CCP (*10*), and small aromatic donors for HRP (*11*). Studies on mutant forms of CCP, where key catalytic residues around the heme (Figure 1a) have been mutated, reveal that the distal His52, and, to a lesser extent, the distal Arg48, control the rate of reaction 1 (*12, 13*); similar studies on HRP mutants have confirmed the catalytic importance of the distal His and Arg (*14*). Although Mb possesses both distal and proximal His residues like the peroxidases (Figure 1b), metMb reacts with H_2O_2 over 10^5-fold more slowly than the peroxidases (*15*). Thus, the high reactivity of peroxidases with H_2O_2 has been ascribed to the effective roles played by both the distal Arg and His. The Arg promotes ionization of H_2O_2 in the heme cavity while the neighboring His accepts the proton, allowing the peroxy anion to bind to the Fe^{III} heme. Heterolytic cleavage of the O–O bond of the peroxy ligand is promoted by back proton transfer (PT) from the distal His to the O_β atom, and (at least in HRP) (*14*) by stabilization of the transient negative charge on O_β by the distal Arg. Ionization of H_2O_2 in the apolar heme pocket of Mb, where Phe43 is a position equivalent to the distal Arg (Figure 1b), is anticipated to be less favorable than in the peroxidases, accounting in part for the 10^5-fold slower rate of reaction 1 in metMb.

The efficiency of heme peroxidase catalysis also depends on the rates of reduction of compounds I and II (equations 2 and 3). For both CCP (*16*) and HRP (*11*), the rate-limiting step under optimal conditions is that of $Fe^{IV}{=}O$ reduction. Details of how heme proteins control the reactivity of $Fe^{IV}{=}O$ catalytic intermediates are poorly understood. The $Fe^{IV}{=}O$ catalytic intermediates of cytochrome P_{450} and cytochrome *c* oxidase are highly unstable (*4, 17*). In fact, substrate must already be bound close to the heme in cytochrome P_{450}(cam) before the catalytic redox cycle begins to ensure that substrate hy-

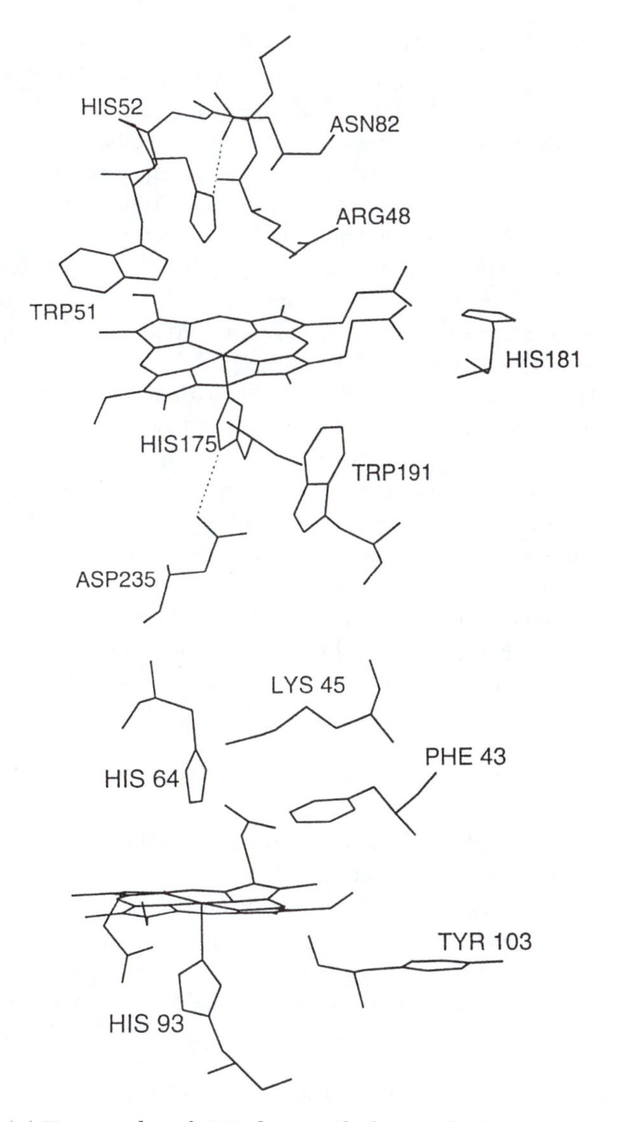

Figure 1. (a) Heme pocket of CCP showing the key catalytic residues. The dashed lines represent H-bonds. This diagram was generated using the X-ray coordinates for the 1.7-Å structure of CCP (44). (b) Diagram of the heme pocket of Mb generated using the X-ray coordinates for the 1.9-Å structure of horse heart met-myoglobin (51).

droxylation is coupled to electron transfer (*17*). In contrast, the $Fe^{IV}=O$ intermediates in heme peroxidases, in particular the compound II species, exhibit half-lives on the order of hours (*1*). Hence, the reduction of the $Fe^{IV}=O$ heme in peroxidases by a variety of redox reagents, including small inorganic complexes (*18*), ferrocytochrome *c* (*1, 19*), and aromatic compounds (*11, 20*), has been studied in detail. For example, the rate of reduction of the $Fe^{IV}=O$ heme in HRP by reagents such as $Fe(CN)_6^{4-}$ was observed to be pH dependent and increase at low pH (*21*), as expected for reductive protonation of the oxene ligand (equation 3). However, in the studies performed to date, pH dependence arising from association of the reagents cannot be distinguished from that due to protonation of the oxene ligand.

To avoid the usual uncertainties associated with bimolecular electron transfer (ET) reactions (*22*), a number of laboratories have covalently bound redox reagents such as a_5Ru^{III} ($a = NH_3$) to surface His residues of a variety of redox proteins (*23–25*). The same approach is under way in our laboratories to compare the ET reactivity of the $Fe^{IV}=O$ heme in both CCP and Mb. This comparison is of interest given that CCP is designed to rapidly turn over H_2O_2, which requires both rapid formation and *decay* of the $Fe^{IV}=O$ center in the presence of reducing substrates (equations 1–3), whereas the O_2-storage function of Mb requires that it reversibly bind dioxygen *without* O–O bond cleavage.

Following surface His ruthenation, the intramolecular ET reaction of interest is the following:

$$a_5Ru^{II}\text{–}Fe^{IV}=O + 2H^+ \rightarrow a_5Ru^{III}\text{–}Fe^{III} + H_2O \qquad \Delta G^\circ \sim 1\ eV \quad (4)$$

Because of the highly favorable driving force for reaction 4, $a_5Ru^{II} \rightarrow Fe^{IV}=O$ heme ET was expected to be rapid (*26, 27*). Hence, the a_5Ru^{II} reductant was generated in situ using both flash photolysis and pulse radiolysis techniques, as outlined in the following sections.

An alternative approach to the generation of suitable protein-bound redox was also investigated. Nitration of surface Tyr residues in CCP was carried out to generate protein-bound reducing $NO_2^{\cdot-}$–Tyr radicals in situ (*28*), and our preliminary results are provided in the section "Pulse Radiolysis Studies of CCP." Finally, the use of flash photolysis and pulse radiolysis techniques in the study of $Fe^{IV}=O$ heme systems is compared.

Flash Photolysis Studies of CCP

Modification of the three surface His residues of CCP (Figure 2a) was attempted as described in detail (*29*). Derivatives containing a_5Ru^{III} groups on His60 and His6 were isolated, and flash photolysis was used to photoinitiate ET (*27, 30*). $Ru(bpy)_3^{2+}$ (bpy = 2,2′-bipyridine) has been used extensively as a photoredox reagent, and its long-lived excited state efficiently reduces a_5Ru^{III} bound to surface His residues of proteins (*31*). For example, $^*Ru(bpy)_3^{2+}$ re-

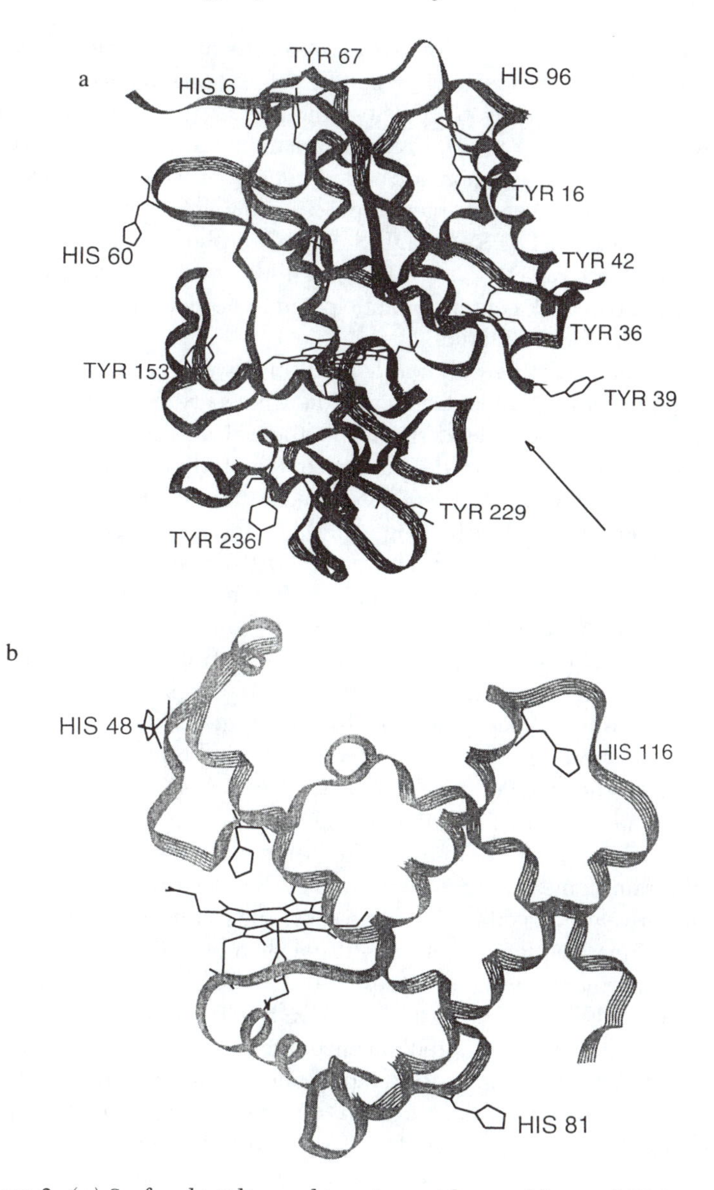

Figure 2. (a) Surface histidine and tyrosine residues and heme of CCP superimposed on the C_α backbone. The cytochrome c *binding domain is centered at the arrow. (b) Surface histidine residues and heme of horse heart myoglobin.*

duces the Ru^{III} and Fe^{III} centers of a_5Ru(His33)cytochrome *c* with rates of ~7 × 10^8 and ~10^8 M^{-1} s^{-1}, respectively, resulting in the formation of excess kinetic over thermodynamic product (*32*). The ground-state $Ru(bpy)_3^{3+}$ complex is a strong oxidant ($E°$ ~ 1.3 V) (*33*), so ethylenediaminetetraacetic acid (EDTA), or other sacrificial electron donors are added to prevent back ET. Unlike cytochrome *c*, CCP is negatively charged at pH 7, so quenching of the negatively charged complexes $Ru(DIPS)_3^{4-}$ (DIPS = 4,7-di(phenyl-4-sulfonate)-1,10-phenanthroline) and $Ru(DIC)_3^{4-}$ (DIC = 4,4′-dicarboxy-2,2′-bipyridine) by the surface-bound a_5Ru^{III} center was anticipated to be electrostatically favored over quenching by the CCP heme.

Studies using the $^*RuL_3$ photoredox reagents were performed with the xenon flash photolysis equipment at Concordia University described previously (*30*). CCP samples were titrated spectrophotometrically with H_2O_2 to form CCP(Fe^{IV}=O) prior to use. For the purpose of this study, compound I of CCP will be designated as CCP(Fe^{IV}=O) because the fate of the protein radical ($P^{\cdot +}$), which is located on Trp191 (Figure 1a) (*1*), following flash photolysis of a_5Ru(His)CCP was not examined. However, Trp191 is not on any direct ET pathway between the Ru and heme centers (Figures 1a and 2a). Anaerobic solutions of 50–100 μM RuL_3, 1–10 mM EDTA, and 5 μM CCP(Fe^{IV}=O) or 5 μM ferricytochrome *c* in 0.1 M phosphate buffer, pH 7.0, were photoexcited with 25-μs pulses from the xenon lamp. When ferricytochrome *c* solutions were flashed, rapid growth of absorbance at 550 nm due to ferrocytochrome *c* formation was observed on the millisecond time scale as reported previously (*32*). When CCP(Fe^{IV}=O) or a_5Ru(His60)CCP(Fe^{IV}=O) solutions were flashed under identical conditions, reduction of the Fe^{IV}=O heme was observed at 564 nm, a maximum in the Fe^{III} minus Fe^{IV}=O difference spectrum, on the same time scale for both samples. Similar results were obtained with the negatively charged RuL_3 complexes, suggesting that quenching of $^*RuL_3$ by the Fe^{IV}=O heme is more efficient than by a_5Ru^{III}. Subsequently, direct k_q measurements revealed values of 1–2 × 10^9 M^{-1} s^{-1} for quenching of $^*RuL_3$ by CCP(Fe^{III}) and CCP(Fe^{IV}=O) (A. English, Concordia University, unpublished results). A further problem encountered, even with 450-nm cutoff filters to eliminate UV light, was rapid photoinduced autoreduction of CCP(Fe^{IV}=O) in the *absence* of RuL_3/EDTA on exposure to the xenon flash.

Deazariboflavin semiquinone (DRFLH$^{\cdot}$) had been shown to exhibit low reactivity (k ~ 10^7 M^{-1} s^{-1}) with the heme of CCP (*34*); thus, DRFLH$^{\cdot}$ was next chosen to selectively reduce the a_5Ru^{III} group on the surface of CCP. Photoexcitation of the flavin quinone (FL) by visible light generates a singlet which decays to the triplet (^{3}FL) in ~10 ns. The triplet can abstract a hydrogen atom from donors such as EDTA to form the flavin semiquinone (FLH$^{\cdot}$) in ≤1 μs, which slowly disproportionates (k = 10^6 M^{-1} s^{-1}) (*35*):

$$FL \xrightarrow{h\nu} {}^1FL \xrightarrow{isc} {}^3FL \qquad (5)$$

$$^{3}\mathrm{FL} + \mathrm{EDTA} \rightarrow \mathrm{FLH}^{\cdot} + \mathrm{EDTA}_{ox} \quad (6)$$

$$\mathrm{FLH}^{\cdot} + \mathrm{FLH}^{\cdot} \rightarrow \mathrm{FLH}_2 + \mathrm{FL} \quad (7)$$

where h is Planck's constant, ν is frequency, and isc is intersystem crossing.

Solutions containing 80–120 μM DRFL, 0.5 mM EDTA, 92 M KCl (added to adjust the ionic strength to 100 mM), and 2.5–40 μM CCP in 4 mM phosphate buffer, pH 7.0, were prepared as previously described (*27, 30*). The nitrogen-pumped dye laser system at the University of Arizona (*36*) provided 1-ns pulses to initiate $DRFLH^{\cdot}$ formation (equations 5 and 6), and photoinduced autoreduction of $CCP(Fe^{IV}{=}O)$ was avoided by excitation at 386 nm 4,4′-bis-(2-butyloctyloxy)-*p*-quarterphenyl (BBQ dye), where DRFL absorption was ~5-fold greater than that of $CCP(Fe^{IV}{=}O)$. Following the laser pulse, slow disproportionation ($k \sim 10^6$ M^{-1} s^{-1}) of the ≤0.6 μM $DRFLH^{\cdot}$ generated per pulse (*27, 30*) was observed in the absence of protein, and slow pseudo-first-order decay ($k \sim 10^7$ M^{-1} s^{-1}) in the presence of 2.5–40 μM native $CCP(Fe^{III})$ at μ = 100 mM. *Rapid* pseudo-first-order decay of $DRFLH^{\cdot}$ was observed, however, in the presence of a_5Ru^{III}(His60)CCP ($k = 1.7 \times 10^9$ M^{-1} s^{-1}) and a_5Ru^{III}(His6)CCP (2.6×10^9 M^{-1} s^{-1}), and the rate constants were the same for both the Fe^{III} and $Fe^{IV}{=}O$ forms of the proteins, consistent with a_5Ru^{III} reduction. The rate of $DRFLH^{\cdot}$ decay was also investigated in the presence of both a_5Ru(His60)CCP and yeast iso-1-cytochrome *c* at μ = 8 mM, conditions where the proteins form a strong complex (*1*). The observed rate constant (1.2×10^9 M^{-1} s^{-1}) is close to that observed for a_5Ru(His60)CCP alone (*27*), which was expected because His60 is remote from the cytochrome binding domain of CCP (Figure 2a).

The ET kinetic data obtained for a_5Ru^{II}(His60)$CCP(Fe^{IV}{=}O)$ have been published in a communication (*27*). The reduction of $CCP(Fe^{IV}{=}O)$ by $DRFLH^{\cdot}$ as monitored at 564 nm was found to be biphasic (Figure 3a). The faster phase, which is clearly concentration dependent, was assigned to intermolecular ET:

$$\begin{aligned} a_5Ru^{II}(\mathrm{His60})\mathrm{CCP}(Fe^{IV}{=}O) &+ a_5Ru^{III}(\mathrm{His60})\mathrm{CCP}(Fe^{IV}{=}O) \\ &+ 2H^+ \rightarrow a_5Ru^{III}(\mathrm{His60})\mathrm{CCP}(Fe^{IV}{=}O) \\ &+ a_5Ru^{III}(\mathrm{His60})\mathrm{CCP}(Fe^{III}) \\ &+ H_2O \end{aligned} \quad (8)$$

The slow, concentration-independent phase was assigned to intramolecular ET:

$$a_5Ru^{II}(\mathrm{His60})\mathrm{CCP}(Fe^{IV}{=}O) + 2H^+ \rightarrow a_5Ru^{III}(\mathrm{His60})\mathrm{CCP}(Fe^{III}) + H_2O \quad (9)$$

Inter- and intramolecular ET was also seen for CCP modified at His6 (Figure 3b).

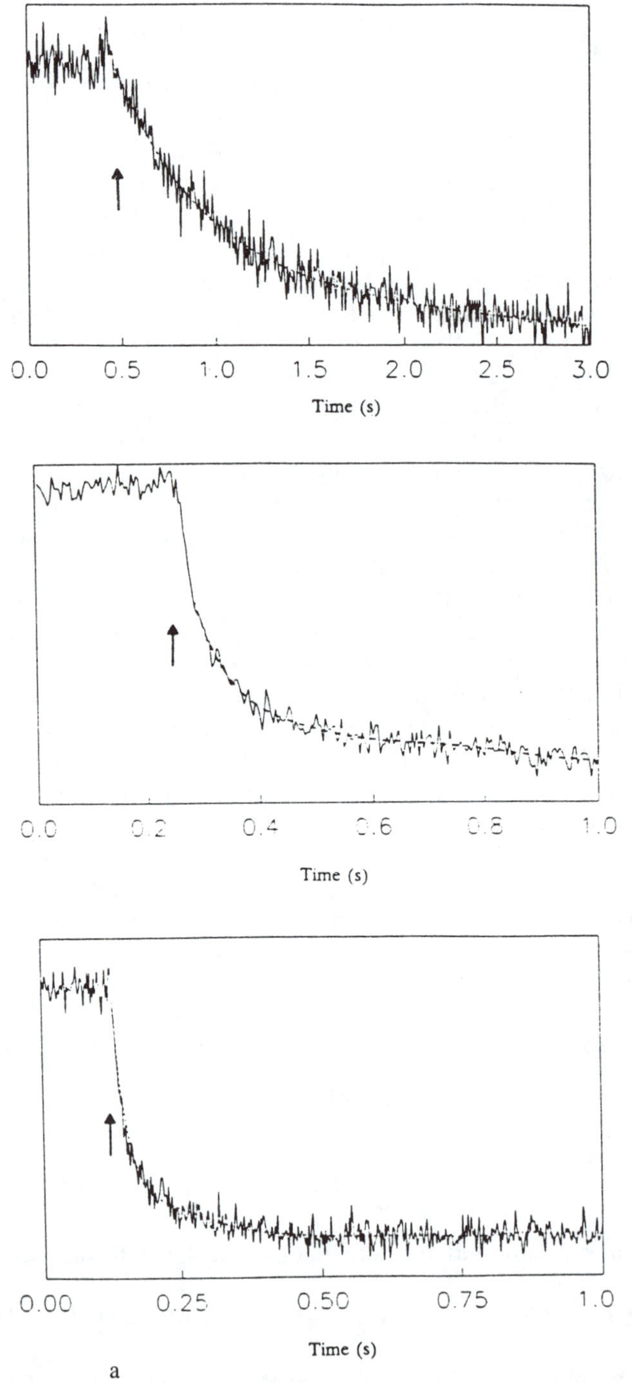
0.0 0.5 1.0 1.5 2.0 2.5 3.0
Time (s)
0.0 0.2 0.4 0.6 0.8 1.0
Time (s)
0.00 0.25 0.50 0.75 1.0
Time (s)
a

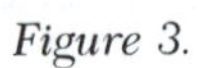

Figure 3.

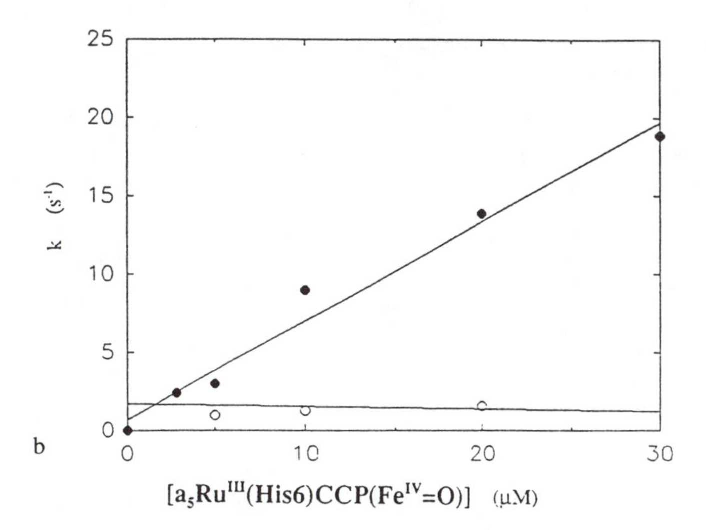

Figure 3. (continued). (a) Absorbance change at 564 nm (y-axes) before and after laser flash photolysis of 2.5 μM (top), 10 μM (middle), and 40 μM a_5Ru^{III}(His60)-CCP($Fe^{IV}{=}O$) (bottom), with 120 μM DRF, 0.5 mM EDTA, and 92 mM KCl in 4 mM phosphate (pH 7.0). The arrow on each trace indicates the time at which the laser flash of <1-ns duration was applied to the sample. The dashed curve though each trace represents the best fit to the data points. Each trace is the sum of the absorbance changes following 4–6 laser flashes, which gave rise to total absorbance at 564 nm of ~0.01–0.03, corresponding to the reduction of 1–4 μM $Fe^{IV}{=}O$ in the laser beam *during each experiment. (Reproduced from reference 27. Copyright 1990 American Chemical Society.) (b) Plot of observed rate constants* (k) *vs. protein concentration for $Fe^{IV}{=}O$ reduction in the His6 derivative. The* k *values were determined from absorbance changes such as those shown in (a). The solid circles represent the* k values for the fast phase and the open circles those for the slow phase.

The observed rate constants for the two CCP derivatives are listed in Table I. The bimolecular rate constants at 100 mM ionic strength for $Fe^{IV}{=}O$ heme reduction by a_5Ru^{II} at His60 ($1.2 \times 10^6\ M^{-1}\ s^{-1}$) and His6 ($6.4 \times 10^5\ M^{-1}\ s^{-1}$) are comparable to that reported for CCP($Fe^{IV}{=}O$) reduction by freely diffusing a_5Ru^{II}pyr ($9.5 \times 10^6\ M^{-1}\ s^{-1}$) (*37*). A smaller diffusion coefficient for CCP-bound a_5Ru^{II}, and electrostatic repulsion, would account in large part for the decrease in the rate constants for reduction by protein-bound versus free a_5Ru^{II}. The X-ray structure of the a_5Ru(His60)CCP derivative reveals that the a_5Ru center is highly solvent-exposed (*27*), consistent with the lack of a large steric effect on the rate of intermolecular ET. The rate constant for reduction by a_5Ru^{II} at His60 doubled on lowering the ionic strength to 8 mM (Table I), which is indicative of an attractive electrostatic interaction between the positively charged Ru center and the negatively charged heme environment of a second CCP molecule. Formation of the a_5Ru(His60)CCP–yeast cytochrome

Table I. Observed Rate Constants for Reduction of $Fe^{IV}=O$ by Surface-Bound a_5Ru^{II} in $a_5Ru(His)CCP(Fe^{IV}=O)$ at pH 7.0[a]

	k ($M^{-1}s^{-1}$)[b]		k (s^{-1})[c]	
μ (mM)	*His60*	*His6*	*His60*	*His6*
100	1.2×10^6	6.4×10^5	3.2	1.6
8	2.5×10^6	—	—	—
8 (+cyt c)[d]	6.4×10^5	—	3.2	—

[a] Reduction of $Fe^{IV}=O$ was followed at 564 nm or 557 nm in the presence of yeast ferricytochrome under the experimental conditions given in Figure 3.
[b] Rate constant for reaction 8 which was obtained from the slopes of plots as shown in Figure 3b.
[c] Rate constant for reaction 9.
[d] In the presence of equimolar yeast iso-1-cytochrome c.

c complex at low ionic strength resulted in a ~4-fold reduction in the bimolecular rate constant for $CCP(Fe^{IV}=O)$ reduction (Table I), which is presumably due to steric hinderance of the CCP heme by the bound cytochrome. Similar results were reported for DRFLH˙ reduction of yeast ferricytochrome *c* on complexation with CCP, which led to a ~10-fold rate decrease (*34*).

The observed rate constants for intramolecular ET over 20–23 Å from the a_5Ru^{II} center at His60 ($k = 3.2 \pm 1.2\ s^{-1}$) and His6 ($k = 1.6 \pm 0.6\ s^{-1}$) to the $Fe^{IV}=O$ heme (Table I) were considered initially to be slow given the large driving force for reaction 9 ($-\Delta G° = 1.04$ eV) (*27*). However, analysis of the ET pathway between the Ru and heme centers in CCP by the tunneling pathway model (*38*) predicted ET rates similar to those observed (A. English and D. Beratan, unpublished results). Therefore, it is necessary to considerably decrease the effective σ-tunneling lengths (σl) (*39*) for ET between the heme and surface-bound redox groups in CCP to observe intramolecular ET rates comparable to the maximal turnover number (1500 s^{-1}) observed for CCP (*40*). Under such conditions it will be possible to ascertain whether ET or PT is rate-limiting in reductive protonation of the oxene ligand (equation 9). It has been suggested previously that the pH dependence of CCP turnover using small reducing substrates such as $Fe(CN)_6^{4-}$ and a_6Ru^{2+} is due to rate-limiting PT to the oxene ligand (*37*). The turnover numbers observed using small-molecule reagents are, however, lower that those observed using the physiological reductant, ferrocytochrome c, suggesting that binding of the latter to the surface of CCP may facilitate ET and/or PT to the $Fe^{IV}=O$ heme. Binding of yeast ferricytochrome *c* did not alter the rate of intramolecular $Ru^{II} \rightarrow Fe^{IV}=O$ ET in the CCP derivative modified at His60 (Table I), indicating that the cytochrome does not act as a gating switch to allow facile ET from nonspecific surface regions of CCP.

The examination of intramolecular ET from surface-bound redox groups within 15 Å of the heme is necessary to determine the effects of pH and cytochrome *c* binding on the redox reactivity of the $Fe^{IV}=O$ heme of CCP.

Suitable derivatives of CCP for such studies are in preparation as outlined in the next section.

Pulse Radiolysis Studies of CCP

The use of pulse radiolysis in ET studies and the Van de Graaff accelerator at Brookhaven National Laboratory are described in detail in another chapter in this book. Briefly, radiolysis of aqueous solutions produces the following major primary products:

$$H_2O \xrightarrow{\rightsquigarrow} e^-_{aq},\ OH^{\cdot},\ H^{\cdot} \qquad (10)$$

A variety of less reactive, negatively or positively charged secondary radicals can be produced from the primary products on addition of suitable reagents (*41*). For example, high concentrations of formate (HCOONa) in N_2O-saturated buffer (to scavenge e^-_{aq}) result in the formation of formate anion radicals (*41*):

$$HCOO^- + OH^{\cdot} \rightarrow CO_2^{\cdot -} + H_2O \qquad (11)$$

Alternatively, addition of metal ions such as Cd^{2+} results in the formation of Cd^+:

$$Cd^{2+} + e^-_{aq} \rightarrow Cd^+ \qquad (12)$$

In this case $OH^{\cdot}$ radicals can be scavenged by the addition of *t*-butanol to generate the unreactive $^{\cdot}CH_2(CH_3)_2COH$ radical (*41*).

Previously, Sykes and co-workers (*28*) have demonstrated that the pulse-radiolytic reduction of plastocyanins NO_2-modified at surface Tyr residues led to rapid $NO_2^{\cdot -} \rightarrow Cu^{II}$ intramolecular ET. Figure 2a reveals that CCP possesses a number of surface exposed Tyr residues close to its heme; thus nitration of CCP was undertaken following the published procedures (*28*). Excess (40–400-fold) tetranitromethane (TNM) was reacted with 100 μM CCP in 50 mM tris(hydroxymethyl)aminomethane (Tris) buffer (pH 7.5) containing 100 mM NaCl for 10 to 30 min. Gel filtration was followed by tryptic digestion of CCP, which was carried out in 0.1 M sodium phosphate buffer (pH 7.0) for 4 h at 37 °C with 50:1 CCP:trypsin. Two nitrated peptides were identified by on-line liquid chromatography–mass spectroscopy (LC–MS) analysis of the tryptic digests of the reaction products. These included peptide 30–48 and peptide 227–243, which contain Tyr36, 39, 42, 229, and 236 (Figure 2a), the most exposed Tyr residues in CCP. The extent of nitration of the peptides depended on the forcing conditions used; products containing ~40% singly nitrated peptide 227–243 could be isolated under less forcing conditions, or in the presence of yeast ferricytochrome c, which protected peptide 30–48 from nitration.

Further purification and characterization of the nitrated derivatives are in progress. However, preliminary pulse radiolysis experiments were carried out on the reaction mixtures nitrated mainly at peptide 227–243. Both $CO_2^{\cdot-}$ and Cd^+ were used as reductants to generate $NO_2^{\cdot-}$–TyrCCP in situ. On pulsing solutions containing native CCP(Fe^{III}), formation of CCP(Fe^{II}) was observed at 438 nm and bimolecular rate constants of ~$2 \times 10^7\ M^{-1}\ s^{-1}$ were observed for direct heme reduction by both Cd^+ and $CO_2^{\cdot-}$. Additional fast phases yielding rate constants of ~$10^8\ M^{-1}\ s^{-1}$ ($CO_2^{\cdot-}$) and $10^9\ M^{-1}\ s^{-1}$ (Cd^+) were observed for heme reduction when solutions of the nitration reaction products were pulsed consistent with the scheme:

$$CO_2^{\cdot-}/Cd^+ + NO_2\text{–TyrCCP}(Fe^{III}) \rightarrow CO_2/Cd^{2+} + NO_2^{\cdot-}\text{–TyrCCP}(Fe^{III}) \rightarrow NO_2\text{–TyrCCP}(Fe^{II}) \quad (13)$$

The absorbance changes accompanying rapid heme reduction revealed that only 5–10% of CCP in the reaction products undergoes reaction 13. Because ~40% of the CCP molecules were nitrated at peptide 227–243, efficient $NO_2^{\cdot-} \rightarrow Fe^{III}$ heme intramolecular ET must occur only from the less abundant NO_2 derivative. This presumably is NO_2–Tyr236CCP because the effective σl for ET (*39*) between this residue and the heme is much smaller than that for Tyr229. Our preliminary results suggest that certain of the NO_2-modified Tyr derivatives may be suitable to investigate the reactivity of the $Fe^{IV}{=}O$ heme in CCP. Purification of the nitrated derivatives to homogeneity is under way, as is the investigation of the stability of CCP($Fe^{IV}{=}O$) in different pulse radiolysis buffer systems. To complement these studies, a number of the exposed Tyr residues in recombinant CCP (*42*) are being mutated to His residues to produce a_5RuHis derivatives with efficient $Ru^{II} \rightarrow$ heme ET pathways.

Pulse Radiolysis Studies of Myoglobin

As mentioned in the introduction, Mb also forms an $Fe^{IV}{=}O$ heme on reaction with peroxides (*7*). Of particular interest is the mechanism of proton delivery to the oxene ligand on reduction because the distal heme pocket of Mb lacks proton donors (Figure 2b) and is isolated from the bulk solvent (*43*), unlike CCP where the hydrated substrate channel connects the distal heme cavity to the solvent (*44*). Thus, it is expected that solvent-assisted PT to the heme of Mb at physiological pH is severely restricted. A spectacular example of solvent-assisted PT to a buried redox center has been highlighted in the X-ray structure of the bacterial reaction centers from *Rhodobacter sphaeroides* (*45*). A narrow hydrated channel extends from the cytoplasmic side of the reaction center to quinone Q_B, which is buried ~23 Å in the L-subunit.

Also of interest is a comparison of the intramolecular $a_5Ru^{II} \rightarrow Fe^{IV}{=}O$ ET rate in Mb with the corresponding rates in Zn-substituted Mb (*46*) over the same ET pathway with similar driving force. PT to the Zn–porphyrin is not required in the ET studies carried out on ZnMb, and because of extended conjugation, the reorganization energy at the Zn–porphyrin center is small (*46*). The reported 0.2-Å difference in the Fe—O bond length in the $Fe^{IV}{=}O$ and $Fe^{III}{-}OH_2$ forms of Mb (*47*) should significantly increase the reorganization energy of the heme center relative to that of the Zn–porphyrin center.

Because horse heart Mb (HHMb) forms a stable $Fe^{IV}{=}O$ heme (*48*), it was chosen for the study of the redox reactivity of the $Fe^{IV}{=}O$ center. A communication (*26*), and a full report (*49*) have been submitted for publication (*49*). Covalent attachment of a_5Ru^{III} to the surface His48 of HHMb was accomplished as previously described (*50*) with some minor modifications (*49*), and the use of on-line LC-MS greatly facilitated the characterization of the HHMb derivatives (*49*). Pulse radiolysis was performed at Brookhaven National Laboratory, using the $CO_2^{\cdot-}$ radical (equation 11) as a reductant. By monitoring the appearance of HHMb(Fe^{II}) at 434 nm, the observed bimolecular rate constants for the reduction of native HHMb(Fe^{III}-OH_2) and a_5Ru^{III}(His48)HHMb(-Fe^{III}–OH_2) by $CO_2^{\cdot-}$ were found to be 2×10^8 and $>10^9$ M^{-1} s^{-1}, respectively, in 40 mM phosphate buffer at pH 7.0 (*26*). Thus, most of the reduction occurred at the a_5Ru^{III} center in the derivatized protein following the 60-ns electron pulse.

Because the reduction potentials for the a_5Ru^{III}(His) and Fe^{III} heme centers are closely matched ($\Delta E^{\circ\prime}$ = 19 mV), the observed rate constant for intramolecular ET over the 12.7 Å from Ru^{II} to the heme followed reversible first-order kinetics (*26*):

$$a_5Ru^{III}(His48)HHMb(Fe^{III}{-}OH_2) \xrightarrow[\text{Fast}]{CO_2^{\cdot-}} a_5Ru^{II}(His48)HHMb(Fe^{III}{-}OH_2) \quad (14)$$

$$a_5Ru^{II}(His48)HHMb(Fe^{III}{-}OH_2) \underset{k_{-1}}{\overset{k_1}{\rightleftarrows}} a_5Ru^{III}(His48)HHMb(Fe^{II}) + H_2O \quad (15)$$

The observed rate constant ($k_{obs} = k_1 + k_{-1}$) was 0.059 ± 0.003 s^{-1}, which is essentially identical to that reported for the a_5Ru(His48) derivative of sperm whale Mb (SWMb) (*50*). As mentioned previously (*26, 49*), the kinetics and thermodynamics of ET in the a_5Ru(His48) derivatives of SWMb and HHMb are very similar, which is not surprising given the similarity in the x-ray structures of the two proteins (*51*).

HHMb$Fe^{IV}{=}O$ was generated prior to use by reaction with 10-fold excess

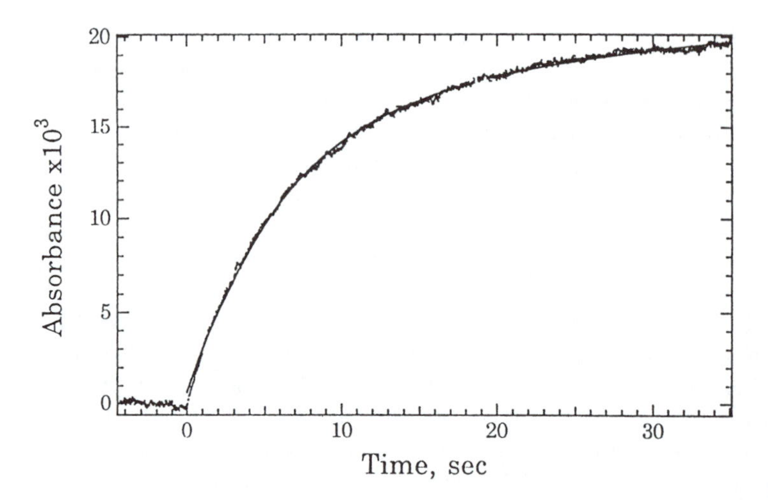

Figure 4. Observed absorbance change at 409 nm vs. time following pulse radiolysis of 2 μM a_5Ru^{III}(His48)HHMb(Fe^{IV}=O) in N_2O-saturated, 40 mM sodium phosphate, 12 mM sodium formate, pH 7.0, I = 0.1 M, 25.2 °C, path length 2.0 cm. The concentration of $CO_2^{\cdot-}$ generated in the pulse was 0.37 μM, and the observed $\Delta\epsilon_{409} \sim 23\ mM^{-1}\ cm^{-1}$ is only 25% of that expected for HHMb(Fe^{IV}=O) reduction due to competition from $CO_2^{\cdot-}$ self-quenching and scavengers. The solid curve shows the fit of the experimental points to first-order kinetics. (Reproduced from reference 26. Copyright 1994 American Chemical Society.)

H_2O_2, and catalase was added to removed unreacted peroxide (*26, 49*). Following rapid reduction of Ru^{III} by $CO_2^{\cdot-}$ (equation 14), slow reduction of the Fe^{IV}=O heme was observed on the second time scale:

$$a_5Ru^{II}(His48)HHMb(Fe^{IV}{=}O) + 2H^+ \xrightarrow{k_{obs}} a_5Ru^{III}(His48)HHMb(Fe^{III}{-}OH_2) \quad (16)$$

The fit of the change in heme absorbance by first-order kinetics is shown in Figure 4, and the average value of k_{obs} was $0.19 \pm 0.02\ s^{-1}$. The k_{obs} for reaction 16 was found to be independent of protein concentration, establishing that bimolecular ET processes are insignificant at the Mb concentrations used (0.5–2.0 μM).

The rate constant for reaction 16 is 5–6 orders of magnitude smaller than those measured on photoexcitation of ZnMb:

$$a_5Ru^{III}(His48)Mb(^3ZnP) \xrightarrow{k_f} a_5Ru^{II}(His48)Mb(ZnP^{\cdot+}) \xrightarrow{k_b} a_5Ru^{III}(His48)Mb(ZnP) \quad (17)$$

where k_f and k_b (the forward and back ET rate constants) are 7×10^4 and $1 \times 10^5\ s^{-1}$ at $-\Delta G°$s of 0.82 and 0.96 eV, respectively (*46*). Thus, at a similar driving force (~1 eV), and over the same pathway, ET to $Fe^{IV}{=}O$ heme of HHMb is ~10^6-fold slower than that to the ZnP^+ center. To ensure that reaction of HHMb(Fe^{III}) with H_2O_2, which also generates a short-lived unidentified radical (*52*), did not alter the polypeptide between the Ru and heme centers, k_{obs} for reaction 15 was remeasured following reduction of the $Fe^{IV}{=}O$ heme to $Fe^{III}{-}OH_2$ and found to be $0.063 \pm 0.016\ s^{-1}$, indicating that radical formation and decay in the polypeptide do not retard ET. Consistent with slow intramolecular reduction of the $Fe^{IV}{=}O$ heme, the bimolecular rate constant for the reduction of *unmodified* HHMb($Fe^{IV}{=}O$) by $CO_2^{\cdot-}$ was observed to be 10^4-fold smaller than that observed for HHMb(Fe^{III}-OH_2) under the same conditions (*26*).

Combined pH and H–D kinetic solvent isotope effects reveal preequilibrium protonation of an acid–base group on Mb (most likely the distal His64) prior to ET (*49*). To determine the driving force dependence of reaction 16, a_4RuLHis48 (L = pyridine and isonicotinamide) derivatives of HHMb were prepared, and rates of intramolecular $Fe^{IV}{=}O$ heme reduction by the surface-bound Ru^{II} were measured as described above for the a_5RuHis48 derivative (*49*). The kinetic data obtained for the three a_4RuL(His48)HHMb derivatives indicate that ET is gated at high–$\Delta G°$. Analysis by Marcus theory of the driving-force dependence of the rate constants extracted for the ET step yielded a reorganization energy (λ) of ~1.8 eV for $Ru^{II} \rightarrow Fe^{IV}{=}O$ ET in the a_4LRu derivatives ($-\Delta G°$ = 0.4–0.5 eV), but λ = 2.1 eV for the a_5Ru derivative ($-\Delta G°$ = 0.75 eV) (*49*). Hence, we assume that ET is gated in the latter since λ for the three derivatives should be similar. Nonetheless, the λ = 1.8 eV obtained for the a_4LRu derivatives is considerably higher than that reported for ZnMb (~1.3 eV) (*46*) and for reaction 15 (~1.5 eV) (*23*), where Mb undergoes a change in coordination number. Thus, we conclude that the study of intramolecular ET over the same pathway from Ru^{II} to either an $Fe^{III}{-}OH_2$, $Fe^{IV}{=}O$, or $ZnP^{\cdot+}$ center in Mb indicates that the rates are determined in large part by the facility with which the prosthetic group can be reduced. Additionally, reduction of $Fe^{IV}{=}O$ in Mb at high driving force is controlled or gated by a conformation that is probably required for H-bonding of a distal residue to the oxene ligand (*49*).

Comparison of Flash Photolysis and Pulse Radiolysis Techniques in the Study of Oxyferryl Heme Reduction

The excitation energy is absorbed by the solvent in pulse radiolysis and by the solutes in flash photolysis. In this respect, pulse radiolysis is more suitable for the study of ET in heme proteins because the absorption of photon energy results in autoreduction of the heme (*53*). When flash photolysis is employed, excitation wavelengths must be chosen to minimize heme absorption. Hence, an ideal photoredox reagent for the study of reaction 4 would (1) absorb strongly at the available excitation wavelength, (2) form a long-lived excited state in high quantum yield that is selectively quenched by ET to the a_5Ru^{III} redox center bound to the protein surface, (3) possess an oxidized form that can be rapidly reduced by a sacrificial donor, and (4) allow heme absorbance changes to be readily monitored.

In the studies on CCP(Fe^{IV}=O), the commonly used Ru polypyridine photoredox reagents were found not to meet criterion (2) because the CCP heme quenched $*RuL_3$ at diffusion-controlled limits. The photogenerated excited flavin precursor of DRFLH$^{\cdot}$ meets the criteria listed above for a suitable photoredox reagent by a different mechanism (equations 5 and 6). Photoinduced autoreduction of CCP(Fe^{IV}=O) was avoided by DRFL absorption of the excitation pulse at 386 nm, and DRFLH$^{\cdot}$ exhibited ~10^2-fold higher reactivity with the surface Ru^{III} group than the CCP heme. However, the absorbance change due to heme reduction per laser flash was very low ($\leq$0.005 AU) because of the low $\Delta\epsilon$ (~8 mM^{-1} cm^{-1}) for CCP(Fe^{III}) minus CCP(Fe^{IV}=O) in the visible region. This required summation of the absorbance changes following 4–6 laser flashes, which in turn necessitated the use of relatively high concentrations (2.5–40 μM) of derivatized CCP (Figure 3a).

Studies on Mb(Fe^{IV} = O) in our laboratory were carried out by pulse radiolysis. This avoided problems associated with photoinduced Fe^{IV}=O heme autoreduction as was observed in the flash photolysis studies on CCP. Furthermore, low concentrations (0.5–2 μM) of a_5Ru(His48)HHMb(Fe^{IV}=O) were sufficient to react with the long-lived $CO_2^{\cdot-}$ radical and for optical detection of heme reduction in the Soret region ($\Delta\epsilon$ ~ 100 mM^{-1} cm^{-1}) (*26, 49*). The use of low protein concentrations promotes intramolecular Ru^{II} → heme ET reactions over intermolecular ET reactions (equations 8 and 9). Further studies on Fe^{IV}=O heme reactivity will be preferentially carried out by pulse radiolysis because of the advantages enumerated here.

Acknowledgments

Research carried out at Concordia University was supported by a grant from the National Sciences and Engineering Research Council (Canada) to A. M. English. Research at Brookhaven National Laboratory was performed under contract DE-AC02-76CH00016 with the U.S. Department of Energy and supported by its Division of Chemical Sciences, Office of Basic Energy Sciences; and National Institutes of Health Grant DK15057 to G. Tsaprailis supported the work carried out at the University of Arizona. Bernard Gibbs (Biotechnology Research Institute, National Research Council, Canada) Montreal) and Stephen Marmor are thanked for their help in the preparation of ruthenated myoglobin.

References

1. English, A. M.; Tsaprailis, G. *Adv. Inorg. Chem.* **1995,** *43,* 79 and references therein.
2. Schonbaum, G.R.; Chance, B. In *The Enzymes;* Boyer, P. D., Ed.; Academic: Orlando, FL, 1976; pp 363–408 and references therein.
3. Gross, Z.; Nimri, S. *J. Am. Chem. Soc.* **1995,** *117,* 8021 and references therein.
4. Babcock, G. T.; Wikström, M. *Nature (London)* **1992,** *356,* 301 and references therein.
5. Springer, B. A.; Sligar, S. G.; Olson, J. S.; Phillips, G. N., Jr. *Chem. Rev.* **1994,** *94,* 699, and references therein.
6. George, P.; Irvine, D. H. *Biochem. J.* **1952,** *52,* 511.
7. Fox, J. R., Jr.; Nicolas, R. A.; Ackerman, S. A.; Swift, C. E. *Biochemistry* **1974,** *13,* 5178.
8. Miller, M. A.; Shaw, A.; Kraut, J. *Struct. Biol.* **1994,** *1,* 524.
9. Baek, H. K.; Van Wart, H. E. *Biochemistry* **1989,** *28,* 5714.
10. Pelletier, H.; Kraut, J. *Science (Washington, D.C.)* **1992,** *258,* 1748.
11. Dunford, H. B. In *Peroxidases in Chemistry and Biology;* Everse, J.; Everse, K. E.; Grisham, M. B., Eds.; CRC Press: Boca Raton, FL, 1991; Vol II, pp 1–24.
12. Erman, J. E.; Vitello, L. B.; Miller, M. A.; Shaw, A.; Brown, K. A.; Kraut, J. *Biochemistry* **1993,** *32,* 9798.
13. Vitello, L. B.; Erman, J. E.; Miller, M. A.; Wang, J.; Kraut, J. *Biochemistry* **1993,** *32,* 9807.
14. Rodriguez-Lopez, J. N.; Smith, A. T.; Thorneley, R. N. F. *J. Biol. Inorg. Chem.* **1996,** *1,* 136.
15. Yonetani, T.; Schleyer, H. *J. Biol. Chem.* **1967,** *242,* 1974.
16. Miller, M. A.; Vitello, L.; Erman, J. E. *Biochemistry* **1995,** *34,* 12048.
17. Raag, R.; Poulos, T. L. *Biochemistry* **1991,** *30,* 2674.
18. Purcell, W. L.; Erman, J. E. *J. Am. Chem. Soc.* **1976,** *98,* 7033.
19. Hazzard, J. T.; Poulos, T. L.; Tollin, G. *Biochemistry* **1987,** *26,* 2863.
20. Huang, J.; Dunford, H. B. *Can. J. Chem.* **1990,** *68,* 1990.
21. Dunford, H. B.; Stillman, J. S. *Coord. Chem. Rev.* **1976,** *19,* 187.
22. Bolton, J. R; Archer, M. D. In *Electron Transfer in Inorganic, Organic, and Biological Systems;* Bolton, J. R.; Mataga, N.; McLendon, G., Eds.; Advances in Chemistry 228; American Chemical Society: Washington, DC, 1991; pp 2–23.
23. Winkler, J. R.; Gray, H. B. *Chem. Rev.* **1992,** *92,* 369.
24. Isied, S. S. In *Electron Transfer in Inorganic, Organic, and Biological Systems;* Bolton, J. R.; Mataga, N.; McLendon, G., Eds.; Advances in Chemistry 228; American Chemical Society: Washington, DC, 1991; pp 229–245.
25. Sykes, A. G. *Chem. Br.* **1988,** June, 551.
26. Fenwick, C.; Marmor, S.; Govindaraju, K.; English, A. M.; Wishart, J. F.; Sun, J. *J. Am. Chem. Soc.* **1994,** *116,* 3169.
27. Fox, T.; Hazzard, J. T.; Edwards, S. L.; English, A. M.; Poulos, T. L.; Tollin, G. *J. Am. Chem. Soc.* **1990,** *112,* 7426.
28. Govindaraju, K.; Christensen, H. E. M.; Lloyd, E.; Olsen, M.; Salmon, G. A.; Tomkinson, N. P.; Sykes, A. G. *Inorg. Chem.* **1993,** *32,* 40.
29. Fox, T.; English, A. M.; Gibbs, B. F. *Bioconjugate Chem.* **1994,** *5,* 14.
30. Fox, T. Ph.D. Dissertation, Concordia University, Montreal, Canada, 1991.
31. Bowler, B. E.; Raphael, A. L.; Gray, H. B. *Prog. Inorg. Chem.* **1990,** *38,* 259.
32. Winkler, J. R.; Nocera, D. G.; Yocam, K. M.; Bordignon, E.; Gray, H. B. *J. Am. Chem. Soc.* **1982,** *104,* 5798.
33. Szentrimay, R.; Yeh, P.; Kuwana, T. *Electrochemical Studies of Biological Systems;* Sawyer, D. T., Ed.; ACS Symposium Series No. 38; American Chemical Society: Washington, DC, 1977; p 162.

34. Hazzard, J. T.; Poulos, T. L.; Tollin, G. *Biochemistry* **1987,** *26,* 2836.
35. Ahmad, L.; Cusanovich, M. A.; Tollin, G. *Biochemistry* **1982,** *21,* 3122.
36. Simondsen, R. P.; Tollin, G. *Biochemistry* **1983,** *22,* 3008.
37. Yandell, J. K.; Yonetani, T. *Biochim. Biophys. Acta* **1983,** *748,* 263.
38. Beratan, D. N.; Betts, J. N.; Onuchic, J. N. *Science (Washington, D.C.)* **1991,** *252,* 1285.
39. Wuttke, D. S.; Bjerrum, M. J.; Winkler, J. R.; Gray, H. B. *Science (Washington, D. C.)* **1992,** *256,* 1007.
40. Kang, C. H.; Ferguson-Miller, S.; Margoliash, E. *J. Biol. Chem.* **1977,** *252,* 919.
41. Buxton, G. V. In *Radiation Chemistry: Principles and Applications;* Rodgers, F.; Rodgers, A. J., Eds.; VCH: New York, 1987; pp 321–349.
42. Fishel, L. A.; Villafranca, J. E.; Mauro, M. J.; Kraut, J. *Biochemistry* **1987,** *26,* 351.
43. Quillin, M. L.; Arduini, R. M.; Olson, J. S.; Phillips, G. N., Jr. *J. Mol. Biol.* **1993,** *234,* 140.
44. Finzel, B. C.; Poulos, T. L.; Kraut, J. *J. Biol. Chem.* **1984,** *259,* 13027.
45. Ermler, U.; Fritzsch, G.; Buchanan, S. K.; Michel, H. *Structure* **1994,** *2,* 925.
46. Casimiro, D. R.; Wong, L.-L.; Colón, J. L.; Zewert, T. E.; Richards, J. H.; Chang, I.-J.; Winkler, J. R.; Gray, H. B. *J. Am. Chem. Soc.* **1993,** *115,* 1485.
47. Chance, M.; Powers, L.; Kumar, C.; Chance, B. *Biochemistry* **1986,** *25,* 1259.
48. Uyeda, M.; Peisach, J. *Biochemistry* **1981,** *20,* 2028.
49. Fenwick, C. W.; English, A. M.; Wishart, J. F. *J. Am. Chem. Soc.,* **1997,** *119, 4758.*
50. Crutchley, R. J.; Ellis, W. R., Jr.; Gray, H. B. J. Am. Chem. Soc. **1985,** *107,* 5002.
51. Evans, S. V.; Brayer, G. D. *J. Mol. Biol.* **1990,** *213,* 885.
52. Davies, M. J. *Biochim. Biophys. Acta* **1991,** *1077,* 86 and references therein.
53. Gu, Y.; Li, P.; Sage, J. T.; Champion, P. M. *J. Am. Chem. Soc.* **1993,** *115,* 4993.

7

Free-Energy Dependence of Electron Transfer in Cytochrome *c* Labeled with Ruthenium(II)–Polypyridine Complexes

J. L. Wright[1], K. Wang[1], L. Geren[1], A. J. Saunders[2], G. J. Pielak[2], B. Durham*,[1], and F. Millett[1]

[1] Department of Chemistry and Biochemistry, University of Arkansas, Fayetteville, AR 72701
[2] Department of Chemistry, University of North Carolina, Chapel Hill, NC 27599–3290

The free-energy dependence of the rate constants for intramolecular electron transfer in yeast cytochrome c *covalently bound to a series of ruthenium(II)–polypyridine complexes has been determined. The H39C;C102T variant of yeast cytochrome* c *was attached to a series of ruthenium complexes through a thioether linkage involving the sulfur atom of Cys39 and a methylene carbon of 4,4′-dimethylbipyridine. The free energies of reaction cover a range from 0.75 to 1.26 V with intramolecular rate constants of* $4–9 \times 10^5\ s^{-1}$*. The observed rate constants are consistent with reactions in which the free energies of reaction are nearly equal to the reorganization energy. In the present case, the reorganization energy for electron transfer between the ruthenium complex and the heme iron is 1 eV.*

Over the past several years, we have developed a technique that has proven extremely valuable in the study of electron transfer between redox sites in metalloproteins. We have reported kinetic studies of the reaction of cytochrome *c* with cytochrome *c* peroxidase (*1–3*), cytochrome oxidase (*4*), cytochrome b_5 (*5, 6*) plastocyanin (*7*), and cytochrome c_1 (*8*). In addition, we have been able to show (*9, 10*) that intramolecular electron transfer in cytochrome b_5 covalently

* Corresponding author.

bound to a series of ruthenium complexes follows the dependence on free energy of reaction predicted by Marcus (*11*). The technique is based on the photoredox chemistry of derivatives of the well characterized Ru(bipyridine)$_3^{2+}$ complex (*12*). When one of these complexes is covalently attached to a metalloprotein, photoexcitation leads to rapid electron transfer quenching of the ruthenium excited state by the iron center of the protein. The Fe(II)–Ru(III) intermediate thus formed rapidly returns to the original oxidation state through a thermal electron transfer reaction as indicated in Scheme I, where h is Planck's constant, ν is the frequency of the exciting light, and k_d describes the rate of excited state decay by processes other than electron transfer.

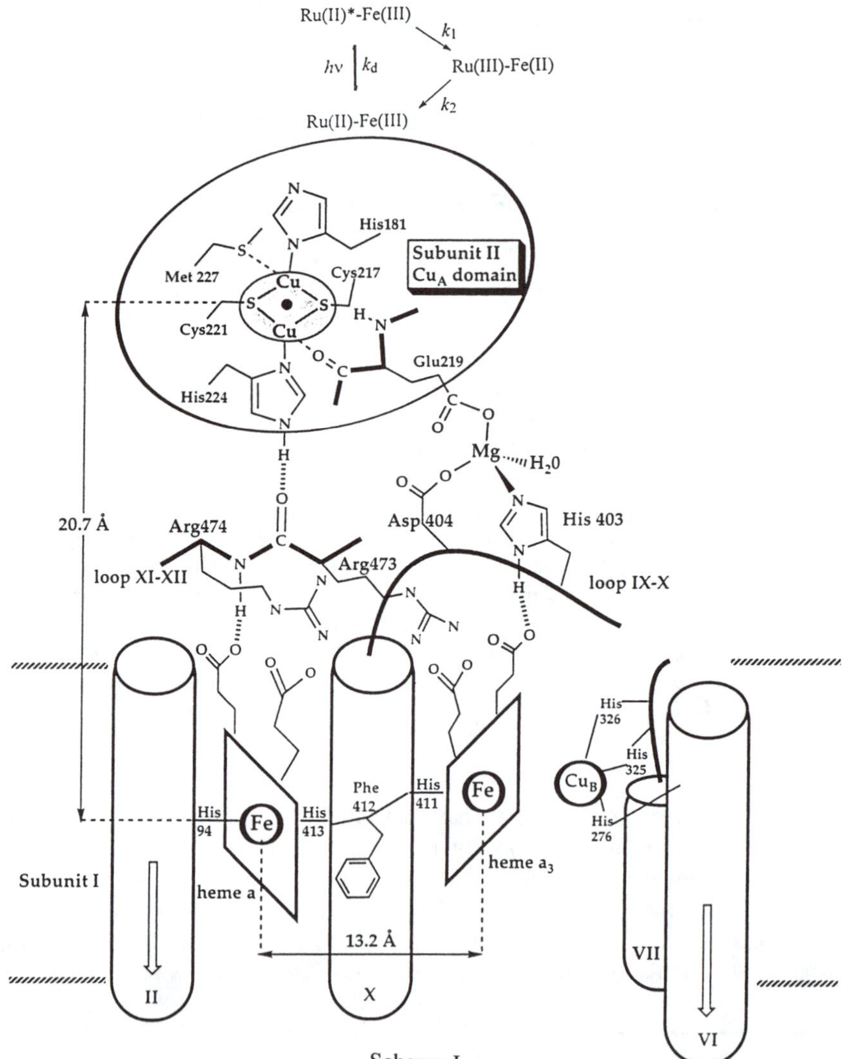

Scheme I

These intramolecular electron transfer processes provide an opportunity to examine electron transfer within the protein environment. Addition of a reductant, such as aniline, results in efficient reaction of the Ru(III) with the reductant to form Ru(II), which leaves the heme iron in the reduced state. If a redox active metalloprotein is present in the solution, electron transfer between the reduced heme and the added protein can be observed. Production of reduced heme iron and removal of the Ru(III) intermediate can be accomplished within a few hundred nanoseconds, which allows the study of extremely rapid interprotein electron transfer reactions.

In recent publications (*5, 6*), we described studies of the electron transfer reaction between horse heart cytochrome *c* and synthetic rat liver cytochrome b_5. We observed a rate constant for electron transfer of $4 \times 10^5\ s^{-1}$ in low ionic strength solutions in which these proteins form a strong 1:1 complex (*13*). The rate constant presumably describes the rate of electron transfer between the two heme centers held together in a specific geometry with a well defined electronic coupling. Alternatively, it may describe electron transfer in a collection of protein complexes in rapid dynamic equilibrium. The binding geometry of the electrostatic complex formed by cytochrome *c* and cytochrome b_5 has been modeled by Salemme and co-workers (*14, 15*) and Northrup et al. (*16*).

If electron transfer between cytochromes *c* and b_5 is viewed in the context of the theoretical description outlined originally by Marcus (*11*), then the rate constant for electron transfer, k_{et}, is given by equation 1.

$$k_{et} = \frac{4\pi^2}{h} H_{AB}^2 (4\pi\lambda k_B T)^{1/2} \exp\left[\frac{-(\lambda + \Delta G°)^2}{4\lambda k_B T}\right] \quad (1)$$

This equation describes a model for electron transfer with two fundamental parameters: H_{AB}, which describes the electronic coupling between the redox centers, and λ, the reorganizational energy, which is a measure of the energy required to alter the solvent and ligands surrounding the redox centers before electron transfer, can take place. The remaining symbols have their usual meaning; T is temperature, h is Planck's constant, k_B is Boltzmann's constant, and $\Delta G°$ is the full energy of reactions. If the rate constant and the reorganizational energy of the reaction are known, then the magnitude of the electronic coupling between the redox centers can be determined. Measurement of the electronic coupling between cytochromes *c* and b_5 will provide valuable information about the geometry of the protein–protein interaction.

This calculation hinges on an accurate measurement of the reorganizational energy. A search of the literature indicates that there is significant uncertainty in the reorganizational energy of the cytochrome *c*–b_5 system. For example, the self-exchange reactions for both cytochrome *c* and cytochrome b_5 have been investigated by Dixon and co-workers (*17*). Intrinsic reorganizational energies of 0.72 and 1.2 eV, respectively, were derived from the self-exchange rate constants. If $\lambda(c\text{–}b_5) = [\lambda(c) + \lambda(b_5)]/2$, then the reorganizational energy for electron transfer between cytochrome *c* and b_5 is 1.0 eV. In contrast, Northrup et al. (*16*) used a value of 0.7 eV for the intracomplex reorganizational energy. Marcus and Sutin (*18*) have calculated a value of 1.04 eV for the reorganizational energy for the cytochrome *c*, which suggests a value of 1.1 eV or larger for the reorganizational energy for the cytochrome c–b_5 reaction. We have been able to show that the reorganizational energy of cytochrome b_5 is 1.3 eV in the electrostatic complex formed with cytochrome *c* at low ionic strength. In these experiments, we assumed that the reorganizational energy of cytochrome b_5 was given by the relation $\lambda(\text{Ru}–b_5) = [\lambda(\text{Ru}) + \lambda(b_5)]/2$, where λ(Ru) is the reorganizational energy for $Ru(bpy)_3^{2+/3+}$, where bpy = 2,2′-bipyridine and $\lambda(\text{Ru}–b_5)$ is the reorganizational energy determined from the intramolecular electron transfer reactions of cytochrome b_5 covalently bonded to a series of ruthenium polypyridine complexes. The reorganizational energy, $\lambda(\text{Ru}–b_5)$, was obtained from a plot of free energy of reaction versus ln (k_{et}) and is equal to $\Delta G^{\circ\prime}$ at the maximum point of the curve. The cytochrome b_5 used in these studies was genetically engineered to contain a unique reactive site for the ruthenium complex that placed it outside the binding domain for cytochrome *c*. The fact that the ruthenium complex does not interfere with binding has been verified (*5, 6*).

In the present study we report preliminary data describing the rates of intramolecular electron transfer of cytochrome *c* covalently linked to several of the same ruthenium complexes used in the cytochrome b_5 study. Data were obtained with a variant of cytochrome *c* genetically engineered to place the ruthenium complexes outside of the binding domain. The goal of these experiments was to determine the reorganizational energy for the intramolecular electron-transfer reactions and calculate the reorganizational energy for cytochrome *c*, $\lambda(c)$, and for the cytochrome c–b_5 reaction, $\lambda(c–b_5)$, as described in the preceding paragraph.

Experimental Details

Materials. The preparation of the H39C;C102T variant of yeast cytochrome *c* has been described by Hilgen and Pielak (*19*). Reaction of this variant with $Ru(bpy)_2(BrCH_2bpyCH_3)^{2+}$ and subsequent purification and characterization of the labeled protein have been described by Geren et al. (*20, 21*). The other derivatives were prepared by analogous methods. Preparation of the ruthenium complexes has been described by Scott et al. (*9*) and Ernst and Kaim (*22*).

Kinetic Measurements. Rate measurements were performed using laser flash photolysis techniques. The excitation source consisted of 10-ns pulses of 355-nm light obtained from a Quanta-Ray Nd:YAG laser. The probe beam, at right angles to the excitation pulse, was obtained from a pulsed Xe arc lamp.

The detector was a photomultiplier tube in a five-dynode configuration with associated electronics of local design. Signals were recorded on a LeCroy 7200 series digital oscilloscope as 2000 point records and transferred to a personal computer for storage and analysis.

The kinetic equations that describe the reactions given in Scheme I have been reported by Pan et al. (7). Emission at wavelengths greater than 580 nm and transient absorbance changes at 434 and 550 nm were recorded and fitted simultaneously to these equations. The emitted light from the ruthenium complexes provided a measure of the excited-state decay rate ($k_1 + k_d$), as did the absorbance changes at 434 nm, which is an isosbestic point for changes in cytochrome *c*. Absorbance changes at 550 nm reflect changes in the redox state of cytochrome *c*. Baseline corrections to the 550-nm transient were made with data collected at 556 nm, which is an isosbestic point for cytochrome c. In cases where the oxidation of the heme iron was slow compared to the lifetime of the excited state, k_2 was obtained from a simple exponential fit of the transient absorbance changes at 417 or 550 nm.

Kinetic measurements were made in solutions containing 1 mM phosphate buffer at pH = 7 at 22 °C. Protein concentrations were 5 to 15 μM. The presence of dissolved oxygen had no measurable effects on the transient absorbance changes or the emission measurements.

Results

Yeast cytochrome *c* has been covalently attached to a series of ruthenium–polypyridine complexes at Cys39. A variant of yeast cytochrome *c* containing cysteine in place of His39 and threonine in place of Cys102 was used in this study. The variant was prepared using genetic engineering techniques. The attachment site was chosen such that binding of cytochrome c to other proteins (i.e., cyt b_5, cyt *c* peroxidase, and cyt *c* oxidase) would not be affected by the presence of the bulky ruthenium complexes. Cys39 and the attached ruthenium complex point toward the lower part of the back side of cytochrome *c*, as illustrated in Figure 1. The binding domain, in general, involves the area around the exposed heme edge and the surrounding lysines. Cysteine 102 was changed to a threonine in this case so that only one surface cysteine residue was available for reaction with the ruthenium complex and to avoid possible disulfide bond formation between two proteins.

The ruthenium complexes were attached to the specified cysteine by formation of a thioether linkage between the sulfur atom of cysteine and the methylene carbon of one of the bipyridine ligands. The reaction makes use of complexes that contain 4-bromomethyl-4′-methylbipyridine, as indicated.

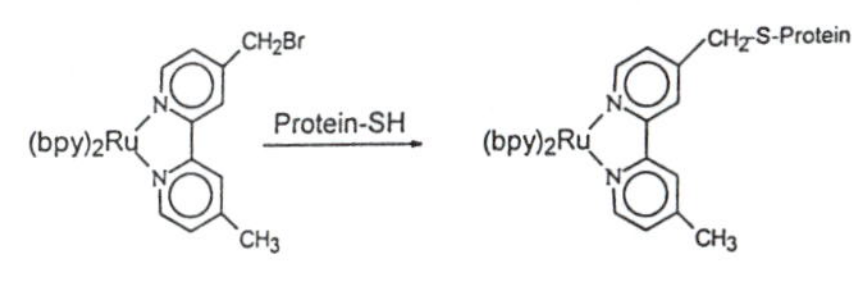

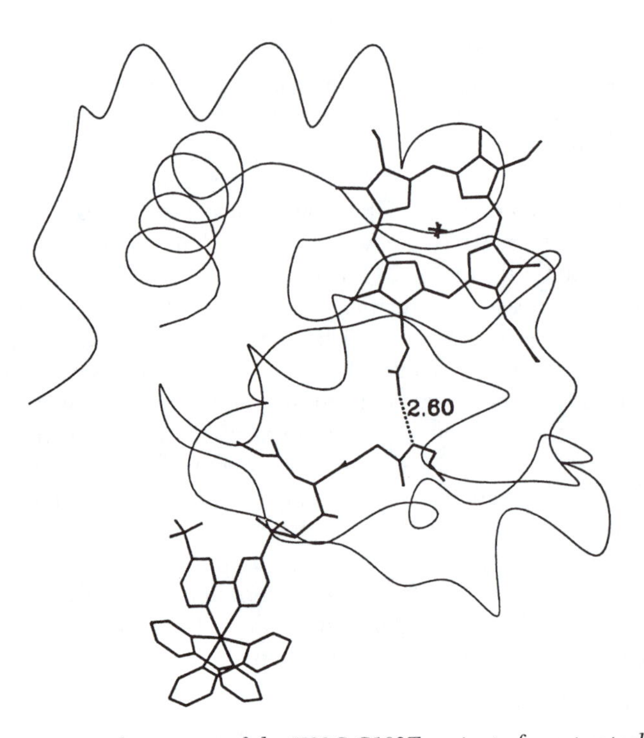

Figure 1. Proposed structure of the H39C;C102T variant of yeast cytochrome c *covalently attached to* $Ru(bpy)_2(CH_3bpyCH_2-)^{2+}$ *at the sulfur atom of Cys39. The pathway for electron transfer proposed by Wuttke et al.* (25) *extends from the ruthenium complex, through Ser40 and Gly41, and includes a hydrogen bond from the amide nitrogen of Gly41 to the heme propionate group. The structure is based on the X-ray structures of yeast iso-1-cytochrome* c *and* $Ru(bpy)_3^{2+}$, *with modifications made using the molecular modeling program Insight II from Biosym, San Diego, CA.*

The remaining bipyridine ligands of the ruthenium complexes can be changed to other ligands to give redox potentials that span a range of free energies of reaction comparable to the expected reorganizational energy. In this study, complexes were prepared that contained 4,4′-dimethylbipyridine and 3,3′-bipyridazine. The overall redox potentials for the electron-transfer reactions investigated are listed in Table I and cover a range from 0.75 to 1.26 V.

Table I. Rate Constants for Intramolecular Electron Transfer in H39C;C102T Variant of Yeast Cytochrome c Covalently Bound to Ruthenium Complexes at Cys39.

	Rate constants (s^{-1})		
Complex	k_1	k_2	$E^{o\prime}$, V[a]
$Ru(bpy)_2(CH_3bpyCH_2-)^{2+}$	6.2×10^5	9.2×10^5	1.09, 1.01
$Ru(CH_3)_2bpy)_2(CH_3bpyCH_2-)^{2+}$	5.2 ts 10^5	6.0×10^5	1.17, 0.91
$Ru(bpdz)_2(CH_3bpyCH_2-)^{2+}$	4.0×10^5	6.0×10^5	0.75, 1.23
$Ru(bpdz)_2(CH_3bpyCH_2-)^{2+}$	k_{et} [Ru(I)–Fe(III)] = 5×10^5		1.26

NOTE: Data were obtained with solutions containing 1 mM phosphate buffer, pH = 7, at 22 °C. Estimated standard deviations in rate constants are 15%.

[a] Overall redox potential for reactions of excited-state Ru(II*) and Ru(III) with heme iron of cytochrome c, respectively.

and 434 nm. The rapid decrease in absorbance observed at 434 nm is due to the formation of the ruthenium-complex excited state. The decrease is followed by conversion of the excited state to Ru(III), which subsequently returns to the Ru(II) form of the complex. The absorption of the excited-state form of the complex is nearly identical to that of the Ru(III) form at this wavelength. The increase and decrease in absorbance at 550 nm correspond to the formation and subsequent oxidation of the Fe(II) heme.

The rate constants for intramolecular electron transfer in the H39C;C102T variant of cytochrome c are listed in Table I. All the rate constants obtained with the H39C;C102T variant fell within a narrow range of $4–9 \times 10^5\ s^{-1}$. Rate constants for intramolecular electron transfer from Fe(II) → Ru(III), Ru(II)* → Fe(III), and Ru(I) → Fe(III) were obtained with the H39C;C102T variant labeled with $Ru(bpdz)_2(CH_3bpyCH_2-)^{2+}$, where bpdz = 3,3′-bipyridazine. The Ru(I) form of the complex was generated by quenching of the excited state with 10 mM dimethylaminobenzoate. Formation of the Ru(I) form of the complex was verified by monitoring the absorbance at 504 nm, which is an isosbestic point for cytochrome *c*.

Discussion

Rate constants for a series of intramolecular electron transfer reactions involving ruthenium polypyridine complexes attached to Cys39 of the H39C;C102T variant of cytochrome *c* have been determined. These reactions cover a range of free energies of reaction from 0.7 to 1.26 V. It is of interest that all the rate constants given in Table I, with one marginal exception, fall within a range of two standard deviations. Thus, no statistically significant variation in rate constants with free energy of reaction was observed.

Two explanations for the independence from free energy of reaction are

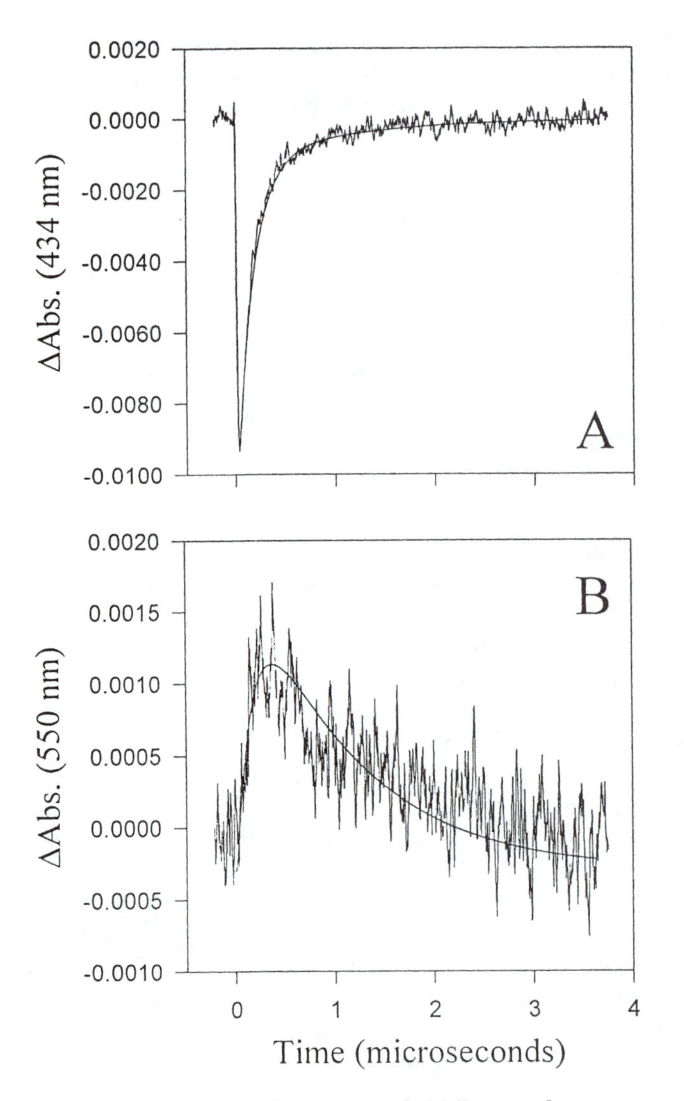

Figure 2. Transient absorbance changes recorded following photoexcitation of the yeast cytochrome c *covalently attached to $Ru(bpy)_2(CH_3bpyCH_2-)^{2+}$ at Cys39. Part A illustrates changes at 434 nm, and Part B shows changes at 550 nm after subtraction of the 556-nm transient absorbance changes. The protein concentration was 10 μM in 1 mM phosphate buffer at pH 7 and 22 °C.*

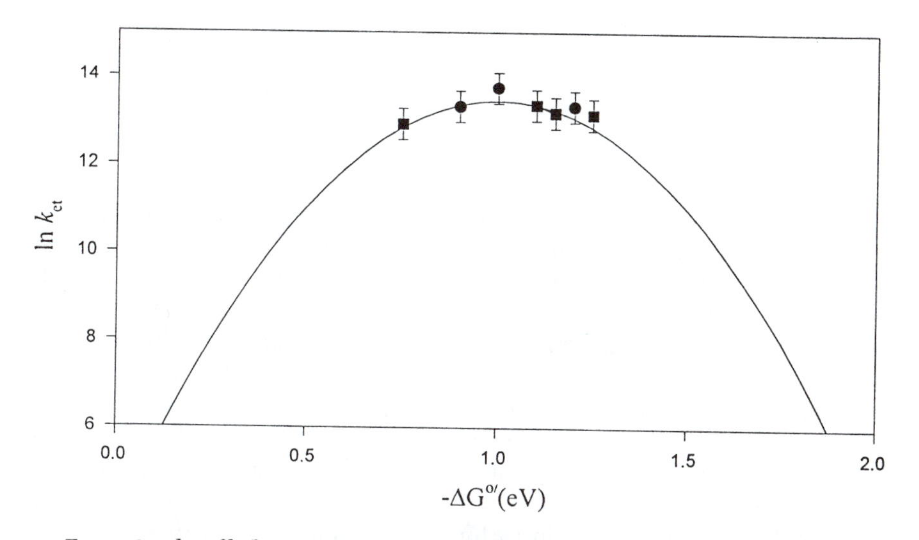

Figure 3. Plot of ln(k_{obs}) vs. the free energy for the intramolecular electron transfer reactions of the H39C;C102T variant of yeast cytochrome c. *The solid curve is a plot of equation 1 with* H_{AB} = *0.06 cm*$^{-1}$ *and* λ = *1 eV. The data points can be identified by using Table I. The rate constant* k_1, are indicated by ■ and k_2 *by* ●.

available. First, if the observed reactions were not limited by electron transfer, then a dependence of the observed rate constants on redox potentials would not necessarily be expected. Second, if the free energies of reaction were such that all the rate measurements fell within a small range around the point where $\Delta G^{o\prime} = \lambda$, then only changes in rate constant comparable to the errors in the determinations would be expected. The latter explanation is illustrated in Figure 3 with a plot of ln(k_{et}) versus the free energy of reaction. The solid curve indicates the best fit of the data to equation 1. The maximum in the plot occurs at $\Delta G^{o\prime} = -1.0$ V ± 0.05 and ln(k_{et}) = 13.6 ± 0.3 (i.e., λ = 1.0 ± 0.05 eV, H_{AB} = 0.06 ± 0.005 cm^{-1}).

Several investigators have indicated that internal vibrational modes of the metal centers can be critical in determining the rate constants for electron transfer, particularly in the region where $\Delta G^{o\prime} > |\lambda|$. Thus equation 1 represents a simplified form of equation 2 in which contributions from an average internal vibration are considered.

$$k_{et} = \frac{4\pi^2}{h} H_{AB}^2 (4\pi\lambda_{out} k_B T)^{-1/2} \sum_{m=0}^{\infty} \left[\frac{e^{-S} S^m}{m!}\right] \exp\left[\frac{-(\lambda_{out} + \Delta G^\circ + mh\nu)^2}{4\lambda_{out} k_B T}\right] \ldots \quad (2)$$

where

$$S = \frac{\lambda_{in}}{h\nu} \quad (3)$$

In this equation, λ_{in} and λ_{out} are the contributions to the reorganizational energy from rearrangement of the ligands coordinated to the metal and reorganization of the solvent, respectively and m integer is defined by summation symbols all values 0 to ∞. The additional terms present in equation 2, however, do not contribute significantly in the region of free energy represented by the complexes used in this study.

Gray and co-workers (*24, 25*) have investigated the rate constants for intramolecular electron transfer in cytochrome *c* covalently bound to $Ru(bpy)_2$ $(imidazole)^{2+}$ at His39. The rate constants have been analyzed in the context of the pathways models suggested by Beratan, Betts, and Onuchic (*26*) using equation 1. The dominant pathway for electron transfer involves 11 covalent bonds and a hydrogen bond between Gly41 and the heme propionate (*see* Figure 1). The electronic coupling term H_{AB} found using a series of ruthenium complexes attached at this location is 0.1 cm^{-1}. The complexes used in the present study share the same through-protein coupling components (independent of the model) with the addition of two covalent bonds between the bipyridine ligand and the protein backbone atoms. In the present study, we obtained H_{AB} = 0.06 cm^{-1}. If we extend the pathway calculated by Gray and co-workers by including two additional covalent bonds, then H_{AB}(calc) = 0.04 cm^{-1}, which agrees well with the observed value. The agreement supports the interpretation of the observed rate constants as valid indicators of the rate of the intramolecular electron transfer reactions indicated in Scheme I.

From the plot shown in Figure 3 and the relation $\lambda(c\text{–Ru}) = [\lambda(c) + \lambda(\text{Ru})]/2$ with $\lambda(\text{Ru})$ = 0.56 eV (*23*), we calculate a reorganizational energy for cytochrome *c* of 1.5 eV. This value is substantially higher than previous investigations have indicated. For example, Dixon and co-workers (*17*) have suggested that $\lambda(c)$ = 0.72 eV, while Marcus and Sutin (*18*) have suggested $\lambda(c)$ = 1.0 eV. Gray and co-workers (*24, 25*) have determined the intramolecular electron transfer rate constants for several ruthenium bipyridine complexes covalently linked to His39. These investigators indicate that the data are best described by $\lambda(c\text{–Ru})$ = 0.8 eV and $\lambda(c)$ = 1.04 eV.

Summary

The observed rate constants for intramolecular electron transfer between the heme center of cytochrome *c* and covalently bonded ruthenium complexes appear to be true measures of rates of intramolecular electron transfer despite the independence of rate from the free energy of reaction. The calculated value of reorganizational energy, however, is in conflict with previous work by Gray and co-workers (*24*), whereas the electronic coupling is consistent with expectations based on the same work. Additional rate measurements obtained using complexes with similar reorganizational energies but lower redox potentials will be required to resolve the conflict. It is of interest that these rate measurements will probably require the use of pulse radiolysis techniques because related ruthenium complexes with low reduction potentials generally do not show the photoredox behavior exhibited by $Ru(bpy)_3^{2+}$. The use of photochemically produced oxidants (typically at concentrations of 10^{-6} M) to oxidize the ruthenium complex and initiate intramolecular electron transfer is probably not feasible. Under these conditions, diffusion of the oxidant would limit the observable electron transfer rate constants to approximately 10^3 s^{-1}.

Acknowledgments

This work was supported in part by National Institutes of Health Grants GM42501 to G. J. Pielak, and GM20488 to Frank Millett and Bill Durham.

References

1. Hahm, S.; Durham, B.; Millett, F. *Biochemistry* **1992,** *31,* 3472.
2. Hahm, S.; Miller, M. A.; Geren, L.; Kraut, J.; Durham, B.; Millett, F. *Biochemistry* **1994,** *33,* 1473.
3. Liu, R-Q.; Hahm, S.; Miller, M.; Durham, B.; Millett, F. *Biochemistry* **1995,** *34,* 973.
4. Pan, L. P.; Hibdon, S.; Liu, R-Q.; Durham, B.; Millett, F. *Biochemistry* **1993,** *32,* 8492.
5. Willie, A.; Stayton, P. S.; Sligar, S. G.; Durham, B.; Millett, F. *Biochemistry* **1992,** *31,* 7237.
6. Willie, A.; McLean, M.; Liu, R-Q.; Hilgen-Willis, S.; Saunders, A. J.; Pielak, G. J.; Sligar, S. G.; Durham, B.; Millett, F. *Biochemistry* **1993,** *32,* 7519.
7. Pan, L. P.; Frame, M.; Durham, B.; Davis, D.; Millett, F. *Biochemistry* **1990,** *29,* 3231.
8. Heacock, D. H.; Liu, R-Q.; Yu, C-A.; Yu, L.; Durham, B.; Millett, F. *J. Biol. Chem.* **1993,** *268,* 27171.
9. Scott, J. R.; Willie, A.; McLean, M.; Stayton, P. S.; Sligar, S. G.; Durham, B.; Millett, F. *J. Am. Chem. Soc.* **1993,** *115,* 6820.
10. Scott, J. R.; McLean, M.; Sligar, S. G.; Durham, B.; Millett, F. *J. Am. Chem. Soc.* **1994,** *116,* 7356.
11. Marcus, R. A. *J. Chem. Phys.* **1956,** *24,* 966.
12. Kalyanasundaram, K. *Photochemistry of Polypyridine and Porphyrin Complexes;* Academic: Orlando, FL, 1992.
13. Mauk, M. R.; Reid, L. S.; Mauk, A. G. *Biochemistry* **1982,** *21,* 1843.
14. Salamme, F. R. *J. Mol. Biol.* **1976,** *102,* 563.
15. Wendoloski, J. J.; Matthew, J. B.; Weber, P. C.; Salemme, F. R. *Science (Washington, D.C.)* **1987,** *238,* 794.
16. Northrup, S. H.; Thomasson, K. A.; Miller, C. M.; Barker, P. D.; Eltis, L. D.; Guillemette, J. G.; Inglis, S. C.; Mauk, A. G. *Biochemistry* **1993,** *32,* 6613.

17. Dixon, D. W.; Hong, X.; Woehler, S. H.; Mauk, A. G.; Sishta, B. P. *J. Am. Chem. Soc.* **1990,** *112,* 1082.
18. Marcus, R. A.; Sutin, N. *Biochim. Biophys. Acta* **1985,** *811,* 265.
19. Hilgen, S. E.; Pielak, G. J. *Protein Eng.* **1991,** *5,* 575.
20. Geren, L.; Hahm, S.; Durham, B; Millett, F. *Biochemistry* **1991,** *30,* 9450.
21. Geren, L.; Beasley, J. R.; Fines, B. R.; Saunders, A. J.; Hibdon, S.; Pielak, G. J.; Durham, B.; Millett, F. *J. Biol. Chem.* **1995,** *270,* 2466.
22. Ernst, S. D.; Kaim, W. *Inorg. Chem.* **1989,** *29,* 1520.
23. Sutin, N. *Acc. Chem Res.* **1982,** *15,* 275.
24. Winkler, J. R.; Gray, H. B. *Chem. Rev.* **1992,** *92,* 369.
25. Wuttke, D. S.; Bjerum, M. J.; Winkler, J. R.; Gray, H. B. *Science (Washington, D.C.)* **1992,** *256,* 1007.
26. Beratan, D. N.; Betts, J. N.; Onuchie, J. N. *J. Phys. Chem.* **1992,** *96,* 2852.

8

Tubular Breakdown of Electron Transfer in Proteins

J. J. Regan, F. K. Chang, and J. N. Onuchic

Department of Physics, University of California at San Diego, 9500 Gilman Drive, La Jolla, CA 92093–0319

We propose and apply a new tubular approach to compute the electron tunneling coupling in a protein. This approach goes beyond the single-pathway view to incorporate multiple-path effects and expose how interference arising from the structure can determine the coupling. An application to recent experiments in Ru-modified azurin is presented. The experimental data are used in a novel way to determine the proper effective electron tunneling energy to use in the model. The data are interpreted in terms of interfering tubes, and hydrogen bonds play a critical role in this interference. As tubes can be blocked or created by mutation, the theory suggests how experimental control of rates can be achieved.

Multiple-site experiments (*1*) have led to a new theoretical approach for electron tunneling (ET) beyond the single-pathway picture (*2–4*). This approach emphasizes tubes—tightly grouped families of pathways—and looks for interactions between these families rather than focusing on individual paths. In some cases, for a given donor D and acceptor A, the electron transfer can be thought of as "pathway-like," wherein the protein bridge can be physically reduced to a tube without changing the overall coupling. In other cases, the transfer is characterized by multiple tubes that can interfere with one another, and a single-path assumption will fail to identify all the structural elements that control the coupling. Reducing the protein to only the relevant parts (tubes) that mediate the tunneling matrix element is a useful tool for understanding ET in a biological medium.

The Electron Transfer Model

The ET model used here arises from the standard non adiabatic expression for the rate constant for electron transfer k_{ET}:

$$k_{\mathrm{ET}} = \frac{2\pi}{\hbar} |T_{\mathrm{DA}}|^2 \, F.C. \tag{1}$$

where $\hbar$ is Planck's constant divided by 2π, *F.C.* is the Franck–Condon density of nuclear states, and T_{DA} is the tunneling matrix element, given by

$$T_{\mathrm{DA}}(E_{\mathrm{tun}}) = \sum_{\mathrm{d,a}} \beta_{\mathrm{Dd}} G_{\mathrm{da}} \, (E_{\mathrm{tun}}) \, \beta_{\mathrm{aA}} \tag{2}$$

$$G = \frac{1}{E_{\mathrm{tun}} - H} \tag{3}$$

where E_{tun} is the energy of the tunneling electron, d and a are the bridge orbitals coupled to D & A, respectively. β's are the couplings and G_{da} the Green's function propagator between d and a. In the couplings discussed here, differences in T_{DA} are expected to be much larger than differences in the *F.C.* term because the rates reported here are k_{max} with an optimized *F.C.* factor. Also, from experiment to experiment the local environment of D and A stays the same while the intervening medium changes. One therefore expects the relative rates to be determined by the tunneling matrix element T_{DA} alone.

All electronic properties of the protein are contained in *H*, a single-electron tight-binding (5) Hamiltonian, representing the protein. A "state" in this system is an electron residing in a particular tight-binding site, and the only sites used are (1) the σ-bonding orbitals and lone-pair orbitals in the protein matrix, and (2) an orbital centered on both the Cu and the Ru. These orbitals are simply localization sites and are not treated in any further detail; *H* is just a large extended-Hückel-like matrix, with a dimension equal to the number of orbitals recognized in the protein. An off-diagonal element in *H* is the coupling between two states and is directly related to the probability that an electron will move between the two sites involved. Two of the sites in the protein are special in that they are associated with the D and A states, whereas the remaining states in the protein are collectively referred to as the bridge (*H* is partitioned into a DA subsystem and a bridge subsystem). The sum in equation 2 is over the bridge entrance and exit states, indexed by d and a, respectively. These are the states with direct coupling to D and/or A. When the energies of D and A are close to each other relative to their distance in energy from the closest bridge state, and when coupling to the bridge is small relative to this distance, then (6) the DA subsystem can be viewed as an effective two-state Hamiltonian with a coupling determined by virtual occupation of the bridge. The so-called tunneling energy, E_{tun}, is an energy parameter indicative of the energies of the D and A states. All bridge states have energy zero on the energy scale used here. The direct coupling between two orbitals which share the same atom (a covalent link) is taken as a constant γ (a pathway-like Hamiltonian).

In this model, a "hydrogen bond" (H bond) is an interaction between a

σ-bonding orbital (between a heavy atom and a hydrogen) and a lone-pair orbital on another heavy atom. If one arranges H so that the diagonal is ordered like the amino acid sequence, then H-bonds (and through-space jumps) are far-off diagonal elements of H. Previous experience has shown that H-bonds are vital for mediating ET in proteins, and for strong H-bonds, such as those involved in protein secondary structure, our current results indicate that their contribution to ET is comparable to that provided by covalent links. H-bond couplings are treated as distance-independent covalent links (providing a direct coupling of γ). Recent experimental results (*7, 8*) support this hypothesis. The Hamiltonian used here has only covalent links and H-bonds; no through-space jumps are important.

The upshot of this is that the theory models the protein bridge with precisely one parameter: the ratio γ/E_{tun}. All the covalent bonds and H-bonds in the bridge are treated as equivalent (i.e., they are all γ), meaning that the Hamiltonian matrix H used with E_{tun} to compute $G = 1/(E_{\text{tun}} - H)$ is just γ times a sparse matrix of 1's. The ratio γ/E_{tun} appears in expansions of G matrix elements. This is a highly simplified picture of the protein. Despite its simplicity, the H used here exhibits the primary features required for this problem. There is a rough exponential decay of coupling with distance (down a tube), the coupling sign changes (*9*) with each step from orbital to orbital (because $\gamma/E_{\text{tun}} < 0$), quantum interference effects (interfering tubes) arise that have a direct connection to the secondary and tertiary structure of the protein, and most important, the computed ratios of T_{DA} are within an order of magnitude of experimental rate ratios (see the section "An Example in Azurin"). More complicated Hamiltonians could and should be used, but they will only be understood in terms of the basic features already present in this Hamiltonian.

G_{da} as a Sum of Pathways. A pathway is a specific sequence of bridge orbitals, starting at a site d (which is directly coupled to the donor D), and ending at a site a (directly coupled to A)—for example, the $-\text{N}-\text{C}_\alpha-\text{C}-\text{N}-\text{C}_\alpha-\text{C}-$ bond sequence of a protein backbone could be a segment of a pathway. The determination of the relevance of a particular pathway is discussed below in the section "The Pathway Approximation"; here paths are simply defined and discussed in general terms. Consider a bridge with only two states, as in Figure 1, where the donor is coupled to only one of the bridge

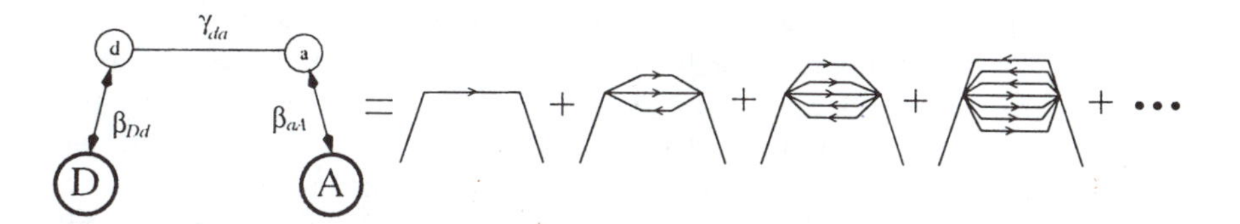

Figure 1. A simple two-state bridge, with its effects in G_{da} expanded as a sum of pathways.

states and the acceptor is coupled to the other. Because of this, the sum in equation 2 has only one term, $T_{DA}(E_{tun}) = \beta_{Dd}\, G_{da}(E_{tun})\, \beta_{aA}$. Recognizing the bridge as a two-state subsystem, for which the desired Green's function matrix element has a simple form, we have

$$G_{da}(E) = \frac{\gamma_{da}}{(E - E_1)(E - E_2) - \gamma_{da}^2} = \frac{\gamma_{da}}{(E - E_1)(E - E_2)} \times \left(1 + \frac{\gamma_{da}^2}{(E - E_1)(E - E_2)} + \frac{\gamma_{da}^4}{[(E - E_1)(E - E_2)]^2} + \cdots\right) \quad (4)$$

The G_{da} matrix element can be written as a sum of terms, each of which corresponds to a specific sequence of states (a pathway). The terms in this sum are depicted in Figure 1. In this simple case of a two-state bridge, each term multiplies in another factor of γ^2, representing a trip from a back to d and back to a again. More complicated bridges are harder to expand this way, but the same idea applies. This "sum of pathways" view shows the potential role of interference in this model. There will be interference effects buried in the calculation of an individual G_{da} (as in this example), and there will be interference effects in the T_{DA} sum, where different G_{da} matrix elements between different bridge exit and entrance points are added. Although these two categories are useful, it is more useful to define two categories of interference effects that are closer to each other than these two extremes.

The Pathway Tube. The first category of interference is called trivial interference, and this is the interference that arises from nearest-neighbor, next-nearest-neighbor, and backscatter effects in propagation down a simple structure, like an ideal linear alkane chain or a protein backbone, centered on a core pathway. The coupling provided by a protein backbone is a much stronger function of backbone length than it is of the types of residues encountered along the backbone. The amide hydrogens, the lone-pairs on the oxygens, even the residues themselves, all provide trivially different alternative pathways that interfere in a way that can easily be renormalized (*10*) into a much simpler set of states with the connectivity of a string of pearls. An example of this is shown in Figure 2, which is discussed in the section "The Pathway Approximation."

A pathway tube is the set of states one finds by first identifying a core pathway (between some d and a) that never visits the same state twice, then adding to this set all nearest neighbors of the core states, then again adding the neighbors of these extra states. This catches all hanging orbitals off the core pathway, and this subset of the bridge is called a pathway tube. To find multiple tubes, a generalization of this method is used that differentiates between different tubes. Such tubes are shown in Figure 6. A tube is a centrally useful concept from an experimental point of view in that it can sometimes be blocked, or created, via molecular replacement.

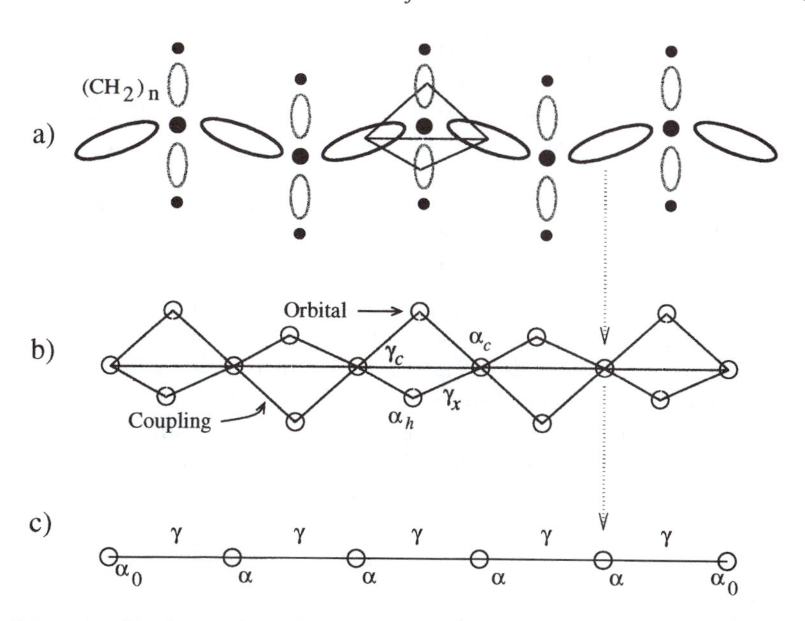

Figure 2. This figure shows how alkane would be converted to a Hamiltonian used in this model (a → b), and then how this periodic Hamiltonian could be further reduced (b → c) to result in the "string of pearls" H *below it. The new energies and couplings in this chain have the effects of the side groups renormalized into them.*

The Sum of Tubes. The second category of interference is that of interfering tubes. Once tubes have been defined, one can ask how they interfere. Whereas the "trivial" interference of the previous section is buried in the calculation of G_{da} for a single tube, this kind of interference can be thought of as sitting at the coarse level of the T_{DA} sum, or just below it. Sometimes there will be only one tube between a given bridge entrance (d) and exit (a) pair, in which case each G_{da} represents a single tube, and the tube interference is explicitly the T_{DA} sum. Sometimes, however, there will be multiple tubes per G_{da} matrix element, when more than one tube share the same D–A pair, and this is the level "just below" the T_{DA} sum.

The Pathway Approximation. The pathways approach (*11*) provides a way to find the virtual route through the protein matrix that contributes the most to the electronic coupling, and easily estimates the coupling provided by this route by turning the Green's function calculation into a simple scalar product along the path rather than an inversion of *H*. If a single tube dominates, then this product of decay factors is a good estimate of the coupling, as described below.

Renormalization. Any bridge can be "renormalized" using matrix partitioning (*12*). The idea is to simply divide up the bridge into the states one wants to keep (one will always want to keep the states at the bridge exit and entrance points, for example), and the states one wants to eliminate. These latter states are removed from the bridge, but their effect on the calculation of any matrix elements between the remaining states is perfectly accounted for by "renormalizing" the energies of these remaining states and the coupling between them. In general, one might want to eliminate states to clarify structural elements in the protein (e.g., reduce a residue to a single state), or to simplify later calculations, as discussed later.

For example, consider the alkane molecule of Figure 2a. Simple systems like these are frequently used in the ET literature (*13–15*). The σ-bonding orbitals between the C–C and C–H pairs have been identified with energies α_c and α_h respectively. Three classes of coupling have also been identified: γ_c, between C–C and C–C bonds; γ_h, between C–H and C–H bonds; and γ_x, between C–C and C–H bonds.

Consider the four-orbital subsystem outlined in Figure 2a, wherein the two C–H bonding orbitals physically located between the two C–C bonding orbitals are coupled to each other and to the C–C orbitals, but are otherwise isolated. From the point of view of the rest of the system, only the C–C bonding orbitals are visible, and one can define new energies and couplings for these orbitals which will include the effects of the C–H bonds implicitly. This permits one to reduce the system as shown in Figure 2b–2c. The original Hamiltonian,

$$H = \begin{bmatrix} \alpha_c & \gamma_x & \gamma_x & \gamma_c & & & & & \cdots \\ \gamma_x & \alpha_h & \gamma_h & \gamma_x & & & & & \\ \gamma_x & \gamma_h & \alpha_h & \gamma_x & & & & & \\ \gamma_c & \gamma_x & \gamma_x & \alpha_c & \gamma_x & \gamma_x & \gamma_c & & \\ & & & \gamma_x & \alpha_h & \gamma_h & \gamma_x & & \\ & & & \gamma_x & \gamma_h & \alpha_h & \gamma_x & & \\ & & & \gamma_c & \gamma_x & \gamma_x & \alpha_c & \gamma_x & \\ \vdots & & & & & & \gamma_x & \ddots & \end{bmatrix} \tag{5}$$

becomes the effective Hamiltonian,

$$H^{\text{eff}} = \begin{bmatrix} \alpha_0 & \gamma & & & & & \\ \gamma & \alpha & \gamma & & & & \\ & \gamma & \alpha & \gamma & & & \\ & & \gamma & \alpha & \gamma & & \\ & & & \gamma & \ddots & \gamma & \\ & & & & \gamma & \alpha & \gamma \\ & & & & & \gamma & \alpha_0 \end{bmatrix} \tag{6}$$

whose matrix elements are functions of E:

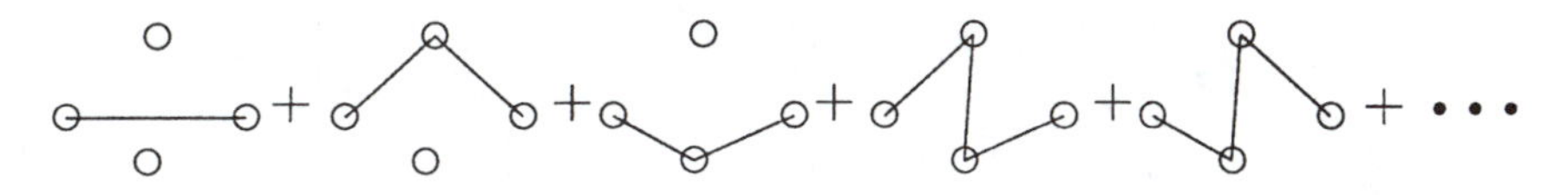

Figure 3. A path breakdown of the effective C–C coupling in the string of pearls renormalized from alkane (see the terms in equation 10).

$$\gamma = \gamma^{\mathrm{eff}}(E) = \gamma_c + 2\gamma_x^2/(E - \alpha_h - \gamma_h) \quad (7)$$

$$\alpha = \alpha^{\mathrm{eff}}(E) = \alpha_c + 4\gamma_c^2/(E - \alpha_h - \gamma_h) \quad (8)$$

$$\alpha_0 = \alpha_0^{\mathrm{eff}}(E) = \alpha_c + 2\gamma_c^2/(E - \alpha_h - \gamma_h) \quad (9)$$

(*see* da Gama (*16*) for a more detailed example of this renormalization procedure). In this simple example, only the end-point energies are different, and the coupling to the end point is the same as the couplings in the middle ($\gamma_0 = \gamma$). The new "Hamiltonian" is a function of E; it is only meaningful in the context of finding $G = 1/(E - H)$ matrix elements. The G matrix elements between the states that were retained (in this case the C–C bonding orbitals in an alkane bridge) in the reduced H will match those computed using the full H.

The C–C to C–C coupling in the reduced alkane (equation 7) can be expanded in terms that correspond to the paths in Figure 3:

$$\gamma^{\mathrm{eff}}(E) = \gamma_c + 2\frac{\gamma_x^2}{E - \alpha_h}\left(1 + \left(\frac{\gamma_h}{E - \alpha_h}\right) + \left(\frac{\gamma_h}{E - \alpha_h}\right)^2 + \cdots\right) \quad (10)$$

The first term is just the simple path straight through γ_c. The first correction to this is $2\gamma_x/E - \alpha_h$, representing the two side trips through the bonds to the H's. The remaining terms involve trips between the H's, with higher powers of γ_h representing multiple rebounds. The factor of 2 is there because the paths come in symmetric pairs, because there are two H's.

What is the effect of these corrections? To arrive at a number for T_{DA}, one must use a specific number E_{tun} in place of E to compute T_{DA}; this is the essence of the two-state approximation (*6*). If the band of bridge states providing the coupling is centered on the zero of the energy scale, then $E = E_{tun}$ (the DA state energy) should have a sign opposite that of any coupling γ (all γ's have the same sign). If $\alpha_h \sim 0$, then the sign of the second term is the sign of $E = E_{tun}$. This sign must oppose that of γ, so the first and second terms have opposite signs. In general, adding a "side" path with one extra step reduces overall coupling because of this localized destructive interference ("trivial renormalization"). The trips with two extra steps interfere constructively with the primary path, but they represent smaller corrections.

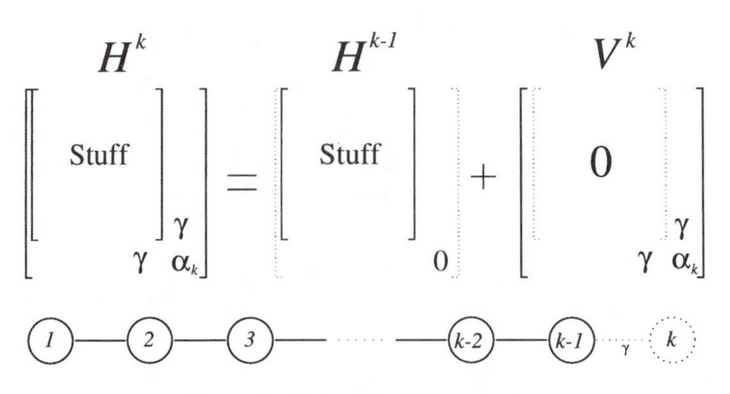

Figure 4. A "growing" Hamiltonian.

Obviously, one cannot address coupling between the C–H bonds in the reduced H, because they no longer exist. This is why, in practice, one must not "normalize away" the states that are directly coupled to D and A. Further, one must not "normalize away" so much of the protein that there is nothing left in the Hamiltonian that makes structural sense; this restraint is exercised in the next section, where a very specific partition and renormalization is done.

Decay per Step. The previous example suggested how a complicated periodic bridge could be reduced to a simple string of pearls. A nonperiodic bridge with some isolated structural elements can also be reduced to a string of pearls, but the site energies and couplings down the chain will not all be the same as they are here. The simplicity of this bridge can be further exploited.

Consider a "growing" Hamiltonian as in Figure 4 and the following equation, where the superscript k indicates the stage of growth:

$$H^k = H^{k-1} + V^k \tag{11}$$

By definition the Green's function can be computed from $(E - H^k)G^k = 1$ (equation 3), so

$$G^{k-1}(E - H^{k-1} - V^k)G^k = G^{k-1} \tag{12}$$

Thus

$$G^k = G^{k-1} + G^{k-1}V^kG^k \tag{13}$$

This is a recurrence relation for $G(E)$, wherein G^k appears on both sides of the equation. Now suppose one is interested only in the matrix elements of G between the end points 1 and k. By taking matrix elements of the above, one finds:

$$G^{k}_{1,k} = f_k G^{k-1}_{1,k-1} \tag{14}$$

where f_k is the bond decay given by

$$f_k \equiv \frac{\gamma}{E} \frac{1}{1 - \frac{\gamma}{E} f_{k-1}} \qquad (f_0 \equiv 0). \tag{15}$$

Expanding f in terms of γ/E, one finds

$$\frac{G^{k}_{1,k}}{G^{k-1}_{1,k-1}} = f_k \approx \begin{cases} f_2 + O\left[\left(\frac{\gamma}{E}\right)^5\right] & k > 2 \\ f_3 + O\left[\left(\frac{\gamma}{E}\right)^7\right] & k > 3 \\ \vdots & \end{cases} \tag{16}$$

That is, if γ/E is small, then once the bridge grows beyond 2 or 3 steps, adding another state to this chain-like bridge just multiplies the G matrix element between the end points by a nearly constant factor of f_2. This number is less than 1, so each additional step means a decay in the coupling. Further, for a physical choice of E_{tun}, this number is negative because γ/E_{tun} is negative. This means that each step results in a sign change in the G matrix element, with obvious implications for pathway interference that a single-path model cannot take into account.

The pathways scheme of Onuchic et al. (*17*) is the following. The protein is converted to a network of sites connected by "ε couplings." This network is basically like the tight-binding Hamiltonians discussed earlier, only the meaning and units of the direct couplings between the tight binding states are different. They define "pathway coupling" as

$$T_{DA} \propto \prod_{N-1} \varepsilon \tag{17}$$

where the ε is exactly the f_2 or f_3 defined above, and N is the number of states in the path with which the coupling is associated. The full pathway model allows for defining different ε's for through-space jumps and H-bonds, but this is covered elsewhere (*17*). The network can be searched for a path which maximizes this pathway coupling, and in a network where all ε's are equal, the maximum coupling path will also have the fewest number of steps because $\varepsilon < 1$. If ε is chosen correctly, the path coupling will exhibit the decay that corresponds to the experimentally measured decay through a simple covalently linked bridge with the effects of hanging orbitals and simple backscatter built in. Likewise, the path coupling should correspond (up to a constant) to the G_{da} one would compute over the tube whose core is the pathway.

Pathways Versus Tubes. The essential idea from the pathway approximation is that the Hamiltonian is converted to a searchable network, and one can compute estimates of T_{DA} that are (1) simple products corresponding to the number of bonds involved, and (2) expected to be correct (up to a prefactor) if the coupling provided by the entire protein bridge is well approximated by only the states included on the pathway tube. The concept of "finding a best path" is dependent on this concept of defining the coupling as a product of constant factors.

The first step in the tube approach is to use the pathways idea to isolate routes through the protein that are physically distinct (see the example in the next section), and then use these paths as tube cores. The analysis then proceeds with an emphasis on partitioning the protein into tubes; the tubes contain trivial interference effects, but they can expose crucial interference between tubes. This is something which one cannot do in the single-pathway approximation.

An Example in Azurin

The blue copper protein azurin plays an important role as an electron carrier (*18–21*) in biological systems. On the basis of extensive studies (*22–24*), it is likely that bimolecular ET reactions between azurin and other redox molecules take place via histidine 117, as it is directly coupled to the copper and close to the surface of the protein. Much less work has been done on long-range ET through the protein, although Farver, Pecht, and other investigators (*25–28*) have made an impressive start by studying ET between the copper atom and a distant disulfide bridge.

We have applied the tube approach to analyze three distant electronic couplings (*1*) in ruthenium-modified derivatives of azurin. Some highlights of this analysis (*29*) are provided in this section. The atomic coordinates used in the calculations of Ru–Cu coupling (Figure 5) come from the Brookhaven Protein Data Bank (PDB) entry "4AZU" (*21, 30*) (chain "A," one of four azurin molecules in the unit cell of this study). The placement of the Ru electron acceptor in these experiments was such that the intervening D–A medium was two β-strands. A best fit to an exponential decay with D–A distance for these results yields a decay constant of 1.09 $Å^{-1}$ (*1*); this is the softest decay that can be expected for σ-bond tunneling through a protein because a β-strand covers the longest distance with the smallest number of bonds. By analyzing ET that goes via more than a single strand, we have been able to show the importance of H-bonds in the coupling across a β-sheet. In addition, we have begun to investigate how the couplings between copper and its different ligands may influence the directionality of long-range ET.

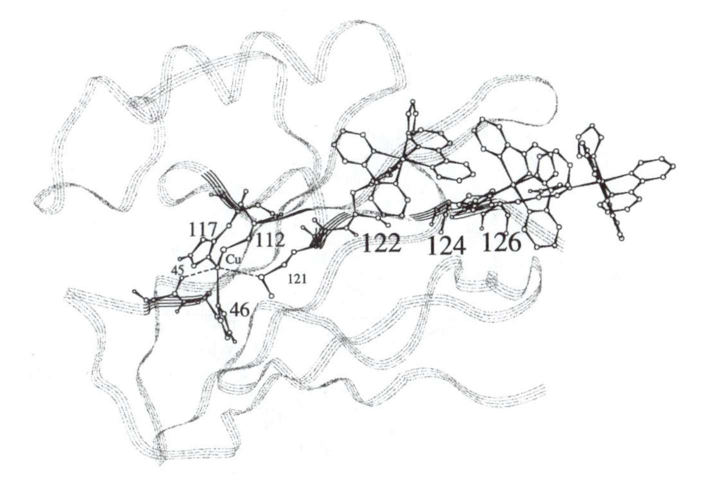

Figure 5. A ribbon model of the azurin molecule showing the copper, its ligands, and Ru–$(bpy)_2$(im)(histidine) groups at positions 122, 124, and 126. Each site represents a different experiment. The chain ends at 128.

The pathway tubes for the case of HIS126 are shown in Figure 6. These paths were found by first finding the best path, and then finding all paths with a path coupling that offered at least 1% of the coupling of the best path. This search would typically generate tens of thousands of paths, each unique in its sequence of states—but most paths would simply be trivial variations of their neighbors. This unmanageable set of paths would then be scanned to extract "tube cores." To do so, a special set of states *S* is initialized to zero. Then the paths are considered in order from highest path coupling to lowest. During consideration, a path is scanned for its smallest continuous segment of states that are not elements of *S*. If this segment is at least *X*% of the entire path (where *X* is typically 10), then the path is retained; otherwise it is deleted from the set of known paths. If it is retained, then all states on the path and within *N* steps of the path are added to the set *S* if they are not in it already (*N* is typically 1). This process continues until all known paths have been considered, and the set of paths has been dramatically reduced, usually to order 10. If one finds more than around 10 paths for a protein the size of azurin, it is probable that these paths are not "unique" enough, and that either *X* or *N* should be increased.

The tubes for 122 and 124 are subsets of those seen for 126 in Figure 6. If one calculates T_{DA} over a bridge that consists of just the states in pathway tubes, and if this coupling is the same as the number one finds over the entire protein, then (*31*) the protein can be reduced (from an ET point of view) to just the tubes, eliminating irrelevant superstructure to expose the important structural features.

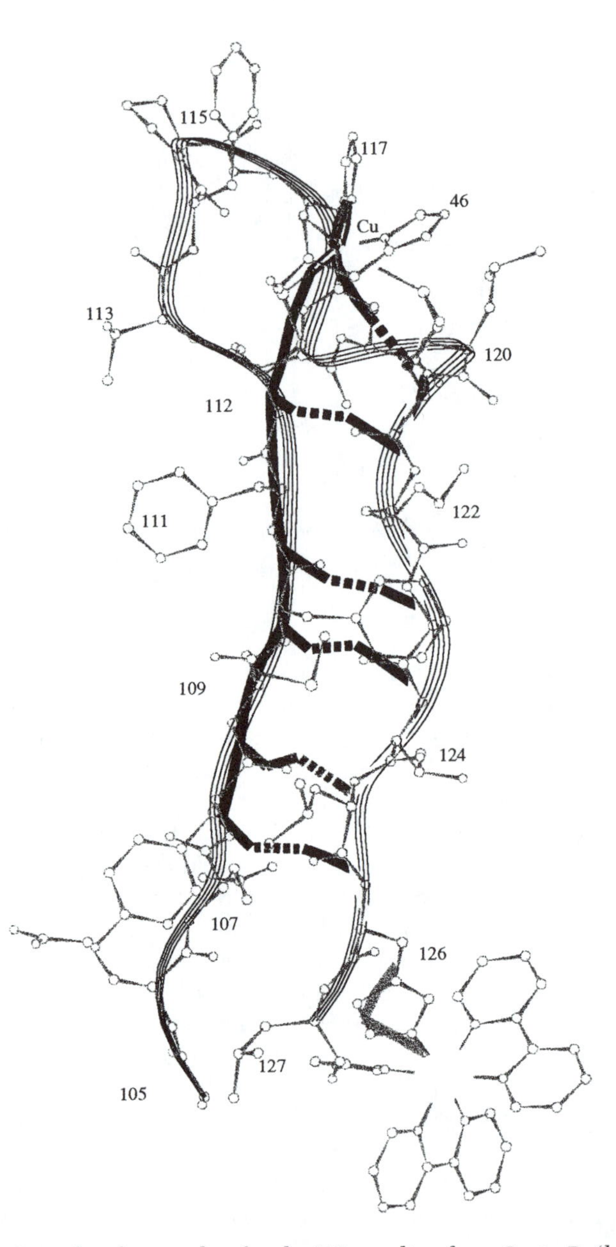

Figure 6. Sets of pathway tubes for the ET coupling from Cu to Ru(bpy)$_2$(im) in HIS 126-modified azurin. The shaded curves (dashed lines are H-bonds) indicate the cores of the tubes that together are responsible for effectively all the electronic coupling of the protein matrix. Nos. 122 and 124 are like 126, but with a subset of the tubes shown here. The copper ligands are 46, 117, and 112, and the coupling is dominated by the 112 β-strand.

The shortest distance, both through-space and through-bond, from the acceptor sites at 122, 124, and 126 back to the copper is directly through the β-strand to the MET at 121. This MET's sulfur is close (3.2 Å in the PDB entry mentioned above) to the copper, but it is not considered a Cu ligand. According to detailed analysis of the electron structure of the copper in this protein (*32*), the strongest coupling from the copper to any of its atomic neighbors is to the sulfur in the CYS at position 112. The coupling to the sulfur of 121 is relatively weak.

β-Strand Calibration of Tunneling Energy. The tunneling matrix element, T_{DA}, in the two-state model is a function of an energy parameter called the tunneling energy, or E_{tun}. This parameter should be indicative of the energies of the DA subsystem (*6*); here it is equal to the "band gap" (the distance from the energies of the DA subsystem to the closest eigen-energy of the bridge subsystem), because the energies of the bridge states have all been centered on zero. The value of E_{tun} is a long-standing question in this model of ET. As already noted, the energy of the DA states (E_{tun}) cannot be too close to the eigen-energies of the bridge, or the effective two-state model breaks down (see above) and the electron would be delocalized over D, A, and the bridge (as in a metal). If it is too far away relative to γ, the contributions to the coupling from nearest-neighbor interactions and other interactions may be improperly weighted. An effective E_{tun} (to be used with our simple Hamiltonian) must be assigned to give an appropriate coupling decay down a tube (with only covalent couplings) that is consistent with experiment.

From experiment to experiment, every D–A pair is associated with a single value of E_{tun}, independent of the intervening bridge. The relevant parameter, when comparing the theoretical model to experiments, is γ/E_{tun}. This "effective" γ itself depends on E_{tun} and on the overlap between orbitals. Therefore, E_{tun} is not a real physical energy since it is highly renormalized. This simple Hamiltonian, however, keeps the main feature of the wave function decay inside of the protein, that is, a single value of γ/E_{tun} is chosen to give an appropriate experimental decay with bond distance, and the wave function alternates sign with every covalent bond.

To compute T_{DA}, a choice must be made for the coupling between the DA subsystem and the bridge—the various β_{Dd} and β_{aA} in equation. 2. The couplings "weight" the different bridge Green's function matrix elements between D and A. In our treatment of azurin in this model, there are five bridge entrances providing routes out of the donor, and one bridge exit to the acceptor. The entrances are the Cu's atomic neighbors, and the weights associated with these entrances will be labeled (as in equation. 2) β_{D46} (2.0 Å), β_{D112} (2.3 Å), β_{D117} (2.1 Å), β_{D121} (3.2 Å), and β_{D45} (2.8 Å). The Cu to ligand atom distances are given in parentheses (*33*); neither 45 nor 121 is strictly a ligand, but each is in close proximity to the Cu. There is only one route into the ruthenium acceptor: via the *X*:HIS:NE2-Ru bond (X = 83, 122, 124, 126). Having only

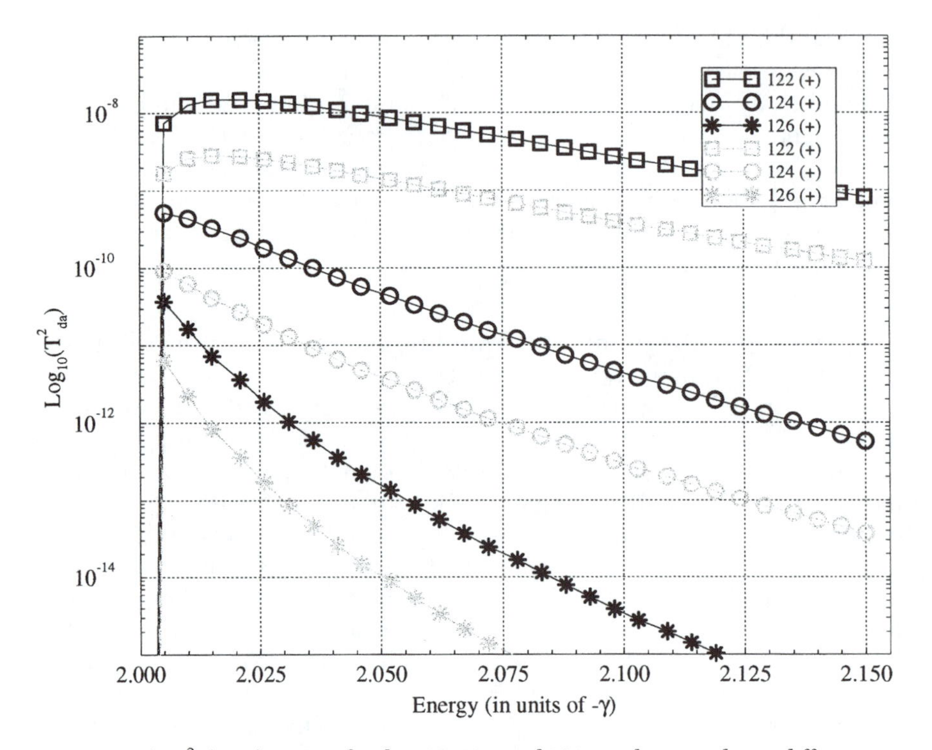

Figure 7. $T^2_{da}(E_{tun})$ *vs.* E_{tun} *for the 122, 124, and 126 couplings, with two different sets of weights on the donor couplings: (1)* $\beta_{D112} = \beta_{D46} = \beta_{D117} = \beta_{D121} = \beta_{D45} = 0.10$ *(gray), and (2)* $\beta_{D112} = 1.00$, $\beta_{D46} = \beta_{D117} = 0.25$, $\beta_{D121} = \beta_{D45} = 0.10$ *(black). The coupling to 122 is greater than the coupling to 124 at all energies because the DA separation is smallest in this case (likewise for 124 compared with 126).*

one bridge exit means having only one β_{aA}, which can then be pulled out of the sum as a T_{DA} scaling factor that goes away in any T_{DA} ratio.

To define the effects of these parameters, two plots will be made involving only the acceptors 122, 124, and 126, as these data are the easiest to interpret. A best fit to an exponential decay with D–A distance for these results yields a decay constant of 1.09 $Å^{-1}$ (*1, 3*); this is the softest decay that can be expected for σ-bond tunneling through a protein because a β-strand covers the longest distance with the smallest number of bonds.

Figure 7 shows a plot of T^2_{DA} as a function of E_{tun} for the 122, 124, and 126 couplings, for two different choices of weights. In both cases, as E_{tun} gets further from the bridge energy, T_{DA} drops off as we would expect from $G = 1/(E - H)$. In this single-electron model (no shielding), G's dependence on E_{tun} may be too strong, but the general trend should be that as expected for hole tunneling. The highest bridge eigenvalue is at -2γ ($\gamma < 0$); if we attempt to use an E_{tun} at this value or lower, we violate the two-state model. This is not a problem because, as described above, E_{tun} is chosen to provide the appropriate distance decay down a covalent chain for a given choice of D and A, and therefore it always falls outside the "bridge band" (the range of eigenvalues of the bridge).

These weights can be given reasonable values. In detailed studies of blue copper proteins, Solomon (*32, 34, 35*) analyzed the electronic structure of the metal and its ligands in an effort to better understand ligand-to-metal charge transfer interactions that, among other things, give these proteins their striking blue color. We only need a rough estimate of these couplings, as our model is only meant to be accurate to an order of magnitude in the bridging matrix element. Solomon found that the strongest coupling is to the SG sulfur in 112, and the weakest is to the SD sulfur in 121. If the absolute coupling to 112 is 1, then the absolute coupling to 121 is roughly 0.1. Any interaction with the lone-pair orbitals on 45:O is likely to be no more than the 121 coupling, so it is also set to 0.1. The coupling to the N's of 46 and 117 is, on the other hand, somewhat stronger; we will take it to be about 0.25 on this absolute scale. This ordering of CYS > HIS > MET will be referred to in what follows as the "rational" set of weights. The 112 interaction is clearly the strongest, and it will be seen to dominate the couplings in the experiments considered here, but as a rule all bridge entrances must be taken into account. Also, although not done here, it would be possible to position an acceptor so that Cu coupling to 112 would be unimportant compared to one of the other bridge entrances.

One set of T_{DA}^2 curves (gray) in Figure 7 was generated with all weights set to 0.1, the smallest weight in the rational set. The second set of curves (black) was generated using the rational weights. Note that the E_{tun} dependence line for 126 in Figure 7 shifts up more than the line for 122 does, with this general increase in coupling weights. This is because 126 has more paths going into it from the strong 112 coupling than 122 does (Figure 6). Because these tubes are all the same length, they interfere coherently (same sign and nearly the same magnitude), so any increase in the 112 coupling is bound to help 126 more than it will help 122. These paths traverse the ladder of H-bonds between the two parallel β-strands seen in Figure 6. The same argument applies to 124, although the effect is not as strong because it has fewer additional paths than 126 does.

Figure 8 shows how experimental data can be used to calibrate E_{tun}. In each of the ET reactions considered here, D and A (and the *F.C.* factors) are assumed to be the same, so we expect that E_{tun} will also be the same for all reactions. But what single value of E_{tun} should be used with our highly renormalized and simplified bridge Hamiltonian? The experimental data determine it: the proper E_{tun} to use is where the theoretical curves intersect the straight experimental lines in this figure: $E_{tun} \sim -2.05\gamma$. Recall that E_{tun} is an effective energy parameter; this value for E_{tun} makes sense when used with the simple Hamiltonian employed in this model and should not be converted to a real energy (see the discussion at beginning of this section).

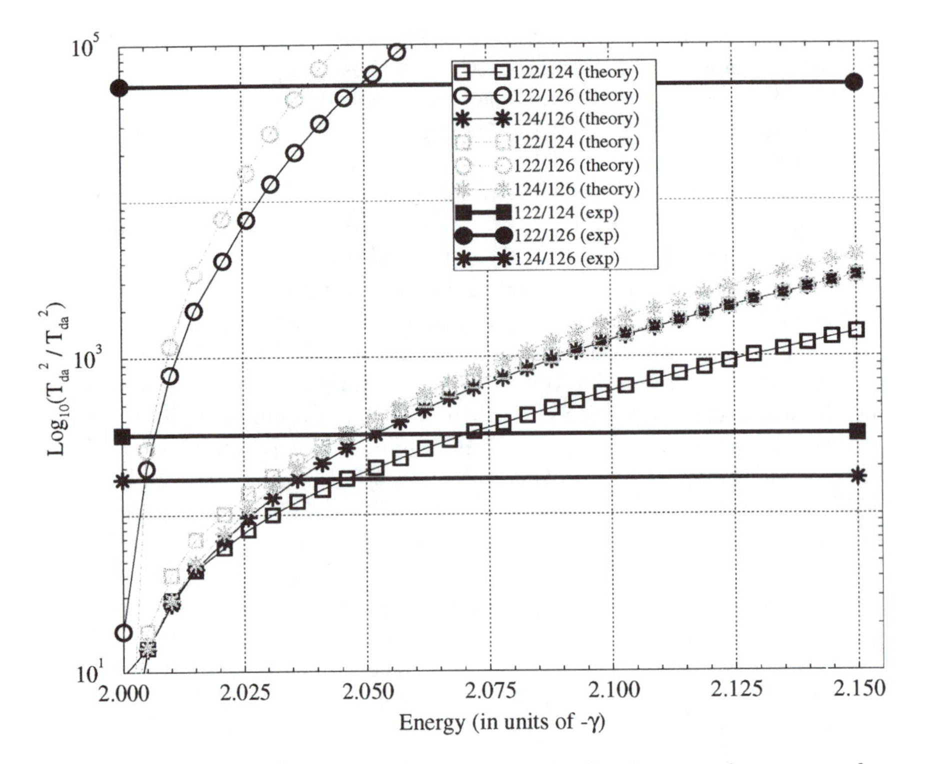

Figure 8. Ratios of T_{DA}^2 (functions of E_{tun}) compared with ratios of experimental k_{max} (straight lines). Theoretical curves cross experimental coupling-ratio lines at roughly the same place, suggesting -2.05γ as an appropriate value for E_{tun}. Two different sets of donor couplings were used (see Figure 7). Three experimental ratios are shown even though only two ratios are uniquely determined.

The tube results suggest that in the cases of 122, 124, and 126, the principal coupling is provided by the tubes from 112 and, to a much lesser extent, 121 (because of weaker copper coupling), and that in any event, these tubes interfere coherently. Because of the dominance of the 112 tubes, ratios of T_{DA}'s (for acceptors on the same β-strand) will not depend on the choice of weights so strongly that shifting the weights slightly will select a substantially different E_{tun}. For the case with all weights the same, the dominant tube leaves the Cu at 121 to feed directly into 122, and so on. For the rational set of weights, the main coupling is through 112. Because the paths down the 112 strand are just a constant three steps longer than those down the 121 strand, the ratios in one strand are similar to those in the other, so both sets of theoretical curves in Figure 8 (for each set of weights) cross experimental lines at roughly the same energy.

As Figure 7 shows, the rational weights shift 126 up more than 124, which

Table I. $G_{da}(-2.05\gamma)$ (in units of $-\gamma^{-1}$), a = X:HIS:NE2–RU, for the full protein and the subset of the protein relevant to the β-strand experiments

d	X = *122*	X = *124*	X = *126*
Full Protein			
112:CYS:SG–CU	–6.1e-05	–5.3e-06	–3.1e-07
121:MET:SD–CU	–3.1e-04	–1.4e-05	–6.6e-07
117:HIS:ND1–CU	–3.0e-06	–1.5e-07	–7.0e-09
46:HIS:ND1–CU	1.1e-06	7.9e-08	4.4e-09
45:GLY:O–LO2	2.1e-06	1.5e-07	7.8e-09
Subset of Protein			
112:CYS:SG–CU	–6.3e-05	–5.5e-06	–3.3e-07
121:MET:SD–CU	–3.2e-04	–1.4e-05	–7.0e-07
117:HIS:ND1–CU	–3.0e-06	–1.6e-07	–7.4e-09
46:HIS:ND1–CU	1.1e-06	7.2e-08	3.5e-09
45:GLY:O–LO2	2.2e-06	1.4e-07	6.8e-09

NOTE: The agreement between these two sets of results shows that the rest of the protein can be neglected in the coupling calculations.

itself shifts up more than 122, and this causes both the ratios 124/126 and 122/124 to drop (Figure 8). Unfortunately, the 122/124 ratio drops more than the 124/126 ratio because 124 is helped more relative to 122 than 126 is relative to 124. Thus, the theoretical 124/126 ratio stays higher than the theoretical 122/124 ratio, in contrast to the experimental results. This minor discrepancy is due to the the rough treatment of H-bonds in this model—they are all treated in exactly the same way. The actual N–O bond length of the 112–121 H-bond feeding 122 is 0.2–0.3 Å longer than the ones feeding 124 and 126, and the coupling to 122 should not improve as much as it does when the 112 weight is increased. If the lengths of these H-bonds were incorporated into the model, it would resolve this problem (*36*). However, a detailed description of the tuning of H-bonds is not the aim of this paper; rather, we seek a qualitative bridge model with only one adjustable parameter (γ/E_{tun}).

The full G_{da} matrix elements taken between bridge entrance and exit points for the full protein bridge are provided for $E_{tun} = -2.05\gamma$ in the upper half of Table I. From this table, and the expression for T_{DA} in equation 2, we can immediately see which bridge entrances are important in which reactions, and what effects the β weights will have, because the weights multiply these numbers directly in the T_{DA} sum. The largest bridge couplings to the β-strand acceptors (122, 124, 126) are of course via 112 and 121 (though the better weight at the 112:SG–Cu coupling will make the 112 tubes far more important). The numbers in the lower half of Table I represent the G_{da} elements taken over a subset of the protein—just the part seen in Figure 6, plus a little more, totaling about 30% of the protein. These numbers have the same order of magnitude (and most important they have the same sign) as the numbers for the full protein, indicating that the subset of the protein used provides essentially all the coupling; the rest of the protein can be dismissed.

The Tubular Breakdown

A straightforward goal of theories describing ET in proteins is to predict the value of the tunneling matrix element T_{DA}. This discussion goes a step further, to define what it is about the structure of the protein that determines this value.

This approach converts a protein, as represented by its atomic coordinates, to a very simple Hamiltonian that has just enough information to retain all essential features of the electron tunneling problem. The coupling derived from this model can be broken down into contributions from individual tubes, each of which is a a family of similar pathways through specific sequences of covalent or H-bonds. The tubes encapsulate trivial interference effects and can expose crucial intertube interferences. The set of all tubes that are important to the coupling can be identified—and the rest of the protein can be dismissed. A purely quantum mechanical effect like tube interference is directly related to the secondary and tertiary structure of the protein, and H-bonds play a central role in this effect, as they are the primary factor distinguishing a protein from what would otherwise be an effective (and uninteresting) one-dimensional chain. As proposed in earlier work, the treatment of H bonds on equal footing with covalent bonds leads to theoretical predictions that are consistent with rate measurements, indicating the critical role that H-bonds can play in ET.

This chapter discusses the particular case of β-strand tubes. However, tube analysis is generally applicable to other motifs in different proteins. Different patterns of tube interferences will exist for different motifs, but the framework for analysis will remain the same. The step from a tube analysis of a protein to experimental design is obvious, as one anticipates that there will be situations where a tube can be created or blocked by an appropriate mutation.

Acknowledgments

We thank David Beratan and Spiros Skourtis for helpful comments about this work. This research was supported by the National Science Foundation (NSF) (Grant MCB–93–16186) and the National Institutes of Health (GM48043). J. J. Regan was also supported by the Berkeley Program in Mathematics in Biology (NSF Grant DMS–94–06348).

References

1. Langen, R.; Chang, I. J.; Germanas, J. R.; Richards, J. H.; Winkler, J. R.; Gray, H. B. *Science (Washington, D. C.)* **1995,** *268,* 1733–1735.

2. Beratan, D. N.; Onuchic, J. N.; Hopfield, J. J. *J. Chem. Phys.* **1987,** *86,* 4488–4498.
3. Beratan, D. N.; Betts, J. N.; Onuchic, J. N. *Science (Washington, D. C.)* **1991,** *252,* 1285.
4. Onuchic, J. N.; Beratan, D. N.; Winkler, J. R.; Gray, H. B. *Annu. Rev. Biophys. Biomol. Struct.* **1992,** *21,* 349–377.
5. Ashcroft, N. W.; Mermin, N. D. *Solid State Physics;* W.B. Saunders: New York, 1976.
6. Skourtis, S. S.; Onuchic, J. N. *Chem. Phys. Lett.* **1993,** *209,* 171–177.
7. Turro, N. J.; Murphy, C. J.; Arkin, M. A.; Jenkins, Y.; Ghatlia, N. D.; Bossmann, S. H.; Barton, J. K. *Science (Washington, D. C.)* **1993,** *262,* 1025–1029.
8. de Rege, P. J. F.; Williams, S. A.; Therien, M. J. *Science (Washington, D. C.)* **1995,** *269,* 1409, 1413.
9. Beratan, D. N. *J. Am. Chem. Soc.* **1986,** *108,* 4321–4326.
10. Onuchic, J. N.; Beratan, D. *J. Chem. Phys.* **1990,** *92,* 722–733.
11. Beratan, D. N.; Onuchic, J. N. *Photosyn. Res.* **1989,** *22,* 173.
12. Löwdin, P. O. *J. Math. Phys.* **1962,** *3,* 969.
13. Liang, C.; Newton, M. D. *J. Phys. Chem.* **1992,** *96,* 2855–2866.
14. Curtiss, L. A.; Naleway, C. A.; Miller, J. R. *J. Phys. Chem.* **1993,** *97,* 4050–4058.
15. Paddon-Row, M. N. *J. Am. Chem. Soc.* **1994,** *116,* 5328–5333.
16. Gama, A. A. S. da *J. Theor. Biol.* **1990,** *142,* 251–260.
17. Onuchic, J. N.; Andrade, P. C. P.; Beratan, D. N. *J. Chem. Phys.* **1991,** *95,* 1131.
18. Adman, E. T.; Jensen, L. H. *Isr. J. Chem.* **1981,** *21,* 8–120.
19. Gray, H. B. *Chem. Soc. Rev.* **1986,** *15,* 17–30.
20. Sykes, A. G. *Advances Inorg. Chem.* **1991,** *36,* 377–408.
21. Nar, H.; Messerschmidt, A.; Huber, R.; Van de Kamp, M.; Canters, G. W. *J. Mol. Biol.* **1991,** *218,* 427–447.
22. Van de Kamp, M.; Floris, R.; Hall, F. C.; Canters, G. W. *J. Am. Chem. Soc.* **1990,** *112,* 907–908.
23. Van de Kamp, M.; Silvestrini, M. C.; Brunori, M.; Beeumen, J. van; Hall, F. C.; Canters, G. W. *Eur. J. Biochem.* **1990,** *194,* 109–118.
24. Mikkelsen, K. V.; Skov, L. K.; Nar, H.; Farver, O. *Proc. Natl. Acad. Sci. U.S.A.* **1993,** *90,* 5443–5445.
25. Farver, O.; Pecht, I. *J. Am. Chem. Soc.* **1992,** *114,* 5764–5767.
26. Farver, O.; Skov, L. K.; Nar, H.; Van de Kamp, M.; Canters, G. W.; Pecht, I. *Eur. J. Biochem.* **1992,** *210,* 399–403.
27. Farver, O.; Skov, L. K.; Pascher, T.; Karlsson, B. G.; Nordling, M.; Lundberg, L. G.; Vänngård, T.; Pecht, I. *Biochemistry* **1993,** *32,* 7317–7322.
28. Broo, A.; Larsson, S. *J. Phys. Chem.* **1991,** *95,* 4925–4928.
29. Regan, J. J.; Bilio, A. J. Di; Langen, R.; Skov, L. K.; Winkler, J. R.; Gray, H. B.; Onuchic, J. N. *Chem. Biol.* **1995,** *2,* 489–496.
30. Nar, H.; Messerschmidt, A.; Huber, R.; Van de Kamp, M.; Canters, G. W. *J. Mol. Biol.* **1991,** *221,* 765–772.
31. Regan, J. J.; Risser, S. M.; Beratan, D. N.; Onuchic, J. N. *J. Phys. Chem.* **1993,** *97,* 13083–13088.
32. Solomon, E. I.; Lowery, M. D. *Science (Washington, D. C.)* **1993,** *259,* 1575–1581.
33. From the coordinates of chain "A," one of four azurin molecules in the unit cell of the Brookhaven Protein Data Bank entry "4AZU" (*21, 30*). Molecular replacement and energy minimization were performed with Enzymix (*36*).
34. Lowery, M. D.; Solomon, E. I. *Inorg. Chim. Acta* **1992,** *200,* 233–243.
35. Solomon, E. I.; Baldwin, M. J.; Lowery, M. D. *Chem. Rev.* **1992,** *92,* 521–542.
36. Lee, F. S.; Chu, Z. T.; Warshel, A. *J. Comput. Chem.* **1993,** *14,* 161–185.

9

Long-Range Electron Transfer Between Proline-Bridged Aromatic Amino Acids

K. Bobrowski,[1,2] J. Poznański,[1] J. Holcman,[3] and K. L. Wierzchowski[1]*

[1] **Institute of Biochemistry and Biophysics, Polish Academy of Sciences, 02–106 Warszawa, Poland**
[2] **Institute of Nuclear Chemistry and Technology, 03–195 Warszawa, Poland**
[3] **Risø National Laboratory, DK–4000 Roskilde, Denmark**

Interpretation of the kinetic pulse radiolysis data for intramolecular Trp˙ → Tyr˙ radical transformation in aqueous solutions of linear H–Trp–(Pro)$_n$–Tyr–OH, n = *0–5, is presented in terms of the Marcus electron transfer theory, taking into account conformational dynamics of the molecules. For this purpose, for each peptide, representative sets of low-energy conformers were selected with the help of experimental methods (^{1}H and ^{13}C NMR, and circular dichroism) and modeling methods (molecular mechanics and dynamics); and relative electron transfer rates averaged over all the conformers were calculated for two assumed competitive electron transfer pathways: through space (TS) and through the peptide backbone (TB). The TS rates were obtained by taking into account the overlap integrals of aromatic ring orbitals calculated quantum mechanically. By fitting the calculated rates to the experimental data for the rate constants for electron transfer,* k_{et}, *with an exponential function appropriate for the two-pathway model, we have demonstrated that in linear short-bridged peptides (*n = *0–2), electron transfer predominantly takes the TS pathway, which consists of van der Waals contacts between the aromatic rings, whereas in longer peptides (*n = *3–5), it occurs exclusively by the TB pathway, which is made of a –(Pro)$_n$–bridge in a helical conformation similar to that of all-*trans *poly-*L*-proline II. This pathway is characterized by a low value of the descriptor of the exponential distance dependence of the electron transfer rate,* $\beta^{TB} = 2.5 \pm 0.1\ nm^{-1}$, *suggesting that helical segments in proteins can function as efficient channels of long-distance electron transfer.*

* Corresponding author.

Long-range electron transfer (LRET) between various oligoproline-bridged redox pairs has been studied over the past 10 years in several laboratories (*1–18*) with the aim of elucidating the parameters of LRET across a single peptide pathway. The choice of oligoprolines for such a study was dictated by the known ability of short H–(Pro)$_n$–OH peptides to attain, in aqueous solution, a stable helical conformation similar to that of the 3_1 left-handed helix of all-*trans* poly-L-proline II (*19–22*). More recently, similar investigations of LRET across α-helical peptide bridges were also begun (*23–25*). These studies complement present challenging attempts to distinguish molecular pathways involved in LRET in proteins (*26–30*). This chapter briefly summarizes the experimental results of our earlier pulse radiolysis investigations on intramolecular LRET accompanying Trp$^\cdot$ → Tyr$^\cdot$ radical transformation in linear H–Trp–(Pro)$_n$–Tyr–OH, n = 0–5, peptides (*1–3*). We also present our current interpretation of the observed dependence of the rate of LRET on the separation distance and spatial disposition of the aromatic side chains in terms of the conformational dynamics of the peptides and the theory of LRET (*31*).

Trp$^\cdot$ → Tyr$^\cdot$ Radical Transformation in Model Peptides

The intramolecular one-electron redox reaction (equation 1) involves electron transfer from the phenol group of the tyrosine side chain to the indolyl radical of the tryptophan side chain (Trp$^\cdot$).

$$\text{Trp}^\cdot\text{–}X\text{–Tyr[OH]} \xrightarrow{k_{et}} \text{H–Trp[H]–}X\text{–Tyr}^\cdot\text{[O]} \qquad (1)$$

where X = (Pro)$_n$. This reaction has been studied by pulse radiolysis in aqueous solution at pH 8 (*1–3*). Reaction 1 was induced by oxidation with the azide radical, $N_3^\cdot$, of the indole side chain of tryptophan to the neutral Trp$^\cdot$ radical (*32*). The use of small electron doses per pulse for generation of $N_3^\cdot$ radicals, and the low concentration of the peptides, eliminated interference from a slow second-order radical decay and intermolecular radical transformation, respectively, and thus allowed determination of the first-order rate constants of intramolecular radical transformation with a reasonable accuracy of about 15% (*3*).

Under neutral solution conditions, the electron transfer is accompanied by a net proton transfer due to breakage of the tyrosine O–H bond: Tyr-[OH]–e^- → Tyr[O]$^\cdot$ + H^+; and formation of the indole N–H bond: Trp$^\cdot$ + e– + H^+ → Trp[H]. Because O–H and N–H groups in water are involved in very fast proton exchange ($k \cong 10^{12}$ s^{-1}), the protonation–deprotonation equilibria accompanying electron transfer cannot be expected to limit the rate of the radical transformation reaction, which occurs on the microsecond time scale (*3, 6*). An additional argument in favor of this conclusion is the recent finding of Mishra et al. (*7*) that mechanisms of reduction in reaction 1 of Trp[H]$^{+\cdot}$ and the N-methylated tryptophan radical cation, MeTrp$^{+\cdot}$, are similar in spite

Table I. Comparison of $\langle k \rangle$, $\langle k^{TS} \rangle$, and $\langle k^{TB} \rangle$ Calculated According to the TB + TS Model, and the Experimental k_{et} (298 K) Rate Constants of LRET in Linear Peptides

Peptide	ω *Trp–Pro*	k_{et} (10^3 s^{-1})	$\langle k \rangle$ (10^3 s^{-1})	$\langle k^{TS} \rangle$ (10^3 s^{-1})	$\langle k^{TB} \rangle$ (10^3 s^{-1})	X^2
Trp–Tyr	—	77.0	48.5	21.6	26.9	1.072
Trp–Pro–Tyr	*cis*	26.0	27.6	14.1	13.5	0.017
Trp–Pro–Tyr	*trans*	26.0	35.2	25.4	9.8	0.462
Trp–$(Pro)_2$–Tyr	*cis*	4.9	9.6	4.3	5.3	2.254
Trp–$(Pro)_2$–Tyr	*trans*	39.0	37.2	33.4	3.8	0.011
Trp–$(Pro)_3$–Tyr	*cis*	1.5	1.8	0	1.8	0.191
Trp–$(Pro)_3$–Tyr	*trans*	1.5	1.5	0	1.5	0.001
Trp–$(Pro)_4$–Tyr	*cis*	0.51	0.69	0	0.69	0.470
Trp–$(Pro)_4$–Tyr	*trans*	0.51	0.56	0	0.56	0.051
Trp–$(Pro)_5$–Tyr	*cis*	0.30	0.27	0	0.27	0.064
Trp–$(Pro)_5$–Tyr	*trans*	0.30	0.22	0	0.22	0.510

of the fact that in the latter case electron transfer is not accompanied by a net proton transfer.

The first-order rate constants, k_{et}, determined by linear least-squares analysis of time-dependent absorbance data for $Trp^{\cdot}$ and/or $Tyr^{\cdot}$ (Table I), thus correspond to the one-electron transfer reaction 1. Note that except for the $n = 2$ linear peptide, where two first-order LRET processes of similar amplitude were resolved, the kinetic data for all the peptides conformed to a single exponential.

The plot of ln k_{et} vs. the number, n, of Pro residues (Figure 1) demonstrates that for longer linear peptides ($n = 3–5$), the rate of electron transfer decreases exponentially with growing n. However, the k_{et} data for shorter ($n = 0–2$) linear peptides fall off considerably from the plot extrapolated to lower n values. This indicates that the rate of LRET in short-bridged peptides is faster than would be expected on the assumption of a common mechanism of electron transfer in the whole group of peptides. In order to rationalize these findings in terms of the theory of the distance dependence of LRET kinetics (*33, 34*), the separation distances and spatial disposition of the aromatic side chains in the linear peptides studied had to be evaluated from their conformational preferences and conformational dynamics.

Conformational Properties of Trp–(Pro)$_n$–Tyr Peptides

The conformational properties of Trp–$(Pro)_n$–Tyr, $n = 0–5$, peptides were deduced from experimental ^1H and ^{13}C NMR (*35*) and circular dichroism (CD) (*36*) investigations, complemented by molecular mechanics and molecular dy-

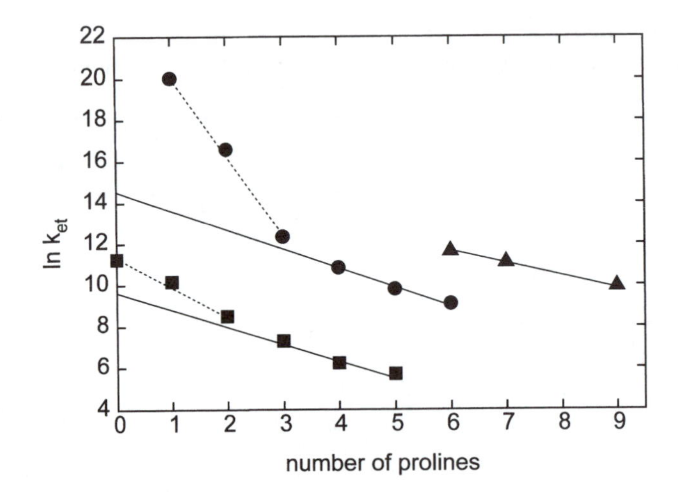

Figure 1. Bridge length dependence of the rate of LRET. Plots are of ln k_{et} vs. number, n, of Pro residues for H–Trp–(Pro)$_n$–Tyr–OH, n = 0–5 (■); [(bpy)$_2$Ru^{II}L(Pro)$_n$Co^{III}(NH$_3$)$_5$]$^{4+}$, n = 0–6 (●); and [(bpy)$_2$Ru^{II}L(Pro)$_n$apyRu^{III}(NH$_3$)$_5$]$^{4+}$, n = 6, 7, 9 (△). Solid lines correspond to linear regression for n = 3–5 in the H–Trp–(Pro)$_n$–Tyr–OH series and for n = 4–6 in the [(bpy)$_2$Ru^{II}L-(Pro)$_n$Co^{III}(NH$_3$)$_5$]$^{4+}$ series. Abbreviations: bpy, 2,2′-bipyridine; 4-amino-pyridine (apy).

namics modeling (*31*) with the help of AMBER 3.0 and 4.0 software (*37*). These properties proved to be governed by a number of interdependent equilibria (cf. Figure 2): (1) *trans* ↔ *cis* isomerization (ω dihedral angle) about the X–Pro peptide bonds, (2) rotation of Trp and Tyr side chains about C$^{\alpha}$–C$^{\beta}$ and C$^{\beta}$–C$^{\gamma}$ bonds (χ_1 and χ_2 dihedrals, respectively), (3) extended helix ↔ all-*trans* poly-L-proline (PLP) II type helical conformation within the –(Pro)$_n$–bridge in peptides with $n \geq 3$, (4) β (~160°) → α (~45°) transition at the ψ(Pro) angle of the Pro–Tyr fragment, and (5) transition between *up* and *down* conformations of the pyrrolidine Pro side chain (χ_3).

Trans ↔ *cis* isomerization occurs most readily at the Trp–Pro bond, so that populations of corresponding major isomers of zwitterionic forms of the peptides are comparable and constitute a 0.85–0.90 molar fraction of the total peptide content. Isomerization about Pro–Pro bonds in peptides with $n > 1$ results in a small population of at least two additional *cis* isomers. The rate constant for interchange between *trans* and *cis* isomers about the Trp–Pro bond at 298 K has been estimated to be close to 10^{-2} s^{-1} (*3, 35, 38*), that is, 4–6 orders of magnitude slower than that observed for the electron transfer reaction 1. Thus, in this reaction the two isomers should be treated as separate species.

Rotations of the Tyr side chain both in *cis* and *trans* isomers proved relatively free, with a marked preference for the $\chi_1(g^-)$ rotamer in longer peptides.

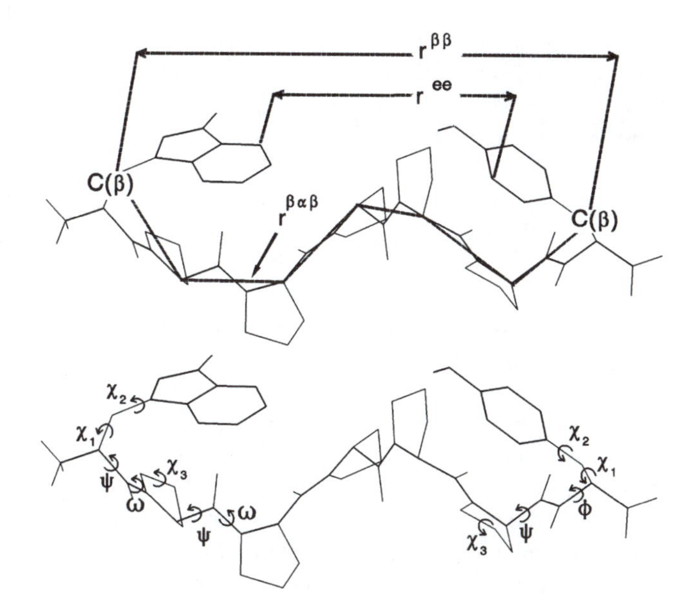

Figure 2. Structural scheme and definition of conformational parameters for H–Trp–(Pro)$_n$–Tyr–OH peptides.

Rotations of Trp, however, were found to be highly dependent on the configuration about the Trp–Pro bond. In the *trans* isomers, the indole ring rotates quite freely, whereas in the *cis* form of all the peptides its rotation is severely restricted, which results in a high population, ~0.85, of the staggered $\chi_1(t)$ rotamer and a $\chi_1(t)$, $\chi_2(-)$ conformation of the whole side chain. All these rotamers, with lifetimes in the time domain of 10^{-9}–10^{-12} s, exchange frequently during the lifetime of the Trp$^{\cdot}$ radical. In the short-bridged peptides (n = 0–2), these side chain rotations, combined with oscillations of the backbone ψ angle, lead to the appearance of multiple low-energy conformers characterized by a close edge-to-edge approach of the indole and phenol rings.

Conformation of the –(Pro)$_n$–fragment varies with the number of adjacent Pro residues and assumes a PLP II–like helical conformation in all-*trans* isomers beginning with n = 3. This conformation also includes the Trp residue. Conformational rigidity of the backbone in the PLP II–like conformation increases with the growing n. In this conformation the ψ(Pro) dihedral angle generally assumes a value in the β-range (+160°).

Recently, Sneddon and Brooks (*39*), on the basis of their CHARMM simulations of conformational dynamics of Pro peptides in aqueous solution, have postulated involvement in electron transfer across the –(Pro)$_n$–bridge of β → α transitions at the ψ angle, as the latter occur more rapidly and bring the donor–acceptor distance to a shorter range than the *trans* ↔ *cis* interconversion at the ω dihedral. We have performed (*31*) similar simulations for Trp–Pro, Pro–Pro, and Pro–Tyr fragments (potential of mean force calculations with explicit H_2O environment in AMBER 4.0 using umbrella sampling along a reaction coordinate in ψ; 36 molecular dynamics runs consisting of 10-ps equilibration and 30-ps evolution time windows). The calculated energies of, and energy barriers between, the β and α ψ(Pro) conformers of the Pro–Pro fragment proved similar to those obtained by Sneddon and Brooks (*39*); it appeared that the ψ(β) conformation is more stable than the ψ(α) one by $\Delta G \cong 3 \pm 1$ kcal mol^{-1} in Pro–Pro and Pro–Tyr, whereas in Trp–Pro, ΔG amounts to as much as 7 ± 1 kcal mol^{-1}, so that the population of Trp–αPro conformers can be expected to be negligibly small.

Using the determined potentials of mean force, we have also evaluated the frequency of $\beta \rightarrow \alpha$ transitions according to the transition state theory: $k_{\beta\rightarrow\alpha} = 1.5 \times 10^3\ s^{-1}$ and $k_{\beta\rightarrow\alpha} = 8 \times 10^5\ s^{-1}$ for Pro in Pro–Pro and Pro–Tyr fragments, respectively. The mean lifetime of the ψ(α)Pro state in Pro–Pro is thus 6×10^{-6} s, and the transient population of corresponding conformers of $-(Pro)_n-$ with one Pro residue in this state can be expected to be insignificantly low on the time scale of the observed electron transfer and was therefore neglected. On the other hand, $\beta \rightarrow \alpha$ transitions at the Pro residue preceding Tyr, which are much faster on the same scale, led us to include this transition in calculations concerning low-energy conformers. The results of these modeling studies find full support in NMR data, which showed that the location and shape of the C^{α} resonances in the ^{13}C spectra of the Trp–$(Pro)_n$–Tyr peptides (*31, 35*) were similar to that of PLP II for all the Pro residues studied except that preceding Tyr.

The cyclic pyrrolidine side chains of Pro residues undergo very fast, picosecond transitions between their *up* and *down* equilibrium conformations at the χ_3 angle, as shown by both NMR and molecular mechanics modeling (*31*).

To sum up, the H–Trp–$(Pro)_n$–Tyr–OH peptides in solution should be represented by ensembles of fast-exchanging, on the observed LRET time scale, side chain and ψ(Pro–Tyr) backbone conformers of the major *cis* and *trans* isomers about the Trp–Pro bond, and $-(Pro)_n-$bridge in a PLP II type conformation beginning with $n = 3$.

Modeling of LRET Pathways in Trp–(Pro)$_n$–Tyr Peptides

In order to interpret the observed dependence of k_{et} on the number of Pro residues in the bridge in terms of the theory of the distance dependence of the LRET rate (*34*), distances between the two aromatic redox centers along some physically possible molecular pathway(s) had to be evaluated for all representative conformers of the two major isomers, all-*trans* and *cis*(Trp–Pro), of each of the peptides. The lowest-energy conformers of zwitterionic forms of the peptides were calculated (*31*) with the use of the AMBER 3.0 program in united atom parametrization and with a dielectric constant of 81 (to mimic an aqueous environment).

For each peptide isomer with a –(Pro)$_n$–bridge in the PLP II structure [adopted from X-ray diffraction data (*40*)], a set of 324 starting conformers, corresponding to all possible combinations of the *N*-terminal and *C*-terminal side chain conformations (with values of the χ_1 and χ_2 dihedral angles varied in 30° steps), and α and β ψ(Pro) conformations of the Pro–Tyr fragment was generated and subjected to further energy minimization. During the latter step, all dihedral angles, including those of the pyrrolidine ring, were allowed to vary. Of the conformers thus obtained, only those with energies $\leq$42 kJ mol^{-1} above the lowest-energy one within a given isomeric ensemble were selected for calculation of separation distances and angles between the aromatic rings. Ensembles of the low energy conformers fairly well reproduced the main conformational features of the peptides derived from the experimental NMR (*35*) and CD data (*36*), especially populations of conformers in particular rotameric side chain states; and the differences between backbone conformations of the n = 0–2 and n = 3–5 groups of peptides were distinct. We were thus justified in assuming that the distributions of calculated distances between various atoms of the terminal side chains in ensembles of the calculated conformers would satisfactorily resemble those prevailing in solution (*3*). Similar distributions were also obtained from molecular dynamics simulations (*31*). The r^{ee} free-energy profiles for n = 2 and n = 3 derived therefrom, shown in Figure 3, are representative for the short- and long-bridged systems, respectively.

Considering the outlined conformational properties of the peptides, two molecular pathways for electron transfer could be reasonably envisaged: (1) through space (TS), viz, directly between the aromatic rings in van der Waals contact and/or mediated by water molecules of the solvation shell, and (2) through the peptide backbone (TB). The corresponding distances for individual conformers (j) within each isomeric ensemble (i), r_{ij}, were consequently calculated for the TS pathway as $r_{ij}(C^{ee})$ between ring carbon, C^{ar}, (or C^{ar} and N/O) atoms of the indole and phenol rings, and for the TB pathway, originally (*3*) as $r_{ij}(C^{\beta\beta})$, the shortest distance between C^{β} atoms of the terminal Trp and Tyr along the peptide backbone. More recently (*31*), r_{ij} was calculated, as $r_{ij}(C^{\beta}C^{\alpha}C^{\beta})$ between the C^{β} atoms of the terminal amino acids along a path joining the backbone C_{α} atoms (cf. Figure 2). The latter parametrization follows a path of highest molecular orbital electronic density and thus better mimics the TB electron transfer pathway.

The distance dependence of the rate of LRET, k, is described (*34*) by equation 2:

$$k = k_0 \exp[-\beta(r - r_0)] \tag{2}$$

where k_0 is the rate constant at $r = r_0$, corresponding to the closest edge-to-edge approach of redox centers, and the descriptor of the exponential distance

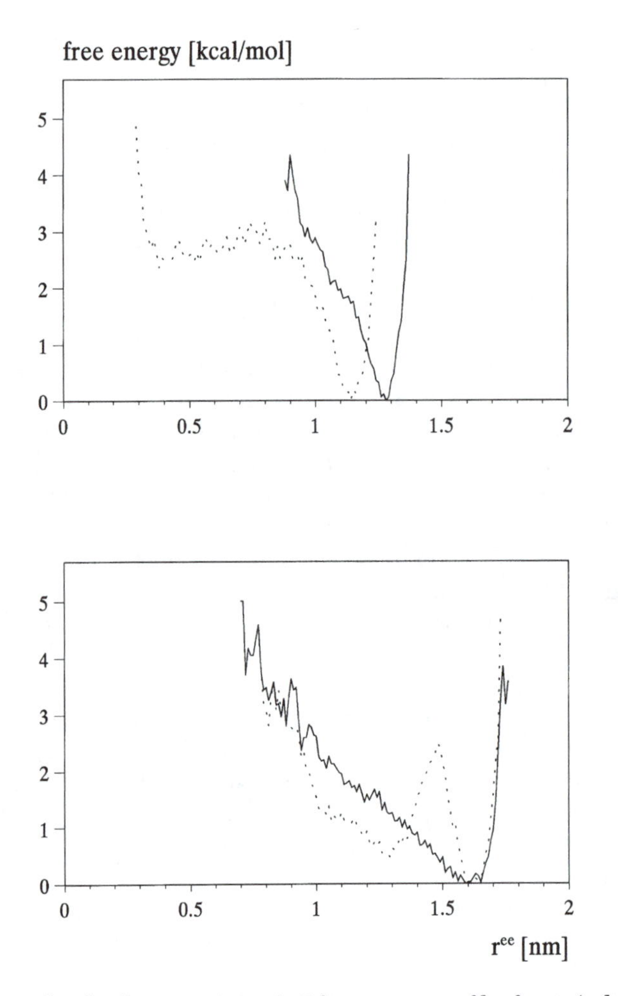

Figure 3. Molecular dynamics (10 ns) r^{ee} free-energy profiles for cis *(solid curves) and* trans *(broken curves) isomers of H–Trp–(Pro)$_2$–Tyr–OH peptide (top) and H–Trp–(Pro)$_3$–Tyr–OH peptide (bottom).*

dependence of k, $\beta = \beta_{el} + \beta_n$, expresses the contribution of electronic (β_{el}) and nuclear (β_n) factors to the overall distance dependence of the rate. Our approach to distinguish which of the possible molecular pathways is actually involved in reaction 1, and to evaluate the corresponding β descriptor, consisted of the following steps: (1) calculation of relative rate constants, k_{ij}, of LRET for individual conformers of a peptide along an assumed pathway, and of average values for each peptide isomer, $\langle k \rangle_i = \Sigma_{ij} k_{ij}\omega_{ij}$ (where ω_{ij} is the Boltzmann probability of occurrence of a jth conformer of the ith isomer), and then (2) fitting these values to the experimental k_{et} data with the use of equation 2, as a fitting function, in an appropriate form for an assumed model of LRET. Originally, four principal LRET models were probed, involving either (1) the TB pathway, (2) the TS pathway, (3) the two pathways TB + TS simultaneously, or (4) TB + TS cos Θ, where Θ is the dihedral angle between the planes of the indole and phenol rings, in which the cos Θ function was introduced to roughly account for the dependence of LRET along the TS pathway on the overlap between the π- and σ-orbitals of the aromatic rings (*3*). The validity of the models was evaluated statistically with the use of the X^2 distribution function, defined as:

$$X^2 = \Sigma_{i=0,n} (\ln \langle k \rangle_i - \ln k_{et,i})^2/\sigma^2 \ln k_{et,i} \quad (3)$$

The best agreement between the experimental and calculated rates has been obtained for the last model (*4*) (documented by the low value of X^2 = 6.23 at the high significance level of 0.513) for the following values of the parameters sought: $\beta^{TB} = 2.8 \pm 0.4\ nm^{-1}$ and $\beta^{TS} = 120 \pm 40\ nm^{-1}$. The high value of the β^{TS} parameter indicates that at $r_{ij}(C^{ar}) \cong 3.65$ Å the contribution to k_{et} from the TS pathway vanishes practically to zero, so that TS–LRET takes place only in conformers with indole and phenol aromatic rings in close van der Waals contact. Because low-energy conformers of this type are dominant in short-bridged (n = 0–2) peptides, the TS pathway proved competitive only in the latter group. In longer peptides (n = 3–5), LRET thus takes place solely through the TB pathway, characterized by unusually weak decay of electronic coupling between the Trp and Tyr redox centers.

We recently calculated (*31*) overlap integrals, I_{AB}, between highest occupied molecular orbitals (HOMOs) of the indolyl radical and phenol rings for all the low-energy conformers of the peptides. (The calculations were actually performed for HOMO orbitals of 3-methylindole and *p*-cresol located in space as indolyl and phenol rings in the low-energy conformers.) Using the $r_{ij}(C^{\beta}\text{-}C^{\alpha}C^{\beta})$ parametrization for the TB pathway and a fitting function in the form

$$k = k_0^{TB} \exp(-\beta^{TB}\, r^{\beta\alpha\beta}) + k_0^{TS}\, I_{AB}^2 \quad (4)$$

all the calculated rates $\langle k \rangle_i$ were again simultaneously fitted to the experimental k_{et} data. The adjustable parameters thus obtained, $\langle k \rangle_i = \langle k^{TB} \rangle_i + \langle k^{TS} \rangle_i$ rates, and values of the X^2 function are summarized in Table I. The plot of $\ln\langle k \rangle_i$ vs. $\ln k_{et}$, shown in Figure 4, illustrates good agreement between the experimental and calculated LRET rates for the following values of the parameters sought: $k_0^{TB} = (10 \pm 2) \times 10^4\ s^{-1}$, $\beta^{TB} = 2.5 \pm 0.1\ nm^{-1}$, and $k_0^{TS} = (14 \pm 2) \times 10^{12}\ s^{-1}$.

Using the same method, we also fitted the calculated rates to the experimental k_{et} data from Klapper and Faraggi's laboratories (*5, 6*) for the same group of peptides. The results were qualitatively similar to those obtained with our k_{et} s; however, the values of the X^2 function were somewhat larger (*31*).

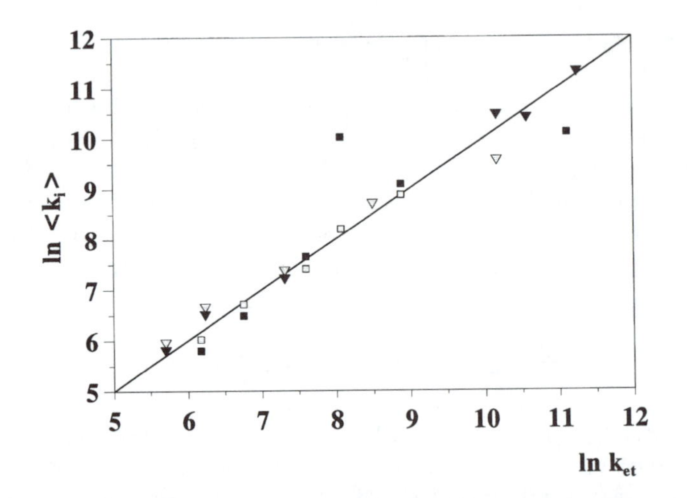

Figure 4. Correlation between experimental k_{et} *and calculated best-fit rate constants of LRET,* $\langle k \rangle_i$*, for* cis *(open data points) and* trans *(filled data points) isomers of H–Trp–(Pro)*$_n$*–Tyr–OH peptides (inverted triangles) and H–Tyr–(Pro)*$_n$*–Trp–OH peptides (squares). The best agreement between the experimental and calculated rates has been obtained for the following values of the parameters sought:* $k_0^{TB} = (10 \pm 2) \times 10^4\ s^{-1}$, $\beta^{TB} = 2.5 \pm 0.1\ nm^{-1}$, *and* $k_0^{TS} = (14 \pm 2) \times 10^{12}\ s^{-1}$.

Moreover, those workers have also studied a series of peptides with reversed order of the terminal amino acids, H–Tyr–(Pro)$_n$–Trp–OH. They found that in this case, the slopes of ln k_{et} versus number, n, of Pro residues in the bridge were somewhat different than for the former series, and they attributed this difference to directional specificity of LRET along the peptide backbone (*6*). We pointed out previously (*3*) that this could be due to somewhat different equilibrium populations of *cis* and *trans* isomers about Trp–Pro and Tyr–Pro bonds and χ_1 rotamers of the *C*- and *N*-terminal indole side chain. Subsequently, we calculated the corresponding low-energy conformers (*31*) using the methods applied for the H–Trp–(Pro)$_n$–Tyr–OH series, and fitted the obtained $\langle k \rangle_i$ values to the published experimental k_{et} data according to equation 4. As shown in Figure 4, we obtained excellent agreement between the calculated and experimental LRET data with the same set of adjustable parameters of the fitting function as used in the case of H–Trp–(Pro)$_n$–Tyr–OH peptides. This shows that in the linear peptide systems, no directional specificity of LRET is observed, provided that the experimental rates are properly interpreted taking into account conformational dynamics of the peptides.

Comparison of LRET Across the –(Pro)$_n$– Bridge Between the Trp$^{\cdot}$–Tyr and Metallic Redox Systems

Comparison of the kinetic LRET data presented for the Trp$^{\cdot}$–Tyr redox pair (cf. Figure 1) with those for intramolecular electron transfer across the same

–(Pro)$_n$– bridge but between different, Ru^{II}–Co^{III} and Ru^{II}–Ru^{III}, donor–acceptor pairs in appropriately metal-derivatized oligoproline peptides with polymerization number ranging up to $n = 9$ (*13–15*) clearly demonstrates a general similarity between the ln k_{et} versus donor–acceptor distance plots for the two systems. They can be approximated by two-component linear plots with different slopes for short-bridged and longer-bridged peptides. While the slopes for short-bridged Trp$^{\cdot}$–Tyr and metallo systems differ considerably, those for longer-bridged peptides are similar (β descriptors ≅ 2.5 and ≅ 3.0 nm^{-1} (*14*), respectively).

Low values and the similarity of the β coefficients for longer peptides provide a strong argument in favor of high electron transferability of the helical PLP II type conformation of –(Pro)$_n$– bridges and independence of the latter from the nature of the attached donor–acceptor pair. On theoretical grounds (*41*), the dependence of β on the nature of the donor–acceptor pair could appear only when energy of the tunneling electron was close to the energy of the intervening bridge atomic orbitals. This is apparently not the case in the systems in question. Because the Trp$^{\cdot}$–Tyr radical and the two charged metal redox pairs, Ru^{II}–Co^{III} and Ru^{II}–Ru^{III}, can be expected to also differ greatly in the outer part of the reorganization energy, λ_{out}, similar values of the observed β descriptors indicate that the distance-dependent nuclear part thereof, β_n, becomes negligibly small at longer donor–acceptor separation distances.

The great difference in slopes for short-bridged peptides between the two systems can be rationalized in terms of differences in the nuclear part of the β descriptor. In the charged metallo systems with $n = 1$–3 prolines, the rates of intra-molecular electron transfer decrease rapidly by about 3 orders of magnitude. For these systems, the LRET mechanism is largely under the control of the nuclear factor ($\beta_n \cong 9.5$ nm^{-1} and $\beta_{el} \cong 6.5$ nm^{-1}) (*10*). The much smaller distance dependence of k_{et} in the Trp$^{\cdot}$–Tyr system for peptides with $n = 0$–2, associated largely with the TS pathway and weak nuclear control, is consistent with the neutral character of the Trp$^{\cdot}$ and Tyr radicals involved in LRET and, thus, the low value of λ_{out} (*3*).

An additional difference between the H–Trp–(Pro)$_n$–Tyr–OH and the two –(Pro)$_n$– bridged metallo systems lies in the number, *n*, of Pro residues at which a change in the slope of the ln k_{et} versus *n* is located (cf. Figure 1). In the first system, it takes place between $n = 2$–3 and is caused by the change in the electron transfer pathway from a largely TS to an exclusively TB one. In the second, it is seen at $n = 3$–4 and can be attributed to a similar change in the LRET pathway. This turnover point is observed in molecules with a longer, –(Pro)$_4$–, bridge than in the Trp$^{\cdot}$–Tyr system because of a larger size of the metallo redox centers compared to that of the indolyl and phenol rings. Indeed, our preliminary modeling of sterically and energetically permissible conformations of the metallo system demonstrated that, at $n = 3$, the two terminal metal complexes may still come into close contact owing to their large van der Waals radii (cf. Figure 5). In the H–Trp–(Pro)$_n$–Tyr–OH system at $n = 3$, the population of conformers with indole and phenol rings with $r^{ee} < 5.0$ nm is negligibly small (cf. Figure 3).

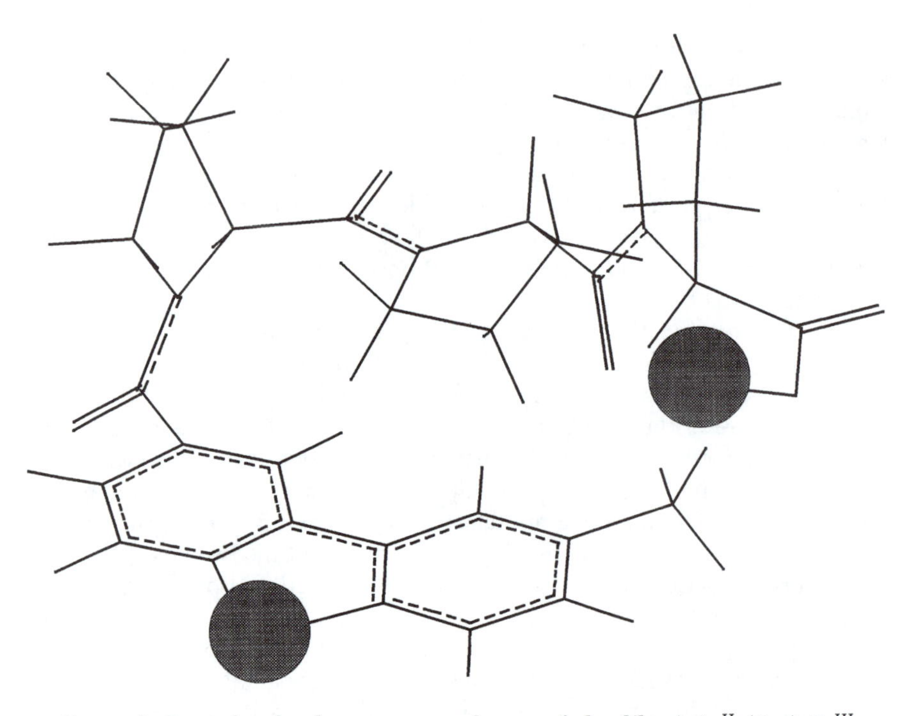

Figure 5. Example of a low-energy conformer of the $[(bpy)_2Ru^{II}L(Pro)_3Co^{III}(NH_3)_5]^{4+}$ peptide, characterized by close contact between the two metal complexes (Ru^{II}, Co^{III}).

References

1. Bobrowski, K.; Wierzchowski, K. L.; Holcman, J.; Ciurak, M. *Stud. Biophys.* **1987,** *122,* 23–28.
2. Bobrowski, K.; Wierzchowski, K. L.; Holcman, J.; Ciurak, M. *Int. J. Radiat. Biol.* **1990,** *57,* 919–932.
3. Bobrowski, K.; Holcman, J.; Poznański, J.; Ciurak, M.; Wierzchowski, K. L. *J. Phys. Chem.* **1992,** *96,* 10036–10043.
4. Bobrowski, K.; Wierzchowski, K. L.; Holcman, J.; Ciurak, M. *Int. J. Radiat. Biol.* **1992,** *62,* 507–516.
5. Faraggi, M.; DeFelippis, M. R.; Klapper, M. H. *J. Am. Chem. Soc.* **1989,** *111,* 5141–5145.
6. DeFelippis, M. R.; Faraggi, M.; Klapper, M. H. *J. Am. Chem. Soc.* **1990,** *112,* 5640–5642.
7. Mishra, A. K.; Chandrasekar, R.; Faraggi, M.; Klapper, M. H. *J. Am. Chem. Soc.* **1994,** *116,* 1414–1422.
8. Isied, S. S.; Vassilian, A. *J. Am. Chem. Soc.* **1984,** *106,* 1732–1736.
9. Isied, S. S.; Vassilian, A.; Magnusson, R. H.; Schwarz, H. A. *J. Am. Chem. Soc.* **1985,** *107,* 7432–7438.
10. Isied, S. S.; Vassilian, A.; Wishart, J. F.; Creutz, C.; Schwarz, H. A.; Sutin, N. *J. Am. Chem. Soc.* **1988,** *110,* 635–637.

11. Vassilian, A.; Wishart, J. F.; van Hemelryck, B.; Schwarz, H. A.; Isied, S. S. *J. Am. Chem. Soc.* **1990,** *112,* 7278–7286.
12. Isied, S. S. In *Electron Transfer in Inorganic, Organic, and Biological Systems;* Bolton, J. R.; Mataga, N.; McLendon, G., Eds.; Advances in Chemistry 228; American Chemical Society: Washington, DC, 1991; Chapter 15.
13. Isied, S. S.; Ogawa, M. Y.; Wishart, J. F. *Chem. Rev.* **1992,** *92,* 381–394.
14. Ogawa, M. Y.; Wishart, J. F.; Young, Z.; Miller, J. R.; Isied, S. S. *J. Phys. Chem.* **1993,** *97,* 11456–11463.
15. Ogawa, M. Y.; Moreira, I.; Wishart, J. F.; Isied, S. S. *Chem. Phys.* **1993,** *176,* 589–600.
16. Schanze, K.; Sauer, K. *J. Am. Chem. Soc.* **1988,** *110,* 1180–1186.
17. Schanze, K.; Cabana, L. *J. Phys. Chem.* **1990,** *94,* 2740–2743.
18. Cabana, L.; Schanze, K. In *Electron Transfer in Biology and the Solid State: Inorganic Compounds with Unusual Properties;* Johnson, M. K.; King, R. B. Kurtz, D. M., Jr.; Kutal, C.; Norton, M. L.; Scott, R. A., Eds; Advances in Chemistry 226; American Chemical Society: Washington, DC, 1990; pp 101–123.
19. Rabanal, F.; Ludevit, M. D.; Pons, M.; Giralt, E. *Biopolymers* **1993,** *33,* 1019–1028.
20. Dukor, R. K.; Keiderling, T. A. *Biopolymers* **1991,** *31,* 1747–1761.
21. Helbecque, N.; Loucheux-Lefebvre, M. H. *Int. J. Pept. Protein Res.* **1982,** *19,* 94–101.
22. Zhang, R.; Madalengoitia, J. S. *Tetrahedron Lett.* **1996,** *37,* 6235–6238.
23. Inai, Y.; Sisido, M.; Imanishi, Y. *J. Phys. Chem.* **1991,** *95,* 3847–3851.
24. Lee, H.; Faraggi, M.; Klapper, M. H. *Biochim. Biophys. Acta* **1992,** *1159,* 286–292.
25. Isied, S. S.; Moreira, I.; Ogawa, M. Y.; Vassilian, A.; Arbo, B.; Sun, J. *J. Photochem. Photobiol. A: Chem.* **1994,** *82,* 203–210.
26. Siddarth, P.; Marcus, R. A. *J. Phys. Chem.* **1993,** *97,* 13078–13082.
27. Regan, J. J.; Risser, S. M.; Beratan, D. N.; Onuchic, J. N. *J. Phys. Chem.* **1993,** *97,* 13083–13091.
28. Winkler J. R.; Gray H. B. *Chem. Rev.* **1992,** *92,* 369–379.
29. Bobrowski, K.; Holcman, J.; Poznanski, J.; Wierzchowski, K. L. *Biophys. Chem.* **1997,** *63,* 153–166.
30. Bjerrum, M. J.; Casimiro, D. R.; Chang, I-Jy; Di Bilio, A. J.; Gray, H. B.; Hill, M. G.; Langen, R.; Mines, G. A.; Skov, L. K.; Winkler, J. R.; Wuttke, D. S. *J. Bioenerg. Biomemb* **1995,** *27,* 295–302.
31. Poznański, J. Ph. D. Thesis, Institute of Biochemistry and Biophysics, Polish Academy of Sciences, 1995.
32. Prütz, W. A.; Land, E. J. *Int. J. Radiat. Biol.* **1979,** *36,* 513–520.
33. Marcus, R. A.; Sutin, N. *Biochim. Biophys. Acta* **1985,** *811,* 265–323.
34. Sutin, N. In *Electron Transfer in Inorganic, Organic, and Biological Systems;* Bolton, J. R.; Mataga, N.; McLendon, G., Eds.; Advances in Chemistry 228; American Chemical Society: Washington, DC, 1991; Chapter 3.
35. Poznański, J.; Ejchart, A.; Wierzchowski, K. L.; Ciurak, M. *Biopolymers* **1993,** *33,* 781–795.
36. Wierzchowski, K. L.; Majcher, K.; Poznański, J. *Acta Biochim. Pol.* **1995,** *42,* 259–268.
37. Weiner, S. J.; Kollman, P. A.; Case, D. A.; Singh, U. C.; Ghio, C.; Alagona, G., Profeta, S.; Weiner, P. *J. Am. Chem. Soc.* **1984,** *106,* 765–784.
38. Grathwohl, Ch.; Wüthrich, K. *Biopolymers* **1981,** *20,* 2623–2633.
39. Sneddon, S. F.; Brooks, C. L. III. *J. Am. Chem. Soc.* **1992,** *114,* 8220–8225.
40. Ramachandran, G. N.; Sasisekharan, V. *Adv. Protein Chem.* **1968,** *23,* 283–437.
41. Evenson, J. W.; Karplus, M. *Science (Washington, D.C.)* **1993,** *262,* 1247–1249.

10

Electron Transfer Kinetics of Bifunctional Redox Protein Maquettes

Mitchell W. Mutz[1], James F. Wishart[2], and George L. McLendon[*,1]

[1] **Department of Chemistry, Princeton University, Princeton, NJ 08544**
[2] **Department of Chemistry, Brookhaven National Laboratory, Upton, NY 11973**

We prepared three bifunctional redox protein maquettes based on 12-, 16-, and 20-mer three-helix bundles. In each case, the helix was capped with a Co(III) tris-bipyridyl electron acceptor and also functionalized with a C-terminal viologen (1-ethyl-1′-ethyl-4,4′-bipyridinium) donor. Electron transfer (ET) was initiated by pulse radiolysis and flash photolysis and followed spectrometrically to determine average, concentration-independent, first-order rates for the 16-mer and 20-mer maquettes. For the 16-mer bundle, the α-helical content was adjusted by the addition of urea or trifluoroethanol to solutions containing the metalloprotein. This conformational flexibility under different solvent conditions was exploited to probe the effects of helical secondary structure on ET rates. In addition to describing experimental results from these helical systems, this chapter discusses several additional metalloprotein models from the recent literature.

In the study of protein electron transfer (ET), radiolytic and photochemical techniques have indeed proven highly complementary. Between them, these techniques provide a range of reaction types and reaction free energies [cf. Zn porphyrin triplets ($F^\circ \sim 0.8$ V) versus Fe porphyrins ($E^\circ \sim 0$ V)]. Of particular interest in the current study is the different dynamic range(s) of the techniques. Photochemistry is subject to a natural "time window" set by the excited state lifetime: only reactions faster than the excited state decay can be observed. Conversely, the bimolecular nature of radiolysis sets an upper limit on the observed rates that is often determined by the rate of electron capture.

An early example of these complementary aspects was provided by studies

* Corresponding author.

of the cytochrome *c*: cytochrome b_5 complex. Initial studies utilized photochemistry to monitor reactions at high free energy, $-\Delta G$, and radiolysis to monitor (slower) reactions at lower $-\Delta G$(*1–2*). An interesting sidelight of these earlier studies has emerged. Using an intramolecular ruthenium bipyridyl (bpy) photosystem, Durham, Millett, and co-workers have reinvestigated the cyt b_5:cyt *c* system, as reported in Chapter 7. With the higher time resolution of their photochemical technique, they observe two populations of reactants: a "slow" population reacts with $k_{obs} = 3 \pm 1 \times 10^4\ s^{-1}$, as reported in the earlier study (*3*). However, they also observe a faster population ($k_{obs} = 4 \pm 1 \times 10^5\ s^{-1}$). This faster phase occurs on a time scale competitive with electron capture and was therefore missed in the radiolysis study. In retrospect, it appears possible to model the earlier data in terms of a limiting rise time of the (fast) signal. Assuming steady-state kinetics where the rate of intracomplex electron transfer is faster than electron capture, the apparent rise time will in fact monitor the rate of electron transfer (and not of electron capture). Such a model indeed appears to reproduce the work of Millett and co-workers (*3*).

A different example of complementarity is explored in this chapter. We report studies of a bifunctional redox protein maquette[1] based on the triple-helix bundle design of Ghadiri, Sasaki, and co-workers (*4–5*). Attempts to monitor electron transfer from an *N*-terminal ruthenium tris-bipyridyl excited state were fruitless, since electron transfer could not compete with the excited state decay of the Ru(II*) tris-bpy. However, as described below, replacement of photoactive ruthenium by a radiolytically accessible $Co^{III/II}$ couple has allowed initial exploration of electron transfer in this synthetic protein couple. In turn, photochemistry initiated at the viologen chromophore helped confirm and extend the radiolytic results.

Electron Transfer in Synthetic Three-Helix Bundles

De novo design of redox proteins represents a significant challenge for biological and biomimetic chemistry (*6–8*). Several maquettes have been designed toward systems in which electrons can be translocated across proteins (*9–11*). A wealth of data now exists for modified natural proteins like cytochrome *c* (*12–16*). Significant data are also available for modified single peptide systems (*17–19*), but conformational equilibria often complicate the interpretation of simple systems (*20*). However, detailed analyses of de novo proteins that adopt well-defined conformations remain rare. Two particularly attractive structural maquettes for the design and study of de novo redox proteins were reported by Ghadiri et al. (*4*) and Lieberman and Sasaki (*5*). Both systems consist of a three-helix bundle whose stoichiometry and topology are defined by the capping metal tris-bipyridyl complex. These maquettes are designed to maximize interhelical interactions, thereby providing a more stable conformation than is gener-

[1] All relevant experimental details on this work can be found in reference 21.

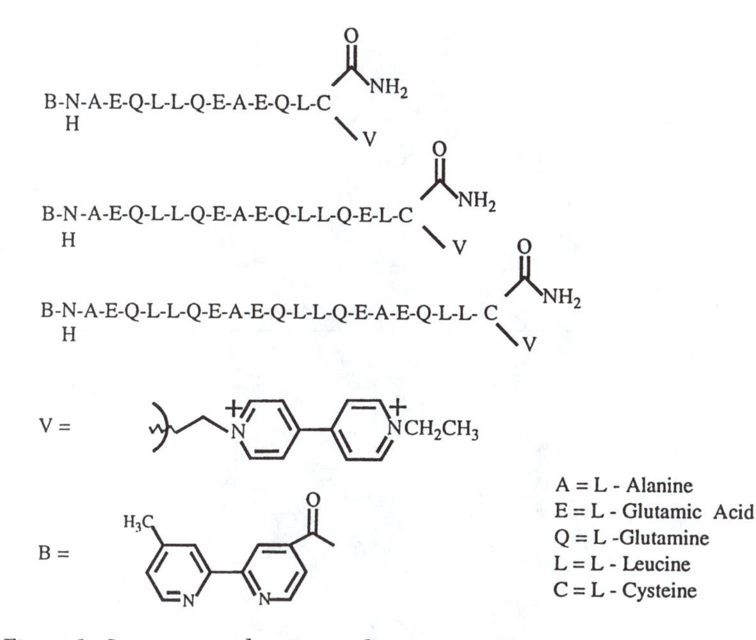

Figure 1. Structures and amino acid sequences of bipyridine–peptide–viologen complexes (bipep–V).

ally available in isolated, single peptides. Additionally, since there are numerous tris-bipyridyl complexes, using this motif to create three-helix bundles allows ready access to the many varied spectroscopic, photophysical, and redox properties offered by these metal compounds.

A minor elaboration on these maquettes provides a model bifunctional redox system, in which the bipyridine-modified peptides, bipep, are covalently modified with a redox-active viologen at the *C*-terminus as shown in Figure 1, and a redox-active metal is incorporated into the *N*-terminus (*see* Figure 2). Details of the synthesis and purification of these compounds can be found elsewhere (*21*). The 16- and 20-mer bundle complexes, M–P_n–V, where M is cobalt, iron, or ruthenium complexed with the bipyridine ligands, P is the peptide bridge, and *n* is the number of residues per single strand, adopt the helical configurations already defined by the parent peptides, as shown in Figure 3 (*22*). However, the helical content of Co^{III}–12-mer was only 30%, even in the presence of trifluoroethanol (TFE), a known helix-inducing reagent. For this reason, the 12-mer bundles were not used as models for helix-mediated ET in this study.

When the metal in these systems is cobalt, the Co^{III} tris-bipyridine and viologen moieties of Co^{III}–P_n–V readily undergo redox reactions at similar potentials to those of the isolated systems (Co^{III} tris-bipep: $E° = +0.15$; viologen: $E° = -0.64$, both versus saturated NaCl calomel electrode (SSCE); pH = 7.0, determined by cyclic voltammetry). When the metal is ruthenium, the Ru(II) tris-bipyridine excited state, Ru(II*), is expected to have $E° \sim -0.8$ V (*17*).

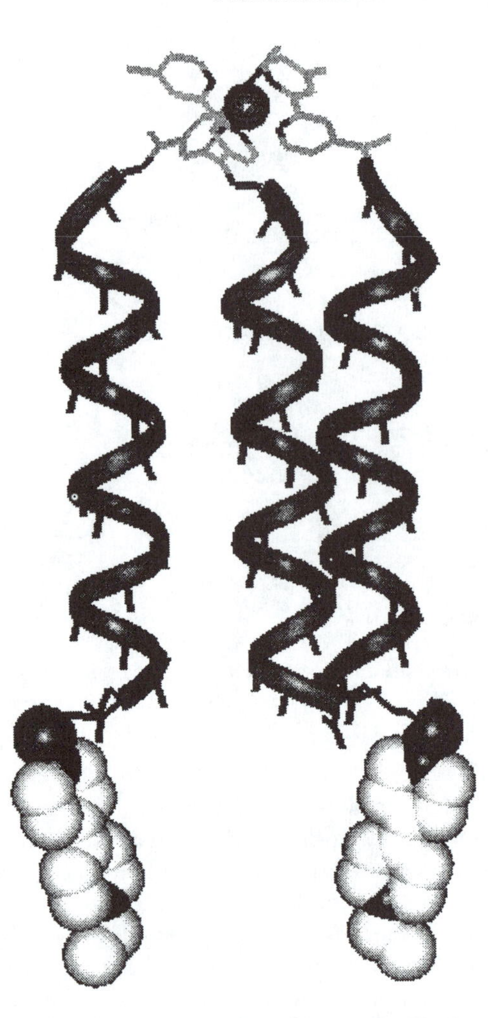

Figure 2. Sketch of the putative structure of an M–P_n–V maquette. The bipyridines are at the top of the page, with the metal center represented as a black sphere. The viologen group is the bicyclic ring at the bottom of the page.

Flash Photolysis of Ruthenium-Modified Helical Maquettes

First attempts to measure electron transfer reactions in these maquettes focused on the Ru^{II}–P_n–V system. All ruthenium-modified complexes were prepared according to standard procedures (*4, 23*). Time-resolved fluorescence measurements of Ru(II*)–P_n revealed lifetimes of about 490 ns, consistent with those of Ru(2,2′-bipyridine)$_3^{+2*}$ (*24*). Based on this lifetime, the only ET-rates that would be directly measurable by this technique are those faster than about 2×10^6 s^{-1}. A possible ET reaction takes place as shown in Scheme I.

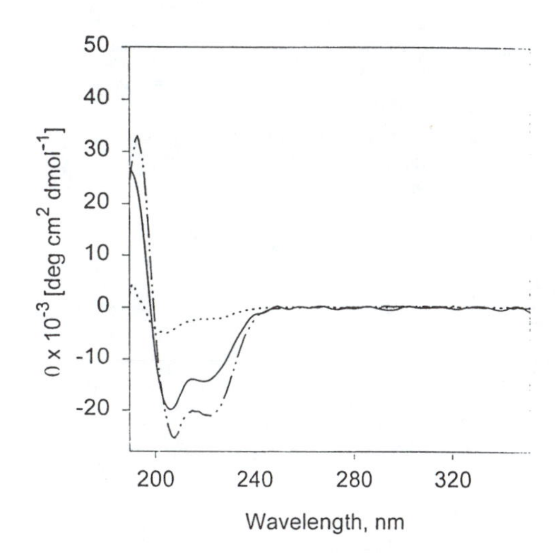

Figure 3. Circular dichroism spectra of Co–P_n–V maquettes, where n = *12, 16, and 20 are represented by dotted, solid, and dotted–dashed curves, respectively. As the number of residues increases, the helicity of the bundles is enhanced, as shown by the increased negative ellipticity at 222 nm. All ellipticity measurements are expressed as mean residue ellipticity. The spectra were obtained in 100 mM formate and 50 mM phosphate, pH = 7.0, at 25 °C. Peptide quantitation was by amino acid analysis in all cases.*

$$\mathrm{Ru^{II}{-}P_n{-}V^{2+}} \xrightarrow{h\nu} \mathrm{Ru^{II*}{-}P_n{-}V^{2+}}$$

$$\mathrm{Ru^{II*}{-}P_n{-}V^{2+}} \xrightarrow{k^{ET}} \mathrm{Ru^{III}{-}P_n{-}V^{+}}.$$

Scheme I.

Following the creation of the Ru(II*) excited state, an electron can, in principle, be transferred from the ruthenium to the viologen moiety. Since there was neither an absorbance change of the viologen moiety (monitored at 600 and 370 nm) nor a change in the fluorescence lifetime of Ru(II*) in the presence of viologen, it was concluded that k_{ET} was less than the rate of Ru(II*) fluorescence decay. As a result of these data, we decided to use pulse radiolysis as a means of accessing slower electron transfer rates which might be evidenced by this system.

Measurements of Cobalt-Modified Systems

As mentioned previously, the upper limit on the measurement of a reaction rate in pulse radiolysis is often the addition rate of the electron or radical to the reactant of interest. To measure the addition rate of CO_2^- to viologen in the bundles, Fe^{II}–P_n–V was used. The Fe^{I} state is inaccessible and the reduction

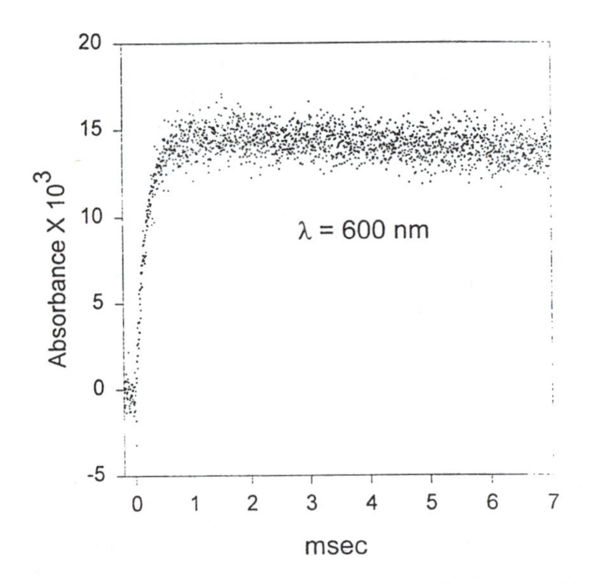

Figure 4. Pulse radiolysis transient absorption data for Fe–P_{12}–V, illustrating the prompt reaction of CO_2^- with the viologen moieties in the maquette. Note that there is a slow (ca. 40 s^{-1}) decay of the viologen absorption, effectively putting a lower limit on the rate of electron transfer that is observable with this technique. One source of this decay could be small amounts of oxygen in the sample.

of the bipyridine moiety is slow, allowing one to monitor the viologen absorbance change without interfering chemistry from another redox-active group and determine a second-order rate constant for the production of reduced viologen. The addition rate of CO_2^- was found to be 4.8×10^3 s^{-1} for the Fe–P_{12}–V complex at a concentration of 5 μM. Since there are, on average, two viologens per complex, the second-order rate constant for viologen reduction is 4.8×10^8 M^{-1} s^{-1}. Kinetic data are shown in Figure 4.

Measurements of electron transfer rates by pulse radiolysis studies of Co^{III}–P_n–V show a diffusion-controlled reduction of the viologen chromophore, followed by exergonic ($\Delta G = -0.80$ eV) electron transfer to the cobalt, with rate constants of 630 and 360 s^{-1} for the 16-mer and 20-mer complexes, respectively (Figure 5). The reaction proceeds according to Scheme II. The

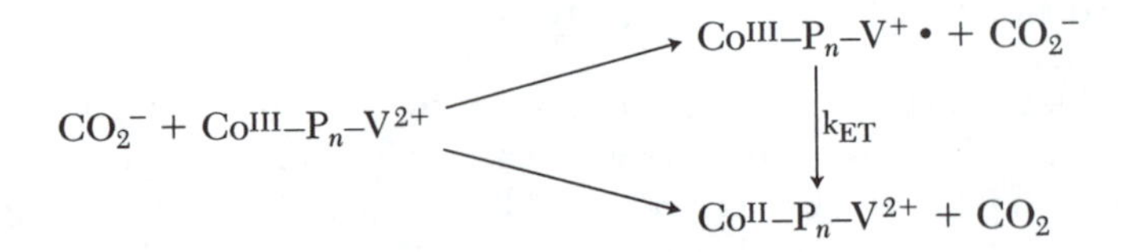

Scheme II.

photochemical experiment proceeds according to, where TEOA is triethanolamine, used here as a reductant (Scheme III) (*25–26*).

$$Co^{III}–P_n–V^{2+} \xrightarrow{h\nu,\ TEOA} Co^{III}–P_n–V^{+\cdot}$$

$$Co^{III}–P_n–V^{+\cdot} \xrightarrow{k_{ET}} Co^{II}–P_n–V^{2+}$$

Scheme III.

The observed rates were independent of peptide concentration, laser pulse energy, and radiation dose, consistent with intraprotein ET along the helix (*21*). Rates of reactions were monitored at 600 and 370 nm for the viologen oxidation and at 320 nm to follow the chemistry of the Co^{III} (bipyridine)$_3$ moiety, respectively. Photolytic control experiments with bipep–viologen show only reduction of the viologen chromophore with no subsequent reaction (Figure 6). Photolytic kinetic data for a $Co^{III}–P_n–V$ complex are also given in Figure 6. Figure 7 shows typical UV–visible spectra for a cobalt-modified maquette.

Effects of Secondary Structure on Electron Transfer Rates

The role of conformational equilibria in peptide-mediated ET was probed by inducing conformational extrema of the peptide using 6 M urea or 25% trifluoroethanol (TFE). Helicities of the maquettes under different solution condi-

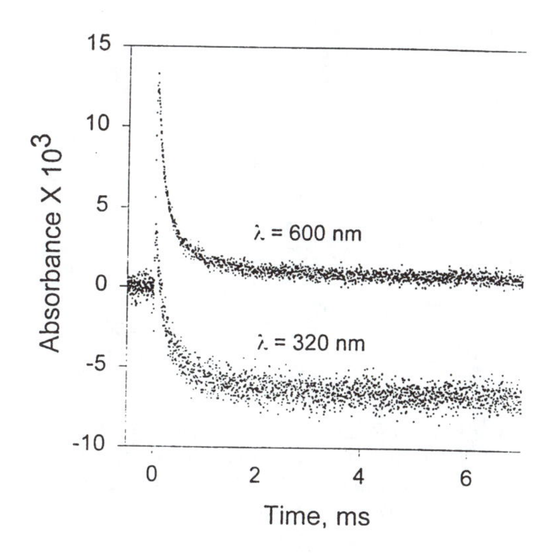

Figure 5. Pulse radiolysis transient absorption data for Co–P_{16}–V, showing the decay at 600 nm due to oxidation of the viologen radical cation moiety and the bleach at 308 nm due to the reduction of the Co^{III}–tris-bipep species. The initial rapid increase at 600 nm is due to the reaction of CO_2^- with viologen.

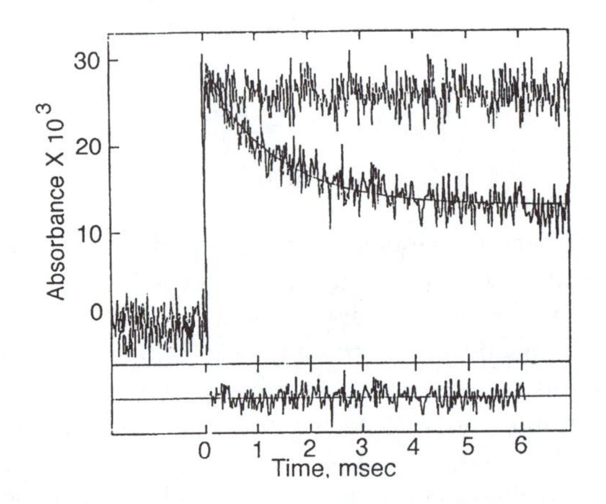

Figure 6. The upper curve shows the persistence of the viologen transient absorption in the absence of cobalt. The lower curve illustrates the photolysis transient absorption data for Co–P_{16}–V, revealing the decay at 600 nm due to oxidation of the viologen radical cation.

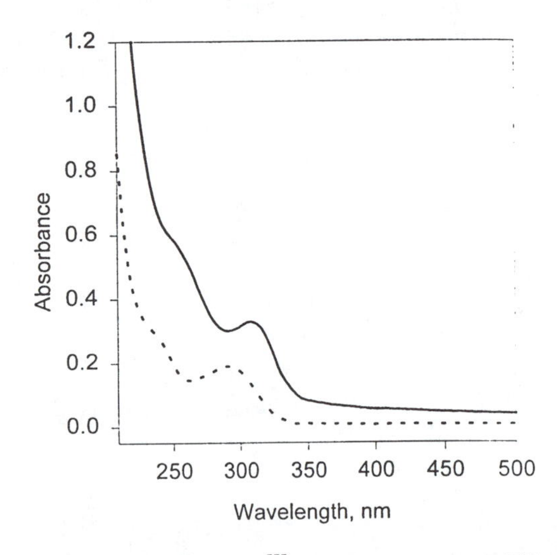

Figure 7. UV–visible spectra for a Co^{III}–P_{16}–V maquette in 100 mM phosphate, pH = 7.0, 8.1 × 10^{-6} M (upper curve), compared with spectra for the unmodified 16-mer–V complex in the same solvent, 19.3 × 10^{-6} M. Note the enhanced absorption at λ = 308 nm.

Table I. Comparison of the Effects of Solution Conditions on Helicity

Complex	*Radiolysis*[a]	*Photolysis*[b]	*25% TFE Added*	*6 M urea Added*
Co–P_{12}–V	6	18	30	0
Co–P_{16}–V	45	28	77	0
Co–P_{20}–V	67	86	94	0

NOTE: Values are % helicity. Helicities were calculated as described in reference 22. CD measurements were collected using a JASCO 710 spectropolarimeter with 0.1- and 0.5-mm cells at 25°C.
[a] Radiolysis solution conditions were 50 mM sodium phosphate, 100 mM sodium formate, pH = 7.0.
[b] Photolysis solution conditions were 100 mM triethanolamine, 100 mM sodium phosphate, and 1 M ethanol.

tions are given in Table I; averaged kinetic data from the pulse radiolysis and flash photolysis experiments are given in Table II. The helical bundles unfold in the presence of 6 M urea to produce random-coil structures (~0% helicity); alternatively, in the presence of 25% TFE, helicity is enhanced (*27*) as shown by circular dichroism (CD) spectroscopy. When the folded and unfolded proteins are examined by pulse radiolysis and flash photolysis, first-order processes are observed. Intriguingly, photolytic studies of the 16-mer in the presence of 25% TFE yield $k_{ET} = 2000 \pm 200\ s^{-1}$, a greater than two-fold increase in the intramolecular ET rate compared with studies of the random coil configuration. As expected, increases in the ET rates are not seen for the 20-mer complexes upon addition of 25% TFE because the helicity is very high in the absence of this reagent. However, the presence of biphasic kinetics in these metalloproteins in the presence of 6 M urea complicates the trend of conformation versus ET rate in this system.

Energy Transfer Studies of Donor–Acceptor Distances

In the presence of 6 M urea, Co^{III}–P_{20}–V exhibits biphasic kinetics, as shown in Table II. These data suggest the presence of different donor–acceptor distances

Table II. Rate Constants for the Co^{III}–P_n–V Systems

Complex	*Radiolysis*	*Photolysis*	*25% TFE Added*	*6 M urea Added*
Co–P_{16}–V	6.3×10^2	7.0×10^2	2.0×10^3	9.0×10^2
Co–P_{20}–V	3.6×10^2	7.0×10^2	4.0×10^2	3.1×10^3; 6.2×10^2

NOTE: Units are s^{-1}. Solution conditions are the same as described in Table I. Measurements were done at sample concentrations ranging from 80 to 0.7 μM. Each measurement is the average of five or more shots at a given concentration. Effects of changing reorganization energies between solvent systems are negligible (<10% change in ET rate) because of the polar nature of all systems used. The driving force for all the reactions is –0.8 eV.

Table III. Comparison of the Effects of Solution Conditions on Helicity, distance, and ET Rates for $Co^{III}(P_{16})$–Viologen

Experiment	*% Helicity*	*Donor–Acceptor Distance*	*Rate* (s^{-1})
Radiolysis[a]	45	18.7	$6.3 \pm 0.5 \times 10^2$
Flash photolysis[b]	28	18.7	$7.0 \pm 0.5 \times 10^2$
In 25% TFE	77	17.8	$2.0 \pm 0.2 \times 10^3$
In 6 M urea	0	18.5	$9.0 \pm 1.0 \times 10^2$

[a] All radiolysis experiments were carried out in 100 mM formate and 50 mM phosphate, pH = 7.0. N_2O was used for degassing.
[b] Laser flash photolysis was performed in 100 mM formate, 50 mM phosphate, pH = 7.0, 100 mM triethanolamine, 1 M ethanol, at 23 °C. Solutions were degassed with argon. Concentration of protein ranged from 80 μM to 0.5 μM.

within these maquettes. Energy transfer conducted on similar maquettes supports this idea: there is clearly more than one donor–acceptor distance because the decay curves cannot be fitted to a single exponential. Energy transfer studies were done with Ru^{II}–P_n–A, where A is 5-((((2-acetyl)amino)ethyl)amino)-naphthalene-1-sulfonic acid (AEDANS). 5-((((2-Iodoacetyl)amino)ethyl)amino)-naphthalene-1-sulfonic acid (IAEDANS, obtained from Molecular Probes, Inc.) was used to specifically modify the *C*-terminal cysteine of bipep. Since the emission of IAEDANS at 482 nm overlaps with Ru^{II}–P_n absorption, Ru^{II}–P_n–A maquettes are good candidates for energy transfer experiments (*28*).

Table III summarizes rate, helicity, and donor–acceptor distance data for the 16-mer bundle. Because of the observable trend in conformation versus rate in the electron transfer studies, it was decided to measure donor–acceptor distances in the 16-mer metalloprotein bundles. In order to study the effects of solution conditions on *R*, the donor–acceptor distance, Förster energy transfer was used as a spectroscopic ruler, according to

$$R = \left\{\frac{(8.8 \times 10^{-28})\kappa^2\phi_{em}}{n^4(\tau_0/\tau - 1)} \int F_D(\lambda)\epsilon_A(\lambda)\lambda^4 d\lambda\right\}^{1/6}$$

where κ is the dipole–dipole orientation factor, ϕ_{em} is the fluorescence quantum yield of the donor, F_D is donor fluorescence, E_A is the extinction coefficient for the acceptors, *n* is the index of refraction of the medium, and τ_0 and τ are the emission lifetimes in the absence and presence of acceptor, respectively (*28*). For this experiment, Ru^{II}–P_{16}–AEDANS was used so that the AEDANS emission would overlap with the absorbance of the ruthenium moiety. The geometry of this bundle is identical to that of the Co^{III}–P_{16}–viologen system as determined by CD measurements. Moreover, the geometries of the Co^{III} and Ru^{II} moieties are known to be similar (*29*). Excitation in all experiments took place at 337 nm, and the emission was monitored at 470 nm. κ^2 was

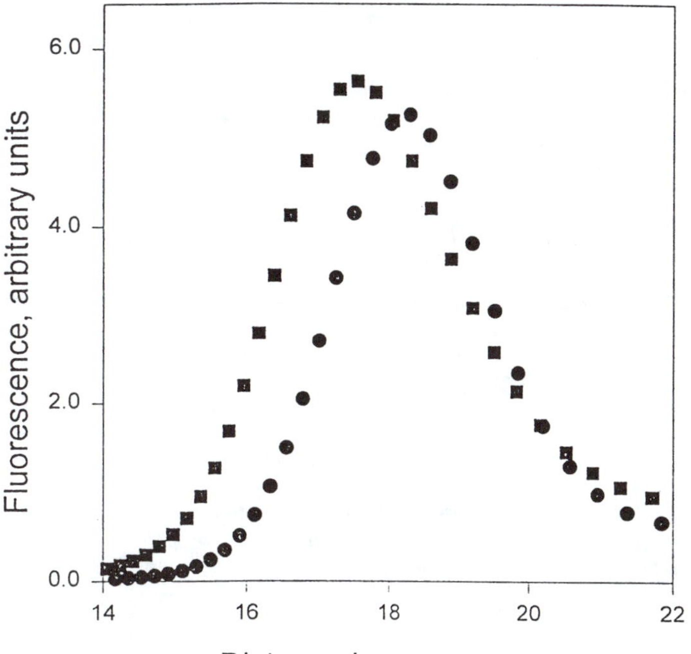

Figure 8. Distributional fluorescence lifetime fit for Ru^{II} (16-mer)$_3$–AEDANS. One hundred exponentials were used, yielding χ^2 values of 0.94 and 0.82 for experiments run in 6 M urea (circles) or trifluoroethanol (squares), respectively. These fits yielded better χ^2 values than attempts using two or three exponentials.

estimated to be 2/3, $n = 1.4$, and $\phi_{em} = 0.58$ and 0.64 in 6 M urea and 25% TFE, respectively.

The resulting data (Figure 8) clearly show that there is a distribution of donor–acceptor distances for the metalloprotein in both 6 M urea or 25% TFE with average R values of 18.5 and 17.8 Å, respectively. These distances are consistent with a conformation in which the AEDANS moiety is oriented parallel to one helical axis of the trimer. The distribution of distances can be explained by the flexible linker used to attach the energy donor to the metalloprotein, but may also be due to flexibility in the bundles themselves. Moreover, the data show that the helical metalloprotein has a higher population of conformers with a smaller R than the same protein under denaturing conditions. The apparent 0.7 Å decrease in donor–acceptor distance in the helical conformation is self-consistent with the observed increase in the ET rate: using the well established exponential dependence of the rate on the donor–acceptor distance, $k_{ET} \propto \exp-(\beta R)$, with a conformationally averaged electronic coupling factor, β, equal to 1.1 Å^{-1} (*30*), and $\Delta R = 0.7$ Å, the predicted $k_{helix}/k_{random\ coil}$ is 2.2, in good agreement with the experimental $k_{helix}/k_{random\ coil}$ of 2.2.

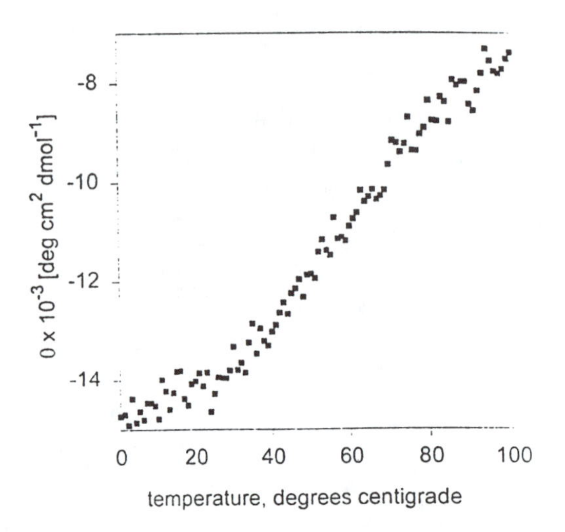

Figure 9. Plot of molar ellipticity at 222 nm versus temperature for a Co^{III}–P_{16}–viologen bundle at 25 °C. The solution conditions were the same as those given for the radiolysis experiments (100 mM formate and 50 mM phosphate, pH = 7.0).

Additional information on the molten globular nature of Ru^{II}–P_{16} and Co^{III}–P_{16}–viologen bundles was obtained by CD thermal melts. In general, naturally occurring proteins exhibit cooperative melting curves as monitored by their molar ellipticity at 222 nm over a large temperature range. The transition from the folded to unfolded state takes place over a temperature range of 5–10°. In contrast, molten globular proteins unfold gradually over a wide temperature range, indicating that they do not possess a unique, folded structure. Thermal melt data for Co^{III}–P_{16}–viologen, presented in Figure 9, reveal a lack of any cooperative unfolding. Similar results were obtained for Ru^{II}–P_{16}–AEDANS. This type of thermal melt is consistent with the behavior of molten globules.

Related Studies of Engineered Metalloproteins and Peptides

In recent years, there have been a number of studies of de novo designed proteins and peptide systems that are based on a metal-binding center. The goals of such studies are diverse, including models in which to study electron transfer, metalloenzyme function, photosynthesis, and transcription-activation factors such as zinc fingers. Several maquettes designed toward the construction of redox active proteins are described below.

To study electron transfer properties of monomeric α-helices, Dahiyat et al. used a 28-mer peptide containing two histidine residues in the sequence

(*31*). This peptide was modified with two redox-active groups, a $Ru^{II}(bpy)_2$ imidazole at one free histidine, and a $Ru(NH_3)_4$(pyridine) at the remaining free histidine (bpy is 2,2′-bipyridine). To avoid frayed ends of the peptide, the histidines are placed at least 6 residues from either terminus. CD spectroscopy confirmed that the addition of these inorganic groups did not perturb the helical structure of the peptide. It will be of interest to compare ET rates from this system with rates in helical bundles to assess whether the presence of interhelix interactions in the de novo metalloprotein exerts an influence on the electronic coupling of donor and acceptor.

A heme-binding four-helix bundle has been designed as a maquette for electron transfer and catalytic studies (*8*). The existence of naturally occurring four-helix bundle cytochromes, such as cytochrome b-562, provided the impetus for the choice of this design. By modulating the sequence of α_2, a peptide that self-assembles to form a four-helix bundle, various heme-binding sites were created. UV–visible spectra show that two peptide sequences with a designed heme-binding pocket, $VAVH_{25}$(S–S), and retro(S–S) exhibit characteristics of bound heme (the Soret band shifted from ~390 nm to 412 nm for oxidized heme), versus no shift for control peptides without a designed heme pocket. Furthermore, redox potentiometry of bound heme in these bundles yields midpoint potentials that are in the same range as midpoint potentials for naturally occurring cytochromes. Because the heme groups in these systems are in a hydrophobic environment, it will be instructive to compare electron transfer rates in such systems to rates in designed systems where the redox-active groups are solvent exposed. A priori, the reorganization energy should be higher for the systems where the redox-active moieties are solvent exposed, but the magnitude of this effect is uncertain.

Another approach to the design of an internal metal center has been explored by Nishino et al. (*32*). A cyclic peptide template was used to bring three 13-mer peptides into close proximity. The artificial amino acid 2,2′-bipyridylalanine was placed at position 7 of each 13-mer to provide a tris-metal binding site. Furthermore, the hydrophobicity of this binding site was enhanced by the inclusion of cyclohexylalanine at positions 3 and 11 and leucine at positions 4 and 10. Nickel was added to the solution, and a titration curve gave the expected 1:1 ratio of nickel to peptide. CD spectroscopy revealed a slight decrease in alpha-helical content after the addition of the metal, consistent with the expected destabilization associated with burying a charged group in a nonpolar environment. Also, the addition of ethylenediaminetetraacetic acid (EDTA) failed to remove the added Ni^{+2}, suggesting that the designed metal binding site is shielded from the solvent.

In contrast to those who study metal binding sites within alpha-helical structures, another group has focused its efforts on the design of a metallopeptide that adopts a parallel β-pleated sheet structure (*33*). The polypyridyl complex $[Ru(bpy)_2L]^{+2}$ is used, where bpy is 2,2′-bipyridine and L = 3,5-dicarboxy-2,2′-bipyridine. The carboxyl groups are used as points of attachment for the short peptide sequence Val-Val-OMe, where OMe is used to indicate that the *C*-terminus of the peptide is capped with a methoxy group instead of a carboxylic acid, presumably to avoid unfavorable charged interactions between two acidic groups. The two diastereomers of this compound were isolated, and ^{1}H NMR was used to assess the solution structure. Vicinal coupling constants from correlation spectroscopy (DQCOSY) experiments are within the range expected for acyclic β-sheet models. Furthermore, nuclear Overhauser enhancement spectroscopy (NOESY) spectra reveal strong inter-residue cross peaks, which are also consistent with the presence of β-sheet structure. A comparison of ET rates through this type of structure with rates over similar distances in an alpha helix could help illustrate the role of hydrogen bonds as mediators of ET.

We conclude that the stable triple helix maquette of Ghadiri and Sasaki can indeed be elaborated to produce bifunctional redox proteins that translocate electrons across significant distances at physically reasonable rates. Moreover, because of their conformational variability under different solution conditions, such systems represent a means of investigating the effects of helical conformation on rates of electron transfer. The current observations set the stage for systematic studies of structural determinants of ET within a single, variable maquette. Additionally, it will be interesting to compare rate data from this study with data from other de novo metalloprotein systems, and see how the results compare with expectations based on current electron transfer and protein design theory.

Acknowledgments

We thank Alan Corin and Elizabeth Gaillard for many helpful discussions and all of their other assistance with this project. The work at Brookhaven National Laboratory was carried out under contract DE–AC02–76H00016 with the U.S. Department of Energy and was supported by its Division of Chemical Sciences, Office of Basic Energy Research. Work at the University of Rochester and Princeton University was supported by the National Institutes of Health (NIH Grant GM50019). Mitchell Mutz is supported by NIH Training Grant GM08309–07.

References

1. McLendon, G.; Miller, J. *J. Am. Chem. Soc.* **1985,** *107,* 7811.
2. McLendon, G.; Winkler, J.; Nocera, D.; Mauk, A. G.; Gray, H. B. *J. Am. Chem. Soc.* **1985,** *107,* 739.
3. Saunders, A. J.; Pielak, G. J.; Sligar, S. G.; Durham, B.; Millett, F. *Biochemistry* **1993,** *32,* 7519.
4. Ghadiri, M. R.; Soares, C.; Choi, C. *J. Am. Chem. Soc.* **1992,** *114,* 825.
5. Lieberman, M.; Sasaki, T. *J. Am. Chem. Soc.* **1991,** *113,* 1470.
6. Kraatz, H. B. *Angew. Chem. Int. Ed. Engl.* **1994,** 33, 2055.

7. Handel, T. M.; Williams, S. A.; DeGrado, W. F. *Science (Washington, D. C.)* **1993,** *261,* 879.
8. Choma, C. T.; Lear, J. D.; Nelson, M. J.; Dutton, P. L.; Robertson, D. E.; DeGrado, W. F. *J. Am. Chem. Soc.* **1994,** *116,* 856.
9. Case, M. A.; Ghadiri, M. R. *Angew. Chem. Int. Ed. Engl.* **1993,** *32,* 1594.
10. Robertson, D. E.; Farid, R. S.; Moser, C. C.; Urbauer, J. L.; Mulholland, S. E.; Piditiki, R.; Lear, J. D.; Wand, A. J.; Degrado, W. F.; Dutton, P. L. *Nature (London)* **1994,** *368,* 425.
11. Vassilian, A; Wishart, J. F.; van Hemeryck, B.; Schwarz, H.; Isied, S. *J. Am. Chem. Soc.* **1990,** *112,* 7278.
12. Therien, M. J.; Bowler, B. E.; Selman, M. A.; Gray, H. B.; Chang, I-J.; Winkler, J. R. In *Electron Transfer in Inorganic, Organic, and Biological Systems;* Bolton, J. R.; Mataga, N.; McLendon, G., Eds.; Advances in Chemistry 228; American Chemical Society: Washington, DC, 1991; p 192.
13. Geren, L.; Hahm, S.; Durham, B.; Millett, F. *Biochemistry* **1991,** *30,* 9450.
14. Meier, M.; Van Eldik, R.; Chang, I-J.; Mines, G.; Wuttke, D. S.; Winkler, J. R.; Gray, H. B. *J. Am. Chem. Soc.* **1994,** *116,* 1577.
15. Cho, K.; Che, C. M.; Chu, W. F.; Choy, C. L. *Biochim. Biophys. Acta* **1988,** *934,* 161.
16. Liang, N.; Hoffman, B. M.; Johnson, J. A.; Smith, M.; Mauk, A. G.; Pielak, G J. *Science (Washington, D.C.)* **1988,** *240,* 311.
17. Isied, S. S.; Ogawa, M. Y.; Wishart, J. F. *Chem. Rev.* **1992,** *92,* 381.
18. Imanishi, Y.; Sisido, M.; Inai, Y. *J. Phys. Chem.* **1990,** *94,* 6237.
19. Mishra, A. K.; Chandrasekar, R.; Faraggi, M.; Klapper, M. H. *J. Am. Chem. Soc.* **1994,** *116,* 1414.
20. Cabana, L. A.; Schanze, K. S. In *Electron Transfer in Biology and the Solid State: Inorganic Compounds with Unusual Properties;* Johnson, M. K.; King, R. B.; Kurtz, D. M., Jr.; Kutal, C.; Norton, M. L.; Scott, R. A., Eds.; Advances in Chemistry 226; American Chemical Society: Washington, DC, 1990; pp 101–124.
21. Mutz, M. W.; Wishart, J. F.; Gaillard, E. R.; Corin, A. F.; McLendon, G. L. *Proc. Natl. Acad. Sci. U.S.A.* **1996,** *93,* 9521.
22. Chen, Y.; Yang, J.; Chau, K. *Biochemistry* **1974,** *13,* 3350.
23. Elliott, C. M., Colorado State University, personal communication, 1994.
24. Van Houten, J.; Watts, R. J. *J. Am. Chem. Soc.* **1976,** *98,* 4853.
25. Hoffman, M. R.; Prasad, D. R.; Jones, G.; Malba, V. *J. Am. Chem. Soc.* **1983,** *105,* 6360.
26. Ebbesen, T.; Ferraudi, G. *J. Phys. Chem.* **1983,** *87,* 3717.
27. Nelson, J. W.; Kallenbach, N. R. *Proteins* **1986,** *1,* 211.
28. Förster, T. *Ann. Phys.* **1948,** *2,* 55.
29. Brunschwig, B. S; Creutz, C.; Macartney, D. H.; Sham, T-K.; Sutin, N. *Faraday Discuss. Chem. Soc.* **1982,** *74,* 113–127.
30. Langen, R.; Chang, I-J.; Germanas, J. P.; Richards, J. H.; Winkler, J. R.; Gray, H. B. *Science (Washington, D.C.)* **1995,** *268,* 1733.
31. Dahiyat, B. I.; Meade, T. J.; Mayo, S. L. *Inorg. Chim. Acta* **1996,** *243,* 207.
32. Nishino, N.; Kato, T.; Murata, T.; Nakayama, H.; Arai, T.; Fujimoto, T.; Yamamoto, H.; Yoshikawa, S. *Chem. Lett.* **1996,** 49.
33. Ogawa, M. Y.; Gretchikhine, A. B.; Soni, S-D.; Davis, S. M. *Inorg. Chem.* **1995,** *34,* 6423.

11

Pulse Radiolysis Measurements of Intramolecular Electron Transfer with Comparisons to Laser Photoexcitation

J. R. Miller, K. Penfield, M. Johnson, G. Closs[†], and N. Green

Chemistry Division, Argonne National Laboratory, Argonne, IL 60439

To understand electron transfer, it is important to separate the effects of the variables that control the rates. Pulse radiolysis offers the advantages of well-known free-energy changes ($\Delta G°$) that are independent of distance and only weakly dependent upon solvent properties. It has the disadvantages of time resolution limited to 10 to 100 ps, the need for high concentrations to enable high time resolution, and concerns about sample degradation. Laser excitation offers much faster time resolution, less sample degradation, and modest needs for concentration. Difficulties are associated with both techniques in that electronic couplings and solvent reorganization energies depend upon distance. The advantages and capabilities of these two complementary techniques are compared. It is shown how pulse radiolysis can yield dependence of electron transfer rates on free energy, distance, temperature, and solvent reorganization energy. Recent results provide information about the distance dependence of the solvent reorganization energy. Examples are also given of the use of pulse radiolysis for measurement of free-energy changes and energetics of ion pairing.

This chapter will discuss the abilities of pulse radiolysis and laser photoexcitation to contribute to an understanding of the fundamental and important process of electron transfer (ET). We begin with a brief discussion of ET reactions.

The scheme in Figure 1 depicts an energy level diagram for energy capture by charge separation, in which a photoexcited molecule transfers an electron to a vacant orbital in a second molecule, which then transfers it to a third.

Many experimenters found how easy it is to implement such a scheme to achieve dismal efficiencies for energy storage. Efficient energy storage can

[†]Deceased.

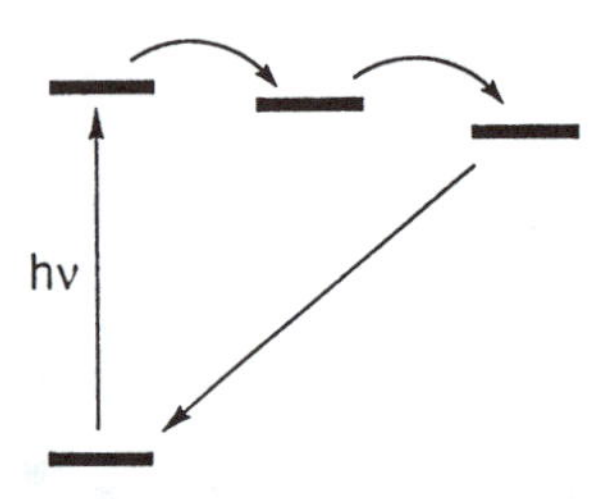

Figure 1. Light absorption by the molecule on the left leads to transfer in the graded set of energy levels to a second and then a third molecule.

occur if the "forward" charge separation reactions are fast, but dissipate little energy, while energy-wasting back reactions are slow, even though they would dissipate large amounts of energy. High efficiencies for such a scheme occur naturally in photosynthetic reaction centers (*1–4*), and can be achieved by rational design. Such rational design requires knowledge of how ET rates depend on several factors. These include:

1. energy ($\Delta G°$)
2. polarity of the medium (solvent) and its dynamics (relaxation times)
3. temperature
4. structural changes in the electron donor (D) and acceptor (A) groups
5. distance
6. electronic properties of the material between D and A
7. angles of orientation between D and A groups

The electron transfer theory of Marcus (*5, 6*) discussed the first four factors. It described an energy specificity that we now know can enhance energy storage. The theory predicted a relationship between kinetics and thermodynamics in which ET rates were expected to be maximized for a moderate free-energy change ($\Delta G°$), but to actually decrease at larger energies. The decrease in rate with decreasing energy, known as the "inverted region", was found not to occur in the famous experiments of Rehm and Weller (*7*), as shown in Figure 2. Quantum mechanical descendants of the Marcus theory (*8–15*) describe the ET rate as a product of the electronic coupling squared and the Franck–Condon weighted density of states (FCWD):

$$k_{\mathrm{ET}} = \frac{2\pi}{\hbar}|V|^2\mathrm{FCWD}$$

$$\mathrm{FCWD} = (4\pi\lambda_S k_B T)^{-1/2}\sum_{w=0}^{\infty}\left(e^{-S}\frac{S^w}{w!}\right)\exp\{-[(\lambda_S + \Delta G° + wh\nu)^2/4\lambda_S k_B T]\}$$

$$S = \lambda_V/h\nu \qquad (1)$$

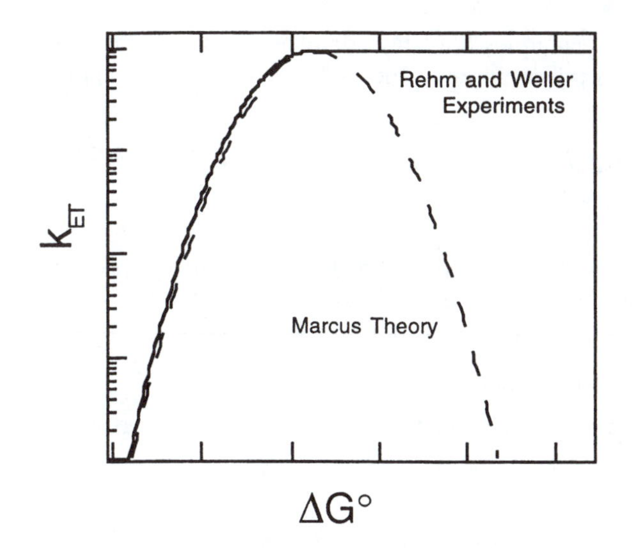

Figure 2. The Marcus theory (dashed curve) predicted that electron transfer rates would decrease with increasing free-energy change, but experiments by Rehm and Weller (solid curve) (7) found no such decrease.

where k_{ET} is the rate constant for electronic transfer, $\hbar$ is Planck's constant/2π, V is the electronic coupling between reactants and products, λ_S is the solvent reorganization energy, k_B is Boltzmann's contant, T is temperature (K), w is the vibrational quantum number in the products. $\Delta G°$ is the free-energy change, h is Planck's constant, ν is frequency, and λ_V is the internal reorganization energy of high-frequency molecular vibrations. The first four factors listed at the beginning of this section affect principally the FCWD, whereas the remaining three affect the electronic coupling, V. In equation 1 the FCWD depends on the free-energy change, $\Delta G°$, the solvent reorganization energy, λ_S, and the internal reorganization energy, λ_V, which couples the ET to high-frequency vibrations of the D and A groups, which are represented by a single mode of frequency ν. Like the classical Marcus theory, equation 1 predicts an "inverted region," although the decrease of rates for highly exoergic reactions may be less pronounced than in the Marcus theory. The quantum mechanical theory also predicts modifications of the effects of temperature and polarity. Some principal features of these predictions have been verified by experiments using both pulse radiolysis and laser photoexcitation.

This chapter will begin by delineating the capabilities of the two techniques, radiolysis and laser photoexcitation, for examination of factors that regulate ET rates. We will begin with a list summarizing the capabilities of the two techniques. Descriptions will be supplemented by reference to published work and examples from pulse radiolysis experiments in our laboratory, along with a brief description of the experimental setup. In this discussion we shall see that the two techniques are truly complementary and that great progress in understanding this important process has come from the application of both.

Photoexcitation and Pulse Radiolysis: Capabilities for Electron Transfer

Photoexcitation can be used to directly investigate the processes involved in photochemical energy storage. In addition, the capabilities of photoexcitation, particularly with lasers, for the study of ET processes include:

1. High time resolution.
2. Selective excitation of particular donor or acceptor groups.
3. Direct study of charge separation and recombination:

$$(D + A)^* \rightarrow D^+ + A^- \rightarrow D + A \quad (2)$$

   which are primary processes in photochemical energy storage.
4. Ability to use very low concentrations of compounds to be studied. For intramolecular ET in donor–spacer–acceptor (DSA) molecules, the ET rate that can be measured is independent of DSA concentration over a wide range.
5. The nondestructive nature of chemistry gives the ability to re-excite the same molecule many times without permanent chemical changes.
6. The ability to readily produce important types of species such as radical ion pairs in solution, coupled with the ability to often obtain such pairs in particular spin states.

Radiolysis, and especially pulse radiolysis, has a different and often complementary set of capabilities:

1. Simple production of radical ions, usually with a desired radical ion being the only species absorbing light in the visible/near-infrared regions.
2. Study of charge-shift reactions, $D^{\pm} + A \rightleftharpoons D + A^{\pm}$, which are important secondary reactions in photochemical energy storage and have several desirable properties as will be noted. In these reactions, an anion, D^-, or a cation, D^+, transfers an electron or hole to a neutral molecule, so there are no long-range forces between the reactants or products. Therefore the free-energy change, $\Delta G°$, is:
   a. rigorously independent of distance
   b. almost independent of solvent polarity and temperature
   c. easily measured when $\Delta G°$ is small

In addition, the same type of reaction may be studied over a wide range of $\Delta G°$.

3. ET is usually the only process occurring in pulse radiolysis, whereas electronic energy transfer may compete with ET in photoexcited reactions, and it is sometimes uncertain which process is responsible for observed rates.
4. Pulse radiolysis readily produces free ions, even in nonpolar solvents.

The next two sections present a more detailed description of the special strengths of both techniques.

Strengths of Laser Photoexcitation

Time Scales. The time scales measurable by the two techniques are illustrated in Figure 3. The time resolution of pulsed lasers is far superior, reaching to as short as 20 fs, with 200-fs measurements becoming routine in many laboratories. Pulse radiolysis measurements achieve rise times of 20–30 ps in only a few laboratories, while 1–10 ns is more common. Pulse radiolysis is slower because the accelerated particles, usually electrons, repel each other, making it difficult to bunch many of them into a very short pulse. Pulses as short as 5 ps have been reported and new accelerators may achieve 1 ps, but it is likely that pulsed lasers will remain the leader in very high time resolution.

For very slow processes, direct initiation by photoexcitation is limited by excited state lifetimes, although very slow subsequent or back reactions may

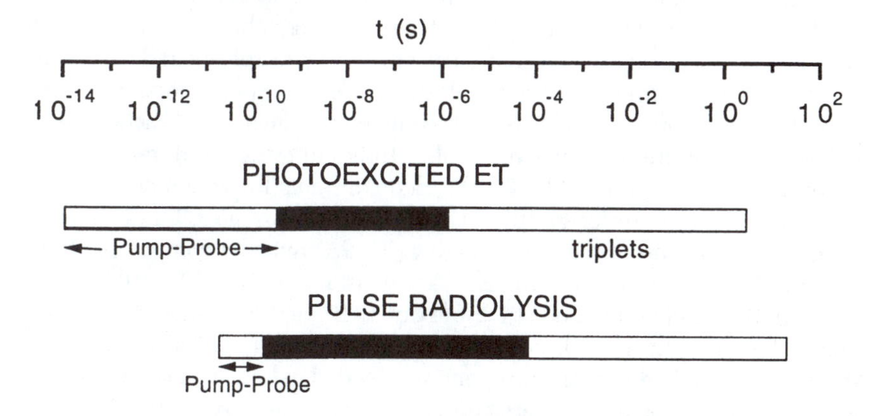

Figure 3. Time scales on which measurements can be made by photoexcitation and pulse radiolysis. For both, the fastest measurements require pump-probe techniques. The dark regions indicate straightforward single-shot measurements. Processes having a very long time scale (light areas on the right) require long-lived excited states in photoexcitation, usually triplets, and for both methods may be applicable in special media, such as solids.

be studied. Pulse radiolysis may have an advantage because the unusual oxidation states it produces in inorganic systems and radical ions in organic ones can in principle live indefinitely in solids (for centuries—this indefinite time scale is not indicated in the figure), although reactive ones will be limited in liquids approximately to the dark region in Figure 3.

Concentrations and Degradation. Photoexcited ET (equation 2) produces ion pairs that often recombine to recover the original molecules, with small quantum yields for processes leading to other products. This means that the same molecules can often be re-excited tens, hundreds, or even thousands of times without significant loss of the starting materials. This, along with high time resolution, makes photoexcitation excellent for pump-probe experiments, which typically require a thousand or more excitation pulses to obtain one kinetic trace. Pulse radiolysis, on the other hand, usually leads to various products, and therefore to the loss of the starting molecules with yields on the order of magnitude of 1. Thus, pulse radiolysis with pump-probe detection is suitable only for easily purchased or prepared molecules, for which one can devote gram quantities to experiments. For valuable molecules synthesized to place donor and acceptor groups at defined distances, practical requirements normally limit experiments to single-shot methods. In our laboratory, this requirement led us to pursue the use of very fast single-shot digitizers.

Even for single-shot experiments, photoexcitation has an advantage. If the starting materials absorb light strongly, experiments can often be performed with concentrations in the 10^{-3}–10^{-6} M region. Because the reactant excited states are produced directly by light absorption, the concentration does not affect the time resolution of the experiment. Pulse radiolysis ionizes the material unselectively to produce ionic species from the most abundant material, the solvent. Usually these are solvated electrons, e_S^-, and solvent radical cations. These must react with the desired molecules (see the experimental discussion) to produce reactants such as $D^{\pm}SA$. The time for producing the reactants scales with the concentration of DSA molecules. In liquid tetrahydrofuran (THF) or 2-methyltetrahydrofuran (MTHF), the rate constants for attachment of $k(e_S^- + DSA)$ are on the order of 10^{11} M^{-1} s^{-1}. To measure an ET rate constant, the rate of formation of the reactants (e.g., D^-SA) must be comparable to the ET rate. Therefore, to measure ET rate constant of 10^8, 10^9, 10^{10} s^{-1}, the concentration [DSA] must be at least about 1, 10, or 100 mM. In highly polar liquids the rate constants for e_S^- attachment are smaller, typically 1–3 $\times$ 10^{10} M^{-1} s^{-1}, so even higher concentrations are needed, whereas in some nonpolar liquids, such as isooctane, e_S^- attachment rate constants are $\approx 10^{13}$ M^{-1} s^{-1}, so smaller concentrations can be used. These concentration requirements are often vexing, and when solubilities of DSA molecules are not large enough, the measurements are not possible.

In pulse radiolysis of DSA molecules, electrons (or holes) are captured statistically by the D and A groups, so that the reactants are typically a 50/50

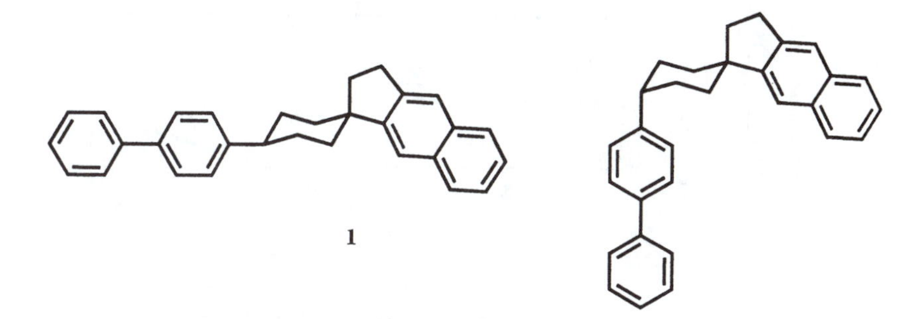

mixture of $D^{\pm}SA$ and $DSA^{\pm}$. In the anion of the *trans* isomer (on the left of the molecule shown in structure **1**), intramolecular ET can be measured. But for the *cis* isomer (right), the equilibrium constant is 1.0 ± 0.1, producing a 50–50 distribution for the electron on biphenyl and naphthalene groups. As a result it was not possible to measure the ET rate. In photoexcited ET the desired group can often be excited almost exclusively. Even if the equilibrium constant were 1.0 for the initial charge transfer, the initial population would be far from equilibrium, so the rate would be measurable.

Strengths of Pulse Radiolysis

Despite the many superior features of laser photoexcitation, pulse radiolysis has played a central and decisive role in developing our understanding of ET reactions. Figures 4 and 5 illustrate two of the reasons for the importance of pulse radiolysis.

Because there are no long-range forces between either the reactants or the products in the charge-shift reaction, $D^- + A \rightleftharpoons D + A^-$, $\Delta G°$ is expected to be independent of distance. Figure 4 demonstrates that this is so. Also plotted in Figure 4 are the calculated estimates of $\Delta G°$ for charge separation–recombination, $D + A \rightleftharpoons D^+ + A^-$, using the Weller equation.

$$\Delta G° = E(D/D^+) - E(A/A^-) - e^2/\varepsilon r - E_{00} \quad (3)$$

where $E(D/D^+)$ is the electrochemical oxidation potential of the donor molecule D, $E(A/A^-)$ is the electrochemical reduction potential of the acceptor molecule A, ε is the static dielectric constant, r is donor–acceptor distance, and E_{00} is the energy of an electronically excited state in its lowest vibrational level. In equation 3, $-e^2/\varepsilon r$ is the coulombic attraction between D^+ and A^-, and E_{00}, the energy of the excited state, is not included for charge recombination. Figure 4 illustrates that for typical ET distances, $\Delta G°$ is strongly dependent on distance. Equation 3 applies to point charges and is inexact for ions having ex-

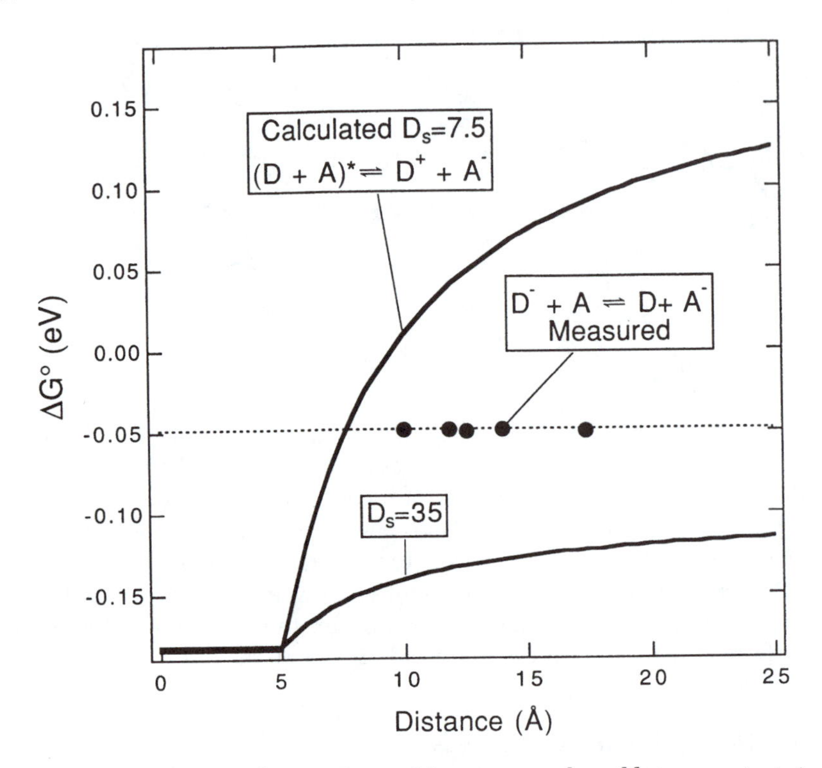

Figure 4. Free-energy changes obtained from measured equilibrium constants in MTHF for the electron transfer reaction $B^-SN \rightleftharpoons BSN^-$, where B = biphenyl, N = naphthalene, and S is a rigid, saturated hydrocarbon spacer group varying from 1,3-cyclohexane to the steroid 3,16-androstane. Also shown are calculated ΔG°_Ss for charge separation and recombination, $D + A \rightleftharpoons D^+ + A^-$, in moderately polar MTHF and a highly polar solvent.

tended charge distributions; it is also inexact if the distance between D^+ and A^- is not precisely known. An additional problem is that the redox potentials, $E(D/D^+)$ and $E(A/A^-)$, are known only in highly polar solvents containing substantial concentrations of inert ions. For other solvents, a correction, $\Delta\Delta G^\circ_S$, must be added to $E(D/D^+)$ and subtracted from $E(A/A^-)$. This is another source of uncertainty. Unfortunately, it is only rarely possible to measure ΔG° for charge separation or recombination.

For the charge shift reactions studied in pulse radiolysis, the situation is simpler.

$$\Delta G^\circ = E(D/D^-) - E(A/A^-) \tag{4}$$

Here there is no distance dependence, so the effect of distance on the ET rate can be studied at constant ΔG°. Furthermore, changes due to solvent polarity,

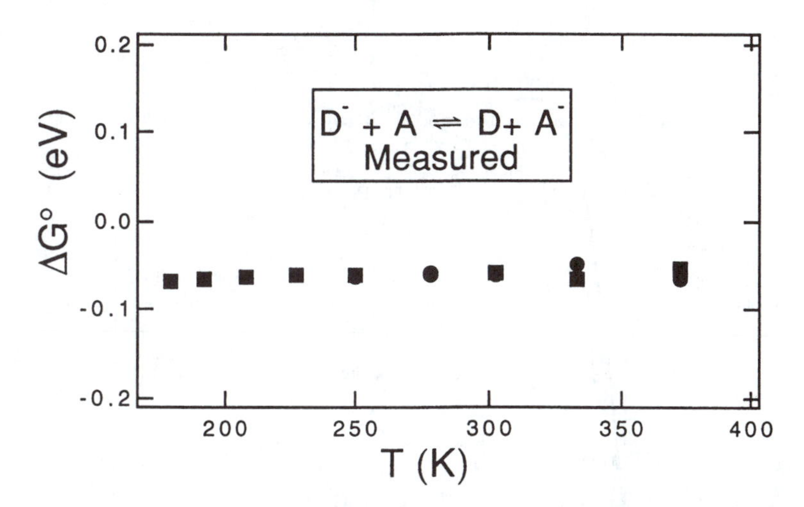

Figure 5. Temperature independence of the free-energy change, ΔG°, obtained from equilibrium constants measured by pulse radiolysis in MTHF, for the electron transfer reaction $B^-SN \rightleftharpoons BSN^-$, where B biphenyl, N naphthelene and S is the steroid 3,16-androstane ■ is measurements by intramolecular ET in BSN; ● is measurements in mixtures of monofunctional molecules BS and NS.

which add for charge separation and recombination, tend to cancel for charge shift reactions. This cancellation leads to the lack of dependence of $\Delta G°$ on temperature seen in Figure 5 and to the small changes with solvent polarity ($-\Delta G°$ changes from 40 to 60 mV, going from completely nonpolar isooctane to highly polar *N*-methylpyrrolidinone) for the same reaction.

Measurement of ET Rates by Pulse Radiolysis

Free-Energy Change, $\Delta G°$. For the charge shift reactions studied by pulse radiolysis, it is relatively easy to assemble a series of donors and acceptors to obtain a range of free energies from very weakly to very strongly exoergic, while maintaining the same type of reaction. This allowed the study of rate vs. $\Delta G°$ in our laboratory to provide confirmation of the Marcus inverted region. This was done first by Beitz and Miller (*16*) in rigid glasses at 77 K with trapped electrons as donors, which produced complex results having a scattered appearance in the inverted region because large $\Delta G°$s could produce excited products, A^{-*}, when the energy of the lowest excited state was less than $\Delta G°$. Similar experiments with B^- as the donor for intermolecular ET in glasses (*17*) (Figure 6), and then for intramolecular ET with DSA molecules in liquids (*18*) (Figure 6, bottom), led to reactions free of the effects of excited state products.

The shape of the rate vs. free-energy curve departed somewhat from that predicted by the classical Marcus theory because of quantum mechanical effects

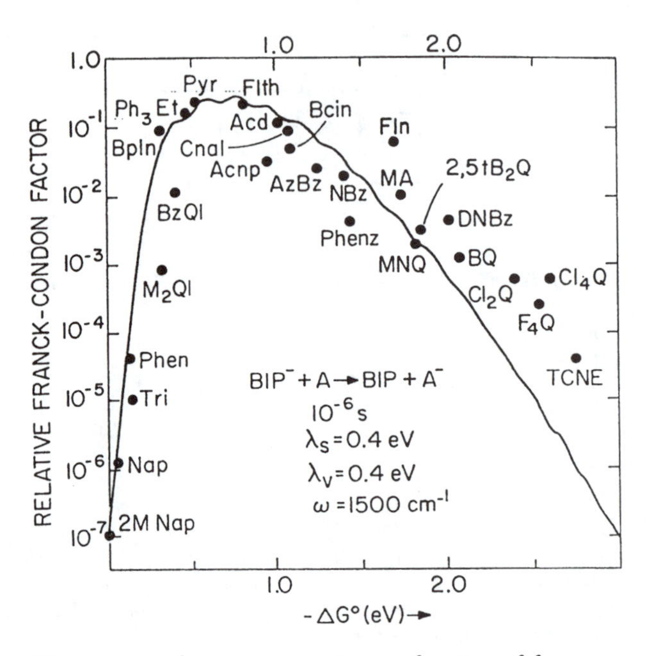

Figure 6 Electron transfer rate constants as a function of free-energy change, $\Delta G°$, by radical anions for intermolecular ET in rigid 2-MTHF glass (top) (17), *and intramolecular ET in 2-MTHF fluid at room temperature (bottom)* (18), *in molecules of the form ASB, where B = biphenyl, S = 3,16-androstane, and A is one of eight acceptor molecules shown. In both parts of the figure, the rate vs. $\Delta G°$ curves are of equation 1 and have identical parameters except for the temperature and solvent reorganization energy.*

from high-frequency molecular vibrations, but when these were included using equation 1, the theory was in good accord with experiment. The experiments clearly supported the theoretical concept that the rate was maximized at a reorganization energy, λ, which was the sum of two parts, λ_S and λ_V, the solvent and internal vibrational reorganization energies. Experiments further showed that the solvent reorganization energy that could be manipulated by changes in solvent polarity produced dramatic effects on ET rates, as shown in Figure 7.

The dependence of ET rates on $\Delta G°$ was also confirmed in experiments involving photoexcited charge separation and recombination (*19–26*). All these results confirm the existence of the inverted region. The pulse radiolysis results for rate vs. $\Delta G°$ have been more suitable for quantitative analysis because $\Delta G°$ is better known and because one type of reaction is used throughout. For photoexcited ET, it is usually possible to study only charge separation in the normal region and only charge recombination in the inverted region. Consistency of the data suggests that electronic couplings did not differ substantially

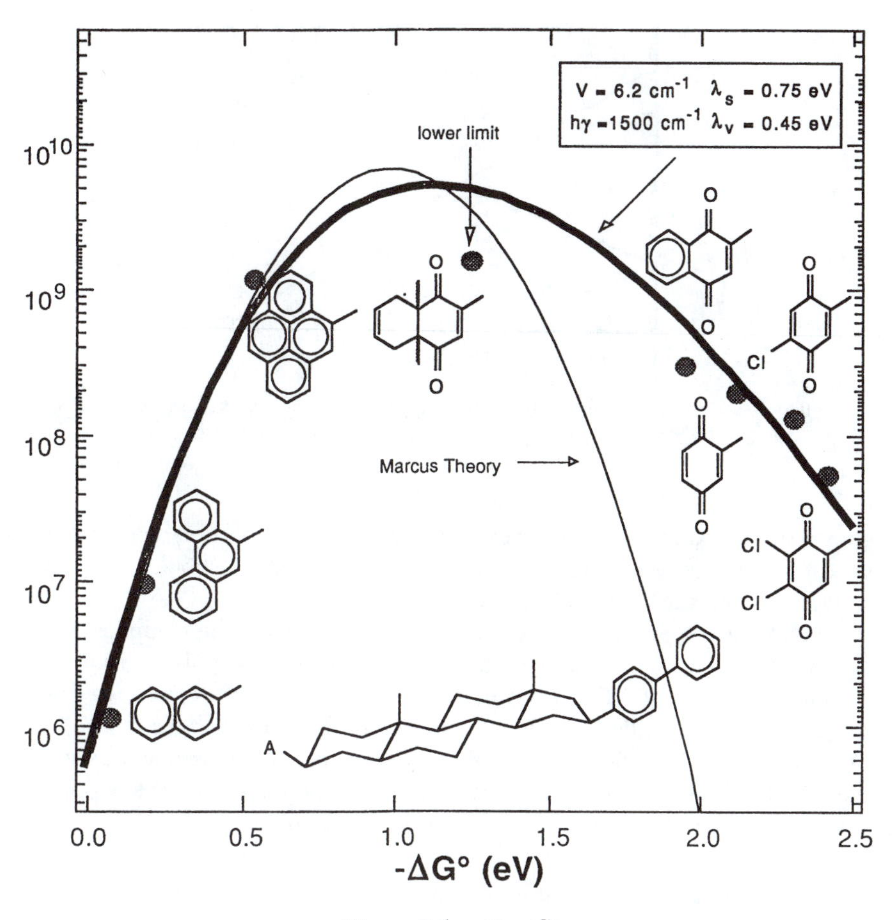

Figure 6 (continued)

for the two types of reactions, but it has been argued that solvent reorganization should be different in separation and recombination (*26–30*).

Polarity. The Marcus theory predicted a substantial dependence of ET rates on solvent polarity. Pulse radiolysis can be used to change solvent polarity while keeping other variables almost constant, whereas large changes in $\Delta G°$ occur for charge separation or recombination. Figure 7 shows the solvent dependence of a weakly and strongly exoergic intramolecular ET reaction. The weakly exoergic reactions are fastest, and the strongly exoergic reaction is the slowest in a nonpolar solvent. In the charge separation scheme of Figure 1, this will make the forward, energy-storing reactions fast and the energy-wasting back reactions slow. So a nonpolar environment enhances the efficiency of

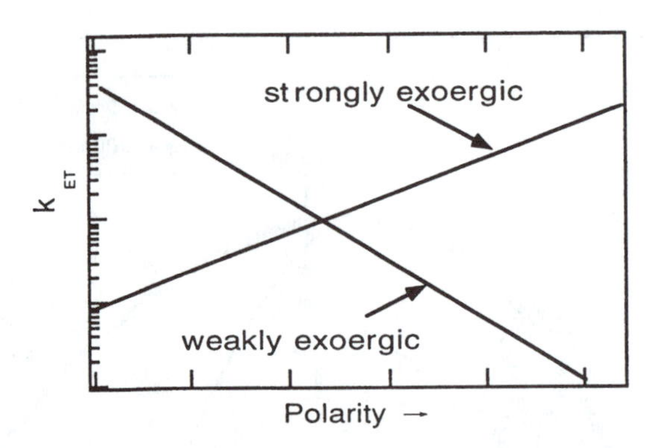

Figure 7. ET rates as a function of polarity for a weakly exoergic and a strongly exoergic ET reaction having the same distance and electronic couplings.

energy storage. Contrary to many early opinions that a highly polar environment is best for charge separation, it is now understood that we should run energy storage reactions in oil or silicon, not water.

Construction of a plot such as that in Figure 7 would be complicated for charge separation or recombination. The effect of solvent polarity cannot be examined at constant $\Delta G°$ unless many different reactions are compared. If a single reaction is studied, the variations in rate due to variations in $\Delta G°$ are likely to exceed those from variations in λ_S. If a reaction is weakly exoergic in the middle of the polarity scale, it will likely become endoergic at low polarity (for charge separation) or at high polarity (for recombination).

Temperature. Measurement of ET rates as a function of temperature by pulse radiolysis provides data that are readily interpretable because of the simplification that $\Delta G°$ is nearly independent of temperature (Figure 5) in many cases. The relative capabilities of pulse radiolysis and photoexcitation for the study of temperature dependence have much in common with those for solvent polarity dependence.

Results have shown that:

1. For a weakly exoergic intramolecular ET reaction, the rate is strongly temperature activated. The observed temperature dependence (*31*) agreed quantitatively with predictions made using the parameters measured earlier for the dependence of rate constants on $\Delta G°$ at constant temperature (i.e., from Figure 6, bottom).
2. For inverted-region ET, the rate was almost completely independent of temperature. This result was in strong contrast to the predictions of classical Marcus theory, but in excellent agreement with quantum mechanical modifications that include the effects of high-frequency molecular vibrations in the donor and acceptor groups.

Particularly the first of these two results is feasible only with charge-shift reactions studied by pulse radiolysis. This is one case where pulse radiolysis is superior for truly quantitative examination of ET mechanisms and testing of theory.

Distance. Both photoexcitation and pulse radiolysis have been used to measure the dependence of ET rates on distance (*17, 32–36*). The approximately exponential decrease of electronic couplings with distance, R,

$$V(R) = V(R_0)\exp(-\beta R/2) \tag{5}$$

is less rapid when a series of connected bonds is present between the donor and acceptor. Evidence for through-bond ET has been seen by both pulse radiolysis and photoexcitation. Measuring the distance dependence of the electronic coupling is complicated for either method because the solvent reorganization energy, λ_S, is also dependent on distance. Therefore, measurement of ET rates as a function of distance does not necessarily give clear measurement of the attenuation coefficient for electronic couplings, β. The advantage that pulse radiolysis can measure the effect of one variable while others are constant is not obtained for measurements as a function of distance, but pulse radiolysis does provide the advantage that two (V and λ_S), rather than three (V, λ_S, and $\Delta G°$) variables depend on distance. For photoexcitation, an additional complication arises because $\Delta G°$ is also distance dependent. Thus, the first reported electronic couplings for the beautiful norborane-based molecules of Paddon-Row were for distance-dependence of optical ET bands measured by pulse radiolysis. Electronic couplings can be deduced from photoexcited ET rates for these compounds. If $\Delta G°$ is nearly optimal, the slope of the curve for rate as a function of $\Delta G°$ is small. Therefore, even though both $\Delta G°$ and λ_S depend on distance, the rate is not very sensitive to either. This method for measurement of β makes use of the superior time resolution of laser photoexcitation to measure the very fast rates for optimally exoergic reactions. Such rates have been measured by pulse radiolysis only at longer distances.

The electronic couplings may be affected by the nature of the bridge or spacer group, independent of the distance. Properties that can be important are conformations and connectedness of the bonds connecting D and A. Examples are *trans* vs. *gauche* or *cis* connections of sigma bonds and the number of through-bond paths. Larger effects on electronic couplings occur when the spacer contains electronically conductive moieties such as π bonds (*37–43*). These effects can be observed both by photoexcitation and pulse radiolysis. If molecular systems can be prepared for which changes of σ vs. π character, or changes in bond angles or connectedness, are accomplished without significant changes in distance, then the two methods are equally effective. Otherwise the comments in the foregoing section on distance will apply.

Molecular orientation can also alter electronic couplings and therefore control ET rates (*44–47*). Experiments by Zeng and Zimmt (*48, 49*) and McLendon and co-workers (*50*) indicate the possibility of substantial orientation effects in ET. Again, both methods should be effective, but photoexcitation has an advantage because high time resolution is often valuable.

Pulse Radiolysis Experiments

This section discusses experimental techniques used in our laboratory for pulse radiolysis, with emphasis on the study of ET reactions and with some general comments on techniques.

Van de Graaff accelerators can be gated to produce pulses to less than 1 ns (*51–55*). Linear accelerators can be gated on for a few nonoseconds, but by nature of the radio frequency (RF) or microwave fields used to accelerate the electrons, these accelerators produce ≈20-ps bunches of electrons spaced at the period of the RF. Hunt and co-workers first used these pulse trains in a pump-probe technique using Cerenkov light generated by the pulses (*56, 57*). At Argonne National Laboratory, "subharmonic bunching" gathered all of the bunches in a ≈4-ns pulse into a single 30-ps pulse (*58, 59*). Jonah developed pump-probe detection using Cerenkov light (*60*). These are repetitive sampling techniques that are useful for solutions using large amounts of material. For synthetic DSA molecules, single-pulse techniques are essential to prevent sample degradation.

The advent of fast digitizing oscilloscopes (IN7000 series, Intertechnique, France, and the 7250 and SCD5000, Tektronix, U.S.) along with biplanar phototubes (e.g., Hamamatsu R1328U) and fast amplifiers (e.g., B&H Electronics) enables ≈100-ps system rise times in a single-shot mode. The biplanar phototubes provide detection in the 200–800-nm range. For near-IR detection, solid-state photodiodes are available, but the user must choose between small areas (fast response time) and large areas (good signal to noise ratios). Devices used in our laboratory are Si for 400–1050 nm (EG&G FOD–100, 2 mm diameter, 2 ns fall time, or Opto Electronics Ltd. CD-10 or PD-10, 0.5 mm, ≈0.15 ns); Ge or InGaAs, 500–1600 nm (Germanium Power Devices GEP–600, 2 mm, 10 ns, or GAP-500L, 0.5 mm, 1.5 ns); and InAs is useful in the range ≈800–3500 nm (Judson J–12, 20 ns). Small flash lamps from EG&G or Hamamatsu, or pulsed dc xenon arc lamps, provide analyzing light. Scattered Cerenkov light must also be measured and subtracted for a time resolution of ≈2 ns or better.

Acknowledgments

This work was performed under the auspices of the Office of Basic Energy Sciences, Division of Chemical Sciences, U.S. Department of Energy, under contract number W–31–109–ENG–38.

References

1. Michel-Beyerle, M. E. *Reaction Centers of Photosynthetic Bacteria: Feldafing II Conference;* Springer-Verlag: New York, 1991; p 469.
2. Michel-Beyerle, M. E. *Antennas and Reaction Centers of Photosynthetic Bacteria: Structure, Interactions and Dynamics;* Springer-Verlag: New York, 1985; p 384.
3. Mathis, P. *Photosynthesis: From Light to Biosphere: Proceedings of the 10th International Congress;* Kluwer Academic: Dordrecht, Netherlands, 1995; p 5168.
4. Devault, D. *Quantum Mechanical Tunnelling in Biological Systems;* Cambridge University, Cambridge, England, 1984; p 207.
5. Marcus, R. A. *J. Chem. Phys.* **1956,** *24,* 966–978.
6. Marcus, R. A. *Discuss. Faraday Soc.* **1960,** *29,* 21–31.
7. Rehm, D.; Weller, A. *Isr. J. Chem.* **1970,** *8,* 259–271.
8. Dogonadze, R. R. *Ber. Bunsen Ges. Phys. Chem.* **1971,** *75,* 628–634.
9. Dogonadze, R. R.; Kuznetsov, A. M.; Vorotyntsev, M. A. *Z. Phys. Chem. (Wiesbaden)* **1976,** *100,* 1–16.
10. Kestner, N. R.; Logan, J.; Jortner, J. *J. Phys. Chem.* **1974,** *78,* 2148–2166.
11. Ulstrup, J.; Jortner, J. *J. Chem. Phys.* **1975,** *63,* 4358–4368.
12. Ulstrup, J. *Charge Transfer Processes in Condensed Media;* Springer-Verlag: Berlin, Germany, 1979; p 419.
13. Van Duyne, R. P.; Fischer, S. F. *Chem. Phys.* **1974,** *5,* 183.
14. Siders, P.; Marcus, R. A. *J. Am. Chem. Soc.* **1981,** *103,* 741–747.
15. Siders, P.; Marcus, R. A. *J. Am. Chem. Soc.* **1981,** *103,* 748–752.
16. Beitz, J. V.; Miller, J. R. *J. Chem. Phys.* **1979,** *71,* 4579.
17. Miller, J. R.; Beitz, J. V.; Huddleston, R. K. *J. Am. Chem. Soc.* **1984,** *106,* 5057–5068.
18. Miller, J. R.; Calcaterra, L. T.; Closs, G. L. *J. Am. Chem. Soc.* **1984,** *106,* 3047.
19. Wasielewski, M. R.; Niemczyk, M. P.; Svec, W. A.; Pewitt, E. B. *J. Am. Chem. Soc.* **1985,** *107,* 1080–1082.
20. Harrison, R. J.; Beddard, G. S.; Cowan, J. A.; Sanders, J. K. M. *Springer Ser. Chem. Phys. 46 (Ultrafast Phenom. 5)* **1986,** 322–325.
21. Harrison, R. J.; Pearce, B.; Beddard, G. S.; Cowan, J. A.; Sanders, J. K. M. *Chem. Phys.* **1987,** *116,* 429–448.
22. Irvine, M. P.; Harrison, R. J.; Beddard, G. S.; Leighton, P.; Sanders, J. K. M. *Chem. Phys.* **1986,** *104,* 315–324.
23. Gould, I. R.; Young, R. H.; Farid, S. In *Photochemical Processes in Organized Molecular Systems;* Tazuke, S.; Honda, K., Eds.; North-Holland: Amsterdam, Netherlands, 1991; pp 19–40.
24. Gould, I. R.; Noukakis, D.; Gomez-Jahn, L. A.; Goodman, J. L.; Farid, S. *J. Am. Chem. Soc.* **1993,** *115,* 4405–4406.
25. Gould, I. R.; Moser, J. E.; Armitage, B.; Farid, S. *Res. Chem. Intermed.* **1995,** *21,* 793–806.
26. Mataga, N. In *Proceedings of the 7th International Conference in Photochemical Conversion of Storage of Solar Energy;* Norris, J. R., Jr.; Meisel, D., Eds.; Elsevier: New York, 1989; pp 32–46.
27. Kakitani, T.; Mataga, N. *J. Phys. Chem.* **1985,** *89,* 4752–4757.
28. Kakitani, T.; Mataga, N. *Chem. Phys.* **1985,** *93,* 381–397.
29. Kakitani, T.; Mataga, N. *J. Phys. Chem.* **1985,** *89,* 8–10.
30. Kakitani, T.; Mataga, N. *J. Phys. Chem.* **1986,** *90,* 993–995.
31. Liang, N.; Miller, J. R.; Closs, G. L. *J. Am. Chem. Soc.* **1989,** *111,* 8740–8741.
32. Closs, G. L.; Calcaterra, L. T.; Green, N. J.; Penfield, K. W.; Miller, J. R. *J. Phys. Chem.* **1986,** *90,* 3673–3683.
33. Paddon-Row, M. N. *Acc. Chem. Res.* **1994,** *27,* 18–25.
34. Jordan, K. D.; Paddon-Row, M. N. *Chem. Rev.* **1992,** *92,* 395–410.

35. Marcus, R. A.; Sutin, N. *Biochim. Biophys. Acta* **1985,** *811,* 265–322.
36. Newton, M. D.; Sutin, N. *Ann. Rev. Phys. Chem.* **1984,** *35,* 437–480.
37. Heitele, H.; Michel-Beyerle, M. E. *J. Am. Chem. Soc.* **1985,** *107,* 8286–8288.
38. Heitele, H.; Michel-Beyerle, M. E.; Finckh, P. *Chem. Phys. Lett.* **1987,** *134,* 273–278.
39. Helms, A.; Heiler, D.; McLendon, G. *J. Am. Chem. Soc.* **1992,** *114,* 6227–6238.
40. Joachim, C.; Launay, J. P.; Woitellier, S. *Chem. Phys.* **1990,** *147,* 131–141.
41. Larsson, S. *Chem. Phys. Lett.* **1982,** *90,* 136–139.
42. Reimers, J. R.; Hush, N. S. *Inorg. Chem.* **1990,** *29,* 3686–3697.
43. Woitellier, S.; Launay, J. P.; Spangler, C. W. *Inorg. Chem.* **1989,** *28,* 758–762.
44. Cave, R. J.; Siders, P.; Marcus, R. A. *J. Phys. Chem.* **1986,** *90,* 1436–1444.
45. Siders, P.; Cave, R. J.; Marcus, R. A. *J. Chem. Phys.* **1984,** *81,* 5613–5624.
46. Oliver, A. M.; Paddon-Row, M. N.; Kroon, J.; Verhoeven, J. W. *Chem. Phys. Lett.* **1992,** *191,* 371–377.
47. Verhoeven, J. W.; Kroon, J.; Paddon-Row, M. N.; Warman, J. M. *NATO ASI Ser., Ser. C 371 (Supramol. Chem.)* **1992,** 181–200.
48. Zeng, Y.; Zimmt, M. B. *J. Phys. Chem.* **1992,** *96,* 8395–8403.
49. Zeng, Y.; Zimmt, M. B. *J. Am. Chem. Soc.* **1991,** *113,* 5107–5109.
50. Helms, A.; Heiler, D.; McLendon, G. *J. Am. Chem. Soc.* **1991,** *113,* 4325–4327.
51. Thomas, J. K.; Johnson, K.; Klippert, T.; Lowers, R. *J. Chem. Phys.* **1968,** *48,* 1608–1612.
52. Frey, H.; Moller, M.; de Haas, M. P.; Zenden, N. J. P.; Schouten, P. G.; van der Laan, G. P.; Warman, J. M. *Macromolecules* **1993,** *26,* 89–93.
53. Schouten, P. G.; Warman, J. M.; de Haas, M. P.; van der Pol, J. F.; Zwikker, J. W. *J. Am. Chem. Soc.* **1992,** *114,* 9028–9034.
54. Luthjens, L. H.; Hom, M. L.; Vermeulen, M. J. W. *Rev. Sci. Instrum.* **1978,** *49,* 230–235.
55. Luthjens, L. H.; Vermeulen, M. J. W.; Hom, M. L. *Rev. Sci. Instrum.* **1980,** *51,* 1183–1189.
56. Bronskill, M. J.; Wolff, R. K.; Hunt, J. W. *J. Chem. Phys.* **1970,** *53,* 4201–4210.
57. Hunt, J. W.; Greenstock, C. L.; Bronskill, M. J. *Int. J. Radiat. Phys. Chem.* **1972,** *4,* 87–106.
58. Ramler, W.; Mavrogenes, G.; Johnson, K. In *Fast Processes in Radiation Chemistry and Biology;* Adams, G. E.; Fielden, E. M.; Michael, B. D., Eds.; Institute of Physics: London, 1975; pp 25–32.
59. Mavrogenes, G. S.; Jonah, C.; Schmidt, K. H.; Gordon, S.; Tripp, G. R.; Coleman, L. W. *Rev. Sci. Instrum.* **1976,** *47,* 187–189.
60. Jonah, C. D. *Rev. Sci. Instrum.* **1975,** *46,* 62–66.

12

Photoinduced Electron and Proton Transfer in a Molecular Triad

Su-Chun Hung, Alisdair N. Macpherson, Su Lin, Paul A. Liddell, Gilbert R. Seely, Ana L. Moore, Thomas A. Moore*, and Devens Gust

Department of Chemistry and Biochemistry, Center for the Study of Early Events in Photosynthesis, Arizona State University, Tempe, AZ 85287–1604

*A series of molecular triads, consisting of a porphyrin (P) covalently linked to a carotenoid polyene (C) and a naphthoquinone moiety (Q), have been prepared. Triad **1** features a quinone with an internally hydrogen-bonded carboxylic acid group. The photochemical properties of these molecules have been studied using steady-state and transient absorption and emission spectroscopies in three solvents: benzonitrile, dichloromethane, and chloroform. Each of the triads undergoes photoinduced electron transfer from the C–^{1}P–Q singlet state to yield the charge-separated state C–$P^{\cdot+}$–$Q^{\cdot-}$. An electron transfer reaction from C to yield $C^{\cdot+}$–P–$Q^{\cdot-}$ competes with fast electron-hole recombination in C–$P^{\cdot+}$–$Q^{\cdot-}$. Triad 1 produces the final $C^{\cdot+}$–P–$Q^{\cdot-}$ state with the highest quantum yield of the series ($\Phi \geq 0.22$), a factor of ca. 2 higher than for reference triads. Following the initial photoinduced electron transfer, a fast (k ~ 10^{12} s^{-1}) proton shift from the carboxylic acid to $Q^{\cdot-}$ generates the semiquinone, increasing the lifetime of $P^{\cdot+}$ and the yield of electron donation by C. In model P–Q dyads, the species $P^{\cdot+}$ is shown to be longer lived for quinones that feature an internal hydrogen bond. A thermodynamic model is proposed in which the increase in the lifetime of the $P^{\cdot+}$ moiety by the proton shift is attributed to the δpK of the Q/$Q^{\cdot-}$ couple, which dramatically lowers the driving force for electron-hole recombination.*

Natural photosynthetic reaction centers, which are responsible for the conversion of light energy into chemical energy in photosynthetic organisms, employ a multistep electron transfer strategy to achieve charge separation across membranes with a total quantum yield near unity. Thus, at each intermediate step

---

* Corresponding author.

the forward electron transfer process dominates charge recombination. It is a significant challenge to design synthetic multicomponent electron transfer molecules in which photoinduced charge separation is followed by forward electron transfer processes that are much faster than electron-hole recombination (*1–5*). Previously, we have used thermodynamic and electronic principles to improve the yield of long-lived charge separation in donor–pigment–acceptor triad molecules, and have designed and synthesized devices having more than three components in which parallel forward electron transfer paths compete with recombination to enhance the yield of the final charge-separated state (*3, 5, 6*). This chapter describes another approach, based on fast proton transfer.

The simplest molecular system that can demonstrate multistep electron transfer is a triad. In the original carotene–porphyrin–quinone (C–P–Q) triad (*7*), the porphyrin first excited singlet state $C{-}^1P{-}Q$ donates an electron to the quinone to form $C{-}P^{\cdot+}{-}Q^{\cdot-}$. A second electron transfer, from the carotenoid, competes with charge recombination to the ground state, yielding a final, long-lived $C^{\cdot+}{-}P{-}Q^{\cdot-}$ charge-separated state. Triad **1**, which features a naphthoquinone moiety with a fused norbornene skeleton bearing a carboxylic acid group at a bridgehead (*8*), has been designed to incorporate a rapid, unimolecular proton transfer reaction (equation 1) immediately after the photoinduced electron transfer step. This proton transfer yields the semiquinone and is driven by a change in p*K* of the quinone upon formation of the anion radical (*9–11*).

$$C{-}P^{\cdot+}{-}Q^{\cdot-}\cdots HOOC \rightarrow C{-}P^{\cdot+}{-}Q^{\cdot}{-}H\cdots{}^{-}OOC \quad (1)$$

For thermodynamic reasons, electron-hole recombination between the porphyrin radical cation and the semiquinone is expected to be slower than recombination involving the quinone radical anion (*vide infra*). Retarding electron-hole recombination should lengthen the lifetime of the porphyrin radical cation and thereby enhance the yield of electron donation to it by the attached carotenoid, and thus increase the yield of the final charge-separated state $C^{\cdot+}{-}P{-}Q^{\cdot}{-}H\cdots{}^{-}OOC$.

The quinone moiety of triad **1** has been designed to accomplish this by positioning a carboxylic acid group so that an internal hydrogen bond can form to a carbonyl group of the quinone (shown schematically as Q···HOOC). This internal hydrogen bond positions the proton for ultrafast transfer.

Triad **1** is also designed to explore one way to couple a photoinduced electron transfer process to a change in proton chemical potential. Equations 2 and 3 illustrate two processes involved in the decay of the final charge-separated state to the ground state.

$$C^{\cdot+}{-}P{-}Q^{\cdot}{-}H\cdots{}^{-}OOC \rightarrow C{-}P{-}Q^{+}{-}H\cdots{}^{-}OOC \quad (2)$$

$$C{-}P{-}Q^{+}{-}H\cdots{}^{-}OOC \rightarrow C{-}P{-}Q\cdots HOOC \quad (3)$$

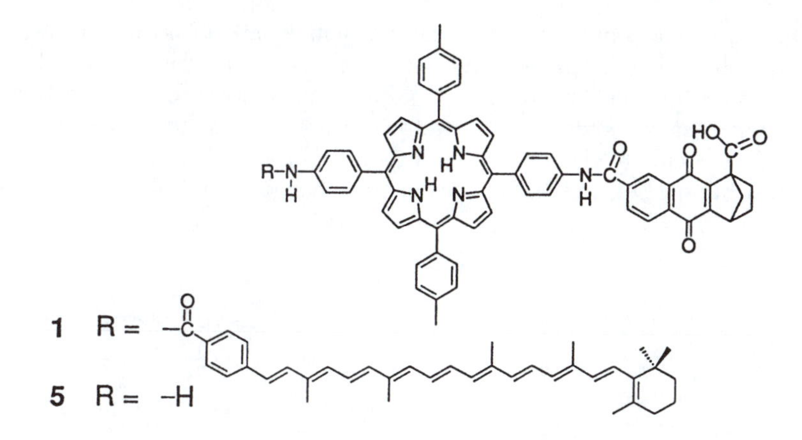
R-N-H
HO
C=O
O
N
H
H
N-C
O
1 R = -C(=O)-
5 R = -H

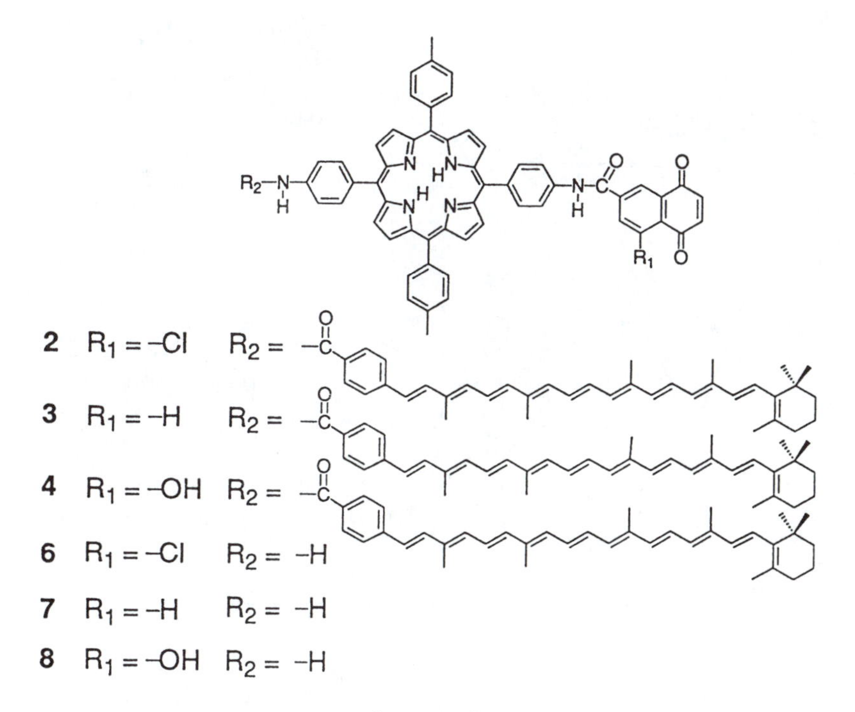
R_2-N-H
N-C
O
R_1
2 R_1 = -Cl R_2 = -C(=O)-
3 R_1 = -H R_2 = -C(=O)-
4 R_1 = -OH R_2 = -C(=O)-
6 R_1 = -Cl R_2 = -H
7 R_1 = -H R_2 = -H
8 R_1 = -OH R_2 = -H

Structure I.

The electron-hole recombination reaction shown in equation 2 is a one- or two-step intramolecular redox process and yields the ground electronic state of the system and the protonated quinone, highly energetic species. This electron transfer step poises the system for the proton transfer process shown in equation 3, in which the energy change is given by the change in pK_a of the relevant proton-bearing species. For triad **1** this energy change is ca. 0.7 eV.

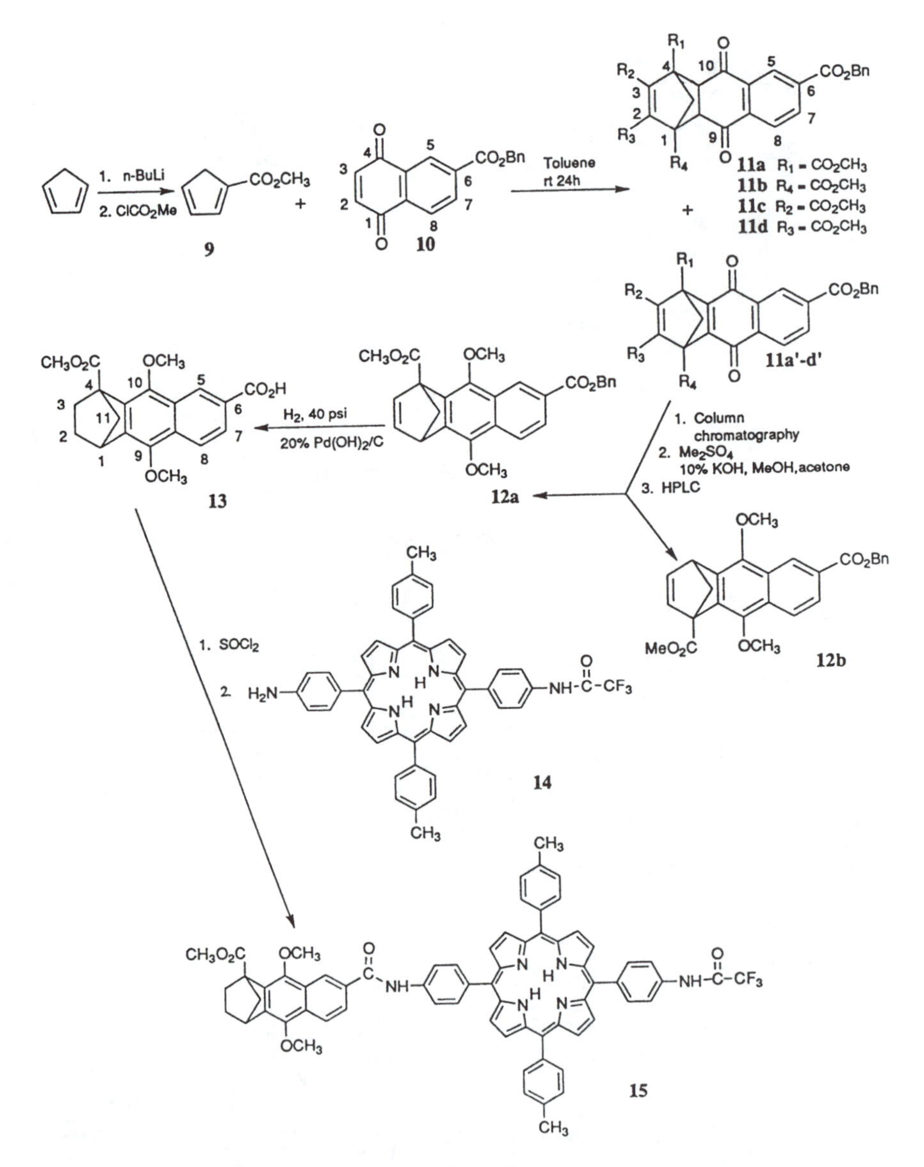

Scheme I.

Results and Discussion

Synthesis. The structural units in the triads and dyads were linked via amide bonds. There are two alternative routes for constructing the triad molecules: coupling the quinone moiety with the porphyrin followed by joining the dyad with the carotenoid polyene, or coupling the porphyrin with the carotenoid polyene and then attaching the quinone. The first approach was used in each of the following examples.

The synthesis of the porphyrin moiety began with 5,15-bis(4-acetamidophenyl)-10,20-bis(4-methylphenyl)porphyrin (**28**) (*12, 32*) which upon treatment with concentrated aqueous HCl gave 5,15-bis(4-aminophenyl)-10,20-bis(4-methylphenyl)porphyrin in 96% yield (*12, 13*). Protection of the diaminoporphyrin with trifluoroacetic anhydride furnished the bis-trifluoroacetylated porphyrin in 98% yield (*6, 13*). Partial hydrolysis of this compound provided monoaminoporphyrin **14** (Scheme I) in 60% yield.

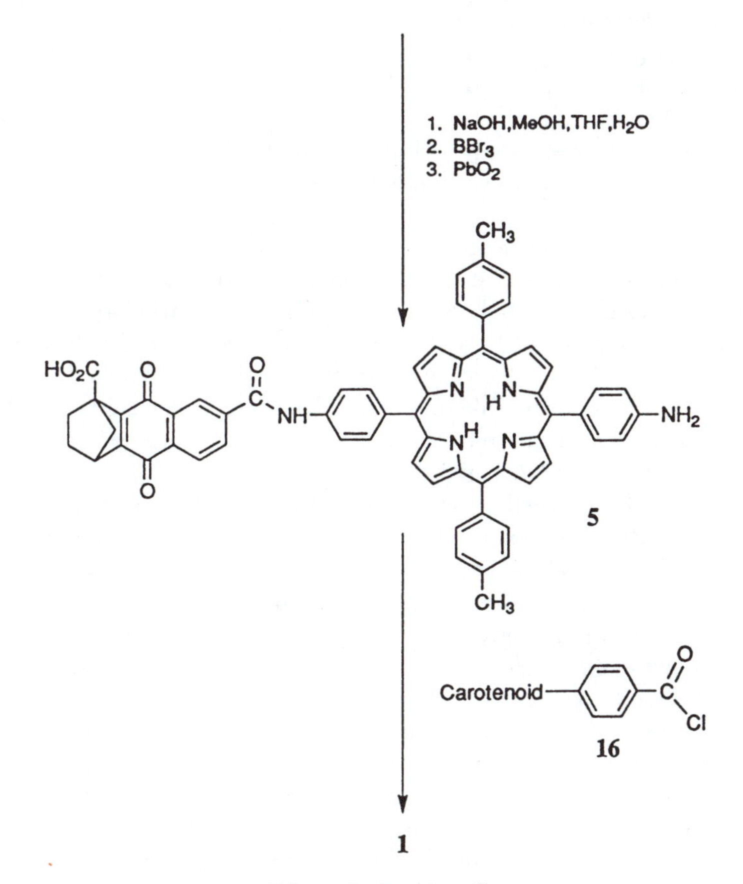

Scheme I. (continued)

Preparation of 7′-apo-7′-(4-carboxyphenyl)-β-carotene (*14, 15*) began with a Wittig reaction of trans-β-apo-8′-carotenal and 4-(carbomethoxy)benzyltriphenylphosphonium bromide, followed by base hydrolysis (*15*). The carotenoid acid was converted to the acid chloride (**16**) with thionyl chloride.

The synthesis of the quinone moiety was initiated by employing 1-methoxycarbonylcyclopenta-1,3-diene (**9**) with naphthoquinone (**10**) as the dienophile in a Diels–Alder reaction (Scheme I). Diene **9** was prepared from commercially available cyclopentadiene using a substitution reaction described by Grunewald and Davis (*16*). The requisite dienophile **10** was prepared from the readily available methyl ester of **25** (Scheme III) (*17*). Transesterification of this ester with benzyl alcohol in the presence of sodium gave the benzyl ester, which was treated with ceric ammonium nitrate to provide **10** in 87% yield. The Diels–Alder reaction of diene **9** with dienophile **10** afforded diketones **11a–d** with a trace of naphthoquinones **11a′–d′** in 90% overall yield. Four sets of regioisomeric Diels–Alder products were produced in this reaction, each with both *endo* and *exo* forms. The bridgehead esters and vinyl esters were generated in approximately equal amounts.

Chromatographic purification of the Diels–Alder products **11a–d** and **11a′–d′** furnished a mixture of four bridgehead esters and a mixture of four vinyl esters. Treatment of the bridgehead esters with dimethyl sulfate provided product **12**, which consisted of a 1:1 mixture of regioisomer **12a** and **12b** (each a racemic mixture). These two isomers were separated by HPLC to give the pure racemic forms.

To prepare the bridgehead carboxylic acid triad **1**, ester **12a** was hydrogenated using a palladium catalyst to deprotect the benzyl ester and give acid **13**, in which the bicyclic double bond is reduced. Next, thionyl chloride was used to generate the corresponding acid chloride, which was coupled with aminoporphyrin **14** to give **15** in 64% yield. Base hydrolysis of porphyrin **15** afforded the carboxylic acid aminoporphyrin (91%), which was deprotected to yield dyad **5**. Finally, coupling this porphyrin quinone with the carotenoid acid chloride **16** yielded the target triad **1** in 53% yield.

To synthesize molecular triad **2**, it was necessary to introduce a chlorine at the *peri* position of the naphthoquinone (Scheme II). Keto ester **17**, which was obtained by literature procedures (*18–21*), was transformed into **18** in 41% yield by using an excess of phosphorus oxychloride and triphenylphosphine dichloride. The resulting ester **18** was conveniently aromatized using 2,3-dichloro-5,6-dicyanobenzoquinone in dry toluene to give 6-carbomethoxy-8-chloro-1,4-dimethoxynaphthalene (**19**) in 53% yield. Hydrolysis of **19** with potassium hydroxide in aqueous methanol–tetrahydrofuran furnished the carboxylic acid in 99% yield. The carboxylic acid was converted into acid chloride **20**, which was coupled with aminoporphyrin **14** to produce porphyrin **21** (40%). The trifluoroacetyl protecting group of **21** was removed using NaOH to form the aminoporphyrin (89%), and the methyl protecting groups were then removed with boron tribromide, affording porphyrin–quinone dyad **6** (78%).

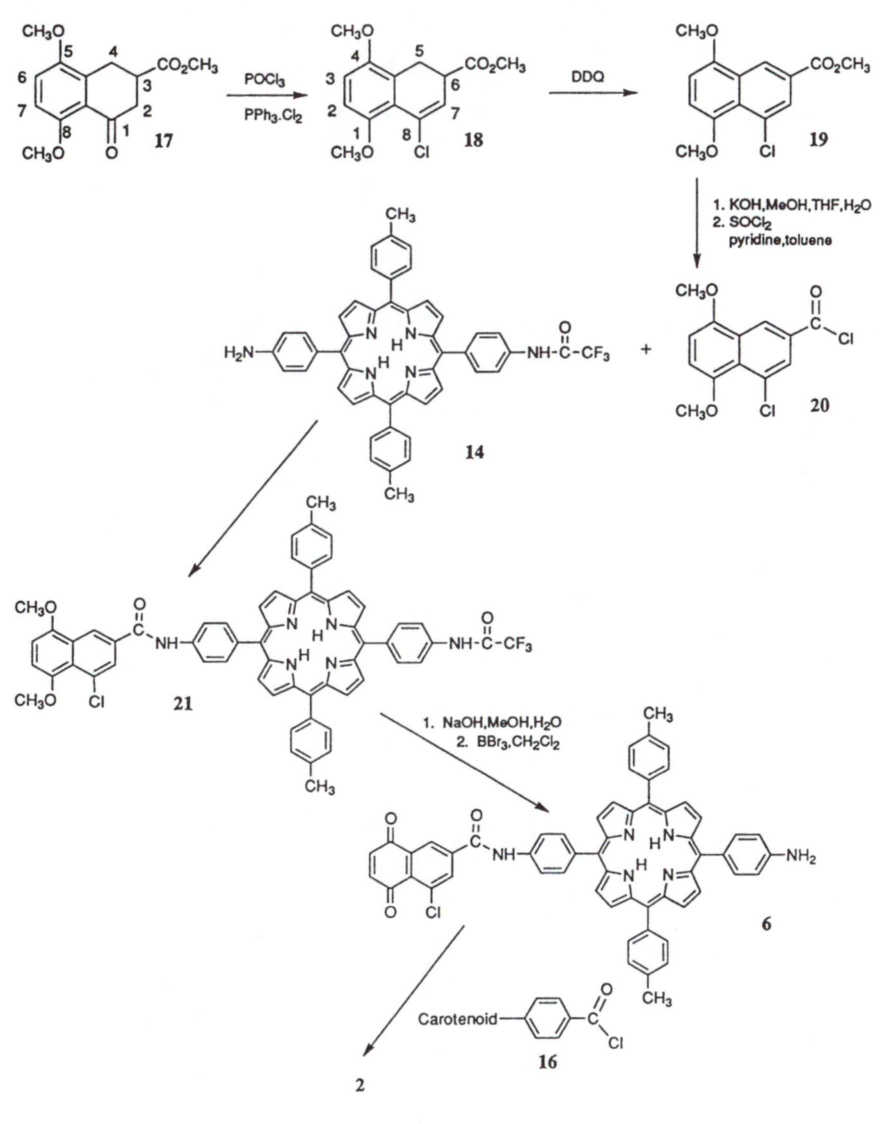

Scheme II.

Finally, triad **2** was synthesized in 44% yield by linking acid chloride **16** with **6**.

The preparation of triad **3** (Scheme III) began with the readily available 6-carboxy-1,4-dimethoxynaphthalene, **25** (*17*). Reaction of the acid chloride **26** with aminoporphyrin **14** gave **27** in 85% yield. Hydrolysis of **27** generated the aminoporphyrin (95%), which was deprotected with boron tribromide and

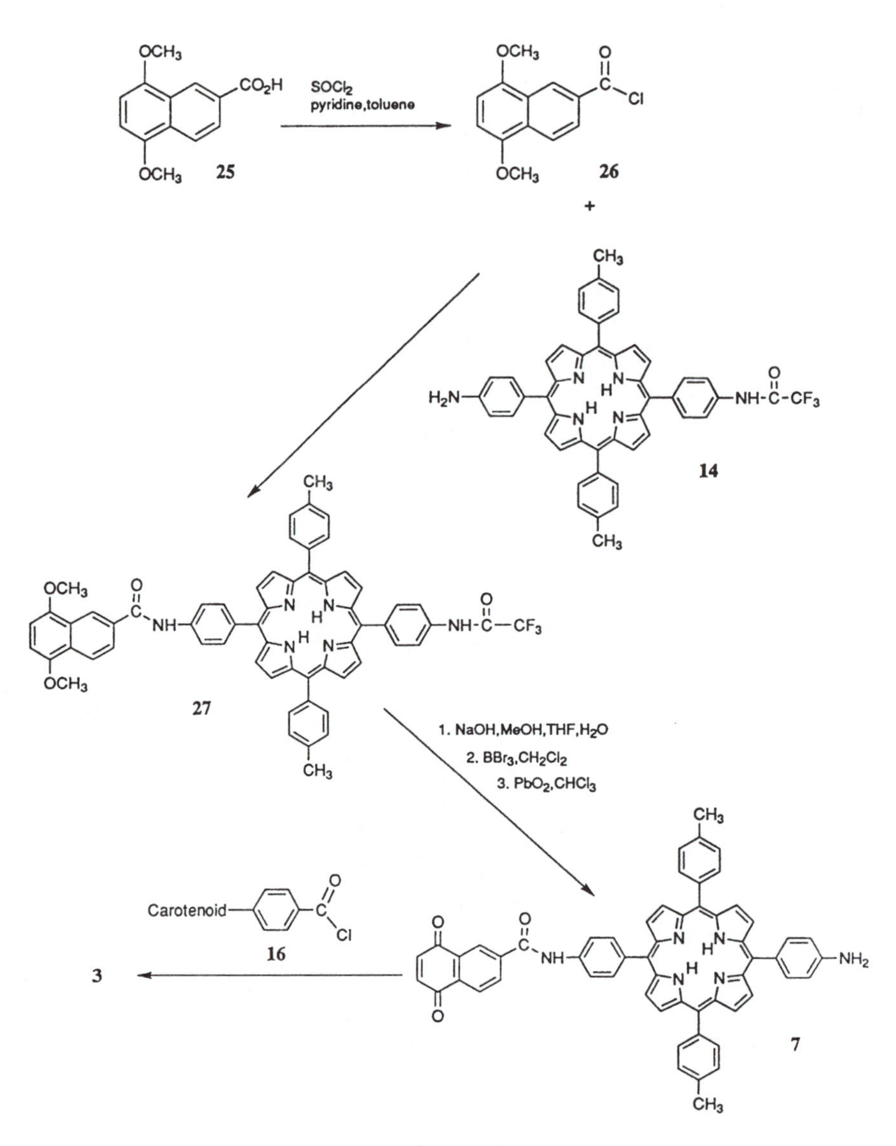

Scheme III.

oxidized with lead dioxide to give dyad **7** (57%). Coupling of **7** with acid chloride **16** provided pure **3** in 36% yield.

The synthesis of triad **4** (Scheme IV) began with aromatization–methylation of compound **17** with 2,3-dichloro-5,6-dicyanobenzoquinone and trimethylorthoformate in methanol. The resulting 6-carbomethoxy-1,4,8-trimethoxynaphthalene (**22**) (81%) (22) was hydrolized to acid **23**, converted to the acid

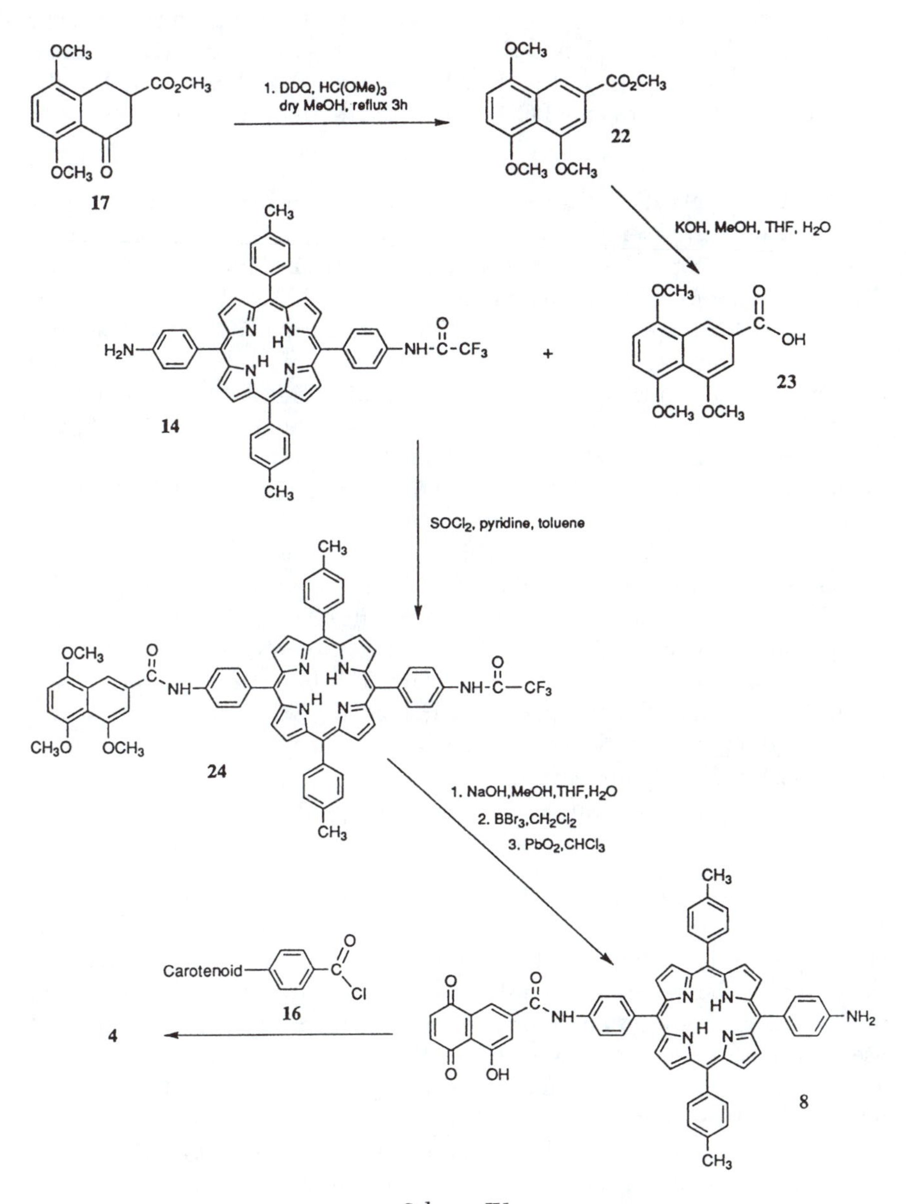
OCH3
CO2CH3
OCH3 O
17
1. DDQ, HC(OMe)3
dry MeOH, reflux 3h
OCH3
CO2CH3
22
OCH3 OCH3
KOH, MeOH, THF, H2O
CH3
H2N
NH-C-CF3
14
+
OCH3
OH
23
OCH3 OCH3
SOCl2, pyridine, toluene
OCH3
NH
CH3O OCH3
24
1. NaOH,MeOH,THF,H2O
2. BBr3,CH2Cl2
3. PbO2,CHCl3
Carotenoid
Cl
16
4
OH
NH2
8

Scheme IV.

chloride by treatment with thionyl chloride, and coupled with aminoporphyrin **14** to give porphyrin **24** in 81% yield. Hydrolysis of **24** using NaOH gave the aminoporphyrin in 99% yield. The removal of the methyl protecting groups with boron tribromide (*23*), in dichloromethane at –78 °C, followed by oxidation with lead dioxide, gave dyad **8** (60%). Finally, dyad **8** was linked to the acid chloride of the carotenoid (**16**) to give the desired triad **4** in 42% yield.

NMR Spectral Assignments. In these studies, NMR techniques [^{1}H, nuclear Overhauser effect (NOE), nuclear Overhauser enhancement spectroscopy (NOESY), and correlation spectroscopy (COSY)] were used to elucidate the structures of the compounds.

The ^{1}H NMR spectra of triads **1–4** were obtained at 500 MHz. The resonances were assigned by comparison with model compounds, known coupling constants for norbornyl systems (*24*), and two-dimensional (2D) COSY experiments. The model compounds included the corresponding dyads (**5, 6, 7**, and **8**), carotenoporphyrins, and aniline amides of the quinone acids. The structures for regioisomers **12a** and **12b** were identified by difference NOE ^{1}H NMR experiments.

Although phenolic proton resonances usually appear in the 4.0–8.0 ppm region, the *peri* hydroxyl group proton of dyad **8** appeared as a sharp singlet at δ 11.64 ppm in $CDCl_3$. This can explained by the effect of the nearby carbonyl group of the quinone, which shifts the phenolic proton absorption downfield because of intramolecular hydrogen bonding. This observed shift is generally a reliable indicator of the formation of H-bonding (*25*). Other systems with hydroxyl groups in the *peri* position of naphthoquinones such as 6-phenylcarbamyl-8-hydroxy-1,4-naphthoquinone and 5-hydroxy-1,4-naphthoquinone (juglone) also show a deshielded peak at about 11.9 ppm.

Absorption Spectra. The absorption spectra in the visible and near-UV of **1–8** are essentially linear combinations of those of their component chromophores, indicating that the interchromophore interactions are weak. In the IR region, the spectra are a sensitive function of hydrogen bonding involving the quinone carbonyl group. Fourier transform IR spectra were recorded and used to evaluate intramolecular hydrogen bonding between the carboxylic acid group and the carbonyl oxygen in the naphthoquinone moiety in triad **1** and dyad **5**. The IR spectrum of a model compound for the quinone moiety of **1** and **5**, 1-carboxy-1,2,3,4-tetrahydro-1,4-methanoanthracene-9,10-dione (**29**), in dilute (0.004 M) dichloromethane solution displayed quinone carbonyl stretching bands at 1665 cm^{-1} and 1630 cm^{-1}. The 1665 cm^{-1} band is assigned to non-hydrogen-bonded quinone carbonyls because it is observed in the methyl ester of **29** (1-carbomethoxy-1,2,3,4-tetrahydro-1,4-methanoanthracene-9,10-dione) at 1665 cm^{-1} and in 1,2,3,4-tetrahydro-1,4-methanoanthracene-9,10-dione at 1661 cm^{-1}. The shift to lower frequency resulting in the 1630 cm^{-1}

band in **29** is characteristic of the shift observed when carbonyl groups are involved in a hydrogen bond (25). For example, the acid carbonyl stretch in **29** occurs at 1748 cm^{-1} in dilute solution. Increasing the concentration to >0.01 M results in substantial dimer formation due to intermolecular hydrogen bond formation and concomitant growth of a band at 1713 cm^{-1}, a 35 cm^{-1} shift to lower frequency. (Neither the band at 1748 cm^{-1} nor the one at 1713 cm^{-1} is observed in dichloromethane solutions of quinones that lack the carboxylic acid function.) The observation of a 35 cm^{-1} shift in the acid carbonyl stretch upon intermolecular hydrogen bond formation in these systems is consistent with the assignment of the band shifted to lower frequency by 35 cm^{-1} (from 1665 cm^{-1} to 1630 cm^{-1}) in **29** to the intramolecular hydrogen bond.

The IR spectra of **29** in dilute dichloromethane solution are interpreted in terms of an equilibrium mixture of internally hydrogen bonded conformers and a population of conformers in which the carboxy group is rotated so that the internal hydrogen bond is not formed. Because the absorption strengths of the two bands are not known, it is not possible to estimate the value of the equilibrium constant from the IR spectra. It is assumed that these salient features of hydrogen bonding in **29** can be extrapolated to triad **1** and dyad **5**.

Fluorescence

The fluorescence of **1–8** derives from the porphyrin moiety and is essentially indistinguishable from that of the porphyrin model except for a reduction in quantum yield. Fluorescence decay measurements of **1–8** were obtained in dichloromethane, chloroform, and benzonitrile with laser excitation at 590 nm. Table I presents the fluorescence lifetimes and initial amplitudes for triads **1–4** in these solvents. The decays were measured at six wavelengths in the 650–750 nm region and analyzed globally. In the case of triad **2**, for example, a satisfac-

Table I. Fluorescence Lifetimes τ_f and Amplitudes of Triads 1–4 in Various Solvents

Triad	*1*	*2*[a]	*3*[b]	*4*[b]
τ_f in $CHCl_3$	9 ps (77%)[c] 49 ps (22%)[e]	25 ps[d]	41 ps[e]	9 ps[c]
τ_f in CH_2Cl_2	17 ps (88%)[c] 76 ps (11%)[e]	30 ps[d]	67 ps[e]	11 ps[c]
τ_f in PhCN	51 ps (69%)[e] 157 ps (23%)[e]	55 ps[e]	109 ps[e]	25 ps[d]

[a] A single exponential component accounted for more than 90% of the decay.
[b] A single exponential component accounted for more than 96% of the decay.
[c] The lifetimes and amplitudes are ±20%.
[d] The lifetimes and amplitudes are ±15%.
[e] The lifetimes and amplitudes are ±10%.

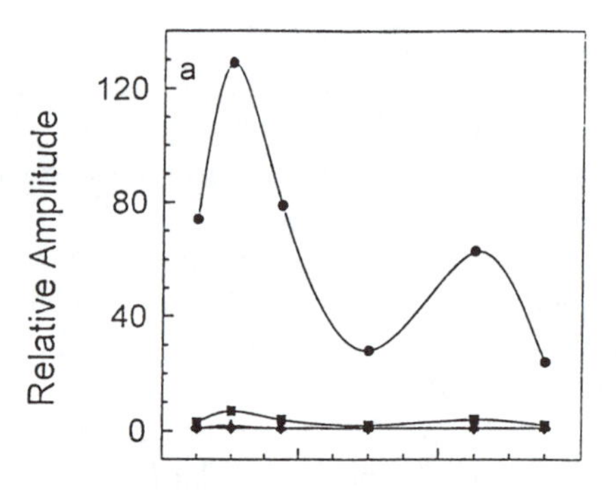

Figure 1. Decay-associated fluorescence emission spectrum for ca. *1* $\times$ *10^{-5} M triad* **2** *in benzonitrile following excitation at 590 nm. The spectrum was obtained from a global analysis of data at the six indicated wavelengths* (χ^2 = *1.14) analyzed as four exponentials. The major component has the lifetime of 55 ps* (●).

tory fit (χ^2 = 1.14) to the data in benzonitrile was obtained with four exponential components (Figure 1). The major component (93%) has a lifetime of 55 ps. This is much shorter than the ca. 10-ns lifetime of the excited singlet state of the porphyrin, or the 2.5-ns lifetime of the excited singlet porphyrin in a carotenoporphyrin, which serves as a model for the linked carotenoid–porphyrin components of triad **2.** Thus, the 55-ps component is one consequence of a new decay pathway in compounds in which a porphyrin and quinone are linked together: electron transfer to the attached quinone acceptor.

In the case of triad **1,** a satisfactory fit to the data (χ^2 = 1.15) was also obtained with four exponential components (Figure 2). However, in this case there are two main exponential components to the decay. The major component, 51 ps (69%), is assigned to the internally hydrogen bonded form of the triad, while the 157 ps (23%) component is assigned to conformers which the carboxylic group is rotated so that hydrogen bonding with the quinone carbonyl does not occur. The component with the shorter lifetime is assigned to the hydrogen-bonded conformer because hydrogen bonding is expected to stabilize the quinone anion radical and thereby increase the driving force for electron transfer, which increases the rate. This assignment is bolstered by considering the rate of photoinduced electron transfer in reference dyads **7** and **8.** Dyad **8** exhibits internal hydrogen bonding between the *peri* hydroxyl group and the naphthoquinone, so that its quinone has a more positive reduction potential [–0.40 V versus standard calomel electrode (SCE)] than that of the quinone moiety of dyad **7** (–0.58 V versus SCE). As expected, photoinduced electron transfer in **8** is faster than in **7** as evidenced by the porphyrin fluorescence lifetime of 28 ps in benzonitrile for **8** versus 113 ps in the same solvent for **7.** Extrapolating the behavior for **7** and **8** to the equilibrium mixture of conformers of triad **1,** those having the internal hyrogen bond are assigned to the 51-ps

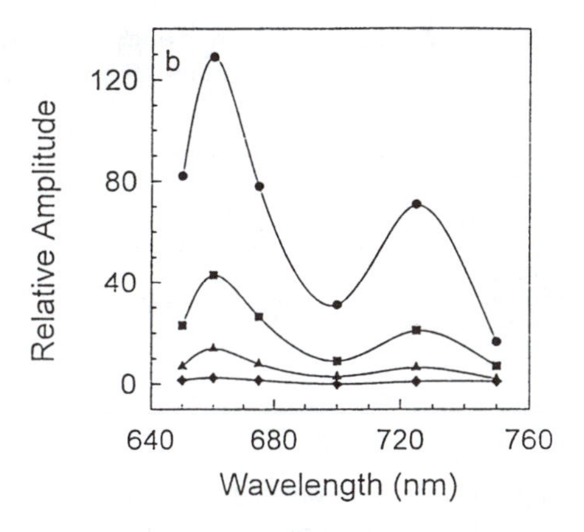

Figure 2. Decay-associated fluorescence emission spectrum for ca. *1 × 10^{-5} M triad **1** in benzonitrile following excitation at 590 nm. The spectrum was obtained from a global analysis of data at the six indicated wavelengths (χ^2 = 1.15) analyzed as four exponentials with lifetimes of 51 ps (●), 157 ps (■), 1.9 ns (▲), and 6.2 ns (◆).*

component of the porphyrin singlet decay, and those not internally hydrogen bonded give rise to the 157-ps component.

Selection of a Reference Triad. In order to assess the role of proton transfer in controlling the yield of long-lived charge-separated species in **1,** it is necessary to design a reference triad in which factors other than the proton transfer step which might affect the yields of the various electron transfer processes are virtually unchanged from those of **1.** These factors comprise structural and thermodynamic features of the triad. Triad **2** meets the required criteria. The chemical linkages between the components are the same in **1** and **2,** and the molecules are therefore expected to have similar conformations. Because the carotenoporphyrin parts of **1** and **2** are identical, the electron transfer step in which the carotenoid donates an electron to the porphyrin radical cation is similar in both compounds. Turning to the porphyrin–quinone part of the triads, identical linkages ensure that the electronic coupling between the porphyrin and quinone is the same in both triads.

In order to match the energy levels of the initial charge-separated species in **1** and the reference triad, and therefore to have the same thermodynamics for charge recombination from this intermediate, it was necessary to consider the electrochemical potential of the process in equation 4.

$$Q\cdots HOOC \rightarrow Q^{\cdot-}\cdots HOOC \qquad (4)$$

Standard electrochemical techniques cannot resolve this process, and they yield only the potential for reduction to the semiquinone species, $Q^{\cdot}$–H⋯$^{-}$OOC. The

Table II. Fluorescence Lifetimes and Decay Times of the Charge-Separated States in Dyads 5–8 in Benzonitrile Solution

Dyad	5	6	7	8
τ_f (ps)	43,[a] 183[a,b]	41[a]	113[a]	28[c]
τ_{decay} (ps)	35[c]	35[c]	121[a]	38[a]

[a] The lifetimes and amplitudes are ±10%.
[b] As was the case for triad **1,** the longer lived component is assigned to a conformer of **5** in which the internal hydrogen bond is not formed.
[c] The lifetimes and amplitues are ±15%.

strategy employed herein to determine the potential of this step (equation 4) was to synthesize a porphyrin–quinone dyad in which the porphyrin singlet lifetime was about the same as that in dyad **5,** but which could not form intramolecular hydrogen bonds. Given that the other factors controlling electron transfer in the two dyads are the same, equal porphyrin fluorescence lifetimes, and therefore equal electron transfer rate constants, would imply similar thermodynamics for the photoinduced electron transfer step. As shown in Table II, dyad **6** satisfies these requirements. Conventional cyclic voltammetric measurements on a model for the quinone moiety of **6,** 8-chloro-6-phenylcarbamyl-1,4-naphthoquinone, established its reduction potential in benzonitrile solution at −0.54 V versus SCE. Therefore, the potential of the process in equation 4 must also be ca. −0.54 V, and energies of the various electron transfer intermediates in triad **2** are closely matched to those of **1,** in which the proton is hydrogen bonded in the quinone carbonyl but not transferred to it.

Time-Resolved Absorption. Unlike time-resolved fluorescence, which unambiguously measures the decay time of the excited singlet state, assignment of the time constants associated with the rise and decay of an intermediate detected by time-resolved absorption is not always straightforward (*26*). It has been shown that in a three-level system composed of an excited state, an intermediate charge-separated state, and the system ground state, the observed rise and decay time constants of the intermediate state are not necessarily the formation and decay time constants, respectively, of the intermediate (*27*). In fact, the smaller of the true formation or decay time constants will always be the observed rise time constant of the intermediate, and the longer of the true formation or decay constants will be the observed decay time of the intermediate. This ambiguity is lifted by fluorescence lifetime measurements. As a general rule, the fluorescence lifetime will match either the observed rise or decay time of the transient absorption associated with the intermediate, and represents the true formation time of the intermediate.

In order to assess the effect of hydrogen bonding and putative proton transfer on the lifetime of the intermediate state $C–P^{\cdot+}–Q^{\cdot}–H\cdots{}^{-}OOC$ in triad

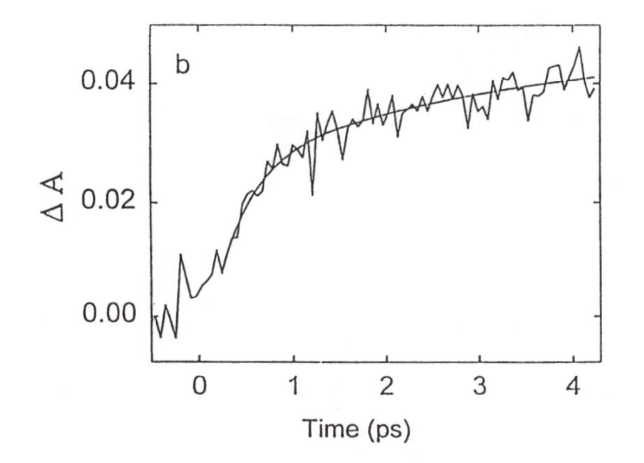

*Figure 3. Rise of the transient absorption at 660–662 nm of dyad **5** in benzonitrile solution following excitation at 590 nm with ~250-fs laser flashes. The fit shown yields rise times of 350 fs and 3.8 ps.*

1, transient absorption studies with excitation by ca. 200-fs pulses at 590 nm have been used to directly probe transient state(s) formed upon photoexcitation of dyads **5–8** in benzonitrile solution. Dyads **5–8** are model compounds for the porphyrin–quinone components of triads **1–4,** respectively. As discussed above, dyad **6,** which lacks the internal hydrogen bond, is the appropriate reference compound for evaluating the effect of the internal hydrogen bond on the dynamics of photoinduced electron transfer and charge recombination in these porphyrin–quinone systems. By comparison with other PQ systems, the spectra obtained between 624 nm and 766 nm are assigned to the absorption of $^1P^*$ and $P^{\cdot +}$ (27). As shown in Figures 3–6 for dyads **5–8,** the spectra in this region rise rapidly and decay more slowly. Global analysis of the decays (Table II) yields time constants of 35, 33, 121, and 38 ps (major component) for **5, 6, 7,** and **8,** respectively. These are similar (within experimental error) to those observed from fluorescence decay (43 (and 183), 41, 113, and 28 ps), and therefore report the true time constants for the formation of the charge-separated intermediates.

Turning to the rise times in Figures 3–6, which actually reflect the prompt formation of $^1P^*$ and the decay of $P^{\cdot +}$, on a short time scale over the 660–664-nm range a single rise time of <300 fs was observed for **6** and **7** (Figures 4 and 5). By contrast, dyads **5** and **8** exhibited rise times having a longer component (Figures 3 and 6). In the case of **6** and **7** the rise time corresponded to the instrument response function, indicating that the decay of $P^{\cdot +}$ is faster than 300 fs. The rise time of the signal for **5** required a two-exponential fit: a 300-fs component and a second component of 3.8 ps (Figure 3). The slow component of the rise of transient absorption in triad **8** was fitted with a time constant of 810 fs (Figure 6). In **5** and **8** the slower component of the rise time is

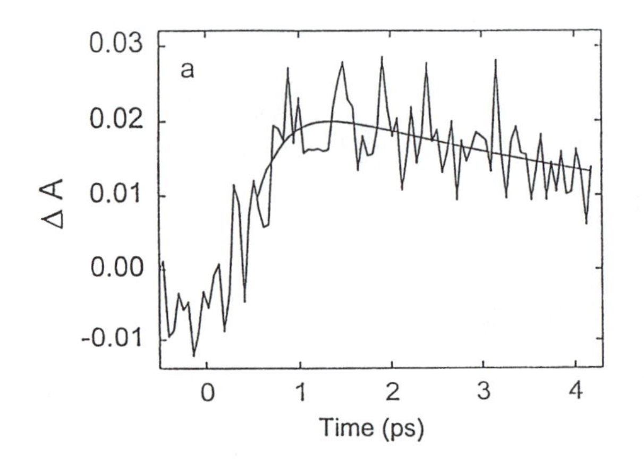

*Figure 4. Transient absorption of dyad **6** under the conditions described in Figure 3. The solid line is the fit to the data with a 250-fs rise-time constant. The decay-time constant presented in Table II was taken from data over a longer time window.*

assigned to the decay of $P^{\cdot +}$. The faster component of the rise in triad **5** is assigned to the prompt formation of $^1P^*$ and the decay of $P^{\cdot +}$ in conformers lacking the internal hydrogen bond.

These results demonstrate that the lifetime of $P^{\cdot +}$ is longer for the dyads in which the quinone bears a proton donor group, and thus electron-hole recombination is slower. Furthermore, $P^{\cdot +}$ is longer lived in dyad **5,** in which the

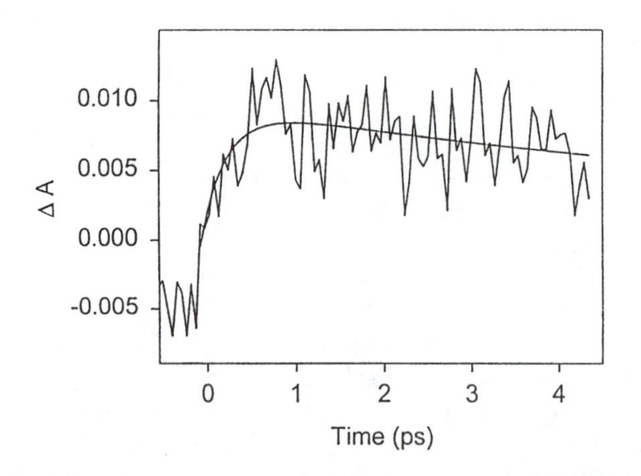

*Figure 5. Transient absorption of dyad **7** under the conditions described in Figure 3. The smooth curve is the fit to the data with a 290-fs rise-time constant. The decay-time constant presented in Table II was taken from data over a longer time window.*

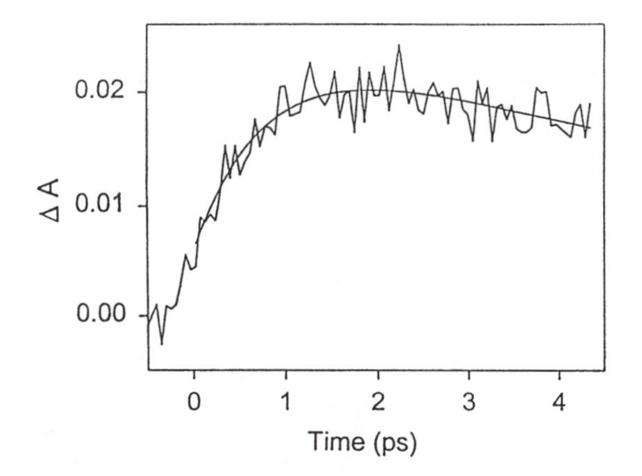

*Figure 6. Transient absorption of dyad **8** under the conditions described in Figure 3. The smooth curve is the fit to the data with an 810-fs rise-time constant. The decay-time constant presented in Table II was taken from data over a longer time window.*

quinone bears the stronger acid. Similar slow electron-hole recombination in the triads should lead to an increased yield of election donation by the attached carotenoid. Therefore, the yield of the final charge-separated species ($C^{\cdot+}$–P–$Q^{\cdot}$–H···–OOC for **1** and $C^{\cdot+}$–P–$Q^{\cdot-}$···H–OR for **4**) was expected to be greater in triads **1** and **4** than in either **2** or **3.**

Energetics. The energies of the species important to this work (Figure 7) are estimates based on the following observations. The energy of C–^{1}P–Q···HOOC in triad **1** was calculated to be 1.90 eV from spectral data and is the same for porphyrin-containing compounds **2–8**. The energy level for C–$P^{\cdot+}$–$Q^{\cdot-}$···HOOC, 1.43 eV, was estimated from the cyclic voltammetric first oxidation potential of 5,15-bis(4-acetamidophenyl)-10,20-bis(4-methylphenyl)porphyrin (**28**) (0.89 V versus SCE in benzonitrile), and the first reduction potential of 8-chloro-6-phenylcarbamyl-1,4-naphthoquinone 0.54 V). As discussed above, this is the appropriate quinone for estimating this energy level. This is also the energy level of the initial charge-separated state in triad **2** and dyads **5** and **6.** In triad **3** and dyad **7** the energy of the initial charge-separated species was 1.47 eV, based on the oxidation potential of **28** and the reduction potential of –0.58 V for 6-phenylcarbamyl-1,4-naphthoquinone. In triad **4** and dyad **8** the energy of the initial charge-separated state is 1.29 eV, based on the reduction potential of –0.40 V for 8-hydroxy-6-phenylcarbamyl-1,4-naphthoquinone and the oxidation potential of **28.**

The energy level for C–$P^{\cdot+}$–$Q^{\cdot}$H···$^{-}$OOC, 1.20 eV, was obtained from the first oxidation potential of **28** and the first reduction potential of 1-carboxy-6-

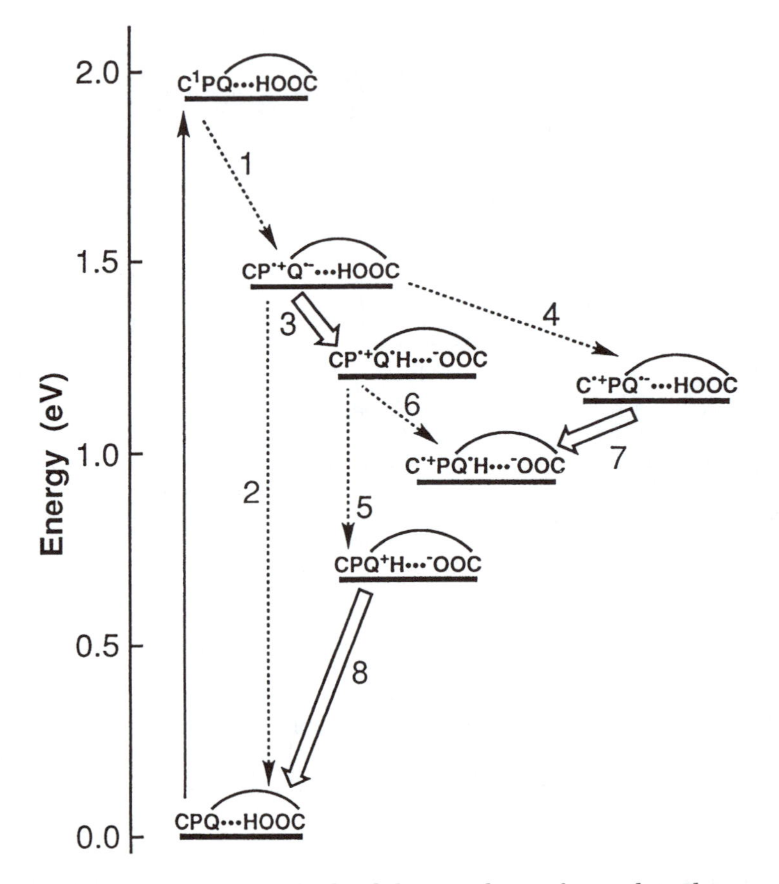

*Figure 7. Transient states and related decay pathways for triad **1**. The energy levels of the various species and the rate constants for the electron and proton transfer processes are discussed in the text. Excitation (⟶), electron transfer (-----⟶), and proton transfer (⇒).*

phenylcarbamyl-1,2,3,4-tetrahydro-1,4-methanoanthracene-9,10-dione (–0.31 V). The energy of C–P–Q$^+$–H···$^-$OOC (0.64 eV) was estimated using a pK_a of –5.6 for the protonated quinone (9) and a pK_a of 5 for the carboxylic acid. The energies of the charged species in triads **1–4** involving C$^{\cdot+}$ were estimated from the data above and the first oxidation potential of a carotenoid model compound (0.59 V). Transfer of the hole from the porphyrin to the carotenoid lowers the energy of the system by 0.30 eV (0.89–0.59). These energy estimates are not corrected for Coulombic effects, which are expected to be relatively minor in benzonitrile.

Quantum Yields. The quantum yields and lifetimes of the final charge-separated species in triads **1–4** were determined by monitoring the

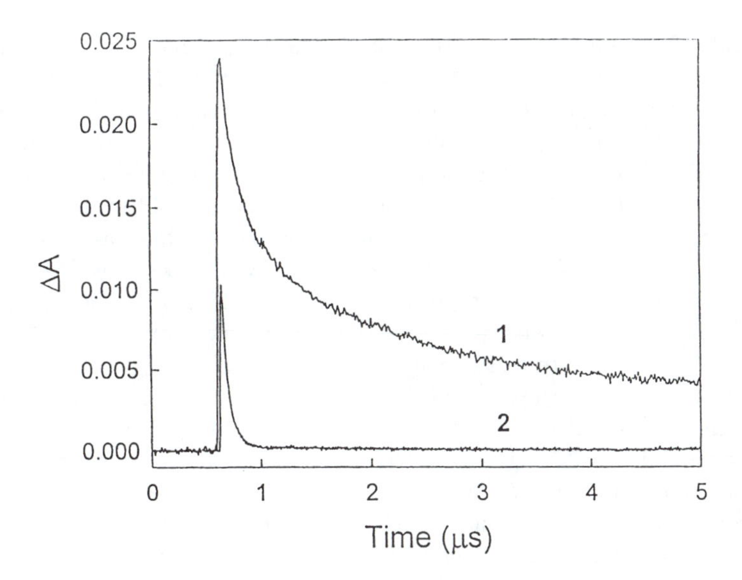

Figure 8. Decays of the transient absorption at 950 nm of $C^{\cdot+}$ *species in triad* **1** *(larger signal) and* **2** *following excitation of* ca. *1* × 10^{-5} *M solutions in benzonitrile with a* ca. *5-ns, 650-nm laser pulse. A sum-of-exponentials fit of the data yields lifetimes of 233 ns and 2.5 μs in the case of* **1** *and 62 ns in the case of* **2.**

carotenoid radical cation absorption at 950 nm following excitation at 650 nm. Figure 8 presents the formation and decay of $C^{\cdot+}$ for triads **1** and **2.** The amplitudes of the transient absorptions at 950 nm are proportional to the relative quantum yield of $C^{\cdot+}$–P–$Q^{\cdot-}$. By the comparative method (*28*), using the triplet–triplet absorption of free base 5,10,15,20-tetra(4-methylphenyl)porphyrin as a standard, the yield of the final charge separation is 22% in the case of triad **1** and 10% in the case of triad **2.** In fact, as shown in Table III, the yield for the formation of the final charge-separated species $C^{\cdot+}$–P–$Q^{\cdot-}$ is larger in

Table III. Quantum Yields and Lifetimes of Long-Lived Charge Separated Species of Triads in Dilute Benzonitrile Solution

Triad	*1*	**2**	*3*	*4*
Quantum yield[a]	0.22	0.10	0.11	0.16
Lifetime (ns)[b]	233 and 2500[c]	62	60	70

[a] The quantum yields relative to each other are ±10%.
[b] The lifetimes are ±10%.
[c] The decay of the transient absorption of the carotenoid radical cation at 960 nm was fitted as the sum of two exponentials; it is likely that the long-lived component involves bimolecular processes and is only approximated as an exponential.

the case of **1** than for any of the triads **2–4**, and the lifetime of the final charge-separated state is longer in the case of **1** than for any of the reference triads.

The increased yield of long-lived charge separation in triad **1** can be interpreted in terms of the chemically relevant species and processes shown in Figure 7. Following excitation, the initial charge-separated state ($C–P^{\cdot+}–Q^{\cdot-}$···HOOC) is generated. Because the lifetime of the charge-$^{\cdot+}–Q^{\cdot-}$ in dyad **6** was less than the instrument response time, the rate constant for electron-hole recombination from $C–P^{\cdot+}–Q^{\cdot-}$···HOOC, k_2, was estimated to be $\geq 4 \times 10^{12}$ s^{-1}. The driving force for step 2, 1.43 eV, is probably greater than the reorganization energy in these systems and therefore beyond the maximum of the Marcus rate versus free-energy curve. Competing with this recombination reaction are steps 3 and 4. Step 3 is the proton transfer process in which the carboxylic acid proton is transferred to the quinone radical anion species yielding $C–P^{\cdot+}–Q^{\cdot}–H$···$^{-}$OOC, the semiquinone and the carboxylate moicties. The rate constant for step 3 (and step 7 which is the same proton transfer process) can be calculated from the quantum yield expression in equation 5, which assumes that the yield of step 7 is unity (vide infra) and yields k_3 ~ 1×10^{12} s^{-1}. Correcting the calculation to allow for the observation that only 70% of **1** is in the hydrogen-bond form yields k_3 ~ 2×10^{12} s^{-1}.

$$\Phi(C^{\cdot+}–P–Q^{\cdot}–H\cdots{}^{-}OOC) = 0.22$$

$$= k_4/(k_4 + k_3 + k_2) + \{k_3/(k_4 + k_3 + k_2)\}\{k_6/(k_6 + k_5)\} \quad (5)$$

Steps 4 and 6 involve electron transfer from the carotenoid to the porphyrin radical cation, and to a first approximation are assumed to be equal. In a closely related system this rate constant has been determined to be ca. 5×10^{11} s^{-1} (27). The electron-hole recombination reaction from $C–P^{\cdot+}–Q^{\cdot}–H$···$^{-}$OOC, step 5, yields $C–P–Q^{+}–H$···$^{-}$OOC, a very energetic species. The driving force for step 5 is 0.56 eV. Because $\Delta G°$ for step 5 is much less than the reorganization energy in these systems (ca. 1 eV), step 5 is low in the Marcus normal region and therefore expected to be much slower than recombination by step 2. Indeed, the rate constant for step 5 can be estimated as 3×10^{11} s^{-1} from the lifetime of $P^{\cdot+}–Q^{\cdot}–H$···$^{-}$OOC in dyad **5.** Thus, proton transfer shifts electron-hole recombination from a high rate slightly in the Marcus inverted region for the species $C–P^{\cdot+}–Q^{\cdot-}$···HOOC to a lower rate in the normal region for the species $C–P^{\cdot+}–Q^{\cdot}–H$···$^{-}$OOC.

From ratios of quantum yield expressions (equation 5) for triads **1** and **2** and the rate constants calculated above, it is possible to compare the yields of key pathways in the two triads. Numerical evaluation of these ratios requires the assumption that the yield of step 7 is unity. This is reasonable as the lifetime of analogous charge-separated states in triads lacking proton transfer is at least 70 ns and proton transfer is subpicosecond in triad **1.** Because the yield of photoinduced electron transfer (step 1) is essentially unity, the yield of electron donation to the porphyrin radical cation by the carotenoid in **2,** which is analogous to step 6 for triad **1,** determines the yield of long-lived charge-separated species in triad **2.** This yield is 0.11 in **2** and 0.66 in **1.** The six-fold increase is attributable to the reduced electron-hole recombination rate resulting from the proton transfer process. The yields for pathways 3 and 4 are 0.2 and 0.09, respectively, indicating that the extremely rapid charge recombination, step 2, accounts for ca. 70% of the decay of the initial charge-separated species. By contrast, in model triad **2** the analogous electron-hole recombination accounts for ca. 90% of the decay.

Conclusions

The coordinated electron transfer/proton transfer process demonstrated by **1** is a viable strategy for increasing the yield of forward electron transfer in multistep systems. Interestingly, in triad **1** a substantial fraction of the intramolecular redox potential from the photoinduced electron transfer process has been translated into a p*K* difference. Of the 0.9 eV of intramolecular redox energy in the species $C^{\cdot+}$–P–$Q^{\cdot}$H···$^{-}$OOC, approximately 0.64 eV is conserved as proton chemical potential in the species C–P–Q^{+}–H···$^{-}$OOC, which formed after the electron-hole recombination process. This system serves as a paradigm for the coupling of electron transfer to a change in proton chemical potential that could ultimately be translated into the generation of proton motive force in a heterogeneous system with appropriate electron and proton transfer relays.

Experimental Details

General Methods. NMR spectra were recorded on a Varian Gemini spectrometer at 300 MHz, or a Varian Unity spectrometer at 500 MHz. Unless otherwise specified, samples were dissolved in deuteriochloroform with tetramethylsilane as an internal reference. The carotenoid and porphyrin resonance assignments are reported using the numbering system previously published (*12*). For simplicity, the quinones, naphthalenes, and methano–anthracenes were numbered as indicated in the schemes. Low-resolution mass spectra were obtained on a Varian MAT 312 mass spectrometer operating in electron ionization mode. Mass spectra of compounds with molecular weights above 1000 were obtained by laser-desorption time-of-flight mass spectrometry (Vestec Laser Tec. Research Instrument).

IR spectra were recorded on a Mattson Model 2020 Galaxy series FT-IR spectrophotometer. The UV–vis spectra were measured on a Shimadzu UV–2100U UV–vis recording spectrophotometer. Thin-layer chromatography was carried out with silica gel GHL or GHLF Uniplates (Analtech, Inc.). Column chromatography was performed using silica gel (70–230 mesh particle size 0.063–0.200 mm), and flash column chromatography was done using silica gel (43 μm average particle diameter). All high-pressure liquid chromatographic (HPLC) separations were performed on a Shimadzu Model SCL–6B using a SIL–6B injector, LC–6A pumps, and a 206 PHD detector, and the columns employed were analytical (4.6 mm i.d. × 250 mm) and semipreparative (10.0 mm i.d. × 250 mm). Both were packed with normal-phase 5-μm silica.

Cyclic voltammetric measurements were carried out with a Pine Instrument Co. Model AFRDE4 potentiostat. All electrochemical measurements were done in benzonitrile at ambient temperature at a glassy carbon working electrode, with a saturated calomel electrode or Ag/Ag^+ (0.01 M) as a reference, and a platinum-wire counterelectrode. The electrolyte was 0.1 M tetra-*n*-butylammonium hexafluorophosphate, and ferrocene was added as an internal reference redox system.

The fluorescence emission spectra were measured using a SPEX Fluorolog-2. Excitation was produced by a 450-W xenon lamp and single-grating monochromator. Fluorescence was detected at a 90° angle to the excitation beam via a single-grating monochromator and an R928 photomultiplier tube having S-20 spectral response operating in the photon counting mode. Fluorescence decay measurements were performed on approximately 1×10^{-5} M solutions using the time-correlated single photon counting method described previously (*29*). All samples were purified either by column chromatography or by TLC prior to being used. The solvents were distilled and stored over anhydrous potassium carbonate to remove any acid. The excitation source was a frequency-doubled Coherent Antares 76s Nd:YAG laser routed through a variable beam splitter to pump a cavity-dumped dye laser. The instrument response function (about 35 ps) was measured at the excitation wavelength (590 or 650 nm) for each decay experiment with a Ludox AS-40 suspension under the same conditions as the sample.

Nanosecond transient absorption measurements were carried out on ~1×10^{-5} M solutions in air-saturated dichloromethane, chloroform, and benzonitrile with excitation at 590 nm or 650 nm. The pump laser was a Continuum Surelite frequency-doubled Nd:YAG laser with a pulse width of about 5 ns; the spectrometer has been described previously (*30*). For each sample, absorption spectra were obtained before and after the experiment to rule out degradation of the compound.

Transient absorption measurements on the subpicosecond time scale were made using the pump-probe technique (*31*). The sample was dissolved in air-saturated benzonitrile to an absorbance of 1.0–1.8 at 590 nm in a 1-cm cell, and the resulting solution was circulated through a flow cuvette of 2-mm path length in order to minimize any possible degradation of the sample by the laser irradiation. Excitation was at 590 nm with 200-fs, 8-μJ pulses at a repetition rate of 540 Hz. The signals from the pump and continuum-generated white-light probe beam were collected by an optical multichannel analyzer with a dual-diode-array detector head.

Solvents and Reagents Commercially available compounds were purchased from Aldrich, Baker, Sigma, and Fluka, and were used without further purification unless otherwise noted. All solvents were redistilled. Tetrahydrofuran (THF) was freshly distilled from lithium aluminum hydride immediately prior to use. Anhydrous chloroform and methylene chloride were distilled from phosphorus pentoxide and stored over anhydrous potassium carbonate and 4–Å molecular sieves. Dry methanol and ethanol were obtained by refluxing with magnesium and iodine followed by distillation, and stored over 3-Å molecular sieves. Acetonitrile, butyronitrile, and benzonitrile (Aldrich HPLC grade) were dried over $MgSO_4$ and warmed at 65 °C for 4 h with DDQ under an argon atmosphere. The resulting red solution was then passed through a dry Al_2O_3 column and distilled (acetonitrile at 760 mmHg, butyronitrile and benzonitrile at 15 mmHg). Only the middle portions were collected and stored over anhydrous K_2CO_3 and 4-Å molecular sieves.

Toluene, dichloromethane, chloroform, and acetone were first distilled and then stored over K_2CO_3 and/or 4-Å molecular sieves. Pyridine, dimethylformamide, and dimethyl sulfoxide were stored over 4-Å molecular sieves for at least one day before use. Evaporation of solvents was performed under reduced pressure on a rotary evaporator and/or a vacuum pump (10^{-4} torr) at room temperature.

Synthesis. The syntheses of 5,15-bis(4-acetamidophenyl)-10,20-bis(4-methylphenyl)porphyrin (**28**) and 7′-apo-7′-(4-carboxyphenyl)-β-carotene (*14*) have been reported previously (*12, 32*).

1-Carbomethoxycyclopenta-1,3-diene (9). Preparation was by the method of Grunewald and Davis (*16*). The crude product was distilled through a 10-cm Vigreux column and collected in a receiver at –78 °C. A clear oily liquid (ester **9**) was obtained (12.6 g, 20%): ^1H NMR (300 MHz, $CDCl_3$) δ 3.29–3.31 (2H, m, CH_2-5), 3.77 (3H, s, CO_2CH_3), 6.56–6.60 (1H, m, H-4), 6.67–6.70 (1H, m, H-3), 7.40-7.43 (1H, m, H-2).

6-Carbobenzyloxy-1,4-dimethoxynaphthalene. To 60 mL of benzyl alcohol was added ca. 0.5 g of sodium under argon. The mixture was stirred at ambient temperature until no more sodium metal could be seen. The resulting colorless solution was added to a 100-mL flask containing the methyl ester of **25** (700 mg, 2.85 mmol). After 21 h of stirring under reduced pressure (~15 mm-Hg), the excess of benzyl alcohol was removed under vacuum. The residue was treated with dilute HCl to adjust the pH to ~8 and then extracted with chloroform (4 × 200 mL). The combined organic extracts were concentrated by rotary evaporation and dried for several hours under vacuum. The resulting yellow solid was purified by column chromatography (CH_2Cl_2, then 15% $MeOH/CH_2Cl_2$) to give 366 mg (40%) of pure benzyl ester as a pale yellow solid, and 241 mg (37%) of pure acid **25** as a white solid. 6-Carbobenzyloxy-1,4-dimethoxynaphthalene: thin-layer chromatography (TLC) R_f = 0.74 (CH_2Cl_2); ^{1}H NMR (300 MHz, $CDCl_3$) δ3.95 (3H, s, OCH_3-1), 3.96 (3H, s, OCH_3 4), 5.44 (2H, s, CH_2Ph), 6.73 (1H, d, J = 8.4 Hz, H-2 or H-3), 6.81 (1H, d, J = 8.4 Hz, H-3 or H-2), 7.33–7.53 (5H, m, PhH), 8.11 (1H, dd, J = 8.8, 1.7 Hz, H-7), 8.24 (1H, d, J = 8.8 Hz, H–8), 9.01 (1H, d, J = 1.7 Hz, H-5).

6-Carbobenzyloxy-1,4-naphthoquinone (10). A solution of ceric ammonium nitrate (CAN) (1.01 g, 1.83 mmol) in 2 mL of water was added dropwise, over 2 min, to a stirred solution of 6-carbobenzyloxy-1,4-dimethoxy-naphthalene (281 mg, 0.87 mmol) in acetonitrile (10 mL). The mixture was stirred for an additional 15 min at room temperature. The reaction was quenched with water (100 mL) and extracted with chloroform (4 × 100 mL). The combined organic extracts were concentrated under reduced pressure to yield a yellow solid, which was purified by column chromatography (silica gel, CH_2Cl_2) to give 222 mg (87%) of quinone **10:** TLC R_f = 0.58 (CH_2Cl_2); ^{1}H NMR (300 MHz, $CDCl_3$) δ 5.42 (2H, s, CH_2Ph), 7.04 (2H, s, H-2 and H-3), 7.33–7.50 (5H, m, PhH), 8.16 (1H, d, J = 8.1 Hz, H-8), 8.42 (1H, dd, J = 8.1, 1.7 Hz, H-7), 8.74 (1H, d, J = 1.7 Hz, H-5).

6-Carbobenzyloxy-4-carbomethoxy-1,4,4a,9a-tetrahydro-1,4-methanoanthracene-9,10-dione (11a), 6-Carbobenzyloxy-1-carbomethoxy-1,4,4a,9a-tetrahydro-1,4-methanoanthracene-9,10-dione (11b), 6-Carbobenzyloxy-3-carbomethoxy-1,4,4a,9a-tetrahydro-1,4-methanoanthracene-9,10-dione (11c), 6-Carbobenzyloxy-2-carbomethoxy-1, 4, 4a, 9a-tetrahydro-1, 4-methanoanthracene-9,10-dione (11d), 6-Carbobenzyloxy-4-carbomethoxy-1,4-dihydro-1,4-methanoanthracene-9,10-dione (11a′), 6-Carbobenzyloxy-1-carbomethoxy-1, 4-dihydro-1, 4-methanoanthracene-9,10-dione (11b′), 6-Carbobenzyloxy-3-carbomethoxy-1,4-dihydro-1,4-methanoanthracene-9,10-dione (11c′), and 6-Carbobenzyloxy-2-carbomethoxy-1, 4-dihydro-1, 4-methanoanthracene-9,10-dione (11d′). To a stirred solution of naphthoquinone **10** (222 mg, 0.76 mmol) in 20 mL of dry toluene that had been saturated with argon were added freshly cracked 1-carbomethoxycyclopenta-1,3-diene **9** (990 mg, 7.98 mmol) and a catalytic amount of hydroquinone. The mixture was stirred for 24 h at ambient temperature and was then quenched by adding 50 mL of water. The solution was extracted with chloroform (3 × 50 mL). The organic extracts were concentrated under reduced pressure to give a yellow oil. TLC analysis of the crude product on silica showed the presence of starting material, diene dimers, isomers **11,** and a small amount of impurities. These components were separated by column chromatography (silica gel, 10% EtOAc/hexanes through 25% EtOAc/hexanes) to give a total of 283 mg (90%) of material, which was collected as four separate samples and analyzed by ^{1}H NMR. The first sample (28.3 mg, 9.3%) was a mixture of **11a′** and **11b′** as a yellowish green solid. The second sample (104 mg, 33%) was a mixture of **11a, 11b, 11a′**, and **11b′** as a yellow oil. The third sample (28 mg, 9%) was a mixture of **11a, 11b, 11c′** and **11d′** as a brown oil. The fourth sample (122 mg, 39%) was a mixture of **11c, 11d, 11c′**, and **11d′** as a brown solid.

6-Carbobenzyloxy-4-carbomethoxy-1,4-dihydro-9,10-dimethoxy-1,4-methanoanthracene (12a) and 6-Carbobenzyloxy-1-carbomethoxy-1, 4-dihydro-9, 10-dimethoxy-1, 4-methanoanthracene (12b). To a 50-mL round-bottomed flask were added the diketones **11a, 11a′, 11b,** and **11b′** (104 mg, 0.25 mmol) and 8 mL of dry acetone under an argon atmosphere. The solution was warmed to reflux and dimethyl sulfate (120 μL, 1.3 mmol) was added slowly by syringe. A 10% methanolic KOH solution (~0.5 mL) was then added dropwise (each drop was accompanied by a transient purple coloration), until no more purple discharges could be seen. The mixture was then refluxed for 5 min. After the removal of the oil bath, a brown suspension formed. The suspension was added to 1 M aqueous HCl (~20 mL) and extracted with chloroform (3 × 50 mL). The organic solvent was removed under reduced pressure to give a brown oil. Column chromatography (silica gel, 1% EtOAc/hexanes to 40% EtOAc/hexanes) yielded two separate samples that were analyzed by TLC and ^{1}H NMR. The first sample (54 mg, 49%) was a mixture of **12a** and **12b** that was purified by HPLC to give pure **12a** and pure **12b** as colorless oils. Both compounds were identified by mass spectra and 1D NOE ^{1}H NMR analysis. The second sample (10 mg, 9%) was a mixture of **12a, 12b, 12c,** and **12d** as a colorless oil. 6-Carbobenzyloxy-4-carbomethoxy-9,10-dimethoxy-1,4-dihydro-1,4-methanoanthracene **12a:** TLC R_f = 0.46 (25% EtOAc/hexanes); ^{1}H NMR (300 MHz, $CDCl_3$) δ 2.49 (1H, dd, J = 7.7, 1.7 Hz, H-11′), 2.62 (1H, d, J = 7.7 Hz, H-11), 3.81 (3H, s, OCH_3-10) 3.92 (3H, s, CO_2CH_3-4), 3.98 (3H, s, OCH_3-9), 4.38 (1H, m, H-1), 5.41 and 5.42 (2H, AB, J = 12.7 Hz, OCH_2Ph), 6.79 (1H, dd, J = 5.3, 3.1 Hz, H-2), 7.02 (1H, d, J = 5.3 Hz, H-3), 7.33–7.50 (5H, m, PhH), 8.08 (2H, s, H-7 and H-8), 8.72 (1H, s, H-5) (proton assignments were made by analysis of the ^{1}H NOE); MS m/z 444 (M^+). 6-Carbobenzyloxy-1-carbomethoxy-9,10-dimethoxy-1,4-dihydro-1,4-methanoanthracene **12b:** TLC R_f = 0.46 (25% EtOAc/hexanes); ^{1}H NMR (300 MHz, $CDCl_3$) δ 2.49 (1H, dd, J = 7.7, 1.7 Hz, H-11′), 2.62 (1H, d, J = 7.7 Hz, H-11), 3.78 (3H, s, OCH_3-9), 3.93 (3H, s, CO_2CH_3-1), 4.02 (3H, s, OCH_3-10), 4.40 (1H, m, H-4), 5.43 (2H, s, OCH_2Ph), 6.81 (1H, dd, J = 5.3, 3.1 Hz, H-3), 7.01 (1H, d, J = 5.3 Hz, H-2), 7.34–7.51 (5H, m, PhH), 7.97 (1H, d, J = 8.8 Hz, H-8), 8.08 (1H, dd, J = 8.8, 1.7 Hz, H-7), 8.83 (1H, d, J = 1.7 Hz, H-5) (proton assignments were made by analysis of the ^{1}H NOE); MS m/z 444 (M^+).

6-Carboxy-4-carbomethoxy-9,10-dimethoxy-1,2,3,4-tetrahydro-1,4-methanoanthracene (13). To a solution of methanoanthracene **12a** (20 mg, 0.045 mmol) in freshly distilled THF (10 mL) that had been thoroughly degassed with argon was added 6 mg of 20% $Pd(OH)_2$ on carbon. The mixture was stirred under a hydrogen atmosphere (40 psi) for 20 h. An additional portion of palladium catalyst (6 mg) was then added and the reaction allowed to continue for another 24 h. The reaction mixture was filtered through glass wool, and the residue was washed several times with chloroform. The filtrate was evaporated under reduced pressure and the residue was purified by column chromatography (silica gel), eluting with a solvent gradient of chloroform through 10% $MeOH/CHCl_3$ to give monoester **13** as a white solid in quantitative yield TLC R_f = 0.08 (10% $MeOH/CHCl_3$); ^1H NMR (300 MHz, 5% $CD_3OD/CDCl_3$) δ 1.30–1.40 (1H m, H-2n), 1.50–1.60 (1H, m, H-3n), 1.81 (1H, d, J = 8.9 Hz, H-11a), 2.02–2.14 (1H, m, H-2x), 2.08 (1H, d, J = 8.9 Hz, H-11s), 2.18–2.28 (1H, m, H-3x), 3.68 (3H, s, CO_2CH_3-4), 3.72 (3H s, OCH_3-10), 3.75 (1H, brs, H-1), 3.84 (3H, s, OCH_3-9), 7.93 (1H, d, J = 8.8 Hz, H-8), 7.98 (1H, d, J = 8.8 Hz, H-7), 8.63 (1H, brs, H-5); MS m/z 356 (M^+).

5,15-Bis(4-trifluoroacetamidophenyl)-10,20-bis(4-methylphenyl)porphyrin Trifluoroacetic anhydride (2 mL) was added dropwise to a solution of 5,15-bis(4-aminophenyl)-10,20-bis(4-methylphenyl)porphyrin (255 mg, 0.38 mmol) in chloroform (20 mL) and pyridine (1 mL). The mixture was stirred at room temperature and the reaction was followed by TLC analysis. After 25 min of stirring, the reaction was quenched with saturated aqueous K_2CO_3 (50 mL) and the mixture was extracted with chloroform (3 × 100 mL). The combined red organic extracts were dried (Na_2SO_4) and concentrated by rotary evaporation. The crude product was purified by column chromatography (silica gel), eluting with a solvent gradient of hexane through 33% EtOAc/hexanes. The product-containing fractions were collected and concentrated to yield 320 mg (98%) of the desired porphyrin as a purple solid: TLC R_f = 0.69 (40% EtOAc/hexanes); ^1H NMR (300 MHz, $CDCl_3$) δ –2.83 (2H, brs, pyrrole–NH), 2.68 (6H, s, 10, 20Ar-CH_3), 7.54 (4H, d, J = 8.2 Hz 10, 20ArH-3, 5), 7.96 (4H, d, J = 8.0 Hz, 5, 15ArH-3, 5), 8.07 (4H, d, J = 8.2 Hz, 10, 20ArH-2, 6), 8.19 (2H, s, 5, 15Ar-NHCO), 8.23 (4H, d, J = 8.0 Hz, 5, 15ArH-2, 6), 8.78 (4H, d, J = 4.6 Hz, pyrrole H-2, 8, 12, 18), 8.87 (4H, d, J = 4.6 Hz, pyrrole H-3, 7, 13, 17).

5-(4-Aminophenyl)-15-(4-trifluoroacetamidophenyl)-10,20-bis(4-methylphenyl)porphyrin (14). A portion of 5,15-bis(4-trifluoroacetamidophenyl)-10,20-bis(4-methylphenyl)porphyrin (320 mg, 0.370 mmol) was dissolved in a mixture of 80 mL of methanol, 80 mL of THF (freshly distilled from $LiAlH_4$), and 27 mL of an aqueous 1.1% solution of KOH. The solution was allowed to stir under argon for 24 h, after which it was poured into a separatory funnel containing water (50 mL). The aqueous phase was extracted with chloroform (3 × 100 mL) and the extracts were dried (Na_2SO_4). The solvent was removed by evaporation at reduced pressure, and the product was purified by flash column chromatography (silica gel), eluting with a solvent gradient of hexanes through 50% EtOAc/hexanes. The product-containing fractions were collected and concentrated to give 170 mg (60%) of **14** as a purple solid: TLC R_f = 0.47 (40% EtOAc/hexanes); ^1H NMR (300 MHz, $CDCl_3$) δ −2.74 (2H, brs, pyrrole–NH), 2.69 (6H, 5, 10, 20Ar-CH_3), 3.97 (2H, brs, 5Ar-NH_2), 7.02 (2H, d, *J* = 8.3 Hz, 5ArH-3, 5), 7.54 (4H, d, *J* = 7.9 Hz, 10, 20ArH-3, 5), 7.94 (2H, d, *J* = 8.5 Hz, 15ArH-3, 5), 7.97 (2H, d, *J* = 8.3 Hz, 5ArH-2, 6), 8.09 (4H, d, *J* = 7.9 Hz, 10, 20ArH-2, 6), 8.24 (2H, d, *J* = 8.5 Hz, 15ArH-2, 6), 8.25 (1H, brs, 15Ar-NHCO), 8.68–8.89 (8H, m, pyrrole-H).

Porphyrin 15. To a stirred solution of compound **13** (14 mg, 0.039 mmol) in dry toluene (8 mL) and dry pyridine (1.5 mL), was added an excess of $SOCl_2$ (2 drops) under argon. After 10 min of stirring, the solvent and excess $SOCl_2$ were evaporated under vacuum, and the acid chloride was redissolved in 4 mL of dry toluene and 0.5 mL of dry pyridine. This solution was added to a solution of porphyrin **14** (18.5 mg, 0.024 mmol) in dry toluene (6 mL) and dry pyridine (1 mL). The reaction was allowed to proceed under argon for 4 h, and this solution was then poured into dilution aqueous $NaHCO_3$ solution. The aqueous phase was extracted with chloroform (3 × 100 mL). The organic solvent was evaporated and the residue was dried for 3 h under vacuum. Flash column chromatography of the crude product on silica gel (1% acetone/CH_2Cl_2) gave 17 mg (63%) of pure porphyrin **15** as a purple solid: TLC R_f = 0.12 (1% acetone/CH_2Cl_2); ^1H NMR (500 MHz, $CDCl_3$) δ −2.78 (2H, s, pyrrole–NH), 1.50–1.58 (1H, m, H-2n), 1.71–1.78 (1H, m, H-3n), 1.98 (1H, d, *J* = 8.8 Hz, H-11a), 2.20–2.28 (1H, m, H-2*x*), 2.27 (1H, d, *J* = 8.8 Hz, H-11s), 2.38–2.46 (1H, m, H-3*x*), 2.68 (6H, s, 10, 20Ar-CH_3), 3.89 (1H, brs, H-1), 3.91 (3H, s, CO_2CH_3–4), 3.94 (3H, s, OCH_3-10), 4.05 (3H, s, OCH_3-9), 7.52 (4H, d, *J* = 7.5 Hz, 10, 20ArH-3, 5), 7.94 (2H, d, *J* = 8.3 Hz, 15ArH-3, 5), 8.02 (2H, d, *J* = 8.3 Hz, 5ArH-3, 5), 8.06 (4H, d, *J* = 7.5 Hz, 10, 20ArH-2, 6), 8.08 (1H, dd, *J* = 9.0, 1.8 Hz, H-7), 8.16 (2H, d, *J* = 8.3 Hz, 5ArH-2, 6), 8.22 (2H, d, *J* = 8.3 Hz, 15ArH-2, 6), 8.29 (1H, s, Ar-NHCO), 8.30 (1H, d, *J* = 9.0 Hz, H-8), 8.33 (1H, s, Ar-NHCO), 8.68 (1H, d, *J* = 1.8 Hz, H-5), 8.76-8.88 (8H, m, pyrrole-H) (proton assignments were made by analysis of the 2D COSY NMR); MS *m/z* 1107 (M^+ + 1).

Dyad 5. Porphyrin **15** (14 mg, 0.013 mmol) was dissolved in MeOH/THF/H_2O (3:3:1 24 mL) and treated with sodium hydroxide (354 mg, 8.86 mmol). The solution was stirred at ambient temperature for 42 h and then warmed to 60 °C for 4 h and quenched by addition of dilute acetic acid. The aqueous layer was extracted with 10% MeOH/$CHCl_3$ (4 × 50 mL), and the combined organic extracts were concentrated to provide a purple solid. The crude product was purified by column chromatography (silica gel), eluting with a solvent gradient of chloroform through 10% MeOH/$CHCl_3$ to give 12 mg (91%) of the pure deprotected amino acid porphyrin: TLC R_f = 0.17 (10% MeOH/$CHCl_3$); ^{1}H NMR (300 MHz, 5% CD_3OD/$CDCl_3$) δ −2.77 (2H, brs pyrrole–NH), 1.38–1.48 (1H, m, H-2n), 1.56–1.66 (1H, m, H-3n), 1.91 (1H, d, *J* = 8.3 Hz, H-11a), 2.10–2.20 (1H, m, H-2x), 2.16 (1H, d, *J* = 8.3 Hz, H-11s), 2.26–2.38 (1H, m, H-3x), 2.59 (6H, s, 10, 20Ar-CH_3), 3.79 (1H, d, *J* = 3.3 Hz, H-1), 3.90 (3H, s, OCH_3-10), 3.96 (3H, s, OCH_3-9), 6.99 (2H, d, *J* = 8.3 Hz, 15ArH-3, 5), 7.45 (4H, d, *J* = 7.8 Hz, 10, 20ArH-3, 5), 7.89 (2H, d, *J* = 8.3 Hz, 15ArH-2, 6), 7.99 (4H, d, *J* = 7.8 Hz, 10, 20ArH-2, 6), 8.01 (1H, dd, *J* = 8.8, 1.7 Hz, H-7), 8.06 (2H, d, *J* = 8.6 Hz, 5ArH-3, 5), 8.13 (2H, d, *J* = 8.6 Hz, 5ArH-2, 6), 8.19 (1H, d, *J* = 8.8 Hz, H-8), 8.65 (1H, d, *J* = 1.7 Hz, H-5), 8.79 (8H, brs, pyrrole–H).

To a purple solution of this amino acid porphyrin (11.5 mg, 0.012 mmol) in dry dichloromethane (4 mL) in a 50-mL round-bottomed flask at −78 °C under argon, a solution of boron tribromide (15 mL, 0.16 mmol) in dry dichloromethane (1 mL) was added dropwise over 2 min. The resulting green solution was kept at −78 °C for 5 min and then allowed to warm. After 100 min at room temperature, the mixture was treated with saturated aqueous sodium bicarbonate until the pH of the water layer was about 7. The aqueous phase was extracted with 10% MeOH/$CHCl_3$ (3 × 100 mL). The combined organic extracts were concentrated to give a purple solid. This porphyrin was dissolved in 5 mL of chloroform and treated with PbO_2 (~1 mg) in darkness for 30 min. After removal of the PbO_2 by centrifugation, the red solution was concentrated and the residue was purified by column chromatography (silica gel, 10% MeOH/$CHCl_3$) to give pure porphyrin-quinone **5** as a purple solid (3 mg, 27%): TLC R_f = 0.14 (10% MeOH/$CHCl_3$); ^{1}H NMR (500 MHz, 15% CD_3OD/$CDCl_3$) δ 1.32–1.39 (1H, m, H-2n), 1.48–1.56 (1H, m, H-3n), 1.79 (1H, d, *J* = 8.8 Hz, H-11a), 2.09 (1H, d, *J* = 8.8 Hz, H-11s), 2.13–2.21 (1H, m, H-2x), 2.30–2.39 (1H, m, H-3x), 2.61 (6H, s, 10, 20Ar-CH_3), 3.64 (1H, brs H-1), 7.03 (2H, d, *J* = 8.0 Hz, 15ArH-3, 5), 7.46 (4H, d, *J* = 7.5 Hz, 10, 20ArH-3, 5), 7.91 (2H, d, *J* = 8.0 Hz, 15ArH-2, 6), 8.00 (4H, d, *J* = 7.5 Hz, 10, 20ArH-2, 6), 8.06 (2H, d, *J* = 8.0 Hz, 5ArH-3, 5), 8.14 (2H, d, *J* = 8.0 Hz, 5ArH-2, 6), 8.16 (1H, d, *J* = 8.0 Hz, H-8), 8.35 (1H, d, *J* = 8.0 Hz, H-7), 8.62 (1H, s, H-5), 8.79 (8H, brs, pyrrole-H) (proton assignments were made by analysis of the 2D COSY NMR); UV–vis (CH_2Cl_2) λ_{max}: 650, 595, 558, 519, and 422 nm.

Triad 1. An excess of $SOCl_2$ (1 drop) was added to a stirred solution of 7′-apo-7′-(4-carboxyphenyl)-β-carotene (14.4 mg, 0.027 mmol) in dry toluene (4 mL) and dry pyridine (2 mL) under argon. After 10 min of stirring, the solvent and excess $SOCl_2$ were evaporated under vacuum, and the acid chloride was redissolved in 3 mL of dry toluene and 0.5 mL of dry pyridine. Porphyrin-quinone **5** (2.6 mg, 0.0027 mmol) was dissolved in dry toluene/pyridine (2:1, 3 mL) and treated with the acid chloride solution. The mixture was kept well stirred under argon for 24 h and then quenched with water. The aqueous phase was extracted with 10% MeOH/$CHCl_3$ (6 × 100 mL) and the combined organic extracts were concentrated to give a red solid. Flash column chromatography on silica gel (5% MeOH/$CHCl_3$) gave 2.1 mg (53%) of the desired triad **1** as a red–purple solid: TLC R_f = 0.25 (10% MeOH/$CHCl_3$); ^{1}H NMR (500 MHz, 20% $CD_3OD/CDCl_3$) δ: 1.04 (6H, s, CH_3-16C and CH_3-17C), 1.37–1.45 (1H, m, H-2nQ), 1.45–1.52 (2H, m, CH_2-2C), 1.52–1.59 (1H, m, H-3nQ), 1.59–1.68 (2H, m, CH_2-3C), 1.73 (3H, s, CH_3, 18C),[56] 1.84 (1H, d, *J* = 9.0 Hz, H-11aQ), 1.99 (3H, s, CH_3-19C), 2.01 (3H, s, CH_3-20C), 2.03 (3H, s, CH_3–20′C), 2.05 (2H, m, CH_2–4C), 2.11 (3H, s, CH_3–19′C), 2.13 (1H, d, *J* = 9.0 Hz, H-11sQ), 2.21–2.28 (1H, m, H-2*x*Q), 2.36–2.43 (1H, m, H-3*x*Q), 2.69 (6H, s, 10, 20Ar-CH_3), 3.69 (1H, brs, H-1Q), 6.15 (1H, d, *J* = 15.5 Hz, H-8C), 6.17 (1H, d, *J* = 12.0 Hz, H-10C), 6.19 (1H, d, *J* = 15.5 Hz, H-7C), 6.26–6.42 (3H, m, H-12C, H-14C, H-14′C), 6.46 (1H d, *J* = 12.0 Hz, H-10′C), 6.50 (1H, d, *J* = 15.0 Hz, H-12′C), 6.63–6.78 (4H, m, H-11C, H-11′C, H-15C and H-15′C), 6.69 (1H, d, *J* = 16.0 Hz, H-7′C), 7.09 (1H, d, *J* = 16.0 Hz, H-8′C) 7.51 (4H, d, *J* = 8.0 Hz, 10, 20ArH-3, 5), 7.63 (2H, d, *J* = 8.0 Hz, H-1′C and H-5′C), 8.04 (2H, d, *J* = 8.0 Hz, H-2′C and H-4′C), 8.06 (4H, d, *J* = 8.0 Hz, 10, 20ArH-2, 6), 8.13 (2H, d, *J* = 8.5 Hz, 15ArH-3, 5), 8.16 (2H, d, *J* = 8.5 Hz, 5ArH-3, 5), 8.23 (4H, d, *J* = 8.5 Hz, 5, 15ArH-2, 6), 8.24 (1H, d, *J* = 8.0 Hz, H-8Q), 8.42 (1H, d, *J* = 8.0 Hz, H-7Q), 8.69 (1H, s, H-5Q) 8.89 (8H, brs, pyrrole–H) (proton assignments were made by analysis of the 2D COSY NMR); UV–vis (CH_2Cl_2) λ_{max}: 650, 591, 515, 481, and 422 nm.

6-Carbomethoxy-8-chloro-5,6-dihydro-1,4-dimethoxynaphthalene (18). To a solution of ketone **17** (110 mg, 0.41 mmol) in dry toluene (12 mL) under argon was added an excess of phosphorus oxychloride (0.32 mL, 3.43 mmol). The mixture was stirred under reflux for 15 days. The reaction progress was monitored throughout this period by TLC, and more $POCl_3$ (4 × 0.5 mL) was added as required. After cooling, the mixture was poured into a separatory funnel containing saturated aqueous $NaHCO_3$ (40 mL). The water layer was extracted with dichloromethane (3 × 100 mL). The combined organic extracts were dried ($MgSO_4$), and the solvent was removed under vacuum to afford a brown oil. Column chromatography of the crude product on silica gel (50% CH_2Cl_2/hexane) gave 37 mg (33%) of starting material and 47 mg (41%) of the desired compound **18** as a colorless oil: TLC R_f = 0.24 ($CHCl_3$); ^{1}H NMR (300 MHz, $CDCl_3$) δ 2.74 (1H, dd, *J* = 17.2, 14.3 Hz, H-4 or H-4′), 3.23–3.33 (2H, m, H-3 and H-4′ or H-4), 3.73 (3H, s, CO_2CH_3), 3.80 (3H, s, OCH_3-5), 3.81 (3H, s, OCH_3-8), 6.32 (1H, dd, *J* = 3.9, 1.1 Hz, H-2), 6.82 (1H, AB, *J* = 9.2 Hz, H-6 or H-7), 6.84 (1H, AB, *J* = 9.2 Hz, H-7 or H-6); MS *m/z* 282 (M^+).

6-Carbomethoxy-8-chloro-1,4-dimethoxynaphthalene (19). To a stirred solution of vinyl chloride **18** (133 mg, 0.47 mmol) in dry toluene (6 mL) that had been flushed with argon was added 2,3-dichloro-5,6-dicyanobenzoquinone (DDQ) (117 mg, 0.52 mmol) in one portion. The mixture was stirred at room temperature for 5 h. The resulting red solution was then poured into a separatory funnel containing water (100 mL) and chloroform (100 mL). The layers were separated and the aqueous layer was extracted with chloroform (3 × 100 mL). The combined organic extracts were concentrated under vacuum to give an orange solid, which was purified by column chromatography (silica gel, 50% CH_2Cl_2/hexanes to 100% CH_2Cl_2) to yield 70 mg (53%) of **19** as a pale yellow solid: TLC R_f = 0.42 (CH_2Cl_2); ^1H NMR (300 MHz, $CDCl_3$) δ 3.92 (3H, s, CO_2CH_3), 3.97 (3H, s, OCH_3-4), 3.98 (3H, s, OCH_3-1), 6.81 (1H, d, J = 8.6 Hz, H-2 or H-3), 6.98 (1H, d, J = 8.6 Hz, H-3 or H-2), 8.12 (1H, J = 1.7 Hz, H-7), 8.92 (1H, d, J = 1.7 Hz, H-5); MS m/z 280 (M^+).

6-Carboxy-8-chloro-1,4-dimethoxynaphthalene. A mixture of ester **19** (2 mg, 0.078 mmol), methanol (6 mL), freshly distilled THF (6 mL), water (2 mL), and potassium hydroxide (0.2 g) was stirred at room temperature for 19 h. Analysis by TLC indicated consumption of the starting material. The reaction mixture was quenched with 2 M aqueous HCl (final pH ~7) and extracted with 10% MeOH/$CHCl_3$ (4 × 100 mL). The combined organic extracts were concentrated by rotary evaporation to give yellow crystals, which were purified by column chromatography (silica gel, 10% MeOH/$CHCl_3$) to afford 20.5 mg (99%) of the desired acid: TLC R_f = 0.06 (10% MeOH/$CHCl_3$); ^1H NMR (300 MHz, 20% $CD_3OD/CDCl_3$) δ 3.85 (3H, s, OCH_3-4), 3.91 (3H, s, OCH_3-1), 6.77 (1H, d, J = 8.6 Hz, H-2 or H-3), 6.95 (1H, d, J = 8.6 Hz, H-3 or H-2), 8.04 (1H, d, J = 1.7 Hz, H-7), 8.87 (1H, d, J = 1.7 Hz, H-5); MS m/z 266 (M^+).

Porphyrin 21. A 19.0 mg (0.07 mmol) portion of 6-carboxy-8-chloro-1,4-dimethoxynaphthalene was dissolved in 5 mL of dry toluene and 2 mL of dry pyridine. The mixture was stirred under argon, and an excess of thionyl chloride (3 drops) was added. After 10 min of stirring, the solvent and excess thionyl chloride were distilled off under reduced pressure and the acid chloride redissolved in 5 mL of dry toluene and 1 mL of dry pyridine. The acid chloride solution was added dropwise over 2 min to a solution of porphyrin **14** (27.4 mg, 0.04 mmol) in dry toluene (5 mL) and dry pyridine (2 mL). The mixture was stirred under argon for 23 h and was then poured into dilute aqueous $NaHCO_3$. The aqueous phase was extracted with chloroform (4 × 100 mL) and the combined organic extracts were concentrated to give a purple solid. The crude product was purified twice on silica gel columns (5% EtOAc/toluene; $CHCl_3$) to afford 14.5 mg (40%) of porphyrin **21:** TLC R_f = 0.39 (40% EtOAc/hexanes); ^1H NMR (500 MHz, 2% CD_3OCDCl_3) δ −2.78 (2H, s, pyrrole–NH), 2.71 (6H, s, 10, 20Ar-CH_3), 3.98 (3H, s, OCH_3-1), 4.06 (3H, s, OCH_3-4), 6.90 (1H, d, J = 8.5 Hz, H-2 or H-3), 7.02 (1H, d, J = 8.5 Hz, H-3 or H-2), 7.56 (4H, d, J = 8.0 Hz, 10, 20ArH-3, 5), 8.03 (2H, d, J = 8.5 Hz, 15ArH-3, 5), 8.09 (4H, d, J = 8.0 Hz, 10, 20ArH-2, 6), 8.13 (2H, d, J = 8.3 Hz, 5ArH-3, 5), 8.21 (1H, d, J = 2.0 Hz, H-7), 8.23 (2H, d, J = 8.5 Hz, 15ArH-2, 6), 8.24 (2H, d, J = 8.3 Hz, 5ArH-2, 6), 8.87 (1H, d, J = 2.0 Hz, H-5), 8.74–8.98 (8H, m, pyrrole–H) (proton assignments were made by analysis of the 2D COSY spectrum).

Dyad 6. Porphyrin **21** (9.1 mg, 0.009 mmol) was dissolved in MeOH (6 mL) and treated with 12.5% aqueous NaOH (2 mL). The reaction mixture was stirred for 30 h. Analysis by TLC indicated consumption of the starting material. The mixture was poured into water (20 mL) and extracted with chloroform (3 × 40 mL). The combined organic extracts were concentrated under reduced pressure to form a purple solid. The product was purified by column chromatography (1.5 × 20 cm silica gel), eluting with a solvent gradient of 30% EtOAc/hexanes through 60% EtOAc/hexanes. The product-containing fractions were collected and concentrated to give 7.3 mg (87%) of the pure aminoporphyrin: TLC R_f = 0.16 (40% EtOAc/hexanes); ^1H NMR (300 MHz, $CDCl_3$) δ −2.74 (2H, brs, pyrrole–NH), 2.70 (6H, s, 10, 20Ar-CH_3), 3.97 (3H, s, OCH_3-1), 4.02 (2H, brs, 15Ar-NH_2), 4.06 (3H, s, OCH_3-4), 6.89 (1H, d, J = 8.6 Hz, H-2 or H-3), 7.01 (1H, d, J = 8.6 Hz, H-3 or H-2), 7.06 (2H, d, J = 8.3 Hz, 15ArH-3, 5), 7.56 (4H, d, J = 7.9 Hz, 10, 20ArH-3, 5), 8.00 (2H, d, J = 8.3 Hz, 15ArH-2, 6), 8.10 (2H, d, J = 8.5 Hz, 5ArH-3, 5), 8.11 (4H, d, J = 7.9 Hz, 10, 20ArH-2, 6), 8.21 (1H, d, J = 1.8 Hz, H-7), 8.25 (2H, d, J = 8.5 Hz, 5ArH-2, 6), 8.34 (1H, brs, 5Ar-NHCO), 8.83–8.95 (9H, m, H-5 and pyrrole-H); MS m/z 920 (M^+).

To a purple solution of the amino porphyrin above (8.5 mg, 0.009 mmol) in dry dichloromethane (2 mL) in a 50-mL round-bottom flask at −78 °C under argon a solution of boron tribromide (13 mL, 0.13 mmol) in dry dichloromethane (1 mL) was added dropwise over 1 min. The resulting green solution was kept at −78 °C for 5 min and then warmed. After 80 min at ambient temperature, the mixture was treated with saturated aqueous sodium bicarbonate until the pH of the water layer was about 7. The aqueous phase was extracted with 10% MeOH–$CHCl_3$ (4×) until the aqueous phase was nearly colorless. The combined organic extracts were concentrated to give a purple solid. The crude product was purified by TLC (silica gel, 1% acetone/CH_2Cl_2) to give 6.4 mg (78%) of dyad **6** as a purple solid: TLC R_f = 0.13 (1% acetone/CH_2Cl_2); ^1H NMR (500 MHz $CDCl_3$) δ −2.84 (2H, s, pyrrole–NH), 2.71 (6H, s, 10, 20Ar-CH_3), 4.04 (2H, brs, 15Ar-NH_2), 6.75 (1H, AB, J = 10.3 Hz, H-2 or H-3), 6.77 (1H, AB, J = 10.3 Hz, H-3 or H-2), 7.08 (2H, d, J = 8.0 Hz, 15ArH-3, 5), 7.56 (4H, d, J = 7.8 Hz, 10, 20ArH-3, 5), 8.02 (2H, d, J = 8.0 Hz, 15ArH-2, 6), 8.06 (2H, d, J = 8.3 Hz, 5ArH-3, 5), 8.11 (4H, d, J = 7.8 Hz, 10, 20ArH-2, 6), 8.16 (1H, s, 5Ar-NHCO), 8.26 (2H, s, H-5 and H-7), 8.28 (2H, d, J =

8.3 Hz, 5ArH-2, 6), 8.83–8.95 (8H, m, pyrrole–H) (proton assignments were made by analysis of the 2D COSY spectrum); UV–vis ($CHCl_3$) λ_{max}: 650, 593, 556, 519, and 420 nm.

Triad 2. To a stirred solution of 7′-apo-7′-(4-carboxyphenyl)-β-carotene (30 mg, 0.056 mmol) in dry toluene (3 mL) and dry pyridine (1.5 mL) under argon was added an excess of $SOCl_2$ (3 drops). After 10 min of stirring, the solvent and excess $SOCl_2$ were evaporated under vacuum, and the acid chloride was redissolved in 2 mL of dry toluene and 1 mL of dry pyridine. Porphyrin-quinone **6** (5 mg, 0.0056 mmol) was dissolved in dry toluene/pyridine (3:1, 4 mL) and treated with the acid chloride solution. The mixture was kept well stirred under argon for 4.5 h and then quenched with water. The aqueous phase was extracted with 10% MeOH–$CHCl_3$ (3 × 100 mL) and the combined organic extracts were concentrated to give a red solid. Column chromatography on silica gel (1% acetone/CH_2Cl_2) gave 3.5 mg (44%) of the desired triad **2** as a red solid: TLC R_f = 0.19 (1% acetone/CH_2Cl_2); ^{1}H NMR (500 MHz, $CDCl_3$) δ −2.82 (2H, brs, pyrrole–NH), 1.04 (6H, s, CH_3-16C and CH_3-17C), 1.46–1.51 (2H, m, CH_2-2C), 1.58–1.66 (2H, m, CH_2-3C), 1.73 (3H, s, CH_3-18C), 1.98 (3H, s, CH_3-19C), 2.00 (3H, s, CH_3-20C), 2.02 (3H, s, CH_3-20′C), 2.04 (2H, m, CH_2-4C), 2.10 (3H, s, CH_3-19′C), 2.72 (6H, s, 10, 20Ar-CH_3), 6.15 (1H, AB, J = 15.8 Hz, H-8C), 6.17 (1H, d, J = 9.5 Hz, H-10C), 6.19 (1H, d, J = 15.8 Hz, H-7C), 6.28 (1H, d, J = 10.5 Hz, H-14C), 6.34 (1H, d, J = 11.0 Hz, H-14′C), 6.37 (1H, d, J = 15.5 Hz, H-12C), 6.45 (1H, d, J = 11.5 Hz, H-10′C), 6.49 (1H, d, J = 14.5 Hz, H-12′C), 6.60–6.74 (5H, m, H-8′C, H-11C, H-11′C, H-15C and H-15′C), 6.98 (2H, brs, H-2Q and H-3Q), 7.06 (1H, d, J = 16.0 Hz, H-8′C), 7.57 (4H, d, J = 8.0 Hz, 10, 20ArH-3, 5), 7.61 (2H, d, J = 8.0 Hz, H-1′C and H-5′C), 7.99 (2H, d, J = 8.0 Hz, H-2′C and H-4′C), 8.07 (2H, d, J = 8.3 Hz, 15ArH-3, 5), 8.09 (2H, d, J = 8.5 Hz, 5ArH-3, 5), 8.10 (4H, d, J = 8.0 Hz, 10, 20ArH-2, 6), 8.16 (1H, s, Ar-NHCO), 8.24 (2H, d, J = 8.3 Hz, 15ArH-2, 6), 8.28 (2H, d, J = 8.5 Hz, 5ArH-2, 6), 8.44 (1H, brs, H-7Q), 8.51 (1H, brs, H-5Q), 8.82–8.90 (8H, brs, pyrrole–H) (proton assignments were made by analysis of the 2D COSY spectrum); UV–vis ($CHCl_3$) λ_{max}: 647, 591, 514, 483, 422, 373, and 302 nm.

6-Carbomethoxy-1,4-dimethoxynaphthalene. Sodium dithionite (4.03 g, 23.1 mmol) in 20-mL water was added dropwise to a solution of 6-carbomethoxy-1,4-naphthoquinone (1.0 g, 4.6 mmol) in freshly distilled THF (60 mL). The mixture was stirred for 60 min and then quenched with dilute aqueous HCl (final pH ~5). The aqueous phase was extracted with chloroform (4 × 100 mL). The combined organic extracts were concentrated by rotary evaporation and dried for 2 h under vacuum. The resulting brown solid, the hydroquinone, was redissolved in 100 mL of dry acetone and treated with K_2CO_3 (6.4 g, 46.3 mmol), followed by dimethyl sulfate (2.2 mL, 23.1 mmol).

The reaction was allowed to proceed for 39 h with stirring under argon. The progress of the reaction was monitored by TLC. One more addition of dimethyl sulfate (2 ml, 21 mmol) was necessary. The workup was effected by extraction ($CHCl_3/H_2O$). Purification by column chromatography (silica gel, hexanes/EtOAc = 6:1) gave the pure desired ester (835 mg, 73%) as white crystals: TLC R_f = 0.59 (CH_2Cl_2); ^1H NMR (300 MHz, $CDCl_3$) δ 3.93 (3H, s, CO_2CH_3), 3.95 (3H, s, OCH_3-1), 3.96 (3H, s, OCH_3-4), 6.69 (1H, d, J = 8.4 Hz, H-2 or H-3) 6.76 (1H, d, J = 8.4 Hz, H-3 or H-2), 8.07 (1H, dd, J = 8.8, 1.8 Hz, H-7), 8.23 (1H, d, J = 8.8 Hz, H-8), 8.95 (1H, d, J = 1.8 Hz, H-5); MS m/z 246 (M^+).

6-Carboxy-1,4-dimethoxynaphthalene (25). A mixture of the above ester (177 mg, 0.720 mmol), methanol (30 mL), freshly distilled THF (30 mL), water (10 mL), and KOH (1.5 g, 26.7 mmol) was stirred at room temperature for 31 h. Analysis by TLC indicated consumption of the starting material. The reaction mixture was quenched with 2 M aqueous HCl (final pH ~1) and extracted with 10% MeOH/$CHCl_3$ (3 × 100 mL). The combined organic extracts were concentrated by rotary evaporation to give 168 mg of white solid **25** in quantitative yield: TLC R_f = 0.14 (10% MeOH/$CHCl_3$); ^1H NMR (300 MHz, 3% $CD_3OD/CDCl_3$) δ 3.88 (3H, s, OCH_3-1), 3.89 (3H, s, OCH_3-4), 6.67 (1H, d, J = 8.4 Hz, H-2 or H-3), 6.74 (1H, d, J = 8.4 Hz, H-3 or H-2), 7.99 (1H, dd, J = 8.8, 1.6 Hz, H-7), 8.14 (1H, d, J = 8.8 Hz, H-8), 8.90 (1H, d, J = 1.6 Hz, H-5); MS m/z 231 (M^+ – 1).

Porphyrin 27. A 24.0-mg (0.10 mmol) portion of 6-carboxy-1,4-dimethoxynaphthalene (**25**) was dissolved in 5 mL of dry toluene and 0.5 mL of dry pyridine. The mixture was stirred under argon and an excess of thionyl chloride (5 drops) was added. After 10 min of stirring the solvent and excess thionyl chloride were distilled under reduced pressure and the acid chloride was redissolved in 5 mL of dry toluene and 1 mL of dry pyridine. The acid chloride solution was added dropwise over 2 min to a solution of porphyrin **14** (53.0 mg, 0.069 mmol) in dry toluene (4 mL) and dry pyridine (0.5 mL). The mixture was stirred under argon for 22 h and was then poured into diluted aqueous $NaHCO_3$. The aqueous phase was extracted with chloroform (4 × 100 mL) and the combined organic extracts were concentrated to give a purple solid. Flash column chromatography on silica gel (CH_2Cl_2) gave 57.3 mg (85%) of porphyrin **27** as a purple solid: TLC R_f = 0.07 (CH_2Cl_2); ^1H NMR (300 MHz, $CDCl_3$) δ –2.77 (2H, s, pyrrole–NH), 2.69 (6H, s, 10, 20Ar-CH_3), 3.99 (3H, s, OCH_3-1), 4.04 (3H, s, OCH_3-4), 6.80 (1H, AB, J = 8.5 Hz, H-2 or H-3), 6.83 (1H, AB, J = 8.5 Hz, H-3 or H-2), 7.52 (4H, d, J = 8.1 Hz, 10, 20ArH-3, 5), 7.97 (2H, d, J = 8.5 Hz, 15 ArH-3, 5), 8.06 (4H, d, J = 8.1 Hz, 10, 20ArH-2, 6), 8.09 (2H, d, J = 8.5 Hz, 5ArH-3, 5), 8.16 (1H, dd, J = 8.8, 1.9 Hz, H-7), 8.22 (2H, d, J = 8.5 Hz, 15ArH-2, 6), 8.24 (2H, d, J = 8.5 Hz,

5ArH-2, 6), 8.33 (1H, brs, Ar-NHCO), 8.41 (1H, d, J = 8.8 Hz, H-8), 8.42 (1H, brs, Ar-NHCO), 8.78–8.92 (9H, m, H-5 and pyrrole–H); MS m/z 982 (M^+).

Dyad 7. Porphyrin **27** (54.6 mg, 0.056 mmol) was dissolved in MeOH/THF (1:1, 60 mL) and treated with 16% aqueous NaOH (10 mL). The reaction mixture was stirred and monitored by TLC; after 24 h, no starting material remained. The mixture was poured into water (100 mL) and extracted with chloroform (3 × 200 mL). The combined organic extracts were concentrated under reduced pressure to yield a purple solid. The product was chromatographed on silica gel eluting with dichloromethane to give 46.9 mg of the desired aminoporphyrin: TLC R_f = 0.21 (40% EtOAc/hexanes); ^{1}H NMR (300 MHz, $CDCl_3$) δ –2.70 (2H, brs, pyrrole–NH), 2.70 (6H, s, 10, 20Ar-CH_3), 3.95 (2H, brs, 15Ar-NH_2), 3.97 (3H, s, OCH_3-1), 4.01 (3H, s, OCH_3-4), 6.75 (1H, AB, J = 8.4 Hz, H-2 or H-3), 6.78 (1H, AB, J = 8.4 Hz, H-3 or H-2), 7.02 (2H, d, J = 8.3 Hz, 15ArH-3, 5), 7.54 (4H, d, J = 7.9 Hz, 10, 20ArH-3, 5), 7.99 (2H, d, J = 8.3 Hz, 15ArH-2, 6), 8.10 (2H, d, J = 8.4 Hz, 5ArH-3, 5), 8.11 (4H, d, J = 7.9 Hz, 10, 20ArH-2, 6), 8.17 (1H, dd, J = 8.8, 1.8 Hz, H-7), 8.24 (2H, d, J = 8.4 Hz, 5ArH-2, 6), 8.39 (1H, d, J = 8.8 Hz, H-8), 8.45 (1H, brs, 5Ar-NHCO), 8.86–8.97 (9H, m, H-5 and pyrrole–H).

A solution of boron tribromide (98 mL, 1.04 mmol) in dry dichloromethane (2 mL) was added dropwise over 9 min to a solution of the aminoporphyrin (45.9 mg, 0.052 mmol) in dry dichloromethane (8 mL) in a 50-mL round-bottom flask at –78°C under argon. The resulting green solution was kept at –78 °C for 60 min and then allowed to warm. After 3 h at room temperature, the mixture was treated with saturated aqueous sodium bicarbonate until the pH of the water layer was about 8. The aqueous phase was extracted with 10% MeOH-$CHCl_3$ (4×) until the aqueous phase was nearly colorless. The combined organic extracts were concentrated to give a purple solid. This porphyrin was then dissolved in 20 mL of chloroform and stirred with PbO_2 (~5 mg) in the dark for 30 min. After removal of the PbO_2 by centrifugation, the red-purple solution was concentrated to give a purple solid. Flash column chromatogaphy of the crude product on silica ($CHCl_3$ to 15% MeOH–$CHCl_3$) gave 25.4 mg (57%) of pure dyad **7**: ^{1}H NMR (500 MHz, $CDCl_3$) δ –2.79 (2H, s, pyrrole–NH), 2.71 (6H, s, 10, 20Ar-CH_3), 4.03 (2H, brs, 15Ar-NH_2), 6.95 (1H, AB, J = 10.3 Hz, H-2 or H-3), 6.96 (1H, AB, J = 10.3 Hz, H-3 or H-2), 7.07 (2H, d, J = 8.0 Hz, 15 ArH-3, 5), 7.56 (4H, d, J = 7.8 Hz, 10, 20ArH-3, 5), 8.00 (2H, d, J = 8.0 Hz, 15ArH-2, 6), 8.07 (2H, d, J = 8.3 Hz, 5ArH-3, 5), 8.11 (4H, d, J = 7.8 Hz, 10, 20ArH-2, 6), 8.12 (1H, d, J = 7.7 Hz, H-8), 8.23 (1H, s, 5Ar-NHCO), 8.27 (2H, d, J = 8.3 Hz, 5ArH-2, 6), 8.37 (1H, dd, J = 7.7, 1.5 Hz, H-7), 8.49 (1H, d, J = 1.5 Hz, H-5), 8.84–8.96 (8H, m, pyrrole-H) (proton assignments were made by analysis of the 2D COSY spectrum).

Triad 3. A 14-mg (0.026 mmol) portion of 7′-apo-7′-(4-carboxyphenyl)-β-carotene was dissolved in 4 mL of dry toluene and 1.0 mL of dry pyridine. The mixture was stirred under argon and an excess of thionyl chloride (1 drop) was added. After 10 min of stirring, the solvent was distilled under reduced pressure and the acid chloride was redissolved in 6 mL of dry toluene and 0.2 mL of dry pyridine. The acid chloride solution was added dropwise over 2 min to a solution porphyrin-quinone **7** (12 mg, 0.014 mmol) in dry toluene/pyridine (4:1, 7.5 mL). The mixture was stirred under argon for 3 h and was then poured into diluted aqueous $NaHCO_3$. The aqueous phase was extracted with chloroform (4 × 100 mL) and the combined organic extracts were concentrated to give a red–purple solid. Flash column chromatography on silica gel (1% acetone/CH_2Cl_2) yielded 7 mg (36%) of the desired triad **3:** TLC R_f = 0.49 (2% acetone/CH_2Cl_2); ^1H NMR (500 MHz, $CDCl_3$) δ −2.89 (2H, s, pyrrole-NH), 1.02 (6H, s, CH_3-16C and CH_3-17C), 1.43–1.49 (2H, m, CH_2-2C), 1.57–1.65 (2H, m, CH_2-3C), 1.71 (3H, s, CH_3-18C), 1.96 (3H, s, CH_3-19C), 1.98 (3H, s, CH_3-20C), 2.00 (3H, s, CH_3-20′C), 2.01 (2H, m, CH_2-4C), 2.07 (3H, s, CH_3-19′C), 2.69 (6H, s, 10, 20Ar-CH_3), 6.13 (1H, AB, J = 15.5 Hz, H-8C), 6.15 (1H, d, J = 9.5 Hz, H-10C), 6.17 (1H, AB, J = 15.5 Hz, H-7C), 6.25 (1H, d, J = 10.5 Hz, H-14C), 6.32 (1H, d, J = 10.0 Hz, H-14′C), 6.35 (1H, d, J = 14.5 Hz, H-12C), 6.42 (1H, d, J = 11.0 Hz, H-10′C), 6.46 (1H, d, J = 14.5 Hz, H-12′C), 6.58–6.72 (4H, m, H-11C, H-11′C, H-15C and 15′C), 6.62 (1H, d, J = 15.3 Hz, H-7′C), 6.82 (1H, AB, J = 10.3 Hz, H-2Q or H-3Q), 6.84 (1H, AB, J = 10.3 Hz, H-3Q or H-2Q), 7.02 (1H, d, J = 15.3 Hz, H-8′C), 7.54 (4H, d, J = 7.5 Hz, 10, 20ArH-3, 5), 7.56 (2H, d, J = 8.0 Hz, H-1′C and H-5′C), 7.94 (2H, d, J = 8.0 Hz, H-2′C and H-4′C), 7.97 (1H, d, J = 7.5 Hz, H-8Q), 8.04 (2H, d, J = 8.3 Hz, 15ArH-3, 5), 8.06 (2H, d, J = 8.3 Hz, 5ArH-3, 5), 8.07 (4H, d, J = 7.5 Hz, 10, 20ArH-2, 6), 8.14 (1H, s, Ar-NHCO), 8.21 (2H, d, J = 8.3 Hz, 15ArH-2, 6), 8.24 (1H, s, Ar-NHCO), 8.25 (2H, d, J = 8.3 Hz, 5ArH-2, 6), 8.29 (1H, d, J = 7.5 Hz, H-7Q), 8.38 (1H, brs, H-5Q) 8.85 (8H, brs, pyrrole-H) (proton assignments were made by analysis of the 2D COSY spectra); UV–vis ($CHCl_3$) λ_{max}: 649, 591, 515, 484, 422, 375, and 305 nm.

2-(2′,5′-Dimethoxybenzyl)butanedioic Anhydride. To a stirred solution of acetic anhydride (500 mL), was added 3-carboxy-3-(2′,5′-dimethoxybenzyl)butanedioic acid (7.8 g, 25.0 mmol) at ambient temperature. The solution was allowed to reflux for 3 h and then cooled, vacuum filtered, and washed with hexane (200 mL). The resulting solution was evaporated at reduced pressure to give a yellow oil, which was crystallized from 50% hexanes/$CHCl_3$ at −5 °C to yield 3.68 g (59%) of the desired product as white crystals: ^1H NMR (300 MHz, $CDCl_3$) δ 2.73 (1H, dd, J = 19.0, 6.7 Hz, CHAr or CH′Ar), 2.86 (1H, dd, J = 19.0, 9.6 Hz, CH′Ar or CHAr), 2.96 (1H, dd, J = 13.9, 8.1 Hz, H-3 or H-3′), 3.26 (1H, dd, J = 13.9, 5.6 Hz, H-3′ or H-3), 3.44–3.54 (1H, m, H-2), 3.75 (3H, s, ArOCH_3-5′), 3.77 (3H, s, ArOCH_3-2′), 6.69 (1H, d, J =

2.3 Hz, ArH-3′ or ArH-4′), 6.78 (1H, d, J = 2.3 Hz, ArH-4′ or ArH-3′), 6.79 (1H, s, ArH-6′); MS m/z 250 (M^+).

3-Carboxy-3,4-dihydro-2*H*-5,8-dimethoxy-1-oxo-naphthalene. To polyphosphoric acid (15 mL) at 85 °C under argon was added the above anhydride (820 mg, 3.28 mmol), and this mixture was stirred at 85–90 °C for 1 h. The resulting brownish-red viscous liquid was then poured into an ice water bath and stirred with a glass rod until no more precipitate formed. The liquid was removed by filtration, leaving a brownish-green solid. The crude solid was purified by chromatography over a 2 × 20 cm silica gel column, eluting with a solvent gradient of 5% MeOH/$CHCl_3$ through 20% MeOH/$CHCl_3$. The product-containing fractions were collected and concentrated to yield 748 mg (91%) of the ketoacid as a pale yellow solid: ^1H NMR (300 MHz, 25% $CD_3OD/CDCl_3$) δ 2.52–2.77 (3H, m, H-2 or H-2′, H4 and H-4′), 2.78–2.90 (1H, m, H-3), 3.18 (1H, dd, J = 16.9, 3.0 Hz, H-2′ or H-2), 3.63 (3H, s, OCH_3-5), 3.64 (3H, s, OCH_3-8), 6.66 (1H, d, J = 9.1 Hz, H-6 or H-7), 6.88 (1H, d, J = 9.1 Hz, H-6 or H-7); MS m/z 205 (M^+–CO_2H).

3-Carbomethoxy-3,4-dihydro-2*H*-5,8-dimethoxy-1-oxo-naphthalene (17). A solution of the keto acid above (748 mg, 2.99 mmol) in THF (10 mL) under argon was treated with diazomethane etherate while stirring at 0 °C. The reaction mixture was stirred for 1 h at 0 °C, warmed up to room temperature, and then quenched with water (20 mL). The aqueous layer was extracted with chloroform (3 × 20 mL). The combined organic extracts were concentrated to provide a yellow solid. The crude product was chromatographed on silica gel eluting with chloroform to give 587 mg (75%) of **17** as a pale yellow solid: TLC R_f = 0.58 (10% acetone/CH_2Cl_2); ^1H NMR (300 MHz, $CDCl_3$) δ 2.66–2.87 (3H, m, H-2 or H-2′, H-4 and H-4′), 2.95–3.06 (1H, m, H-3), 3.31 (1H, ddd, J = 17.0, 4.3, 1.6 Hz, H-2′ or H-2), 3.67 (3H, s, CO_2CH_3), 3.77 (3H, s, OCH_3-5), 3.80 (3H, s, OCH_3-8), 6.77 (1H, d, J = 9.1 Hz, H-6 or H-7), 6.96 (1H, d, J = 9.1 Hz, H-6 or H-7); MS m/z 264 (M^+).

6-Carbomethoxy-1,4,8-trimethoxynaphthalene (22). To a stirred solution of keto ester **17** (162 mg, 0.16 mmol) in dry methanol (10 mL) that had been saturated with argon were added DDQ (175 mg, 0.77 mmol) and trimethylorthoformate ($HC(OMe)_3$) (136 mg, 1.28 mmol), and the mixture was then warmed to reflux for 3 h. The reaction progress was monitored through this period by TLC, and more DDQ and $HC(OMe)_3$ were added as required. The resulting solution was cooled to room temperature and then poured into a separatory funnel containing water (40 mL) and chloroform (50 mL). The layers were separated and the aqueous layer was extracted with chloroform (3 × 50 mL). The combined organic extracts were concentrated under vacuum to give a light brown solid, which was purified twice on flash silica gel columns

($CHCl_3$; CH_2Cl_2) to yield 137 mg (81%) of the desired ester as a pale yellow solid: TLC R_f = 0.26 ($CHCl_3$); ^{1}H NMR (300 MHz, $CDCl_3$) δ 3.89 (3H, s, OCH_3-1), 3.93 (3H, s, CO_2CH_3), 3.94 (3H, s, OCH_3-4), 4.01 (3H, s, OCH_3-8), 6.73 (1H, d, J = 8.6 Hz, H-2 or H-3), 6.86 (1H, d, J = 8.6 Hz, H-3 or H-2), 7.45 (1H, J = 1.6 Hz, H-7), 8.60 (1H, d, J = 1.6 Hz, H-5); MS *m/z* 276 (M^+).

6-Carboxy-1,4,8-trimethoxynaphthalene (23). A mixture of **22** (787 mg, 2.85 mmol), methanol (30 mL), freshly distilled THF (75 mL), water (15 mL), and potassium hydroxide (480 mg, 8.55 mmol) was stirred at room temperature for 30 h. Analysis by TLC indicated consumption of the starting material. The reaction mixture was quenched with 2 M aqueous HCl (final pH ~ 2) and extracted with 5% MeOH/$CHCl_3$ (3 × 100 mL). The combined organic extract were concentrated by rotary evaporation to give a peach-colored solid that was washed with 15% $CHCl_3$/hexanes (200 mL) to afford 733 mg (98%) of **23:** ^{1}H NMR (300 MHz, $CDCl_3$/CD_3OD) δ 3.83 (3H, s, OCH_3-1), 3.89 (3H, s, OCH_3-4), 3.94 (3H, s, OCH_3-8), 6.72 (1H, d, J = 8.6 Hz, H-2 or H-3), 6.85 (1H, d, J = 8.6 Hz, H-3 or H-2), 7.40 (1H, s, H-7), 8.57 (1H, s, H-5); MS *m/z* 262 (M^+).

Porphyrin 24. A 25.2 mg (0.10 mmol) portion of acid **23** was dissolved in 6 mL of dry toluene and 1.0 mL of dry pyridine. The mixture was stirred under argon and an excess of thionyl chloride (4 drops) was added. The formation of the acid chloride was instantaneous. The solvent and excess thionyl chloride were distilled under reduced pressure and the acid chloride was redissolved in 4 mL of dry toluene. The acid chloride solution was added dropwise over 2 min to a solution of porphyrin **14** (49.3 mg, 0.064 mmol) in dry toluene (4 mL) and dry pyridine (1.5 mL). The mixture was kept well stirred under argon for 2 h and was then poured into diluted aqueous $NaHCO_3$. The aqueous phase was extracted with chloroform (4 × 100 mL) and the combined organic extracts were concentrated to give a purple solid. Flash column chromatography on silica gel (hexanes to 40% EtOAc/hexanes) gave 52.8 mg (81%) of porphyrin **24** as the major product: TLC R_f = 0.32 (40% EtOAc/hexanes); ^{1}H NMR (300 MHz, $CDCl_3$) δ –2.76 (2H, s, pyrrole–NH), 2.69 (6H, s, 10, 20Ar-CH_3), 3.97 (3H, s, OCH_3-1), 4.04 (3H, s, OCH_3-4), 4.14 (3H, s, OCH-8), 6.85 (1H, AB, J = 8.6 Hz, H-2 or H-3), 6.93 (1H, AB, J = 8.6 Hz, H-3 or H-2), 7.53 (4H, d, J = 7.8 Hz, 10, 20ArH-3,5), 7.63 (1H, d, J = 1.6 Hz, H-7), 7.97 (2H, d, J = 8.4 Hz, 15ArH-3, 5), 8.07 (4H, d, J = 7.8 Hz, 10, 20ArH-2, 6), 8.09 (2H, d, J = 8.3 Hz, 5ArH-3, 5), 8.22 (2H, d, J = 8.4 Hz, 15ArH-2, 6), 8.24 (2H, d, J = 8.3 Hz, 5ArH-2, 6), 8.37 (1H, brs, Ar-NHCO), 8.47 (1H, d, J = 1.6 Hz, H-5), 8.49 (1H, brs, Ar-NHCO), 8.77–8.94 (8H, m, pyrrole–H); MS *m/z* 1012 (M^+).

Dyad 8. Porphyrin **24** (29.5 mg, 0.03 mmol) was dissolved in MeOH/THF (1:1, 30 mL) and treated with 17% aqueous NaOH (5 mL). The reaction mixture was stirred at ambient temperature and was monitored by TLC; after 23 h no starting material remained. The mixture was poured into water (50 mL) and extracted with chloroform (3 × 100 mL). The combined organic extracts were concentrated under reduced pressure to give a purple solid. Flash column chromatography of the crude product on silica gel (hexane ~50% EtOAc/hexane) gave 26.5 mg (99%) of the pure aminoporphyrin: TLC R_f = 0.23 (50% EtOAc/hexanes); ^{1}H NMR (300 MHz, $CDCl_3$) δ –2.75 (2H, brs, pyrrole–NH), 2.70 (6H, s, 10, 20Ar-CH_3), 3.97 (3H, s, OCH_3-1), 4.02 (2H, s, 15Ar-NH_2), 4.06 (3H, s, OCH_3-4), 4.14 (3H, s, OCH_3-8), 6.87 (1H, AB, J = 8.6 Hz, H-2 or H-3), 6.95 (1H, AB, J = 8.6 Hz, H-3 or H-2), 7.06 (2H, d, J = 8.2 Hz, 15ArH-3, 5), 7.56 (4H, d, J = 8.0 Hz, 10, 20ArH-3, 5), 7.63 (1H, d, J = 1.6 Hz, H-7), 8.00 (2H, d, J = 8.2 Hz, 15ArH-2, 6), 8.10 (4H, d, J = 8.0 Hz, 10, 20ArH-2, 6), 8.12 (2H, d, J = 8.4 Hz, 5ArH-3, 5), 8.25 (2H, d, J = 8.4 Hz, 5ArH-2, 6), 8.45 (1H, brs, 5Ar-NHCO), 8.46 (1H, d, J = 1.6 Hz, H-5), 8.84–8.95 (8H, m, pyrrole–H); MS m/z 916 (M^+).

A solution of boron tribromide (111 μL, 1.17 mmol) in dry dichloromethane (5 mL) was added dropwise over 9 min to a purple solution of the aminoporphyrin (53.6 mg, 0.059 mmol) in dry dichloromethane (5 mL) at –78 °C under argon. The resulting green solution was kept at –78 °C for 10 min and then warmed. After 80 min at room temperature, the mixture was treated with saturated aqueous sodium bicarbonate until the pH of the water layer was about 7. The aqueous phase was nearly colorless. The organic phase was concentrated and dried for several hours under vacuum to give a brown-red solid. This porphyrin was then dissolved in 20 mL of chloroform and treated with PbO_2 (~5 mg) in the dark for 40 min. After removal of the PbO_2 by centrifugation, the red solution was concentrated to give a brown–red solid. Flash column chromatography of the crude product on silica gel ($CHCl_3$ to 15% MeOH/$CHCl_3$) gave 30.4 mg (60%) of pure porphyrin quinone **8** as a brown-purple solid: ^{1}H NMR (300 MHz, $CDCl_3$) δ –2.84 (2H, s, pyrrole–NH), 2.70 (6H, s, 10, 20Ar-CH_3), 4.04 (2H, brs, 15Ar-NH_2), 6.80 (1H, AB, J = 10.4 Hz, H-2 or H-3), 6.84 (1H, AB, J = 10.4 Hz, H-3 or H-2), 7.08 (2H, d, J = 8.3 Hz, 15ArH-3, 5), 7.56 (4H, d, J = 7.9 Hz, 10, 20ArH-3, 5), 7.80 (1H, s, H-7), 7.89 (1H, s, H-5), 8.01 (2H, d, J = 8.3, Hz, 15ArH-2, 6), 8.06 (2H, d, J = 8.4 Hz, 5ArH-3, 5), 8.12 (4H, d, J = 7.9 Hz, 10, 20ArH-2, 6), 8.13 (1H, brs, 5Ar-NHCO), 8.27 (2H, d, J = 8.4 Hz, 5ArH-2, 6), 8.84–8.96 (8H, m, pyrrole–H), 11.64 (1H, s, OH-8) (proton assignments were made by analysis of the 2D COSY spectra).

Triad 4. An excess of $SOCl_2$ (1 drop) was added to a stirred solution of 7′-apo-7′-(4-carboxyphenyl)-β-carotene (10.0 mg, 0.019 mmol) in dry toluene (3 mL) and dry pyridine (1.5 mL), under argon. After 6 min of stirring, the solvent and excess $SOCl_2$ were evaporated under vacuum, and the acid chloride was redissolved in 2 mL of dry toluene. Porphyrin-quinone **8** (11.3 mg, 0.013 mmol) was dissolved in dry toluene/pyridine (4:1, 5 mL) and treated with the acid chloride solution. The mixture was kept well stirred under argon for 12 min and then quenched with saturated aqueous $NaHCO_3$. The aqueous phase was extracted with chloroform (4 × 100 mL) and the combined organic extracts were concentrated to give a red solid. Flash column chromatography on silica gel ($CHCl_3$) gave 7.5 mg (42%) of the desired triad **4** as a red–purple solid: TLC R_f = 0.15 (0.5% MeOH/$CHCl_3$); ^{1}H NMR (500 MHz, 15% $CD_3OD/CDCl_3$) δ 1.04 (6H, s, CH_3-16C and CH_3-17C), 1.46–1.51 (2H, m, CH_2-2C), 1.58–1.66 (2H, m, CH_2-3C), 1.73 (3H, s, CH_3-18C), 1.99 (3H, s, CH_3-19C), 2.00 (3H, s, CH_3-20C), 2.02 (3H, s, CH_3-20′C), 2.03 (2H, m, CH_2-4C), 2.11 (3H, s, CH_3-19′C), 2.72 (6H, s, 10, 20Ar-CH_3), 6.15 (1H, d, *J* = 16.5 Hz, H-8C), 6.17 (1H, d, *J* = 10.5 Hz, H-10C), 6.20 (1H, d, *J* = 16.5 Hz, H-7C), 6.28 (1H, d, *J* = 10.5 Hz, H-14C), 6.35 (1H, d, *J* = 12.5 Hz, H-14′C), 6.38 (1H, d, *J* = 15.0 Hz, H-12C), 6.45 (1H, d, *J* = 12.0 Hz, H-10′C), 6.49 (1H, d, *J* = 15.0 Hz, H-12′C), 6.61–6.75 (4H, m, H-11C, H-11′C, H-15C and H-15′C), 6.68 (1H, d, *J* = 16.0 Hz, H-7′C), 6.99 (2H, brs, H-2Q and H-3Q), 7.07 (1H, d, *J* = 16.0 Hz, H-8′C), 7.58 (4H, d, *J* = 7.8 Hz, 10, 20ArH-3, 5), 7.63 (2H, d, *J* = 8.0 Hz, H-1′C and H-5′C), 7.96 (2H, brs, H-5Q and H-7Q), 8.04 (2H, d, *J* = 8.0 H, H-2′C and H-4′C), 8.11 (4H, d, *J* = 7.8 Hz, 10, 20ArH-2, 6), 8.11 (2H, d, *J* = 8.3 Hz, 15ArH-3, 5), 8.15 (2H, d, *J* = 8.5 Hz, 5ArH-3, 5), 8.24 (2H, d, *J* = 8.3 Hz, 15ArH-2, 6), 8.25 (2H, d, *J* = 8.5 Hz, 5ArH-2, 6), 8.90 (8H, brs, pyrrole–H) (proton assignments were made by analysis of the 2D COSY spectra); UV–vis ($CHCl_3$) $\lambda_{max:}$ 649, 592, 515, 483, 422, 375, and 303 nm.

1-Carboxy-1,2,3,4-tetrahydro-1,4-methanoanthracene-9,10-dione (29). Diene **9** (15 g, 0.12 mmol), prepared as described above, was allowed to drip into a solution of 30 g (0.19 mmol) naphthoquinone in 500 mL of dichloromethane. The resulting green solution was stirred at room temperature under nitrogen for 48 h. Flash column chromatography of the crude product on silica gel (dichloromethane to 3% acetone-CH_2Cl_2) gave two fractions. The less polar fraction was further purified by chromatography (silica gel, 20–25% EtOAc/hexanes) to yield 5 g (15%) of the desired product: 1-carbomethoxy-1,4,4a,9a-tetrahydro-1,4-methanoanthracene-9,10-dione.

To a 200-mL round-bottom flask equipped with a stirring bar and an addition funnel was added 5 g (17.7 mmol) of the diketone prepared above dissolved in 60 mL of acetone and 12 mL of dimethyl sulfate (DMS). The mixture was stirred while a solution of 20% KOH in methanol was added at the rate of 1–2 drops/10 s. TLC indicated that as the starting material was being consumed, both a more polar intermediate and a less polar product began to appear. After the addition of an additional 10 mL DMS, the reaction mixture was poured into ether and washed with water (2 ×). The organic extracts were concentrated and residual DMS was removed under high vacuum. The dark oily residue that remained was chromatographed on silica gel (flash column

10% EtOAc/ hexanes) to afford 4.60 g (84%) of the desired 1-carbomethoxy-1,4-dihydro-9,10-dimethoxy-1,4-methanoanthracene. This yellow material slowly solidified upon standing.

To a 250-mL hydrogenation bottle was added 4 g (12.9 mmol) of the olefin prepared above 80 mL of methanol, and 0.4 g of 10% Pd on carbon. The system was flushed with hydrogen before shaking the suspension under a hydrogen pressure of 40 psi for 4 h. TLC indicated that the reaction was complete. The solution was filtered through Celite and the solvent was evaporated. Flash column chromatography (silica gel, 10% hexanes/CH_2Cl_2) afforded 4.0 g of the desired ester (1-carbomethoxy-1,2,3,4-tetrahydro-9,10-dimethoxy-1,4-methanoanthracene) in 99% yield.

The ester above (0.8 g, 2.6 mmol), 20 mL of THF, 5 mL of methanol, 3 mL of water, and 0.32 g (7.7 mmol) of LiOH monohydrate were stirred under nitrogen at 40 °C for 48 h. The reaction product was poured into a mixture of chloroform and diluted HCl. The aqueous phase was washed several times with chloroform, the organic extracts were back-washed with brine, and the solvent removed under vacuum. The product was of high enough purity to be used in the next step without further purification.

To a 200-mL round-bottom flask equipped with a stirring bar and an addition funnel were added the crude product of the reaction above (*ca.* 2.5 mmol) and 15 mL of acetonitrile. The mixture was stirred under nitrogen and 4.1 g (7.5 mmol) of ceric ammonium nitrate (CAN) dissolved in 15 mL of water was added dropwise during approximately 7 min. Initially, the white precipitate dissolved, forming a yellow solution. However, by the time all the oxidant was added, an intense yellow crystalline solid was formed. The stirring was continued for an additional 15 min. The reaction product was poured into a mixture of chloroform and dilute HCl. The aqueous layer was washed with chloroform (4 × 60 mL). The combined extracts were back-washed with water, and the solvent was evaporated to yield 0.68 g (98%) of pure **29** as a yellow powder. TLC $R_f = 0.5$ (6% MeOH/CH_2Cl_2, trace of acetic acid); ^1H NMR (300 MHz, 5% $CD_3OD/CDCl_3$) δ 1.32 (1H, m, H-3n), 1.57 (1H, m, H-2n), 1.81(1H, m. H-11a), 2.06 (1H, m, H-11s), 2.15 (1H, m, H-3x), 2.33 (1H, m, H-2x), 3.64 (1H, brs, H-4), 7.66 (2H, m, H-5,8), 8.01 (2H, d, H-6,7).

Acknowledgment

This work was supported by a grant from the U.S. Department of Energy (DE–FG03–93ER14404) and the Petroleum Research Fund (23911 AC4).

References

1. Gust, D.; Moore, T. A.; Moore, A. L. *Acc. Chem. Res.* **1993,** *26,* 198–205.
2. Wasielewski, M. R. *Chem. Rev.* **1992,** 92, 435–461.

3. Moore, T. A.; Gust, D.; Hatlevig, S.; Moore, A. L.; Makings, L. R.; Pessiki, P. J.; DeSchryver, F. C.; Van der Auweraer, M.; Lexa, D.; Bensasson, R. V.; Rougée, M. *Isr. J. Chem.* **1988,** *28,* 87–95.
4. Gust, D.; Moore, T. A.; Moore, A. L.; Barrett, D.; Harding, L. O.; Makings, L. R.; Liddell, P. A.; DeSchryver, F. C.; Van der Auweraer, M.; Bensasson, R. V.; Rougée, M. *J. Am. Chem. Soc.* **1988,** *110,* 321–323.
5. Gust, D.; Moore, T. A.; Moore, A. L.; Lee, S.-J.; Bittersmann, E.; Luttrull, D. K.; Rehms, A. A.; DeGraziano, J. M.; Ma, X. C.; Gao, F.; Belford, R. E.; Trier, T. T. *Science (Washington D.C.),* **1990,** *248,* 199–201.
6. Gust, D.; Moore, T. A.; Moore, A. L.; Macpherson, A. N.; Lopez, A.; DeGraziano, J. M.; Gouni, I.; Bittersmann, E.; Seely, G. R.; Gao, F.; Nieman, R. A.; Ma, X.-C.; Demanche, L. J.; Hung, S.-C.; Luttrull, D. K.; Lee, S.-J.; Kerrigan, P. K. *J. Am. Chem. Soc.* **1993,** *115,* 11141–11152.
7. Moore, T. A.; Gust, D.; Mathis, P.; Mialocq, J. C.; Chachaty, C.; Bensasson, R. V.; Land, E. J.; Doizi, D.; Liddell, P. A.; Lehman, W. R.; Nemeth, G. A; Moore, A. L. *Nature (London),* **1984,** *307,* 630–632.
8. Hung, S.-C.; Macpherson, A. N.; Lin, S.; Liddell, P. A.; Seely, G. R.; Moore, A. L.; Moore, T. A.; Gust, D. *J. Am. Chem. Soc.* **1995,** *117,* 1657–1658.
9. Beauchamp, A.; Benoit, R. L. *Can. J. Chem.* **1966,** *44,* 1607–1613.
10. Neta, P. *The Chemistry of Quinonoid Compounds;* Patai, S.; Rappoport, Z., Eds.; John Wiley & Sons, 1988; Vol. 2, Part 2, p. 886.
11. Rao, P. S.; Hayon, E. *J. Phys. Chem.* **1973,** *77,* 2274–2276.
12. Gust, D.; Moore, T. A.; Liddell, P. A.; Nemeth, G. A.; Making, L. R.; Moore, A. L.; Barrett, D.; Pessiki, P. J.; Bensasson, R. V.; Rougée, M.; Chachaty, C.; De Schryver, F. C.; Van der Auweraer, M.; Holzwarth, A. R.; Connolly, J. S. *J. Am. Chem. Soc.* **1987,** *109,* 846–856.
13. Lopez, A., Masters Thesis, Arizona State University, Tempe, AZ, 1992.
14. Gust, D.; Moore, T. A.; Moore, A. L.; Liddell, P. A. In *Methods in Enzymology;* Packer, L., Ed.; Academic: Orlando, FL, 1992; Vol. 213, pp 87–100.
15. Gust, D.; Moore, T. A.; Bensasson, R. V.; Mathis, P. A.; Land, E. J.; Chachaty, C.; Moore, A. L.; Liddell, P. A.; Nemeth, G. A. *J. Am. Chem. Soc.* **1985,** *107,* 3631–3640.
16. Grunewald, G. L.; Davis, D. P. *J. Org. Chem.* **1978,** *43,* 3074–3076.
17. Gust, D.; Moore, T. A.; Moore, A. L.; Seely, G.; Liddell, P. A.; Barrett, D.; Harding, L. O.; Ma, X.-C.; Lee, S.-J.; Gao, F. *Tetrahedron* **1989,** *45,* 4867–4891.
18. Swenton, J. S.; Raynolds, P. W. *J. Am. Chem. Soc.* **1978,** *100,* 6188–6195.
19. Wong, C. M.; Schwenk, R.; Popien, D.; Ho, T.-L. *Can. J. Chem.* **1973,** *51,* 466–467.
20. Wong, C. M.; Popien, D.; Schwenk, R.; Te Raa, J. *Can. J. Chem.* **1971,** *49,* 2712–2718.
21. Langlois, M.; Gaudy, F. *Synth. Commun.* **1992,** *22,* 1723–1734.
22. Kende, A. S.; Gesson, J.-P.; Demuth, T. P. *Tetrahedron Lett.* **1981,** *22,* 1667–1670.
23. Semmelhack, M. F.; Zask, A. *J. Am. Chem. Soc.* **1983,** *105,* 2034–2043.
24. Marchand, A. P. In *Stereochemical Applications of NMR in Rigid Bicyclic Systems, Methods in Stereochemical Analysis;* Verlag Chemie International: FL, 1982; Vol. 1.
25. Silverstein, R. M.; Bassler, G. C.; Morrill, T. C. In *Spectrometric Identification of Organic Compounds;* Wiley: New York, 1991; pp 117, 184.
26. Hermant, R. M.; Liddell, P. A.; Lin, S.; Alden, R. G.; Kang, H. K.; Moore, A. L.; Moore, T. A.; Gust, D. *J. Am. Chem. Soc.* **1993,** *115,* 2080–2081.
27. Hung, S.-C.; Lin, S.; Macpherson, A. N.; DeGraziano, J. M.; Kerrigan, P. K.; Liddell, P. A.; Moore, A. L.; Moore, T. A.; Gust, *D. J. Photochem. Photobiol. A: Chem.* **1994,** *77,* 207–216.
28. Bensasson, R.; Land, E. J. *Photochem. Photobiol. Rev.* **1978,** *3,* 163–191.

29. Gust, D.; Moore, T. A.; Luttrull, D. K.; Seely, G. R.; Bittersmann, E.; Bensasson, R. V.; Rougée, M.; Land, E. J.; De Schryver, F. C.; Van der Auweraer, M. *Photochem. Photobiol.* **1990,** *51,* 419–426.
30. Davis, F. S.; Nemeth, G. A.; Anjo, D. M.; Makings, L. R.; Gust, D.; Moore, T. A. *Rev. Sci. Instrum.* **1987,** *58,* 1629–1631.
31. Taguchi, A. K. W.; Stocker, J. W.; Alden, R. G.; Causgrove, T. P.; Peloquin, J. M.; Boxer, S. G.; Woodbury, N. W. *Biochemistry* **1992,** *31,* 10345–10355.
32. Liddell, P. A. Ph. D. Thesis, Arizona State University, Tempe, AZ, 1985.

13

Electrolyte Effects in Intramolecular Electron Transfer

Piotr Piotrowiak

Department of Chemistry, Rutgers University, Newark, NJ 07102

Pulse radiolysis and laser photolysis were used to study intramolecular electron transfer (ET) reactions in solutions of electrolytes. The presence of counterions has been found to slow the rate of weakly exoergic (ΔG = 100 meV) electron transfer by as much as 3 orders of magnitude. The larger counterions resulted in a larger decrease of the ET rate, suggesting that the transfer rate is influenced by the mobility of ions. In a separate study, transient triplet charge-transfer (CT) absorption bands of p-*aminonitroterphenyl were used as probes of ion-pairing dynamics and energetics. It has been found that the pairing process is diffusion controlled, and that in weakly polar media, association with ions can stabilize the photoinduced charge-separated state by up to 1 eV.*

Specific ion pairing and ionic atmosphere relaxation can play an important role in determining the energetics and dynamics of intramolecular electron transfer. Such effects can be expected not only when an electrolyte is intentionally introduced into the system, but also whenever counterions are present in the system, such as in electron transfer (ET) in transition metal complexes, in biological assemblies, and at electrolyte–electrode interfaces. Although the dynamics and energetics of dipolar reorganization have been exhaustively studied in practically all solvents and for a wide variety of donor–acceptor molecules (*1, 2*), relatively little is known about the analogous behavior of electrolytes (*3–6*). In addition, the majority of existing work on electrolyte solutions has dealt with intermolecular ET (*7, 8*).

Because electrolyte relaxation, that is, the translational diffusion of ions, occurs at a slower time scale and is physically distinct from the dipolar reorientation, it is convenient to view it as a separate coordinate of the van der

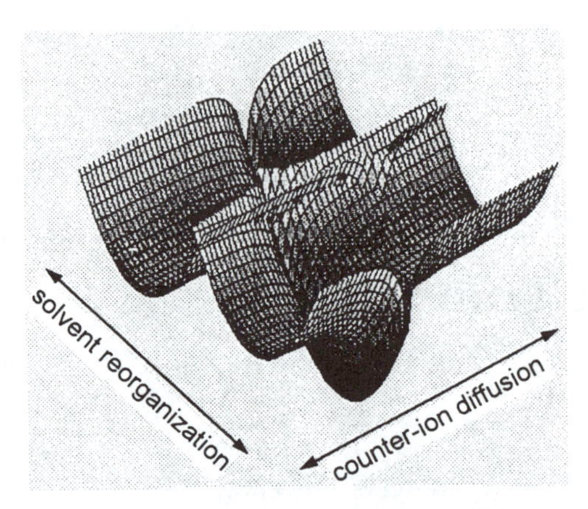

Figure 1. Schematic representation of the adaptation of the Sumi–Marcus two-dimensional free energy surface concept for intramolecular electron transfer in solutions of electrolytes: parabolic potential along the solvent reorganization coordinate and Coulomb potential along the counterion translational motion coordinate. The scaling of the two components, Λ_{polar} and Λ_{ionic}, is arbitrary.

Zwan–Hynes (9) or Sumi–Marcus (*10*) theory (Figure 1). Depending on the relative rate of the ionic relaxation, the electrolyte effects can be divided into two limiting cases (*11*): (1) the ionic reorganization is considerably faster than the intrinsic electron transfer rate, in which case the influence of the electrolyte is fully described by the additional contribution, λ_{ionic}, to the overall reorganization energy (this is the "slow reaction limit" of the Sumi–Marcus theory), and (2) the ionic relaxation is slower than the intrinsic electron transfer rate (this is the "fast reaction limit" of the Sumi–Marcus theory). In the second case, two behaviors are possible: (1) if the ET reaction is weakly exoergic, and particularly when $\lambda_{ionic} > \Delta G°$, the transfer rate will be determined by the ion dynamics, and a transition from a nonadiabatic to adiabatic ET will be observed and (2) when the ET is strongly exothermic, and particularly when it is in the "inverted region," the reaction becomes a two-step process, with a fast electron transfer followed by slow relaxation of the electrolyte.

The work presented in this chapter was performed in moderately polar media [primarily tetrahydrofuran (THF)], in which electrostatic interactions between ions are considerably larger than the thermal energy, kT. Therefore, all results are interpreted in terms of specific ion-pairing effects rather than the bulk ionic atmosphere relaxation.

Experiments

As in many other investigations, the combination of pulse radiolysis and laser photolysis afforded a much more complete exploration of the problem than would have been possible with just one of these techniques. There are important differences between the sequences of events in pulse radiolysis and laser photolysis ET experiments in solutions of electrolytes, which make one or the other method particularly suited for certain aspects of the study. These differences are briefly outlined in this section.

Pulse Radiolysis. Under typical salt concentrations (1–100 mM), the primary event in a pulse radiolysis experiment on a solution of an electrolyte in a moderately polar solvent is the rapid association of the solvated electron with the cation of an electrolyte (a pseudo-unimolecular rate of ~8×10^{11} M^{-1} s^{-1}). These special "ion pairs" are surprisingly stable. For example, the {e^-, Na^+} in THF lives for ~0.1 ms before collapsing into a sodium atom (*12*), whereas the pairs involving tetraalkylammonium cations slowly decompose following the reaction {e^-, $^+NR_4$} → NR_3 + ·R (*13*). In the presence of electron scavengers, such as the donor–acceptor model compounds, the {e^-, cation$^+$} pair reacts with them at a nearly diffusion-controlled rate to form a new {scavenger$^{(-)}$, cation$^+$} ion pair. Because the excess electron is "delivered" to the donor–acceptor assembly together with the cation of the salt, pulse radiolysis allows one to prepare a system in which at time, $t = 0$ generated donor$^{(-)}$–acceptor radical anions have a counterion in the vicinity of the donor$^{(-)}$ unit.

$$\{e^-, Na^+\} + D\text{–}A \rightarrow \{Na^+, D^-\text{–}A\} \xrightarrow{ET} \{D\text{–}A^-, Na^+\} \quad (1)$$

In this fashion, the complicated competition between the dynamics of ion pairing, the ET in free donor$^{(-)}$–acceptor radical anions, and the ET in donor$^{(-)}$–acceptor radical anions paired with cations is circumvented, and only the last process is being studied. For a more detailed discussion, see reference 14.

All pulse radiolysis experiments were conducted at Argonne National Laboratory with the 20-MeV linac producing 30-ps pulses.

Laser Photolysis. In most donor–acceptor compounds, the rate of the photoinduced forward ET is faster than the translational diffusion of a dilute electrolyte. Therefore, it can be assumed that in the absence of pre-association with the ground state of the probe, the free charge-separated species is formed

$$D\text{–}A \xrightarrow{h\nu} D^+\text{–}A^- \xrightarrow{\text{Electrolyte relaxation}} \{\text{anion}^-, D^+\text{–}A^-\} + \{D^+\text{–}A^-, \text{cation}^+\} \quad (2)$$

instantaneously at $t = 0$, and the dynamics of the subsequent response of the randomly distributed electrolyte can be conveniently studied. The systems described in this contribution are well within the limits of the above assumption.

The third harmonic of a Nd:YAG laser [355 nm, 5 ns FWHM (full width at half-maximum), ~30 mJ] was used as the excitation source in all reported laser photolysis experiments.

Ion-Pairing Control of Weakly Exothermic Intramolecular Electron Transfer

The rate of intramolecular charge shift reaction from biphenyl to naphthalene in the well characterized *trans*-1-(4-biphenylyl)-4-(2-naphthylcyclohexane) (*15, 16*) was investigated in THF solutions containing 10–30 mM of tetraphenylboron (TPB^-) salts of several cations: Li^+, Na^+, tetrabutylammonium (TBA^+), cetyl(trimethylammonium) ($C(TMA)^+$), tetraoctylammonium (TOA^+), and tetraoctadecylammonium ($TOdA^+$). The salts were selected on the basis of solubility in THF and compatibility with the pulse radiolysis experiments. In all instances, dramatic reductions of the intramolecular ET rate were recorded. The observed dependence on the size of the counter-ion was unexpected: the small alkali metal cations with high charge density reduced the transfer rate by a factor of ~100, whereas the large, relatively weakly charged tetraalkylammonium cations caused a 10 times larger ET rate reduction (Table I and Figure 2).

Clearly, the results cannot be explained on the basis of the magnitude of electrostatic interactions between the donor-acceptor radical anion and the cation of the salt, as these have been shown to follow the expected trend, that is, to increase with decreasing cation size *(14)*. Therefore, the counter-ion dynamics need to be considered. It is important to note that as long as the counter-ion remains in the vicinity of the donor unit, the ET reaction is effectively endothermic because its $\Delta G°$ is only ~100 meV, while the stabilization due to ion association can easily be as large as 0.5 eV (*6, 14*). This suggests that the counter-ion must diffuse away from the donor if the ET is to take place. It is true that thanks to a particularly large solvent fluctuation the barrier

Table I. Electron Transfer Rates in 1,4-ee-BCN in THF in the Presence of Various Tetraphenylboron Salts[a]

Cation of the Salt	r_{cation} *(Å)*	k_{ET} *(s^{-1})*	k_{ET}/k_0
Neat THF	—	1.6×10^9	1.00
Li^+	0.65	2.3×10^7	1.4×10^{-2}
Na^+	1.05	1.2×10^7	7.5×10^{-3}
TBA^+	4.13	1.2×10^6	7.5×10^{-4}
$C(TMA)^+$	4.36	1.9×10^6	1.2×10^{-3}
TOA^+	5.54	8.0×10^5	5.0×10^{-4}
$TOdA^+$	7.20	7.6×10^5	4.8×10^{-4}

[a] The rates are accurate to within ±20%.

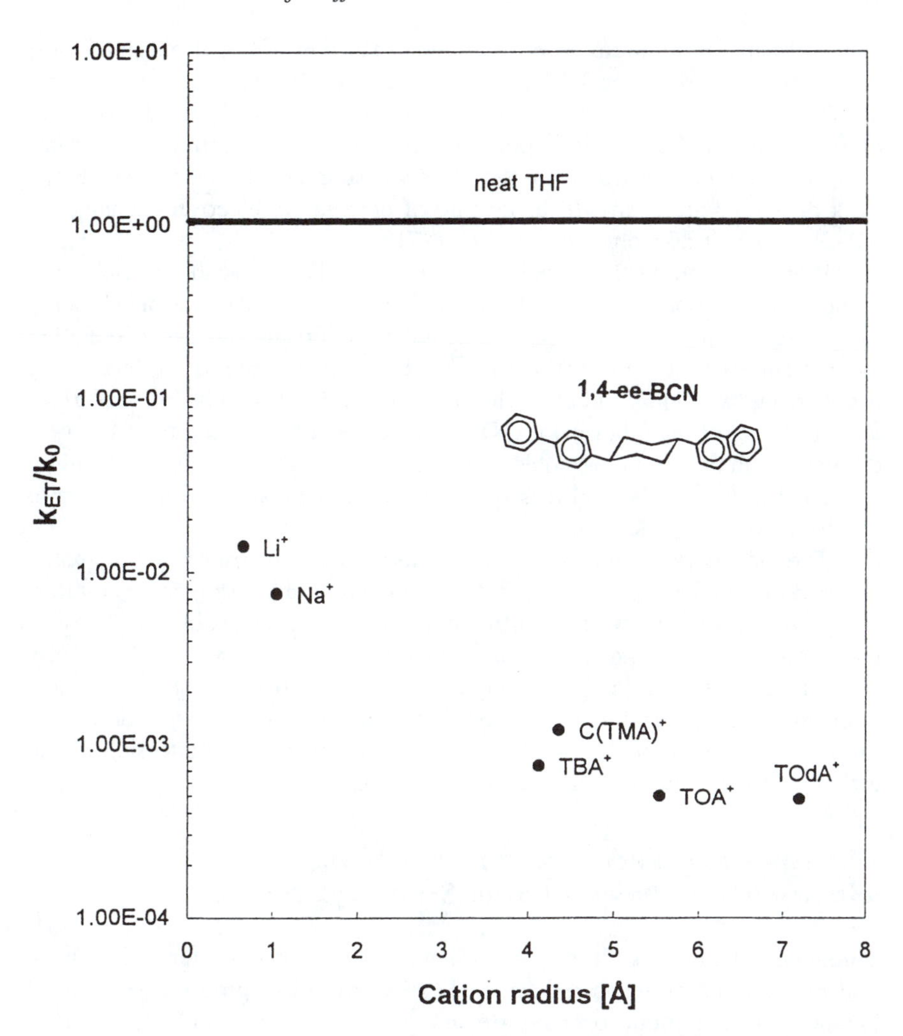

Figure 2. The ratio of the electron transfer rate in 1,4-ee-BCN (trans-1 (4-biphenyl)-4-(2-naphthyl)cyclohexane) in the presence of counter-ions to the rate in neat THF, plotted against the radius of the counter-ion. The concentrations of the salts were in the 10–30-mM range, and the anion of the salt was tetraphenylboron in all cases.

crossing can occur following evolution purely along the fast polar coordinate (Figure 1). However, with the counter-ion still close to the donor, a rapid back-transfer (exothermic by ~400 meV) will always be preferred over the slow evolution toward product equilibrium, and no net reaction will be observed. Therefore, the counter-ion diffusion is indeed the rate-determining step, and it effectively controls the intramolecular electron transfer. The diffusion of the bulky tetraalkylammonium cations is slower than that of the alkali metal ions; hence the decrease of the ET rate with an increase in cation size.

The observation that the motion of counter-ions can determine the rate of intramolecular ET in weakly polar media is not without a precedent. It has been elegantly demonstrated by the electron paramagnetic resonance (EPR) work of Mazur et al., in which the rate of degenerate electron exchange in radical anions of diketones was studied (*17*).

While the counter-ion mobility argument explains well the overall trend of the data, a satisfactory quantitative model of the observed behavior is lacking. An attempt to reproduce the experimental results using the Eigen equation was not successful (*14*). Most likely, it is at least partly due to the fact that it is not trivial to assign a radius to the studied cations. The alkali metal cations form tight solvation spheres in THF, and their hydrodynamic radius is very different from the crystallographic or gas phase value. The tetraalkylammonium ions are highly flexible, and it is questionable how meaningful it is to try to approximate them as spheres.

The most important achievement of this work is the observation of greatly reduced intramolecular ET rates in the presence of chemically inert counter-ions. The dynamic counter-ion control of the ET rate could be observed because the studied electron transfer reaction was very weakly exothermic. Therefore, even in the case of the large organic cations and relatively weak electrostatic interactions, the condition $\lambda_{ionic} > \Delta G°$ was satisfied, and the evolution along the slow coordinate of the Sumi–Marcus theory, that is, the diffusion of the cation, could become the rate-determining step.

Dynamics and Energetics of Ion Pairing with a Photoinduced Charge-Separated Species

Time-dependent blue shifts of the transient triplet charge-transfer (CT) absorption of *p*-aminonitroterphenyl (*p*-ANTP) were used to probe association of ions with a photoinduced charge-separated species (*6*). Thanks to the long lifetime of the triplet excited state of *p*-ANTP ($\tau > 3$ μs), the study of ion-pairing dynamics can be extended to much lower electrolyte concentrations, and consequently to much less polar media, than is possible with the short-lived fluorescent probes ($\tau \leq 3$ ns) (*3, 4*). The ability to probe weakly polar and nonpolar solvents is important because the electrostatic interactions are much larger in these media than in strongly polar solvents.

After photoexcitation into a locally excited singlet state, *p*-ANTP undergoes subnanosecond ET from the amino- to the nitro- end of the molecule. The resulting singlet charge-separated state intersystem crosses to the triplet charge-separated state within < 3 ns (*18*). These events can be viewed as instantaneous on the time scale of the experiments. The triplet charge-separated state of *p*-ANTP is characterized by a strong CT absorption band, $\varepsilon \approx 40{,}000$, peaking in neat THF at 990 nm. Immediately after the laser pulse, the transient

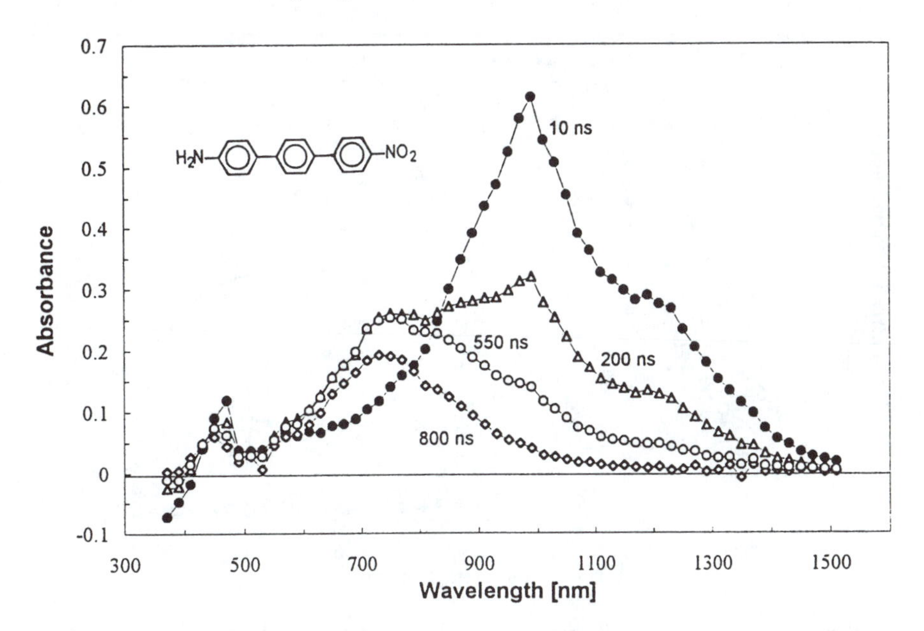

Figure 3. Transient triplet absorption spectra of 0.1 mM p-ANTP *in THF containing 1.0 mM of TBABr, taken at increasing delays after a 5-ns, 355-nm excitation pulse.*

absorption spectrum of *p*-ANTP in a 1 mM solution of TBABr is identical to the spectrum in neat THF (compare the 10-ns trace in Figure 3 and the THF reference in Figure 4. As the association of ions with the probe takes place, the energy of the CT triplet state is lowered, and the spectrum evolves toward a new, very strongly blue-shifted maximum, λ_{max} = 730 nm (Figure 3). This behavior is strictly analogous to the dynamic red shifts in the emission spectra of fluorescent probes of solvation dynamics. It is important to note that millimolar salt concentrations have no influence on the position and intensity of the fluorescence spectrum of *p*-ANTP because the emissive state is too short-lived in comparison with the slow electrolyte dynamics.

A systematic study of the ion-size dependence of the electrolyte-induced spectral shifts of the *p*-ANTP triplet absorption band was performed. The work focused on anions because it is possible to select a relatively large set of rigid, approximately spherical negative ions. In addition, unlike the cations, they do not form specific coordination spheres in ethereal solvents. THF solutions of the fluoride, chloride, bromide, tetrafluoroborate, perchlorate, and hexafluorophosphate salts of the tetrabutylammonium (TBA^+) cation were investigated.

The recorded fully equilibrated spectra follow the expected trend, with the smaller anions leading to a larger blue shift and, therefore, larger stabilization energies (Figure 4 and Table II). The largest blue shift of 6850 cm^{-1} was

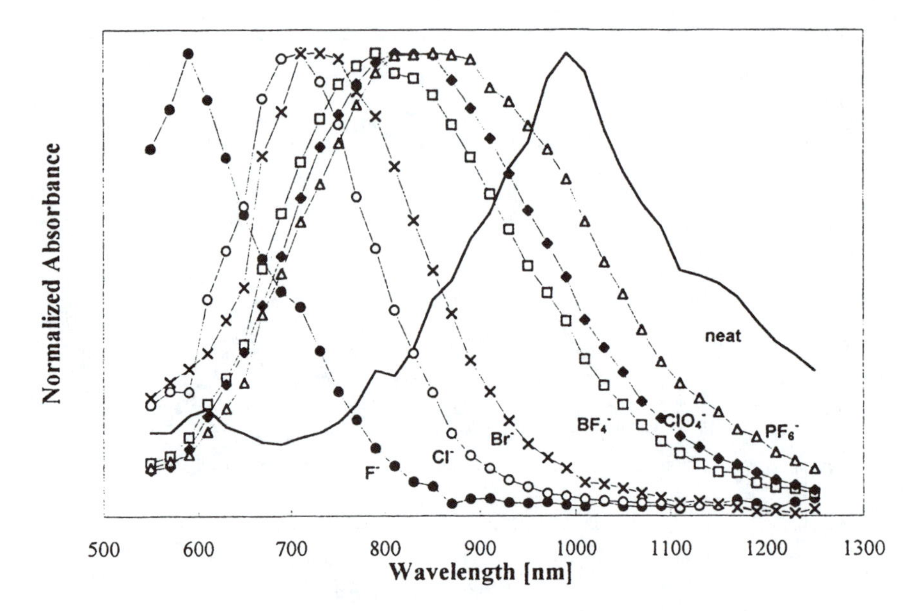

Figure 4. Normalized, fully equilibrated, transient triplet absorption spectra of 0.1 mM p-ANTP *in THF containing 5.0 mM of various tetrabutylammonium salts: TBAF, TBACl, TBABr, $TBABF_4$, $TBAClO_4$, and $TBAPF_6$. The spectrum in neat THF (heavy curve) is provided as a reference.*

obtained in the presence of the F^- salt. It is more than 3 times greater than the largest electrolyte-induced shift reported for fluorescent probes (*3, 4*).

It is impossible to conclude unequivocally on the basis of the presented data whether the association takes place primarily at the H_2N^+– side or the $–NO_2^-$ side of the probe, and whether it is dominated by the anion or the

Table II. Spectral Shifts of the Triplet Charge-transfer Absorption of *p*-ANTP in THF in the Presence of 5 mM of Various Tetrabutylammonium Salts

Anion of the Salt	r_{anion} (Å)	λ_{max} (nm)	$\Delta\nu$ (cm^{-1})
neat THF		990	0
F^-	1.19	590	6850 ± 250
Cl^-	1.67	710	3980 ± 200
Br^-	1.82	730	3600 ± 200
BF_4^-	2.18	790	2560 ± 150
ClO_4^-	2.26	810	2240 ± 150
PF_6^-	2.96	850	1670 ± 150

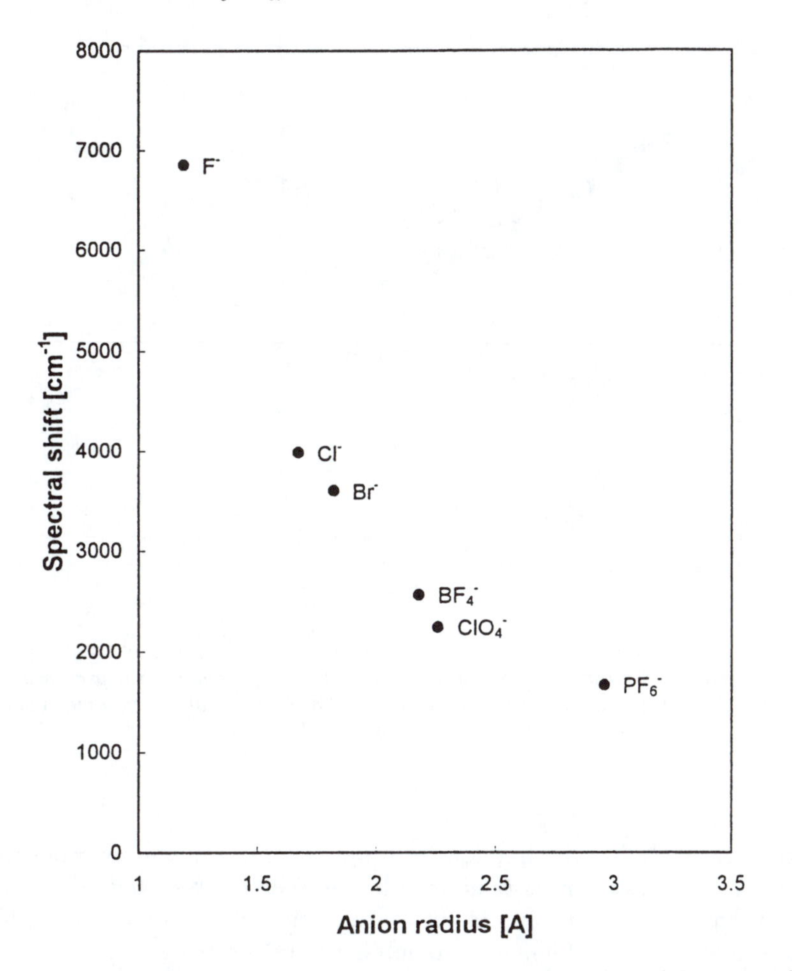

Figure 5. Spectral shifts from Table I plotted against the radius of the anion of the salt.

cation of the salt. However, it appears from Figure 5 that as the anion radius is extrapolated to ∞, the magnitude of the shift approaches an asymptotic value of ~1500 cm^{-1}, which can be interpreted as the constant contribution of the TBA^+ cation. These questions are being addressed in detail by time-resolved resonant Raman experiments.

The dynamics of the evolution of the blue shift and of the association process exhibit linear dependence on the overall concentration of the salt rather than on the concentration of the free ions (*6*), and in the case of TBABr yield a pseudo-unimolecular rate of 7.7×10^9 M^{-1} s^{-1}. This finding is not surprising considering the low dissociation constants of salts in THF (from ~1×10^{-6} to ~1×10^{-5}). Even at a 1 mM concentration, the salts are less than 10% disso-

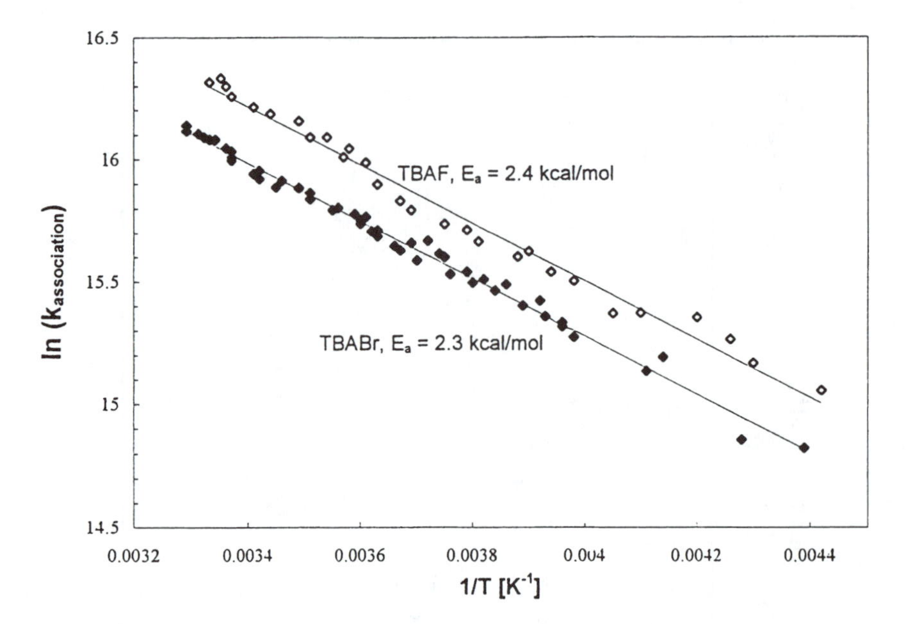

Figure 6. The rate vs. temperature dependence of the association between the triplet charge-separated species of p-*ANTP and the electrolyte in the presence of 1.0 mM of TBAF and TBABr. Data were monitored at the absorption maximum of the free probe,* λ_{max} = *990 nm.*

ciated, and therefore the free ions can provide only a minor contribution to the overall process. This is reinforced by the spectra presented in Figure 4, which show that at a millimolar salt concentration, the equilibrium is shifted completely toward the ion-paired complex, with no residual free probe absorption. It is expected that only at salt concentrations below 1×10^{-4} M will the free ions begin to dominate the dynamics. It is interesting that the preliminary electrolyte concentration dependence data obtained in nonpolar solvents (toluene and benzene) indicate that in these media, salts are present primarily in the form of clusters consisting of several ions rather than as simple ion pairs (T. Schatz, R. Kobetic, and P. Piotrowiak, manuscript in preparation).

Temperature dependence studies were undertaken in order to evaluate the activation energy of the association process. The association rates obtained by monitoring the disappearance of the free probe absorption, λ_{max} = 990 nm, yielded remarkably linear Arrhenius plots (Figure 6) with the activation energies of 2.4 ± 0.2 kcal/mol, that is, the activation energy for diffusion of a small molecule in THF. Therefore, it can be concluded that the primary event of the association process is a barrierless, diffusion-controlled encounter between the probe molecule and an ion pair of the salt. The association rates for TBAF were consistently ~20% higher than for the somewhat more bulky TBABr. The

Arrhenius plots obtained by monitoring the buildup of the fully associated complex, λ_{max}(TBABr) = 730 nm and λ_{max}(TBAF) = 590 nm, yield somewhat higher activation energies, 2.7 kcal/mol for TBABr and 3.2 kcal/mol for TBAF. This suggests that after the initial encounter with an ion pair of the salt, the association complex undergoes a rearrangement, the limiting case of which would be the dissociation of one of the ions of the salt.

Finally, it is of interest to examine how the ion-pairing influences the lifetime of the charge-separated state of *p*-ANTP. The CT state → ground state charge recombination in *p*-ANTP is strongly exoergic (1.6–2.0 eV). Therefore, one would expect it to be considerably accelerated in the presence of counter-ions, in accordance with the "inverted region" behavior. Rather surprisingly, the preliminary results indicate a weak dependence of the CT state lifetime on the presence of an electrolyte. No clear correlation between the magnitude of the salt-induced spectral shift and the lifetime of the charge-separated state of *p*-ANTP has been found. Because the monitored charge-separated state is a triplet, its nonradiative decay is most likely controlled primarily by the intersystem crossing (ISC) rate. In this case, association with a counter-ion would influence the lifetime of the CT state by modifying the ISC rate through the "heavy atom" effect, rather than by tuning the energetics of charge recombination in the standard Marcus theory sense. As a consequence, systems with singlet CT states will be better suited for studying the influence of ion-pairing on the rate of charge recombination.

Summary

We have demonstrated in two independent studies that electrolyte effects can have a strong influence on the energetics of a charge-separated species and on the dynamics of intramolecular electron transfer in moderately polar media. In the case of weakly exoergic ET, the ion dynamics can effectively control the transfer rate. The developed spectroscopic probes indicate that association with ions can stabilize a photoinduced charge-separated species by as much as 1 eV. Further investigation of electrolyte effects in photoinduced electron transfer, particularly in the "inverted region," is needed.

Acknowledgments

The support by the Office of Basic Energy Sciences, Division of Chemical Sciences, U.S. Department of Energy, under grant No. FG–05–92–ER14310, is gratefully acknowledged. All pulse radiolysis experiments described here were performed in collaboration with John R. Miller at the Argonne National Laboratory Linac facility. Renata Kobetic and Timothy R. Schatz are thanked for their work on the laser photolysis experiments.

References

1. Barbara, P. F.; Walker, G. C.; Smith, T. P. *Science (Washington, D.C.)* **1992,** *256,* 975 and references therein.
2. Chapman, C. F.; Fee, R. S.; Maroncelli, M. *J. Phys. Chem.* **1995,** *99,* 4811 and references therein.
3. Huppert, D.; Ittah, V.; Kosower, E. M. *Chem. Phys. Lett.* **1989,** *159,* 267.
4. Chapman, C. F.; Maroncelli, M. *J. Phys. Chem.* **1991,** *95,* 9095.
5. Thompson, P. A.; Simon, J. D. *J. Chem. Phys.* **1992,** *97,* 4792.
6. Piotrowiak, P.; Kobetic, R.; Schatz, T.; Strati, G. *J. Phys. Chem.* **1995,** *99,* 2250.
7. Chiorboli, C.; Indelli, M. T.; Rampi-Scandola, M. A.; Scandola, F. *J. Phys. Chem.* **1988,** *92,* 156.
8. Kawanishi, Y.; Kitamura, N.; Tazuke, S. *J. Phys. Chem.* **1986,** *90,* 2469.
9. Zwan, G. v. d.; Hynes, J. T. *J. Chem. Phys.* **1983,** *78,* 4174.
10. Sumi, H.; Marcus, R. A. *J. Chem. Phys.* **1985,** *84,* 4894.
11. Piotrowiak, P. *Inorg. Chim. Acta* **1994,** *225,* 269.
12. Bockrath, B.; Dorfman, L. M. *J. Phys. Chem.* **1973,** *77,* 1002.
13. Konovalov, V. V.; Bilkis, I. I.; Selivanov, B. A.; Shteingarts, V. D.; Tsvetkov, Y. D. *J. Chem. Soc., Perkin Trans.* **1993,** *2,* 1707.
14. Piotrowiak, P.; Miller, J. R. *J. Phys. Chem.* **1993,** *97,* 13052.
15. Closs, G. L.; Johnson, M. D.; Miller, J. R.; Piotrowiak, P. *J. Am. Chem. Soc.* **1989,** *111,* 3751.
16. Closs, G. L.; Calcaterra, L. T.; Green, N. J.; Penfield, K. W.; Miller, J. R. *J. Am. Chem. Soc.* **1986,** *90,* 3673.
17. Furderer, P.; Gerson, F.; Heinzer, J.; Mazur, S.; Ohya-Nishiguchi, H.; Schroeder, A. H. *J. Am. Chem. Soc.* **1979,** *101,* 2275.
18. O'Connor, D. B.; Scott, G. W.; Tran, K.; Coulter, D. R.; Miskowski, V. M.; Stiegman, A. E.; Wnek, G. E. *J. Chem. Phys.* **1992,** *97,* 4018.

14

Mechanisms of Photochemical Reactions of Transition-Metal Complexes Elucidated by Pulse Radiolysis Experiments

Carol Creutz, Harold A. Schwarz, and Norman Sutin

Chemistry Department, Brookhaven National Laboratory, Upton, NY 11973–5000

This chapter illustrates the complementarity of photochemical and radiation chemical techniques to elucidate elementary pathways in mechanistically rich systems. Some of the mechanistic conclusions that have resulted from these studies in aqueous media are presented. Extreme (both high and low) oxidation states of transition-metal complexes are included. Reactivity with respect to electron transfer reactions and small-molecule activation are addressed.

Since the early 1970s, the work of our group at Brookhaven National Laboratory has focused on basic research directed toward the conversion of the energy from sunlight into other forms, useful on a practical human time scale. Our approach (illustrated in Figure 1) has involved the use of transition-metal complexes with good light absorption properties and relatively long excited state lifetimes as sensitizers.

Typically, these excited state species possess ~2 eV of excitation energy, and accordingly (depending on ground-state properties, of course), they are powerful reducing and/or oxidizing agents. The energy, captured originally in the light-absorption step, is largely retained through charge-separating, electron transfer reactions of the excited state, which produce ground-state, highly reactive electron transfer products. In principle, the energy stored (albeit fleetingly) in the charge-separated species can be captured and stored in several ways, such as through electron transfer to appropriate electrodes installed in a suitably designed circuit to either yield a photogalvanic cell or charge a battery. The direction we have taken is mainly a more chemical one, in which the energy

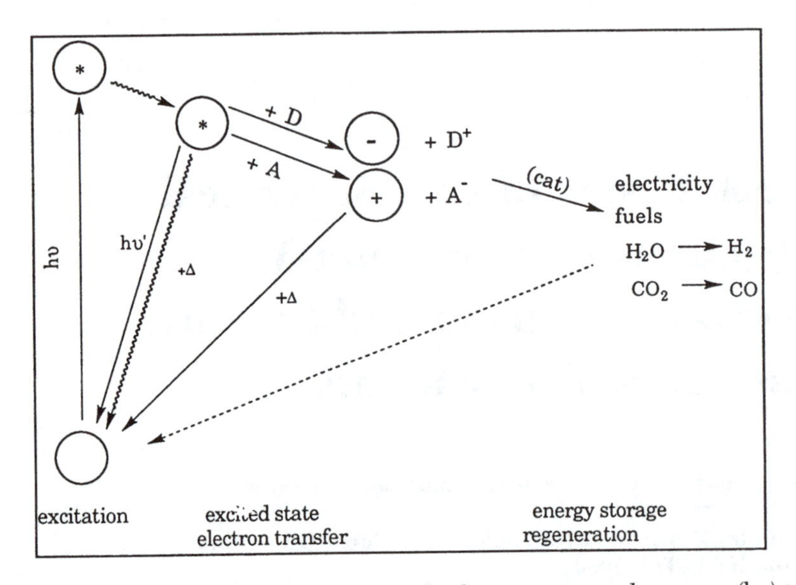

Figure 1. Cartoon outlining the steps involved in converting solar energy (hν) to stored chemical energy or electricity. Electron transfer reactions of the excited state (denoted by an asterisk) with donors (D) or acceptors (A) to yield charge-separated species (D^+ or A–) take place in competition with nonproductive emission (hν′), nonradiative excited-state decay, and back electron transfer, which produce heat.

from the light is used to drive uphill chemical reactions, such as dissociation of water into its elements or reduction of carbon dioxide. The chemicals produced in these reactions could then be used as fuels (energy sources) or as raw materials for other processes. Critical to an efficient overall photoconversion process are high efficiency at each juncture and, in particular, for high yields in the chemical-forming reactions, efficient catalysis by species that can be generated by one-electron transfer reactions. One- and two-electron reduction potentials for hydrogen ion and for carbon dioxide vs. normal hydrogen electrode (NHE) are compared in Figure 2.

One role of the catalyst is to stabilize the one-electron reduction (or oxidation) products, thereby lowering the kinetic barrier for the net two (or more) electron transfer process. Transition-metal catalysts serve to stabilize the one-electron reduction products and to promote rapid formation of the desired products.

$$M^{I} + H^{+} \rightarrow M^{II}\,H\cdot \rightarrow M^{III}\,H^{-} \qquad (1)$$

$$M^{I} + CO_2 \rightarrow M^{II}\cdot CO_2^{-} \rightarrow M^{III}:CO_2^{2-} \qquad (2)$$

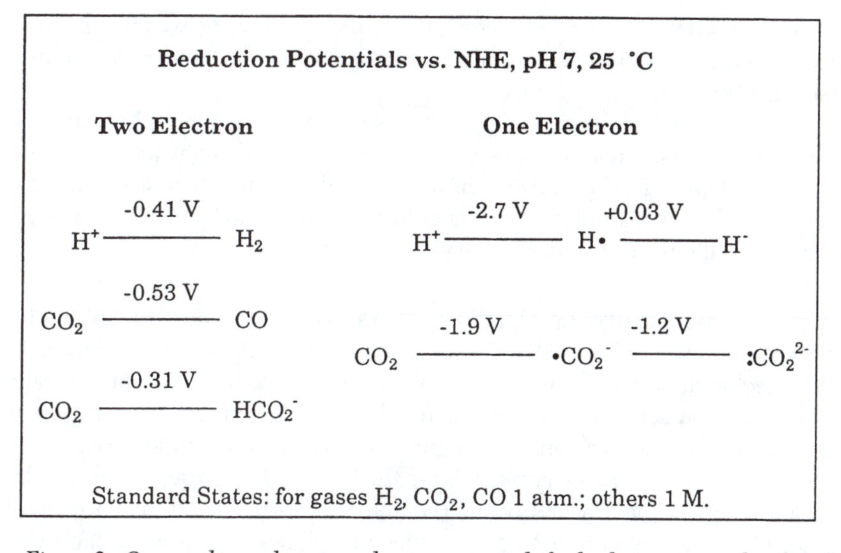

Figure 2. One- and two-electron reduction potentials for hydrogen ion and carbon dioxide vs. NHE (pH 7, 25 °C). Standard states: 1 atm for gases H_2, CO_2, and CO; 1 M for others.

This chapter focuses on our mechanistic studies of some of these chemical systems.

Flash Photolysis versus Pulse Radiolysis Techniques

Clearly, excited state reactions, the true photochemical reactions, are best studied directly through photochemical methods (although there are systems in which an excited state of interest may not be attainable through direct light absorption but might be attained through a radiation chemical route). What then is the role of radiation chemistry in the mechanistic study of complex photochemical systems? It is a versatile, powerful tool, particularly the pulse radiolysis technique, for the study of the thermal reactions that may be induced by photochemical events. In our work, these follow-up reactions have involved organic radicals and metal ions in unusually high or low oxidation states. We have used UV–vis spectroscopy to identify these species and probe their fates. First we outline the strengths of pulse radiolysis in these studies; then we return to them in photochemical case studies that illustrate their value. Note that all of these comments pertain to aqueous solutions near room temperature; without a large and reliable database on the kinetics and products of such solutions, the planning and interpretation of these complex experiments would not be possible.

Sensitivity. Pulse radiolysis has high sensitivity for weakly absorbing species, which is especially useful for monitoring ligand-field absorptions of first-transition-series metal ions.

There is no inner filter (or at least less of one): functional photochemical systems of necessity contain species that strongly absorb light in some region of the spectrum. The sensitizer absorption makes it much more difficult to detect signal changes arising from chemical reactions not involving the sensitizer, especially when the desired signals are small.

Synthetic Control. The ionization of water produces hydrated electron, e^-_{aq}, and hydroxyl radical, ·OH, species with a remarkable range of oxidizing and reducing ability, and these are stable for a remarkably long time in water.

By manipulation of scavenger and substrate concentrations, the primary reagents can be readily changed. To give a simple example, in an Ar-saturated aqueous solution, the primary radicals are the hydrated electron, e^-_{aq}, and hydroxyl radical, ·OH, but saturation of the solution with nitrous oxide converts e^-_{aq} to ·OH, thereby eliminating the reducing species and giving a high yield of the oxidizing ·OH. By contrast, with use of a solution containing CO_2 and formate, a very high yield of the strongly reducing CO_2^- radical is produced.

Flexibility. There is ready control of reagent concentrations through dose variations. Furthermore, in the experiments conducted at Brookhaven National Laboratory (BNL), great flexibility is provided by the cell design. The cell consists of a large reservoir of solution which is used to fill and refill (2 mL per change) the working cell area in the side arm. The working area is irradiated with 2-MeV electrons from one side through 0.5-mm-thick quartz. The solution is monitored at 90° to the electron pulse through a 2-cm path length (which can be increased to 6 cm, effectively through use of mirrors in the sample holder). The solution pH is easily monitored and adjusted through the top port, so that a single stock solution can be used for a pH dependence study. The "purge" gas is easily changed, so that a single solution can be used for a range of, for example, CO_2 concentrations.

The current BNL setup is also readily adapted for work on time scales ranging from microseconds to minutes, with the long time scale capability being particularly useful.

Applications

We turn now to concrete examples of the interplay between photochemical and radiation chemical techniques in the form of case histories taken mainly from our work at Brookhaven National Laboratory.

Excited State Characterization. Charge-transfer excited states of transition-metal complexes fulfill many important photochemical roles, and an understanding of their properties has been the focus of many kinds of studies. Metal-to-ligand charge transfer (MLCT) excited states have long been of keen interest, and pulse radiolysis techniques have enriched our understanding of these states and aided in their assignments. Typically, MLCT excitation involves transfer of a metal electron to an aromatic ligand bound to the metal, i.e.,

$$\text{M–L} \xrightarrow{h\nu_{CT}} \text{M}^{+}\text{–L}^{-} \qquad (3)$$

where h is Planck's constant and ν_{CT} is frequency. The transfer produces an oxidized metal center, M^+, and a reduced ligand, L^-. Since many aromatic anion radicals absorb at relatively low energy, the MLCT excited state might be expected to exhibit low-energy absorptions characteristic of the L^- moiety. Thus, the MLCT state of $Ru^{II}(bpy)_3{}^{2+}$, $*\ Ru(bpy)_3{}^{2+}$, or $Ru^{III}(bpy)_2(bpy^-)^{2+}$ where bpy is 2, 2′-bipyridine exhibits bands at 390 and 520 nm, reminiscent of the spectrum of bpy^-, produced by reduction with sodium metal in tetrahydrofuran (THF) (*1*). When finally determined by picosecond absorption spectroscopic methods, the spectra of the MLCT states of a series of $Ru(NH_3)_5L^{2+}$ complexes (*2*) also proved to exhibit characteristic L^- chromophores by comparison with radical spectra determined earlier in pulse radiolysis measurements.

In photochemical systems, highly reactive photoproducts such as $Ru(bpm)_3{}^+$, which is produced by reduction of the MLCT state $*\ Ru(bpm)_3{}^{2+}$ are common (bpm = 2,2′-bipyrimidine). In the photochemical system, a sacrificial donor, D, such as ethylenediaminetetraacetic acid (EDTA), TEOA (triethanolamine), or oxalate, is used to reduce the excited state ("reductive quenching"). Such sacrificial reagents owe their high efficiencies to the irreversible character of their two-electron couples. Their one-electron oxidized forms, D^+, generally rearrange and ultimately provide a second reducing equivalent to the system. However, the complexity of their chemistry generally makes it more difficult to characterize the authentic chemistry of the reactive transition-metal complexes when the two are produced together. Thus, the definitive characterization of $Ru(bpm)_3{}^+$ was carried out by pulse radiolysis (*3*). Protonation of $Ru(bpm)_3{}^+$ to yield $Ru^{II}(bpm)_2(bpmH)^{2+}$ ($pK_a = 6.3$) was established, and the chemistries of these two Ru(II)-bound ligand radicals were fully explored. Indeed, Hoffman, Mulazzani, and colleagues, through comparable photochemical and radiation chemical studies, mapped out the chemistry of the $Ru^{II}(L)_2(L^-)^+$ series for $Ru(bpy)_3{}^+$ (*4*) and $Ru(bpz)_3{}^+$ (*5*).

MLCT excited states can also be produced via radiation chemical techniques. The reduction of $Ru(bpy)_3{}^{3+}$ with e^-_{aq} yields the MLCT excited state of $Ru(bpy)_3{}^{2+}$ in high yield (*6, 7*):

$$e^-_{aq} + Rh(bpy)_3{}^{3+} \longrightarrow *\ Ru(bpy)_3{}^{2+} \qquad (4)$$

Characterization of "Invisible" Photoproducts. In the late 1970s, our attention was drawn to the work of Lehn and Sauvage, who reported photoproduction of H_2 from a system containing $Ru(bpy)_3^{2+}$, $Rh(bpy)_3^{3+}$, triethanolamine, and tetrachloroplatinate (*8*). We conducted in-depth mechanistic studies of this system to learn how the H_2 was formed (*9*). From continuous photolyses, we learned that omission of the platinum salt resulted in completely different products: no H_2 was produced, but a purple-to-brown colored product, ultimately identified as the low-spin d^8 rhodium(I) complex $Rh(bpy)_2^+$, formed instead. Prepared independently, the latter does not reduce water to dihydrogen in the presence of the platinum salts or colloidal platinum, consistent with the Rh^{III}/Rh^{I} and H^+/H_2 potentials (–0.25 and –0.47 V versus NHE, respectively, at pH 8).

$$Rh(bpy)_2^+ + 2H_2O \xrightarrow{[Pt^0]} \!\!\!\!\!\!\!\!\!/ \;\;\; H_2 + Rh(bpy)_2(OH)_2^+, \; K \sim 10^{-4} \text{ atm at pH 8} \quad (5)$$

(The line through the arrow indicates that the reaction does not take place.) Emission quenching and flash photolysis studies revealed that the primary photoreaction involved oxidation of the MLCT excited state $*Ru(bpy)_3^{2+}$ by $Rh(bpy)_3^{3+}$ ($k = 6.2 \times 10^8\ M^{-1}\ s^{-1}$):

$$*Ru(bpy)_3^{2+} + Rh(bpy)_3^{3+} \rightarrow Ru(bpy)_3^{3+} + Rh(bpy)_3^{2+} \quad (6)$$

In the transient absorption studies, bleaching of the distinctive $Ru(bpy)_3^{2+}$ ground-state and excited absorption spectra were readily observed, consistent with formation of $Ru(bpy)_3^{3+}$ and $Rh(bpy)_3^{2+}$. However, no intense new absorptions were observed. When triethanolamine was added to the $Ru(bpy)_3^{2+}$ + $Rh(bpy)_3^{3+}$ solution, the re-formation of $Ru(bpy)_3^{2+}$ via reduction of $Ru(bpy)_3^{3+}$ by TEOA could be observed in the microsecond-to-millisecond regime, depending upon conditions.

$$Ru(bpy)_3^{3+} + TEOA \rightarrow Ru(bpy)_3^{2+} + TEOA^+ \quad (7)$$

Typically, however, no other color changes occurred in these solutions for times up to a second. On this very long time scale, the growth of the Rh(I) absorption could be detected, but reliable kinetic data could not be obtained because of diffusion; only a small volume element of the solution was irradiated, and on the time scale of seconds, that volume began to mix with unirradiated solution, giving the appearance that the Rh(I) was being consumed.

We also tried to use electrochemical methods to characterize $Rh(bpy)_3^{2+}$, but the latter is reduced more readily than the $Rh(bpy)_3^{3+}$ complex, so that only the Rh(I) complex can be prepared by the electrochemical technique.

Finally, to characterize the elusive $Rh(bpy)_3^{2+}$ species, we turned to pulse radiolysis methods, using the electron to reduce Rh(III) and TEOA as OH-scavenger (although the chemistry of TEOA differs in photo- and radiation chemistry systems (*10*)).

$$Rh(bpy)_3^{3+} + e^-_{aq} \rightarrow Rh(bpy)_3^{2+} \tag{8}$$

$$N(CH_2CH_2OH)_3 + \cdot OH \rightarrow N(CH_2CHOH)(CH_2CH_2OH)_2 + H_2O \tag{9}$$

The $Rh(bpy)_3^{2+}$ species exhibited only weak absorption features in the visible–near-UV region of the spectrum ($\varepsilon_{350} 4000$, $\varepsilon_{490} 1000$ M^{-1} cm^{-1}). These low intensities suggest that the species is actually an authentic 19-electron, Rh(II) complex. The alternative, $Rh^{III}(bpy)_2(bpy^-)$, should exhibit much more intense absorption. The disproportionation of $Rh(bpy)_3^{2+}$ occurs by a highly unusual mechanism, rate-determining bpy-loss.

$$Rh(bpy)_3^{2+} \rightleftharpoons Rh(bpy)_2(H_2O)_2^{2+} + bpy, \qquad k_f = 2\ s^{-1} \tag{10}$$

where f is forward.

$$Rh(bpy)_3^{2+} + Rh(bpy)_2(H_2O)_2^{2+} \rightarrow Rh(bpy)_3^{3+} + Rh(bpy)_2^{+} \tag{11}$$

The most surprising outcome of this study was the conclusion that the 19-electron species is so stable with respect to loss of ligand. This conclusion is derived from the kinetic data for bpy loss and addition (*11*).

$$Rh(bpy)_3^{2+} \rightleftharpoons Rh(bpy)_2(H_2O)_2^{2+} + bpy \tag{12}$$

$$k_f = 2\ s^{-1};\ k_r = 0.2 \times 10^9\ M^{-1}\ s^{-1};\ K = k_f/k_r = 10^{-8}\ M$$

where r is reverse.

At the time of this discovery, 19-electron species beyond the first transition series were not regarded as electronically viable. Subsequent work by David Tyler and his group demonstrated the existence of many other 19-electron species and the important role such species can play in organometallic systems.

Characterization of an Unstable Photoproduct. Pulse radiolysis provides a unique experimental tool for the characterization of an unstable photoproduct. The highly oxidizing, ligand-field excited state of $^*Os^{VI}(TMC)O_2^{2+}$, where TMC is tetramethyl-cyclam, has now been studied

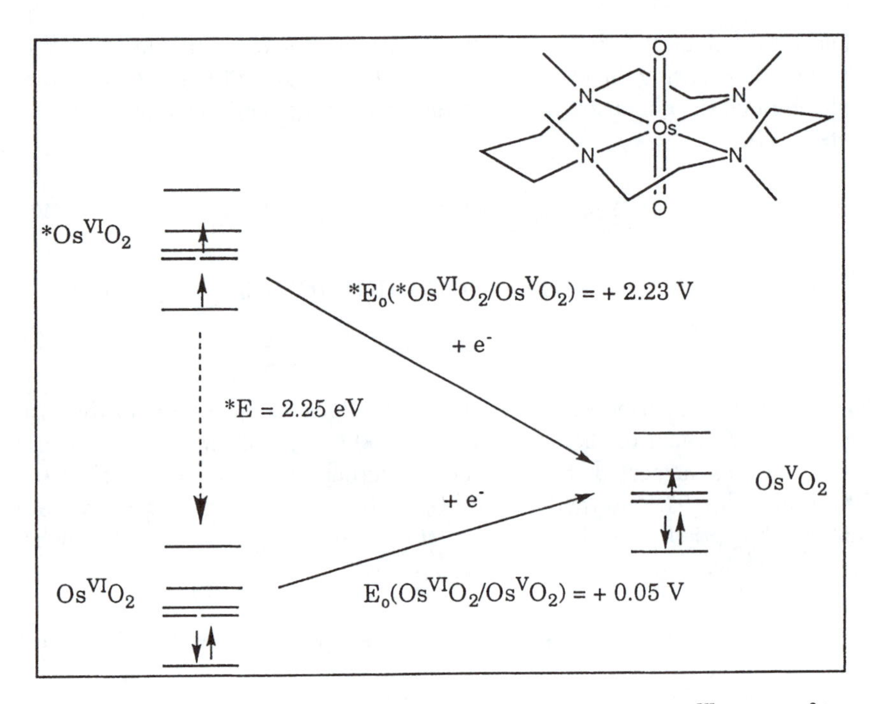

Figure 3. Energetics of the ground- and excited-state $Os^{VI}(TMC)O_2^{2+}$–$Os^{V}(TMC)O_2^{+}$ couples.

fairly extensively (*12*). As shown in Figure 3, it is a powerful outer-sphere oxidant, capable of being reduced by such poor reductants as the halide ions and hydroxide ion (*13*). However, relatively little attention has been paid to its Os(V) reduction product. In photochemical experiments, Os(V) can be produced by excited state quenching:

$$* Os^{VI}(TMC)O_2^{2+} + Q \rightarrow Os^{V}(TMC)O_2^{+} + Q^{+} \qquad \Delta G^\circ \sim 0 \text{ eV} \quad (13)$$

However, photochemical systems are not suitable for characterizing the ground-state Os^{VI}/Os^{V} couple, because the "back" reaction is extremely exergonic.

$$Os^{V}(TMC)O_2^{+} + Q^{+} \rightarrow Os^{VI}(TMC)O_2^{2+} + Q, \qquad \Delta G^\circ \sim -2 \text{ eV} \quad (14)$$

In general, a self-exchange rate is best estimated from very low driving force reactions involving similar reaction partners. Furthermore, the Os(V) is unstable toward Os(III), which renders application of more conventional techniques, such as stopped flow, difficult. Accordingly, we used pulse radiolysis to produce Os(V),

$$Os^{VI}(TMC)O_2^{2+} + e^{-}_{aq} \rightarrow Os^{V}(TMC)O_2^{+} \quad (15)$$

and studied its reaction,

$$Os^{V}(TMC)O_2^{+} + BQ \rightleftharpoons Os^{VI}(TMC)O_2^{2+} + BQ^{-}, \qquad K = 0.315 \quad (16)$$

with the characterized reaction partner benzoquinone (BQ),

$$BQ + e_{aq}^{-} \rightleftharpoons BQ^{-} \qquad E_{1/2} = 0.078 \text{ V vs. NHE} \quad (17)$$

$$BQ + BQ^{-} \rightleftharpoons BQ^{-} + BQ \qquad k_{ex} = 6.2 \times 10^{7}\ M^{-1}\ s^{-1} \quad (18)$$

in order to estimate the self-exchange rate, $k_{11} = 1.1 \times 10^{6}\ M^{-1}\text{–}s^{-1}$, of the ground state couple (*14*).

$$Os^{V}(TMC)O_2^{+} + Os^{VI}(TMC)O_2^{2+} \rightarrow Os^{VI}(TMC)O_2^{2+} + Os^{V}(TMC)O_2^{+} \quad (19)$$

In both the above rhodium and osmium systems, the oxidation state studied by pulse radiolysis is at a thermodynamic maximum for the system, so that it is very difficult to utilize more conventional techniques such as electrochemical methods or chemical reductants to produce them for study. The difficulties of studying these unstable oxidation states are further exacerbated by their relatively low molar absorptivities and the overlap of their spectra with those of the higher oxidation states from which they are produced.

Synthetic Control as a Probe of Mechanism. In certain systems, synthetic control of very subtle power is possible. The oxidation of water to dioxygen by $Ru(bpy)_3^{3+}$ has long (*15–17*) been of interest because this process is one putative step in the use of $Ru(bpy)_3^{3+}$ to mediate the photodecomposition of water into its elements. In the reaction of $Ru(bpy)_3^{3+}$ with water/hydroxide ion, electron transfer to produce $Ru(bpy)_3^{2+}$ + ·OH can be postulated as an elementary step. The ·OH would be expected to add rapidly to the aromatic ring to yield $Ru(bpy)_2(bpyOH)^{2+}$. Studies of the $Ru(bpy)_3^{3+} + OH^{-}$ reaction by stopped-flow techniques reveal an intermediate absorbing at long wavelengths. To test whether this species might be the ·OH-radical adduct, pulse radiolysis was used to study the reaction of ·OH with $Ru(bpy)_3^{2+}$.

$$Ru(bpy)_3^{3+} + OH^{-} \rightarrow Ru(bpy)_2(bpyOH)^{2+} \quad (20)$$

$$Ru(bpy)_3^{2+} + \cdot OH \rightarrow Ru(bpy)_2(bpyOH)^{2+} \quad (21)$$

Evidence that common adducts are formed in the two reactions provided substantiation for $Ru(bpy)_2(bpyOH)^{2+}$ species. Recently, these have been observed in the direct (uncatalyzed) oxidation of water in zeolites into which the $Ru(bpy)_3^{3+}$ has been incorporated (*18*).

Carbon Dioxide Complexes. In contrast to some of the unstable systems described above, macrocyclic cobalt(I) complexes can be generated by electrochemical or chemical methods in organic solvents, but when mixed with water react with it rapidly. We have conducted extensive studies of tetraazamacrocyclic cobalt complexes. The cobalt(I) complex was first described by Vaselevskis and Olson (*19*); its protonation in water was reported by Tait et al. (*20*); and its binding to CO_2 was discovered by Fisher and Eisenberg (*21*). Fujita carried out elegant work in acetonitrile to characterize the binding of CO_2 to the macrocycle and its reduction in protic media (*22, 23*). We wished to carry out parallel studies in aqueous media, but recognized that protonation of the low oxidation state would complicate the work. Pulse radiolysis ultimately proved extremely useful (*24, 25*).

One theme in the CO_2-binding systems is an "electron isomerism" reminiscent of the $Ru(bpy)_3^{3+}/OH^-$ and $Ru(bpy)_3^{2+}/OH\cdot$ systems described above. In principle, there are two routes to the cobalt(I)–CO_2 adduct:

$$CoL^{2+} + e^-_{aq} \rightarrow CoL^+ \qquad CoL^+ + CO_2 \rightarrow CoL(CO_2)^+ \tag{22}$$

$$CO_2 + e^-_{aq} \rightarrow CO_2^- \qquad CoL^{2+} + CO_2^- \rightarrow CoL(CO_2)^+ \tag{23}$$

Depending upon the complex, both may be observed. For the case of *rac* CoL, the two routes yield different isomers; addition of CO_2 to Co(I) produces primary *rac* $CoL(CO_2)^+$, whereas reaction of CO_2^- with CoL^{2+} results in secondary *rac* $CoL(CO_2)^+$. The isomers are depicted in Figure 4.

Primary *rac* $CoL(CO_2)^+$ also exhibits coordination isomerism (*22*); at low temperatures it is six-coordinate, with a solvent molecule occupying the position *trans* to CO_2, but at high temperatures a purple five-coordinate form predominates.

$$\underset{purple}{\text{primary } rac \text{ } CoLCO_2^+} + S \rightleftharpoons \underset{yellow}{\text{primary } rac \text{ } CoL(CO_2)(S)^+} \tag{24}$$

For S = CH_3CN (*22*), the enthalpy change, $\Delta H° = -6.2$ kcal mol^{-1}, and the entropy change, $\Delta S° = -26$ cal K^{-1} mol^{-1}, and for S = H_2O (*25*), $\Delta H° = -6.1$ kcal mol^{-1}, and $\Delta S° = -19$ cal K^{-1} mol^{-1}.

Ultimately, we learned to adequately control the starting isomer of the cobalt(II) macrocycle and the kinetics of formation of the H^+, CO, and CO_2 adducts so that a full comparison of the thermodynamics and kinetics of these adducts could be made for both stereoisomers of the macrocycle (*25*). Ironically, the binding constants for both CO and CO_2 to the *rac* isomer in water are so great that it would be very difficult to determine them in the absence

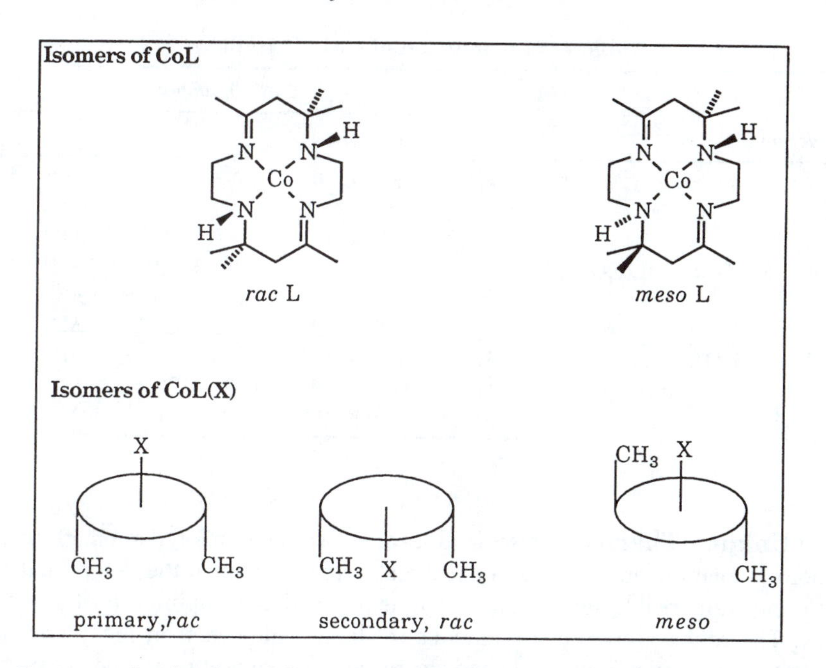

*Figure 4. The chirality of the amine nitrogen in the macrocyclic complex (*rac L *and* meso L*) leads to three potential isomers of the axially substituted complexes.*

of the protonation processes originally of great concern. Thus, data for the equilibrium

$$CoLCO_2{}^{+} + H^{+} \rightleftharpoons \text{primary } rac\ CoL(H)^{2+} + CO_2 \qquad (25)$$

were determined as a function of partial CO_2 (or CO) pressure and pH for the most accurate data. Combining our results with those of earlier work (*20*) led to the comparisons shown in Table I. As first reported by Tait et al. (*20*), *rac* CoL^{+} reacts with Brönsted acids, HA, with a rate constant that depends on the acidity of HA.

$$\text{primary } rac\ CoL^{+} + HA \rightleftharpoons \text{primary } rac\ CoL(H)^{2+} + A^{-}\ ,\ k_{HA},\ K_{eq} \qquad (26)$$

Indeed, a plot of log k_{HA} versus log K_{eq} for the proton transfer (log($K_{CoL(H)^{2+}}$) = 11.3) is linear, with a slope of ca. 0.5. Most remarkably, the rate constants for both CO and CO_2 addition to the macrocycle fall on the same plot. Protonation and formation of the metal–carbon bond appear to lend themselves to a description in terms of an associative reaction (S_N2), in which the low-spin d^8 cobalt(I) metal center serves as the nucleophile.

Table I. Comparison of CoL(X) Isomers

		k *(298 K), water*		
Reaction		*Primary* rac	meso	
$CoL^+ + CO_2 \rightleftharpoons CoL(CO_2)^+$	k_f	1.7×10^8	1.5×10^7	$M^{-1} s^{-1}$
	k_r	0.38	2.5	s^{-1}
	k_f/k_r	4.5×10^8	6.0×10^6	M^{-1}
$CoL^+ + CO \rightleftharpoons CoL(CO^+$	k_f	5.0×10^8	8.0×10^8	$M^{-1} s^{-1}$
	k_r	3.1	10	s^{-1}
	k_f/k_r	1.6×10^8	0.8×10^8	M^{-1}
$CoL^+ + H_3O^+ \rightleftharpoons CoL(H)^+$	k_f	3.1×10^9	2.4×10^9	$M^{-1} s^{-1}$
	k_r	1.2×10^{-2}	$<10^{-4}$	s^{-1}
	k_f/k_r	2.5×10^{11}	$>8 \times 10^{13}$	M^{-1}

Unique Thermodynamic Data. Exploitation of combinations of photo-, radiation, and electrochemical techniques has led to the determination of rather remarkable sets of data for aqueous media exemplified for those for bpy. The redox properties in aqueous media are important in many systems because bpy is such a useful ligand for ruthenium, rhodium, and cobalt metal centers. The full reduction scheme shown in Scheme I was determined through a combination of methods. The left-hand branch was characterized through studies of quenching the emission of polypyridyl–ruthenium(II) complexes as a function of pH. The bipyridine radical protonation equilibria and equilibrations (right-hand branch) were studied by pulse radiolysis methods (*26*). Although the bpy^-, produced through reduction of bpy by e^-_{aq}, protonates very rapidly

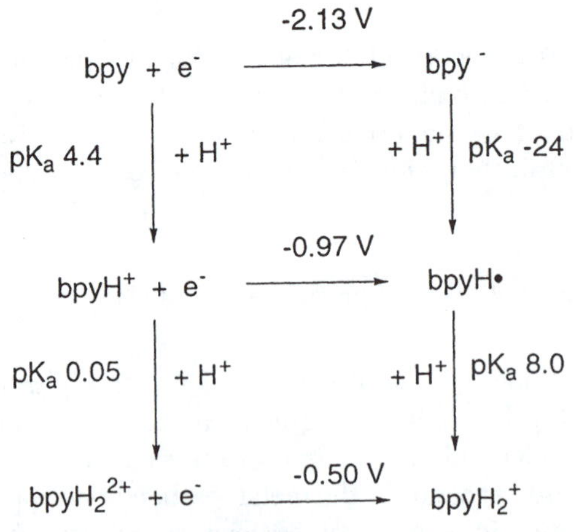

$$bpy + e^- \xrightarrow{-2.13\ V} bpy^-$$

pK$_a$ 4.4 (left, $+ H^+$) ↓   ↓ ($+ H^+$, right) pK$_a$ -24

$$bpyH^+ + e^- \xrightarrow{-0.97\ V} bpyH\bullet$$

pK$_a$ 0.05 (left, $+ H^+$) ↓   ↓ ($+ H^+$, right) pK$_a$ 8.0

$$bpyH_2^{2+} + e^- \xrightarrow{-0.50\ V} bpyH_2^+$$

Scheme I.

(27) to yield bpyH, equilibration of bpyH with phosphate buffers could be studied on the millisecond time scale. Furthermore, equilibration of $bpyH^+$–bpyH with $Co(bpy)_3{}^+$–$Co(bpy)_3{}^{2+}$ was utilized to define the reduction potentials for the protonated species. With the reduction potentials established, the quenching rate constants could be interpreted in terms of intrinsic and thermodynamic components to the rates.

Mechanistic Surprises

Both our photochemical and radiation studies have focused on the chemistry of very reactive species in aqueous solution. Indeed, it is because the photochemical work involved aqueous media that radiation chemistry techniques could be so useful to us. Our pulse radiolysis work has led to a number of highly unusual mechanistic conclusions. In the area of low-oxidation-state chemistry, several of the systems violate standard organometallic dogma. We investigated the rate of hydride formation in another cobalt(I) system, that derived from the high-spin d^8 polypyridyl–cobalt(I) complexes (28). Remarkably, electron transfer was found to be the rate-determining step for formation of the hydride complex, and contributions from Brönsted acid pathways contribute negligibly to the rate. Rather, the hydride formation appears to involve H-atom transfer from the protonated bpy radical. The "H-atom receptor" may be either $Co(bpy)_2{}^{2+}$ or $Co(bpy)^{2+}$ as shown in Scheme II.

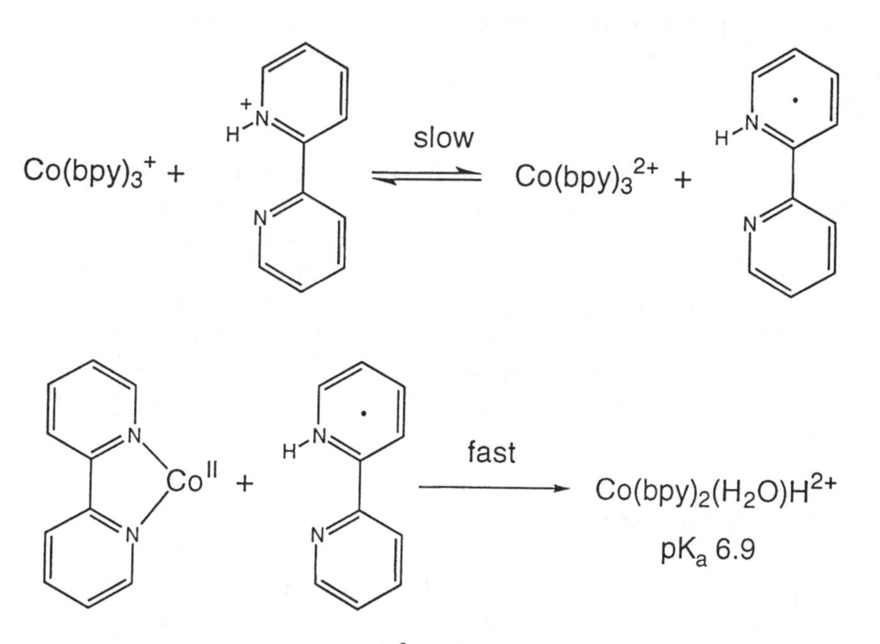

Scheme II.

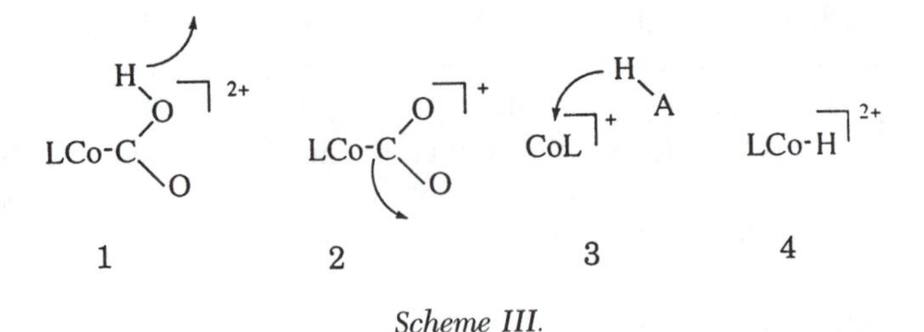

Scheme III.

Conversion of a metallocarboxylic acid to a hydride complex is supposed to occur via a hydrogen transfer to the metal as CO_2 is eliminated:

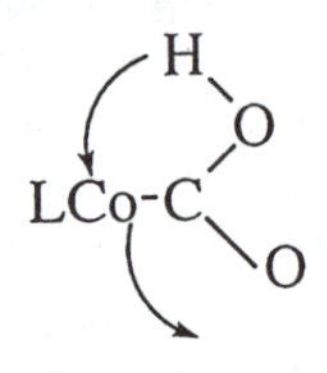

Instead, we found a strikingly different route for the macrocyclic cobalt systems. As shown in Scheme III, the metallocarboxylic acid (**1**) is essentially unreactive, but its conjugate base, the metallocarboxylate (**2**), can eliminate CO_2 at ~1 s^{-1} to form the Co(I) complex (**3**). The Co(I) complex can then undergo protonation by available proton sources to form the cobalt(III) hydride (**4**) (*24*).

Concluding Remarks

We and others have used pulse radiolysis methods to clarify a number of complex photochemical mechanisms. In the course of these studies we have also been able to learn a great deal of new chemistry, including the electronic absorption spectra, thermodynamics, and reaction mechanisms of highly reactive transition-metal centers in both unusually high and low oxidation states. As these data pertain to aqueous media, they contribute in an important way to future work on solar photoconversion in water (the ideal medium from both economic and environmental points of view) and to catalysis in aqueous media in general.

Acknowledgments

This research was carried out at Brookhaven National Laboratory under contract DE–AC02–76CH00016 with the U.S. Department of Energy and was supported by its Division of Chemical Sciences, Office of Basic Energy Sciences. We thank our many collaborators for their invaluable contributions to the work presented here.

References

1. Creutz, C. *Comments Inorg. Chem.* **1982,** *1,* 293–311.
2. Winkler, J. R.; Netzel, T. L.; Creutz, C.; Sutin, N. *J. Am. Chem. Soc.* **1987,** *109,* 2381–2392.
3. Neshvad, G.; Hoffman, M. Z.; Mulazzani, Q. G.; Venturi, M.; Ciano, M.; D'Angelantonio, M. *J. Phys. Chem.* **1989,** *93,* 6080–6088.
4. Mulazzani, Q. G.; Emmi, S.; Fuochi, P. G.; Hoffman, M. Z.; Venturi, M. *J. Am. Chem. Soc.* **1978,** *100,* 981–983.
5. Venturi, M.; Mulazzani, Q. G.; Ciano, M.; Hoffman, M. Z. *Inorg. Chem.* **1986,** *25,* 4493–4498.
6. Martin, J. E.; Hart, E. J.; Adamson, A. W.; Gafney, H.; Halpern, J. *J. Am. Chem. Soc.* **1972,** *94,* 9238–9240.
7. Jonah, C. D.; Matheson, M. S.; Meisel, D. *J. Am. Chem. Soc.* **1978,** *100,* 1449–1456.
8. Lehn, J.-M.; Sauvage, J.-P. *Nouv. J. Chim.* **1977,** *1,* 449.
9. Brown, G. M.; Chan, S. F.; Creutz, C.; Schwarz, H. A.; Sutin, N. *J. Am. Chem. Soc.* **1979,** *101,* 7638–7640.
10. Schwarz, H. A. *J. Phys. Chem.* **1982,** *86,* 3431–3435.
11. Schwarz, H. A.; Creutz, C. *Inorg. Chem.* **1983,** *22,* 707–713.
12. Che, C.-M.; Yam, V. W.-W.; Cho, K.-C.; Gray, H. B. *J. Chem. Soc. Chem. Commun.* **1987,** 948–949; Yam, V. W.-W.; Che, C.-M. *J. Chem. Soc. Dalton Trans.* **1990,** 3741–3746; Che, C.-M.; Cheng, W.-K.; Yam, V. W.-W. *J. Chem. Soc. Dalton Trans.* **1990,** 3095–3100.
13. Schindler, S.; Castner, E. W., Jr.; Creutz, C.; Sutin, N. *Inorg. Chem.* **1993,** *32,* 4200–4208.
14. Kelly, C.; Szalda, D. J.; Creutz, C.; Schwarz, H. A.; Sutin, N. *Inorg. Chim. Acta* **1996,** *243,* 39–45.
15. Creutz, C.; Sutin, N. *Proc. Natl. Acad. Sci. U.S.A.* **1975,** *72,* 2858–2862.
16. Ghosh, P. K.; Brunschwig, B. S.; Chou, M.; Creutz, C.; Sutin, N. *J. Am. Chem. Soc.* **1984,** *106,* 4772–4783.
17. Brunschwig, B. S.; Chou, M. H.; Creutz, C.; Ghosh, P.; Sutin, N. *J. Am. Chem. Soc.* **1983,** *105,* 4832–4833.
18. Ledney, M.; Dutta, P. K. *J. Am. Chem. Soc.* **1995,** *117,* 7687–7695.
19. Vaselevskis, J.; Olson, D. *Inorg. Chem.* **1971,** *10,* 1228.
20. Tait, A. M.; Hoffman, M. Z.; Hayon, E. *J. Am. Chem. Soc.* **1976,** *98,* 86.
21. Fisher, B.; Eisenberg, R. *J. Am. Chem. Soc.* **1980,** *102,* 7363–7366.
22. Fujita, E.; Szalda, D. J.; Creutz, C.; Sutin, N. *J. Am. Chem. Soc.* **1988,** *110,* 4870–4871.
23. Fujita, E.; Creutz, C.; Sutin, N.; Szalda, D. J. *J. Am. Chem. Soc.* **1991,** *113,* 343–353.
24. Creutz, C.; Schwarz, H. A.; Wishart, J. F.; Fujita, E.; Sutin, N. *J. Am. Chem. Soc.* **1989,** *111,* 1153–1154.
25. Creutz, C.; Schwarz, H. A.; Wishart, J. F.; Fujita, E.; Sutin, N. *J. Am. Chem. Soc.* **1991,** *113,* 3361–3371.
26. Krishnan, C. V.; Creutz, C.; Schwarz, H. A.; Sutin, N. *J. Am. Chem. Soc.* **1983,** *105,* 5617–5623.
27. Mulazzani, Q. G.; Emmi, S.; Fuochi, P. G.; Venturi, M.; Hoffman, M. Z.; Simic, M. G. *J. Phys. Chem.* **1979,** *83,* 1582.
28. Creutz, C.; Schwarz, H. A.; Sutin, N. *J. Am. Chem. Soc.* **1984,** *106,* 3036–3037.

15

Studies of Superoxide with Manganese Complexes and Manganese Superoxide Dismutase from *Escherichia coli*

Diane Esther Cabelli

Chemistry Department, Brookhaven National Laboratory, Upton, NY 11973–5000

The reactions of Mn^{II} TTHA (Mn^{II}–triethylenetetraminehexaacetate) complexes with HO_2–O_2^- radicals were studied (pH 2.5–9.5), and a mechanism was suggested that involves the formation of a transient Mn^{II}tthaH$(O_2^-)^{3-}$ complex. At low pH, this complex is protonated, with the release of H_2O_2. At higher pH, the dismutation of O_2^- from the equilibrium complex ($Mn^{II}TTHA(O_2^-)^{3-} \rightleftharpoons Mn^{II}TTHA + O_2^-$) is competitive with protonation. At low pH, the results indicate that there is a rapid first-order process that may be an isomerization from end-bound to side-bound of the attached superoxide radical. In contrast, the kinetics of the dismutation of superoxide radical by Escherichia coli *MnSOD (manganese superoxide dismutase) were measured and shown to fit a mechanism involving the rapid reduction of Mn^{3+}SOD by superoxide followed by both the direct reoxidation of Mn^{2+}SOD by superoxide and the formation of a $Mn^{II}SOD(O_2^-)$ complex. The differences in the mechanisms are discussed.*

The reactions of superoxide radicals with metal complexes have received much attention as O_2^- can play a role in cycling oxidation states in metal complexes (*1*). This feature has been implicated in catalytic processes such as the Fenton-type chemistry ($M^{n+} + H_2O_2 \rightarrow M^{(n+1)+} + HO + OH^-$; $M^{(n+1)+} + O_2^- \rightarrow M^{n+} + O_2$), which is thought to be involved in hydroxylation processes (*2*). Reactions involving O_2^- radicals have also been tied to the deleterious effects of oxygen upon aerobic organisms (*1, 3*), probably as a result of similar Fenton-type chemical reactions.

Superoxide dismutases (SODs) are thought to have evolved as nature's way of disposing of the superoxide produced in aerobic cells. They are metal-containing enzymes that catalyze the dismutation of superoxide to dioxygen and hydrogen peroxide (reaction 1) (*3, 4*).

$$2O_2^- + 2H^+ \xrightarrow{\text{SOD}} O_2 + H_2O_2 \quad (1)$$

SODs are differentiated mainly by the redox-active metal in the active site: copper, manganese, or iron. The iron and manganese SODs are structurally similar (*5–11*) and are structurally distinct from the Cu,Zn SOD (*12*). The dramatic features of these enzymes are that they catalytically dismutate superoxide at rates that are not only diffusion controlled but have been shown to be electrostatically facilitated (*13*). In these systems, modifications of amino acid residues near the active site have been shown to alter the enzymatic activity, indicating that superoxide is electrostatically drawn into the active site channel (*14*). In addition, in contrast to the spontaneous dismutation rate of O_2^- and the dismutation rates of O_2^- by many metal complexes, all of which are pH dependent, the enzymatic dismutation rate is largely pH independent over the pH range (*5–10*).

The mechanisms by which manganese complexes and manganese superoxide dismutase react with superoxide radicals are of interest as knowledge of the kinetic parameters and the reaction pathways may allow the synthesis of model compounds with specific chemical features. These compounds may then have clinical application or may allow the control of specific redox chemistry in catalytic processes.

This chapter describes the reactions between mononuclear Mn^{II}TTHA complexes and HO_2–O_2^- radicals over a broad pH range. The polyaminocarboxylate ligand TTHA (triethylenetetraminehexaacetate) was chosen as the ligand in this study because the manganese(II) complexes in equilibrium at a particular pH are well established (*15, 16*), soluble in water, and stable over a wide pH range (pH 2 to 10). TTHA binds manganous ion tightly and can form both mononuclear and binuclear complexes (*15, 16*). The experiments described here were carried out under conditions in which only mononuclear complexes were formed. We discuss the overall mechanism in light of a reaction scheme suggested earlier from studies involving the reactions of simple manganous complexes (Mn^{II}–sulfate, Mn^{II}–formate, Mn^{II}–pyrophosphate, and Mn^{II}–phosphate) with HO_2–O_2^- radicals (*17, 18*). Finally, we examine the reaction between MnSOD from *E. coli* and HO_2–O_2^- radicals and contrast the enzymatic system with systems involving simple mononuclear complexes.

Materials and Methods

Materials. All solutions were prepared using water which, after distillation, had been passed through a Millipore ultrapurification system. TTHA (triethylenetetramine hexaacetate), EDTA (ethylenediaminetetraacetate), and sodium formate were purchased from Sigma Chemical Co. Very pure manganous sulfate was used as purchased (Alfa/Ventron Corp. Puratronic, 99.998%). The pHs of all solutions were adjusted by the addition of H_2SO_4 (double distilled from Vycor, GFS Chemical Co.) and NaOH (Puratronic, JT Baker Chemical Co.). Monobasic phosphate (Ultrex, JT Baker Chemical Co.) was used as a buffer in the MnSOD experiments; the excess TTHA and/or formate provided sufficient buffering capacity to maintain constant pH in the MnTTHA experiments. *Escherichia coli* MnSOD was used as purchased (Sigma Chemical Co.). The O_2 used was ultrahigh-purity (UHP) grade (99.999%).

Methods. The pulse radiolysis experiments were carried out using the 2-MeV van de Graaff accelerator at Brookhaven National Laboratory. Dosimetry was established using the KSCN dosimeter, assuming that $(SCN)_2^-$ has a G value (the number of atoms or molecules formed per 100 eV of energy dissipated in the solution) of 6.13 and a molar absorptivity of 7950 M^{-1} cm^{-1} at 472 nm. All UV–Vis spectra were recorded on a Cary 210 spectrophotometer thermostated at 25 °C. The actual concentration of manganese in the *E. coli* MnSOD was determined by atomic absorption (AA) using a Pye-Unicam AA instrument. The reported rate constants for the MnSOD studies are based on manganese concentration and not MnSOD concentration, with the assumption that all of the metal is bound and active and behaves independently from the other metal centers.

Superoxide radicals were generated upon pulse radiolysis of an aqueous, air–O_2 saturated solution containing sodium formate according to the following mechanism (*19*):

$$H_2O^- \rightsquigarrow OH\ (2.75),\ e^-_{aq}\ (2.65),\ H\ (0.65),\ H_2\ (0.45),\ H_2O_2\ (0.70) \quad (1)$$

where the values in parentheses are G values, and

$$OH + HCO_2^- \rightarrow H_2O + CO_2^- \quad (2)$$

$$CO_2^- + O_2 \rightarrow CO_2 + O_2^- \quad (3)$$

$$e^-_{aq} + O_2 \rightarrow HO_2^- \quad (4)$$

$$H + O_2 \rightarrow HO_2 \quad (5)$$

where

$$HO_2 \rightleftharpoons O_2^- + H^+ \text{ and } pK_6 = 4.8\ (13). \quad (6, -6)$$

Under our conditions, the formation of HO_2–O_2^- radicals is more than 90% complete by the first microsecond after the pulse.

Results and Discussion

Manganese(II)TTHA Complexes. Between pH 2 and 9, manganous ions are known to form three complexes with $ttha^{6-}$ as given in equilibria 7 and 8, where $pK_7 = 8.75$ and $pK_8 = 3.5$ (*15, 16*).

$$Mn^{II}ttha^{4-} + H^+ \rightleftharpoons Mn^{II}tthaH^{3-} \qquad (7, -7)$$

$$Mn^{II}tthaH^{3-} + H^+ \rightleftharpoons Mn^{II}tthaH_2^{2-} \qquad (8, -8)$$

When it is impossible to establish the exact protonation state of the complex(es) in a specific reaction, or when all complexes in equilibrium over a broad pH range are under discussion, they will be written here as Mn^{II}TTHA, Mn^{III}TTHA, etc. The reactions of Mn^{II}TTHA complexes with HO_2–O_2^- radicals were studied in solutions containing at least a twofold excess of TTHA to $MnSO_4$ and a tenfold excess of formate over TTHA, in order to ensure scavenging of the OH radical by formate and not TTHA. Under these conditions, only the mononuclear complexes described above were formed in significant yield. Studies were not carried out at pH < 2 because preliminary results indicated a complex reaction mechanism, likely due to many different manganous complexes (various manganous formate complexes, aquo Mn^{2+}) in equilibrium.

Upon reaction of HO_2–O_2^- with Mn^{II}TTHA (oxygen-saturated solutions containing 0.1 M formate, 10 mM TTHA, and 0.1–5.0 mM $MnSO_4$, pH 2–9), a transient is formed with an absorption maximum at 350 nm (ε_{350} nm $= 3200$ M^{-1} cm^{-1}), a shoulder at 400 nm, and an absorbance in the UV (*see* Figure 1). Although the spectrum does not change over this pH range, the kinetics of its formation and disappearance vary with pH.

At pH 6–9, the observed rate of formation of the transient is first-order in [Mn^{II}TTHA], and a plot of the observed rate as a function of [Mn^{II}TTHA] yields a slope and intercept (Figure 2). This kinetic behavior indicates an equilibrium between O_2^- and $Mn^{II}tthaH(O_2^-)^{3-}$. Equilibria such as this were observed previously (*17, 18, 20, 21*) in the reactions of Mn^{2+} complexes with O_2^-.

$$Mn^{II}tthaH^{3-} + O_2^- \rightleftharpoons Mn^{II}tthaH(O_2^-)^{3-} \qquad (9, -9)$$

The rate constants, k_9 and k_{-9}, and the equilibrium constant, K_9, in this pH range are given in Table I. The relatively constant value of k_9 over this broad pH range, coupled with the known values for K_6, K_7, and K_8, suggests that the monoprotonated complex reacts with O_2^- between pH 6 and 9. The protonation state of the transient is unknown and is written as $Mn^{II}tthaH(O_2^-)^{3-}$ as a convenience. Although the spectral characteristics of this transient are invariant

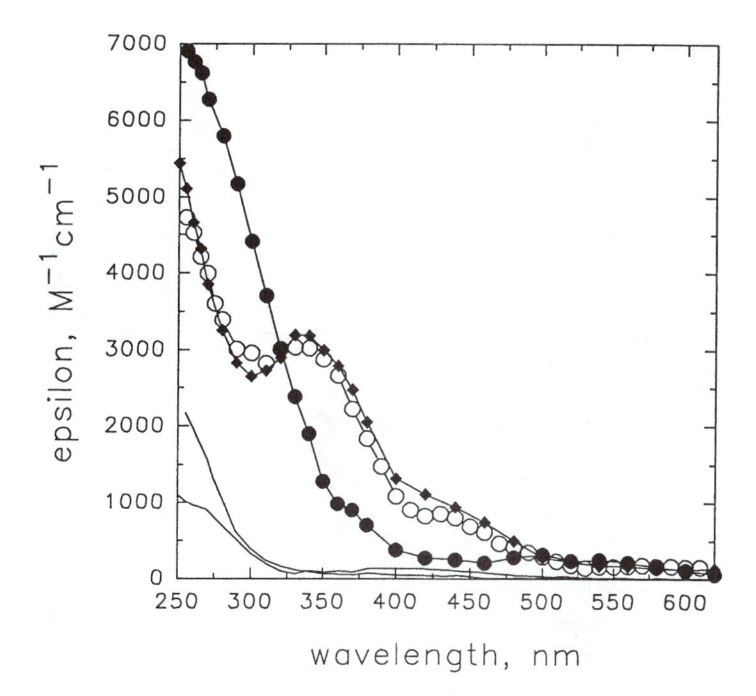

Figure 1. Spectra of species produced upon reaction of HO_2–O_2^- with Mn^{II}TTHA complexes: O_2^- (——) and Mn^{II}tthaH(O_2^-)$^{3-}$ (○) as measured at pH 7.0 (0.1 M formate, 0.01 M TTHA, and 5 mM $MnSO_4$); HO_2 (——) and Mn^{II}tthaH(O_2^-)$^{3-}$ (◆) and Mn^{III}TTHA (●) as measured at pH 2.75 (0.1 M formate, 0.01 M TTHA, and 5 mM $MnSO_4$).

from pH 2.5–9.0, the kinetic data in Table II suggest that there might be a protonation or an isomerization of the transient.

At pH 6–9, the transient Mn^{II}tthaH(O_2^-)$^{3-}$ complex disappears by a second-order process that varies nonlinearly with [Mn^{2+}TTHA] and almost linearly with [H^+]. The equilibrium constant, K_9', is also calculated from a study of [Mn^{2+}TTHA] and pH dependence on the observed rate of disappearance of Mn^{II}tthaH(O_2^-)$^{3-}$, using equation II. This kinetic equation is easily derived assuming a mechanism involving equilibria 6 and 9 and reaction 10. Equilibrium 6 is necessary to the overall mechanism, as the observed spontaneous disappearance of O_2^- is pH dependent (22).

$$O_2^- + HO_2 \xrightarrow{H^+} H_2O_2 + O_2 \quad (10)$$

$$k_{obs},\ M^{-1}\ s^{-1} = \frac{k_{10}[H^+]}{K_6(1 + K_9[Mn^{II}TTHA])^2} \quad (II)$$

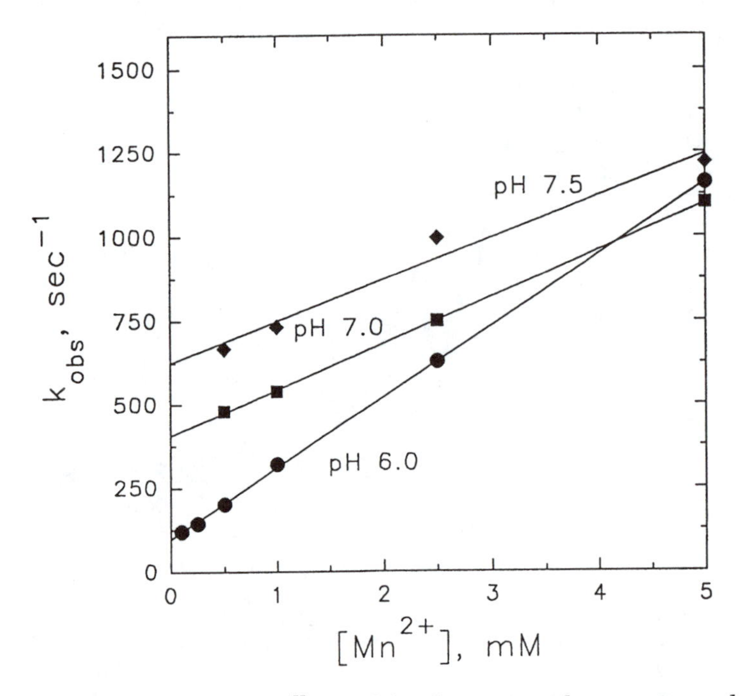

Figure 2. Plots of k_{obs} *vs.* $[Mn^{II}TTHA]$ *for the reactions between* O_2^- *and the* $Mn^{II}TTHA$ *complexes at pH 6.0 (●), 7.0 (■), and 7.5 (◆) (0.1 M formate, 0.01 M TTHA, and 1.2 mM* O_2*). The lines are least-squares fits to the data points.*

Table I. Measured Forward and Reverse Rate Constants for the Reaction of $Mn^{II}tthaH^{3-}$ with O_2^- (pH 6–9) and the Corresponding Equilibrium Constants

pH	k_9 $(M^{-1}s^{-1})$	k_{-9} (s^{-1})	K_9 (M^{-1})	K_9' (M^{-1})
6.0	2.1×10^5	97.4	2.2×10^3	3.0×10^3
6.5	1.8×10^5	1.4×10^2	1.3×10^3	1.1×10^3
7.0	1.5×10^5	3.9×10^2	3.4×10^2	4.3×10^2
7.5	1.4×10^5	6.1×10^2	2.0×10^2	3.4×10^2
8.3	2.0×10^5	6.0×10^2	3.3×10^2	
9.0	9.3×10^4	6.0×10^2	1.6×10^2	

NOTE: $K_9 = k_9/k_{-9}$, and K_9' was calculated using equation II and the observed rate of disappearance of $Mn^{II}tthaH(O_2^-)^{3-}$ at varying $[Mn^{II}tthaH^{3-}]$.

Using the known values (*22*) of $k_{10} = 1 \times 10^8$ M^{-1} s^{-1} and $pK_6 = 4.8$, K_9 was calculated at different pHs. These calculated equilibrium constants, K'_9, are in reasonable agreement with the equilibrium constants K_9, obtained by dividing k_9 by k_{-9} (*see* Table I). Our results do not preclude the possibility that $Mn^{II}tthaH(O_2^-)^{3-}$ may itself react with O_2^-, as was observed earlier (*17*) in the reaction between $Mn^{II}sulfate(O_2^-)$ and O_2^-, or that $Mn^{II}tthaH(O_2^-)^{3-}$ may disproportionate, but they indicate that these are minor pathways here.

At lower pH, the mechanism is somewhat more complicated. Upon pulse radiolysis of aqueous oxygen-saturated solutions of $Mn^{II}TTHA$ at pH 2.5–4.0 (the same conditions as described above), two processes are observed in the spectral range of 250–400 nm. The initial reaction leads to the formation of a transient spectrally identical to that formed at higher pH. This transient disappears with the concomitant formation of a species with an absorbance in the UV similar to the spectra of other manganese(III) complexes (*see* Figure 1). The kinetic traces corresponding to both of these processes are fitted to first-order kinetics. A plot of the reciprocal of the rate of formation of Mn^{II}-$tthaH(O_2^-)^{3-}$ versus the reciprocal of the total [MnTTHA] at different pHs, yields a set of straight lines; *see* Figure 3 and Table II for the slope and intercept

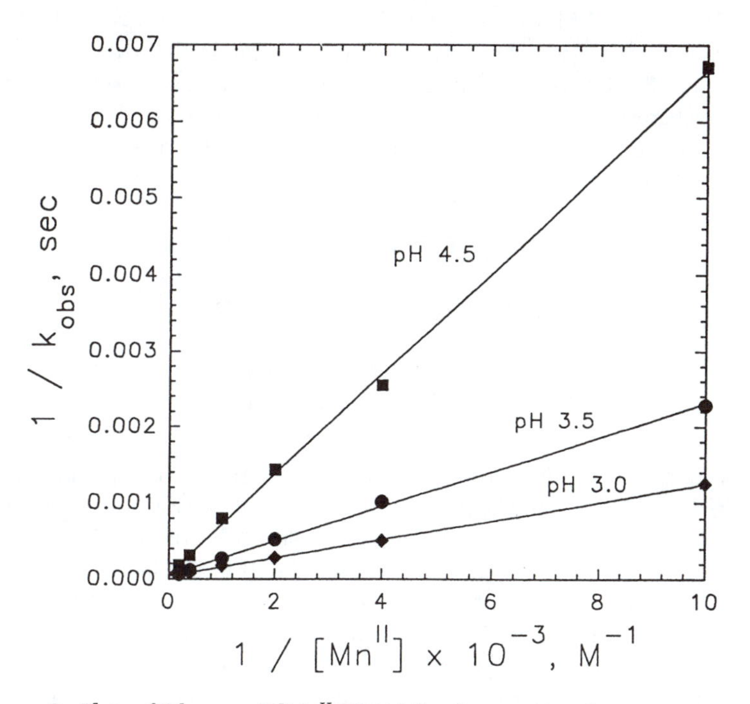

Figure 3. Plots of $1/k_{obs}$ *vs.* $1/[Mn^{II}TTHA]$ *for the reactions between* HO_2–O_2^- *and the* $Mn^{II}TTHA$ *complexes in equilibrium at pH 3.0, 3.5, and 4.5 (0.1 M formate, 0.01 M TTHA, and 1.2 mM* O_2*). The lines are least-squares fits to the experimental points.*

Table II. Equilibrium Constants for the Reactions of O_2^-/HO_2 Radicals with $Mn^{II}TTHA$ Complexes (pH 3–4.5) and Rate Constants for the Subsequent Process at Various pHs

pH	K_{11}, M^{-1}	k_{12}, s^{-1}
3.0	3.37×10^2	2.46×10^4
3.5	2.34×10^2	1.9×10^4
4.5	9.25×10^1	1.64×10^4

NOTE: Numbers were calculated from the data in Figure 3, using equation III.

at different pHs. This behavior suggests that either different ratios of manganous complexes were in equilibrium in the reaction mixture at the different $Mn^{II}TTHA$ concentrations or that the mechanism consists of an equilibrium followed by a fast unidirectional process. The former possibility was eliminated by two sets of experiments, both carried out at pH 3.0. In one experiment, [TTHA] was maintained at 0.01 M while the $[MnSO_4]$ was varied from 0.005–0.0001M. In the other experiment, the ratio $[MnSO_4]$: [TTHA] was kept constant at 1 : 2 and the total concentration was varied from 0.01 M TTHA/0.005 M $MnSO_4$ to 0.0002 M TTHA/0.0001 M $MnSO_4$. No difference in the kinetic behavior was observed in these two systems, suggesting that the ratio of complexes in equilibrium is constant as the $MnSO_4$ concentration is changed.

A mechanism consistent with this process is, therefore, equilibrium 11 followed by reaction 12. A kinetic equation to describe this mechanism is given by equation III.

$$HO_2 + Mn^{II}TTHA \rightleftharpoons Mn^{II}TTHA(HO_2) \qquad (11, -11)$$

$$Mn^{II}TTHA(HO_2) \rightarrow Mn^{II}tthaH(O_2^-)^{3-} \qquad (12)$$

$$k_{obs}, s^{-1} = \frac{k_{12}K_{11}[Mn^{II}TTHA]}{(1 + K_{11}[Mn^{II}TTHA])} \qquad (III)$$

Reaction 12 could plausibly be attributed to an isomerization of the bound HO_2. A likely isomerization pathway that has precedent in Fe–EDTA complexes (*23*) is that of isomerization of an end-bound O_2^- moiety to a side-bound O_2^- moiety.

In contrast to the second-order process for the disappearance of Mn^{II}-$tthaH(O_2^-)^{3-}$ observed at pH 6–9, in this pH range the disappearance of Mn^{II}-$tthaH(O_2^-)^{3-}$ occurs by a first-order process. The rate is independent of $[Mn^{2+}TTHA]$, [TTHA], and $[HCO_2^-]$ but varies linearly with $[H^+]$ (*see* Figure 4). The rate constant for the reaction between $Mn^{II}tthaH(O_2^-)^{3-}$ and H^+ in 1.0 M $NaClO_4$ is $k_{13} = 9.5 \times 10^4\ M^{-1}\ s^{-1}$.

$$Mn^{II}tthaH(O_2^-)^{3-} + H^+ \rightarrow Mn^{III}TTHA + HO_2^- \qquad (13)$$

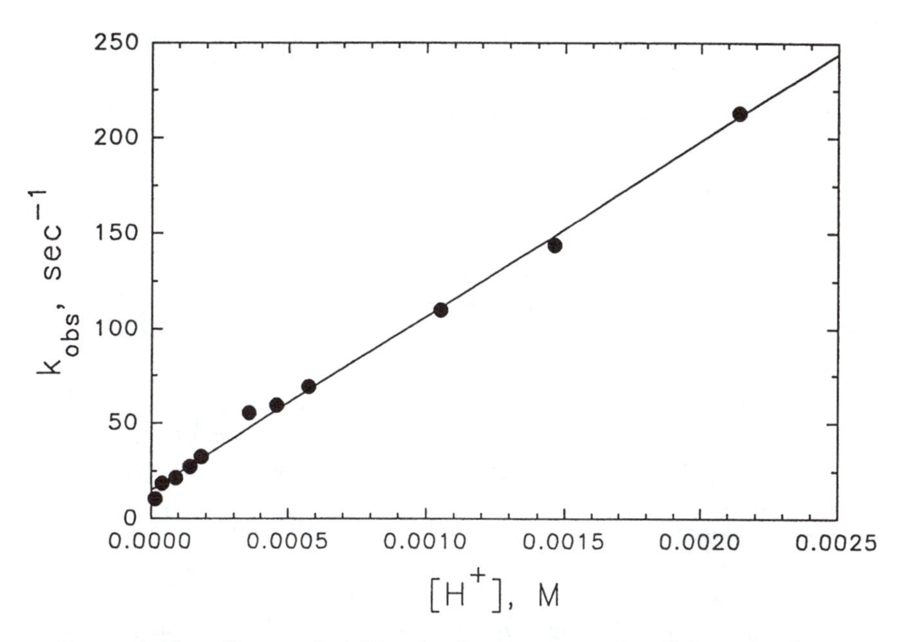

Figure 4. Plot of k_{obs} *vs.* $[H^+]$ *for the disappearance of* $Mn^{II}tthaH(O_2^-)^{3-}$ *and concomitant formation of* $Mn^{III}TTHA$ *(1.0 M* $NaClO_4$*, 0.1 M formate, 0.01 M TTHA, 2 mM* $Mn^{II}TTHA$*, and 1.2 mM* O_2*). The line represents a least-squares fit to the experimental points.*

A significant ionic strength dependence was observed here, with the reaction becoming more rapid at lower ionic strength. This is in accord with a negatively charged $Mn^{II}tthaH(O_2^-)^{3-}$ complex reacting with a positively charged H^+.

This reaction is observed in competition with the dismutation of O_2^- at pH 5.5–7.0; at low concentrations of O_2^-, where reaction 10 has a longer half-life, the protonation reaction (reaction 13) dominates, whereas at high $[O_2^-]$, reaction 10 is the predominant pathway. The spectrum of $Mn^{III}TTHA$ (Figure 1) has a large absorption spectrum in the UV with a maximum that is near or slightly less than 250 nm (ε_{250nm} = 7000 M^- 1 cm^{-1}), consistent with spectral characteristics observed previously for Mn^{3+} complexes in aqueous solution (*17, 18*).

The reactions of Mn^{2+} complexes with HO_2–O_2^- have been studied for several Mn^{2+} complexes, and an overall mechanism has been postulated that involves a number of equilibria and reactions, with rate constants that are ligand-dependent (*see* Scheme I). The mechanism shown here can generally be accommodated in this overall mechanism. Rate constants for the reaction of O_2^- with various Mn^{II} complexes and for the disappearance of the $Mn^{II}(O_2^-)$ complexes are given in Table III. A pH-dependent disappearance of the transient $Mn^{II}(O_2^-)$ complexes was observed in the reactions of EDTA and nitrilotriace-

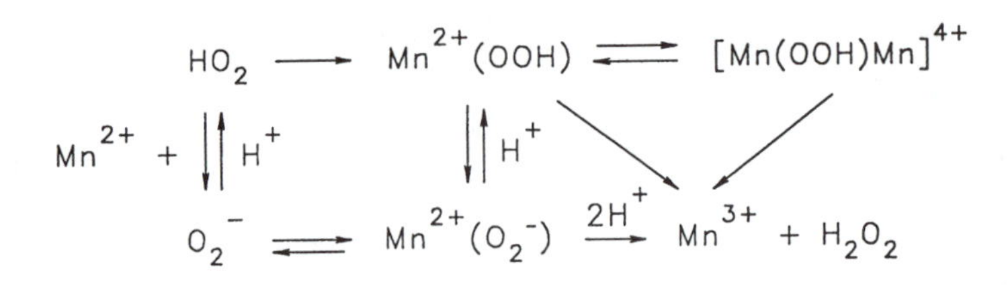

Scheme I.

tate (NTA) (*15*) complexes of Mn^{2+} with O_2^-, and the pathway likely involves the same protonation of the bound O_2^- to peroxide that is observed in this system (reaction 13). The isomerization reaction (reaction 12) that we postulate at lower pH has not been observed in other manganous systems (*17, 18*), but this step may be pH dependent and only a few of these complexes were studied at low pH. An isomerization such as this was postulated earlier for the mechanism of the reaction of O_2^- with MnSOD (*see* the next section). The relatively slow reaction between Mn^{2+} and O_2^- observed at neutral to alkaline pH is also quite reasonable in a system where the Mn(II) ion is bound by a chelating agent that not only has the ability to coordinate to all available binding sites of the metal but may also sterically hinder the access of O_2^- to the metal. The

Table III. Rate Constants for the Reaction of Various Mn^{2+} Complexes with Superoxide Radicals and the Disappearance of Mn^{2+} (O_2^-) Complexes

Ligand	k $(Mn^{2+} + O_2^-)$ $(M^{-1}\,s^{-1})$	k $(Mn^{II}(O_2^-) \rightarrow Mn^{3+})$ (s^{-1})	*Ref.*
$P_2O_7^{4-}$	1.3×10^7 (pH 7.3)		32
	2.6×10^7 (pH 6.5)	$>2 \times 10^4$	17
PO_4^{3-}	5×10^7 (pH 2–6)	2×10^3	17
NTA[a]	4×10^8 (pH 4.5)	3×10^3	21
EDTA[b]	3×10^7 (pH 4.5)	1.8×10^3	21
	7.5×10^6 (pH 5.5)	9×10^1	21
	$<1 \times 10^6$ (pH 6.0)		33
EDDA[c]	3×10^7 (pH 7.1)		34
DTPA[d]	$<1 \times 10^6$ (pH 6.0)		33
Formate	4.3×10^7 (pH 5.7–7.1)	<5	17
Sulfate	5.4×10^7	<1.4	17
TTHA	2×10^5	9×10^4[e]	This work

[a] NTA = nitrilotriacetate, $N(CH_2CO_2^-)_3$.
[b] EDTA = ethylenediaminetetraacetate, $(^-O_2CCH_2)_2N(CH_2)_2N(CH_2CO_2^-)_2$.
[c] EDDA = ethylenediaminediacetate, $(CH_2NH(CH_2CO_2^-))_2$.
[d] DTPA = diethylenetriaminepentaacetate.
[e] Reaction with H^+ (*see* the text).

interesting feature of all of these systems, however, is that although the rate constants for reaction between Mn^{2+} complexes and O_2^- can range from 10^8 M^{-1} s^{-1} to the somewhat slower rates observed here, the reaction between Mn^{3+} and O_2^- is uniformly slower, as reflected in the measured catalytic dismutation rates of simple Mn^{2+} complexes with superoxide. Direct measurements of the rate constants for the reaction of both $Mn^{III}EDTA$ (*24*) and $Mn^{III}NTA$ (*25*) with O_2^- also yielded a relatively slow rate constant: k ($Mn^{III}EDTA$ + O_2^-) ≈ 5 × 10^4 M^{-1} s^{-1} at pH 10.0, and k ($Mn^{III}NTA$ + O_2^-) = 1.2 × 10^7 M^{-1} s^{-1} at pH 6.0.

***E. coli* MnSOD.** Manganese superoxide dismutases are either dimeric or tetrameric enzymes containing monomeric units of approximately 22,000 g/mol (*5*). In general, eucaryotic MnSODs are dimers while prokaryotic MnSODs are tetramers. The structure has been established for a variety of manganese superoxide dismutases (*6–8*); however, the structure of the MnSOD from *E. coli* has not been determined. Therefore, the rate constants reported here are given per manganese concentration and not per enzyme concentration, where the manganese concentration of the enzyme was determined by atomic absorption, as discussed above.

Air–oxygen saturated solutions of varying concentrations of *E. coli* MnSOD (0.214–10.7 μM manganese as determined by atomic absorption spectroscopy) containing 10 mM formate, 50 mM phosphate, and 10–100 μM EDTA were pulse irradiated. Kinetic traces of the disappearance of the absorbance at 260 nm were recorded at different concentrations of MnSOD (Figure 5). When the ratio of superoxide to manganese was small, the absorbance due to superoxide radical disappeared by a first-order process (Figure 5B). A subsequent growth of an absorbance in the UV also occurred by a first-order process (Figure 5C). However, with increased superoxide concentration relative to the enzyme concentration, the kinetic trace clearly indicated complex kinetics (Figure 5A). Our results are consistent with studies reported (*26–29*) earlier, and our data could be fitted (using PRKIN, written by H. A. Schwarz, Brookhaven National Laboratory, Upton, NY) by the mechanism given by McAdam et al. in an earlier study of *T. thermophilus* MnSOD (Scheme II) (*26, 27*). The observed differences in the kinetic traces as the MnSOD concentration is varied can be explained by assuming that reactions 14 and 15 are substantially faster than reaction 16 and that reaction 17 is relatively slow. Reactions 14 and 15 constitute a catalytic cycle that is analogous to that observed in the mechanism of the reaction between Cu,Zn superoxide dismutase and O_2^-. The data in Figure 5B represent conditions where only this fast catalytic cycle (reactions 14 and 15) is operative to any extent, and thus the disappearance of the O_2^- absorbance can be fitted to a first-order process. The regrowth of an absorbance at 260 nm as seen in Figure 5C can be explained as the growth of the absorbance of $Mn^{III}SOD$. The resting state of the enzyme has been shown to be the Mn(III) state (*30*), and $Mn^{III}SOD$ has a small absorbance at 260 nm while $Mn^{II}SOD$

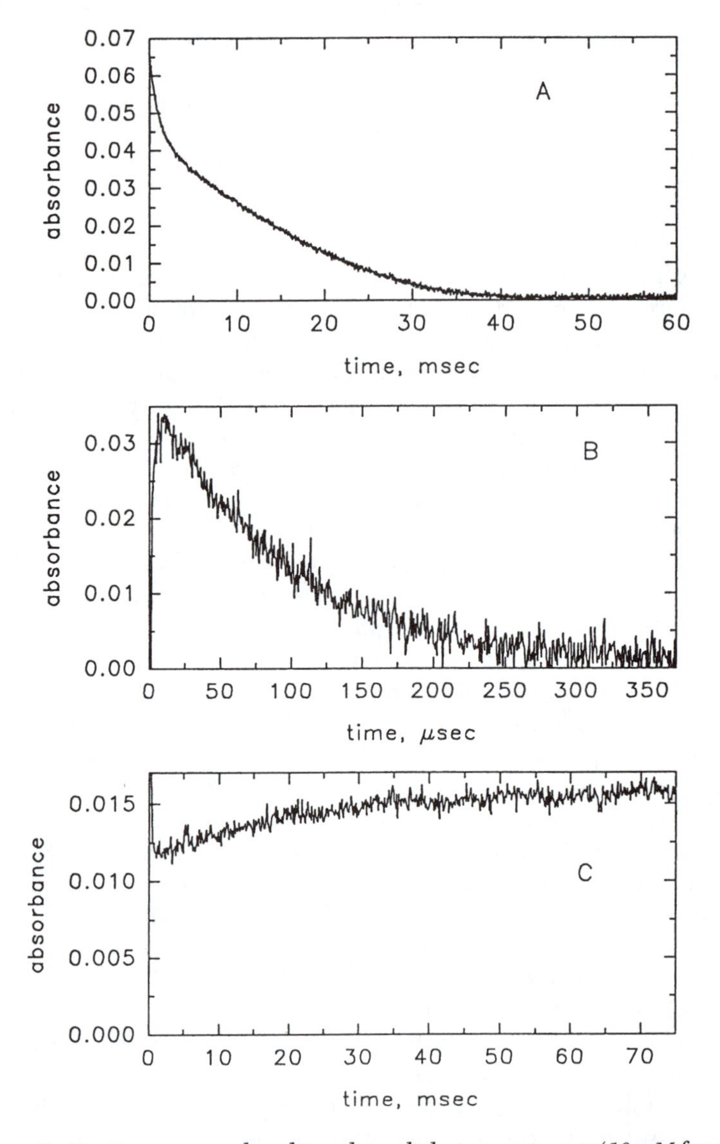

Figure 5. Kinetic traces produced in pulse radiolysis experiments (10 mM formate, 50 mM phosphate, and 260 nm). Total manganese concentration of the E. coli *MnSOD: (A) 0.41 μM manganese, 16 μM* O_2^-*; (B) 8.1 μM manganese, 8 μM* O_2^-*; and (C) 8.1 μM manganese, 8 μM* O_2^-*.*

$$Mn^{3+}SOD + O_2^- \xrightarrow{k_{14}} Mn^{2+}SOD + O_2$$

$$Mn^{2+}SOD + O_2^- \xrightarrow{k_{15},\ 2H^+} Mn^{3+}SOD + H_2O_2$$

$$Mn^{2+}SOD + O_2^- \xrightarrow{k_{16}} [Mn^{2+}SOD(O_2^-)]$$

$$[Mn^{2+}SOD(O_2^-)] \xrightarrow{k_{17},\ 2H^+} Mn^{3+}SOD + H_2O_2$$

Scheme II.

absorbs even less here. Under conditions where the concentration of O_2^- is roughly equivalent to the concentration of manganese in the MnSOD, the enzyme is not in turnover, and the ratio $[Mn^{II}]/[Mn^{III}]$ varies depending upon the initial ratio in the enzyme and the amount of O_2^- that is generated relative to the manganese concentration. This leads to the net change in absorbance at 260 nm observed here, corresponding to reaction 17; $k_{17} = 70\ s^{-1}$.

The rate constant for reaction 16 is obtained using the data generated at low manganese concentration relative to the superoxide concentration and using PRKIN to fit theoretical curve of absorbance versus time to this data, using the mechanism given in Scheme II. This calculated rate constant is roughly 10-fold slower than the catalytic rate (reactions 14 and 15) measured under conditions described in Figure 5A. The rate constant for reaction 17 is also extracted from the data in Figure 5A using PRKIN; $k_{17} \approx 70\ s^{-1}$ independent of [MnSOD], $[O_2^-]$, and ionic strength. This value is in good agreement with that measured from the regrowth at 260 nm (Figure 5C).

In Figure 6, the rate constants for reactions 14 and 16 are plotted as a function of pH. As can be seen, the rate constants are consistent over a wide range of manganese and O_2^- concentrations. In the mechanism suggested by McAdam et. al., the alternate pathways for the reaction between $Mn^{II}SOD$ and O_2^- were suggested to be controlled by a conformational change in the enzyme. The ionic strength dependence of the rate constants supports this mechanism in that significant decreases in the rate constants k_{14}, k_{15}, and k_{16} with increasing ionic strength have been observed (*14*), while no ionic strength effect upon k_{17} was observed (*14*). The latter effect is consistent with dissociation of a bound O_2^-, which should be ionic strength independent.

An alternate mechanism to that proposed above by McAdam et al. (*26, 27*) has been postulated by Fee et al. (*29*) (Scheme III). In this mechanism, the interaction of MnSOD with O_2^- was suggested to involve equilibria and, instead of a mechanism involving dual reoxidation pathways separated by a conformational change, an isomerization of the superoxide radical from an end-

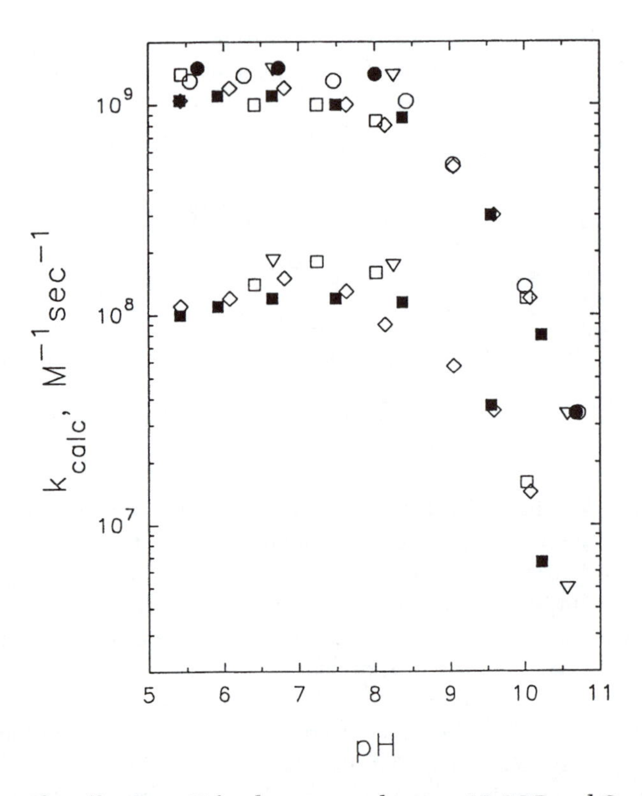

Figure 6. Plot of k_{calc} vs. pH for the reactions between MnSOD and O_2^- (10 mM formate, 50 mM phosphate, 0.25 or 1.2 mM O_2, 2–20 μM O_2^-, 260 nm). Key: (○) 10.7 μM manganese; (●) 8.1 μM manganese; (▽) 1.92 μM manganese; (◇) 0.714 μM manganese; (□) 0.411 μM manganese; and (■) 0.214 μM manganese. The rate constants were calculated by modeling the kinetic traces using the mechanism of McAdam et. al. and the modeling software PRKIN.

$$Mn^{3+}SOD + O_2^- \rightleftharpoons [Mn^{3+}SOD(O_2^-)] \rightarrow Mn^{2+}SOD + O_2$$

$$Mn^{2+}SOD + O_2^- \rightleftharpoons [Mn^{2+}SOD(O_2^-)] \xrightarrow{2H^+} Mn^{3+}SOD + H_2O_2$$

$$\upharpoonleft\downharpoonright$$

$$Mn^{2+}SOD(O{-}O^-)]$$

Scheme III.

on to a side-on configuration was suggested. This mechanism is attractive in that it parallels chemical behavior observed in Fe–EDTA complexes (*21*) and the known behavior of many metal complexes (*31*). However, on the basis of our experiments and the published data of previous experiments, it is impossible to distinguish between these two mechanistic pathways.

In MnSOD, as in many manganous complexes, no significant reaction between the reduced metal center and peroxide has been observed. This makes it an interesting enzyme in terms of a potential therapeutic agent, as no adverse effect from a Fenton-type reaction ($M^{n+} + H_2O_2 \rightarrow M^{(n+1)} + OH^{\cdot} + OH^-$) is likely.

Conclusions

The contrast between the reactions of manganous complexes and MnSOD with superoxide is striking. The reactions between the lower oxidation state manganous complexes and O_2^- are similar in both the enzymatic and nonenzymatic systems, where $Mn^{II}(O_2^-)$ transient adducts are formed. The mechanism observed in MnSOD is somewhat more involved than those of simple Mn^{II} complexes in that the kinetics are complex in the former and indicate either dual pathways that are "conformationally gated" or an isomerization of the $Mn^{II}(O_2^-)$ adduct to an unreactive manganous superoxide adduct. However, the resting state of MnSOD has been shown to be the Mn(III) form, and the reduction of the Mn^{3+} to Mn^{2+} is facile ($>1 \times 10^9\ M^{-1}\ s^{-1}$). In contrast, the rate constants for the reduction of Mn^{III} complexes by O_2^- in the complexes discussed in this chapter are significantly slower. This suggests that the design of a MnSOD model compound or any manganese complex that is catalytically active should involve activation of the Mn^{3+} oxidation state.

Acknowledgments

This research was carried out at Brookhaven National Laboratory under contract DE–AC02–76CH000016 with the U.S. Department of Energy and supported by its Division of Chemical Sciences, Office of Basic Energy Sciences. Discussion of this work with H. A. Schwarz was extremely helpful.

References

1. Aust, S. D.; Morehouse, L. A.; Thomas, C. E. *J. Free Rad. Biol. Med.* **1985,** *1,* 3.
2. Bielski, B. H. J. In *Oxygen Radicals in Biology and Medicine;* Simic, M. G.; Taylor, K. A.; Ward, J. F.; von Sonntag, C., Eds.; Plenum: New York, 1989; pp 123–130.
3. *Chelating Agents in Oxidation–Reduction Reactions;* Technical Information-Literature Code G-1; Organic Chemical Division, W. R. Grace & Co.; 1985.
4. Fridovich, I. *Ann. Rev. Biochem.* **1995,** *64,* 97–112.
5. Beyer, W.; Imlay, J.; Fridovich, I. *Prog. Nucleic Acid Res. Mol. Biol.* **1991,** *40,* 221.

6. Ludwig, M. L.; Metzger, A. L.; Pattridge, K. A.; Stallings, W. C. *J. Mol. Biol.* **1991,** *219,* 335.
7. Parker, M. W.; Blake, C. C. F. *J. Mol. Biol.* **1988,** *199,* 649.
8. Borgstahl, G. E. O.; Parge, H. E.; Hickey, M. J.; Beyer, W. F.; Hallewell, R. A.; Tainer, J. A. *Cell* **1992,** *71,* 107.
9. Stallings, W. C.; Powers, T. B.; Pattridge, K. A.; Fee, J. A.; Ludwig, M. L. *Proc. Natl. Acad. Sci. U.S.A.* **1983,** *80,* 3884.
10. Ringe, D.; Petsko, G. A.; Yamakura, F.; Suzuki, K.; Ohmori, D. *Proc. Natl. Acad. Sci. U.S.A.* **1983,** *80,* 3879.
11. Morris, D. C.; Searey, D. G.; Edwards, B. F. *J. Mol. Biol.* **1985,** *186,* 213.
12. Getzoff, E. D.; Tainer, J. A.; Weiner, P. K.; Kollman, P. A.; Richardson, J. S.; Richardson, D. C. *Nature (London)* **1992,** *358,* 347–351.
13. Cudd, A.; Fridovich, I. *J. Biol. Chem.* **1982,** *257,* 11443–11447.
14. Benovic, I.; Tillman, T.; Cudd, A.; Fridovich. I. *Archives Biochem. Biophys.* **1983,** *221,* 329–332.
15. Harju, L. *Anal. Chim. Acta* **1970,** *50,* 475.
16. Harju, L.; Ringbom, A. *Anal. Chim. Acta* **1970,** *49,* 205, 221.
17. Cabelli, D. E.; Bielski, B. H. J. *J. Phys. Chem.* **1984,** *88,* 3111–3115.
18. Cabelli, D. E.; Bielski, B. H. J. *J. Phys. Chem.* **1984,** *88,* 6291–6294.
19. Schwarz, H. A. *J. Chem. Ed.* **1981,** *58,* 101.
20. Pick-Kaplan, M.; Rabani, J. *J. Phys. Chem.* **1976,** *80,* 1840.
21. Lati, J.; Meyerstein, D. *J. Chem. Soc., Dalton Trans.* **1978,** 1105–1118.
22. Bielski, B. H. J.; Cabelli, D. E.; Arudi, R. L.; Ross, A. B. *J. Phys. Chem. Ref. Data* **1985,** *14,* 1041–1100.
23. Bull, C.; McClune, G. J.; Fee, J. A. *J. Am. Chem. Soc.* **1983,** *105,* 5290–5293.
24. Stein, J.; Fackler, J. P., Jr.; McClune, G. J.; Fee, J. A.; Chan, L. T. *Inorg. Chem.* **1979,** *18,* 3511–3519.
25. Koppenol, W. H.; Levine, F.; Hatmaker, T. L.; Epp, J.; Rush, J. D. *Arch. Biochem. Biophys.* **1986,** *251,* 594–599.
26. McAdam, M. E.; Fox, R. A.; Lavelle, F.; Fielden, E. M. *Biochem. J.* **1977,** *165,* 71–79.
27. McAdam, M. E.; Fox, R. A.; Lavelle, F.; Fielden, E. M. *Biochem. J.* **1977,** *165,* 81–87.
28. Pick, M.; Rabani, J.; Yost, F.; Fridovich, I. *J. Am. Chem. Soc.* **1974,** *96,* 7329–7333.
29. Bull, C.; Niederhoffer, E. C.; Yoshida, T.; Fee, J. A. *J. Am. Chem. Soc.* **1991,** *113,* 4069–4076.
30. Fee, J. A.; Shapiro, E. R.; Moss, T. H. *J. Biol. Chem.* **1976,** *251,* 6157.
31. Valentine, J. S. *Chem. Rev.* **1973,** *73,* 235.
32. Goetz, F.; Lengfelder, E. In *Oxy-Radicals and Their Scavenger Systems, Vol. 1, Molecular Aspects;* Cohen, G.; Greenwald, R. A., Eds.; Elsevier Biomedical: New York, **1983;** pp 228–233.
33. Epp, J.; Fairchild, S.; Erickson, G.; Koppenol, W. H. In *Superoxide and Superoxide Dismutase in Chemistry, Biology and Medicine;* Rotilio, G., Ed.; Elsevier: Amsterdam, Netherlands, **1981;** pp 76–78.
34. Rush, J. D.; Maskos, Z. *Inorg. Chem.* **1990,** *29,* 897–905.

16

The Primary Process in Photo-oxidation of $Fe^{2+}(H_2O)_6$ in Water

N. S. Hush[1,2], J. Zeng[1], J. R. Reimers[1], and J. S. Craw[1,3]

[1] **Departments of Physical and Theoretical Chemistry and[2] Biochemistry, University of Sydney, NSW 2006, Australia**
[2] **Department of Biochemistry, University of Sydney, NSW 2006, Australia**

When aqueous solutions containing Fe^{2+} ions are irradiated at <250 nm, photooxidation to Fe^{3+} occurs and molecular hydrogen is generated. This photoprocess has been studied extensively for over 60 years, but without agreement being reached about the nature of the primary step. Possible initial steps include metal-to-ligand charge transfer (MLCT), internal Fe^{2+} 3d → 4s absorption, direct electron photodetachment producing a partially solvated electron in a pre-existing solvent cavity, and polaron-type charge transfer to solvent (CTTS) absorption. We consider the energetics and solvent shift of the first three of these processes, concluding that the MLCT band is too high in energy, the 3d → 4s excitation could participate, and the direct photodetachment band is at the correct energy and intensity to account for all that is (as yet) observed of the absorption band. In general, a rather complicated picture of this process in inorganic complexes emerges. In this work, we apply a general method we have developed for estimating the effects of solvents on transitions of species that have strong specific interactions (e.g., hydrogen bonding) with the solvent molecules.

Oxidation–reduction processes involving the Fe^{2+}–Fe^{3+} couple have been intensively studied, and the elementary electron exchange, taking place either in solution or at an electrode interface, has served for a long time as a test case for theoretical interpretations of electron transfer kinetics. For reactions involving ground-state ions, the mechanisms of the reactions are now well understood: for the homogeneous systems, for example, electron transfer can occur, as demonstrated by Taube, via either a simple outer-sphere electron

[3] Current address: Department of Chemistry, University of Manchester, Oxford Road, Manchester, England.

transfer step, via an inner-sphere reaction in which electron transfer is accompanied by ligand exchange, or via a transition state in which the ions maintain their water coordination shells intact but are bridged by an anion such as a halide. In contrast to this, the photochemical process in which Fe^{2+} in aqueous solution is oxidized to Fe^{3+} when irradiated at <250 nm, although extensively studied (*1–13*) since the 1930s, is still not understood: in particular, the nature of the primary step has not yet been established. The main reason for the early interest in this reaction is that it leads to the photoproduction of molecular hydrogen, according to the overall equation

$$2Fe^{2+} + 2H_2O \xrightarrow{h\nu} 2Fe^{3+} + 2OH^- + H_2 \qquad (1)$$

where h is Planck's constant and ν is frequency, indicating that absorption of one quantum leads to the oxidation of two ferrous ions together with one molecule of H_2. It may well have been the possibility of generation of hydrogen from water in a solar energy device that inspired the early work in this area, for example that of Farkas and Farkas in Israel (*3*), although the low extinction coefficient (ca. 28 mol^{-1} cm^{-1} at 250 nm), the low quantum yield (ca. 8%), and high energy (50,000 cm^{-1}) would not seem to favor this as a practical process.

The earliest suggestion for a mechanism for the photoprocess appears to have been made by Weiss (*14*) in 1935. The primary step was proposed to be

$$Fe^{2+}(H_2O) \xrightarrow{h\nu} Fe^{3+} + OH^- + H \qquad (2)$$

that is, photoreduction of a water molecule in the first hydration shell to yield a hydrogen atom. Variants of this persisted into the 1960s: thus Jortner and Stein (*15*) in 1962 [following Rigg and Weiss (*4*)] took the initial light absorption step to be formation of an excited ferrous ion, $Fe_{aq}^{2+}*$, followed by

$$Fe_{aq}^{2+}* \rightarrow Fe_{aq}^{3+} + OH^- + H \qquad (3)$$

According to Jortner and Stein, the evidence from radiation chemistry and from experiments on the action of hydrogen atoms on acid solutions of ferrous sulfate in favor of the above mechanism was "conclusive".

The relevance of the discovery of the solvated electron, which revolutionized radiation chemistry (*16*), to the photooxidation of ferrous ion was appreciated only rather slowly. However, in 1966, Airey and Dainton (*12*) proposed the alternative mechanism

$$Fe^{2+}H_2O \xrightarrow{h\nu} Fe^{3+} + e_{aq}^-, \qquad (4)$$

"possibly via $Fe_{aq}^{2+}*$", where e_{aq}^- refers to the aquated electron, adducing evidence from a number of sources. This was not universally accepted: in 1969 Weiss (*17*) stated that the Airey and Dainton proposal could not be confirmed, and in 1975 Fox concluded (*18*) that a solvated electron ("in the normally accepted definition") was not an intermediate in the Fe^{2+} photooxidation. That the solvated electron does indeed have a role was shown conclusively by Sloper et al. (*13*), who in 1983 observed a broad transient absorption peaking between 700 and 800 nm characteristic of the solvated electron following a 25-ns pulse: light intensity and scavenger experiments established that this absorption corresponded to monophotonic production of e^-_{aq}. However, the exact nature of this role has not yet been established, and reaction pathways in which it does not participate may also contribute to the overall photoprocess, according to several suggested mechanisms.

We can classify the proposals for the nature of the primary step in photooxidation of aqueous ferrous ion under four different mechanisms. These are:

1. direct photodetachment to produce a free electron (i.e., an electron lying beyond the first coordination shell (*19, 20*) partially but not fully solvated (*11, 12*)) followed by reactions of the free electron with water and other species,
2. a charge transfer to solvent (CTTS) transition (*21*) in which the electron is excited into a hydrogenic-like orbital over the inner solvent shells, analogous to that observed (*22*) for halide and other anions, followed by electron ejection and further chemistry,
3. charge-transfer absorption of a metal-to-ligand (MLCT) state in which the transferring electron is localized on one of the ligands (*3–5*), followed by a subsequent ligand decomposition reaction (the Farkas and Farkas mechanism),
4. internal metal 3d → 4s absorption producing (*8–10*) a ^5S state of Fe^{2+} with configuration $3d^54s^1$, traditionally thought to be followed by nonradiative transfer of the excitation into an MLCT state and hence to ligand decomposition.

It should be noted that the expression "charge transfer to solvent" has come to have a colloquial meaning as an umbrella term covering all mechanisms by which an electron is transferred from a chromophore into the solvent's realm. We use this term here explicitly with its original and rather technical meaning as described by polaron theory: this process does not change the expectation value of the position of the electron, and the electron is simply placed in a hydrogenic chromophore-centered orbital that permeates out and through the solvent. Motion of the electron's probability center into the solvent happens in a second step after the primary absorption process is complete. This is in contrast to the direct photodetachment mechanism, in which the electron's position expectation value moves away from the chromophore into the solution as a result of the primary absorption process, much as happens in gas-phase photoelectron spectroscopy.

The situation is clearly quite complex, as the absorption band (*1, 4, 6*) shows structure, suggesting that more than one of the above processes may be occurring simultaneously. Subsequent to these initial steps, a complex system of chemical reactions ensues (e.g., *see* references 4, 5, 10, 15, and 23), finally resulting in the evolution of hydrogen gas at ca. 8% quantum yield (*10, 13*). One clue as to the nature of the process is given by the observation of solvated electrons as reaction intermediates (*11, 13*), which indicates that mechanisms 3 and 4, as originally formulated, are incorrect: light absorption leads first to electron ejection, not ligand decomposition. However, the primary steps postulated in each of these mechanisms, charge-transfer absorption of a metal-to-ligand (MLCT) state localized on one of the water molecules and 3d → 4s absorption, are still appropriate as they could lead to subsequent electron loss.

Somewhat surprisingly, no clear resolution of this fundamental question has been reached; this has been perhaps partly due to the realization that photooxidation of aqueous and other ions leading to hydrogen gas production is a general phenomenon, not a specific property of the Fe^{2+} system (*6, 12, 22, 24*), and a consequent turning of attention to other systems. There has only been one (*13*) major paper on this system in the past 30 years. Over the past few years, advances in femtosecond experimental and quantum simulation techniques have led to the unambiguous characterization of the primary absorption process in a number of systems, and examples of mechanisms 1, 2, and 3 have been characterized. Liquid water, for example, undergoes (*25*) a two-photon photodetachment (mechanism 1) at energies above 6.5 eV: a 2p electron leaves the chromophore extremely rapidly, localizes retaining p angular momentum (as required for a two-photon allowed process) in a pre-existing solvent cavity not far from its source, and there relaxes to produce a solvated electron in its ground s state (*19, 20, 26–29*). Other molecular liquids such as neat alkanes display similar effects (*30*). Alternatively, halide ions (*22, 31–36*) undergo a one-photon allowed p → s CTTS transition (mechanism 3) to produce some short-lived excited complexes, which is followed by electron release and capture. This basic mechanism, somewhat modified, has also been observed to apply to neutral chromophores (*37*).

The spectra of few inorganic complexes exhibiting photooxidation have been studied. One example is that of the $[Fe(CN)_6]^{4-}$ and related complexes (*12, 38–40*). These have an intense MLCT band with a long low-frequency shoulder which is believed to be a CTTS transition; absorption at both the band center and shoulder gives rise to electron ejection, suggesting that considerable interaction between the two bands may occur, or possibly that more than one mechanism for electron release may be involved. Another example is the $Ru^{2+}(NH_3)_6$ complex (*21*), which displays a moderately intense, isolated band which is believed to represent a CTTS transition. The final electronic state is suggested to involve (*21*) a hydrogenic 2s orbital, but this is unlikely as this would require a Laporte-forbidden g → g electronic transition. An interesting feature of the photooxidation of inorganic complexes is that the observed absorption intensities appear to vary over 5 orders of magnitude (e.g., *see* references 6 and 12), suggesting that a variety of mechanisms are involved. Unfortunately, in these examples only part of the absorption spectrum is recorded, and estimation of the total intensity may be very difficult: it could well be that these bands in general have considerable structure, with each substructure originating from a different absorption mechanism. Indeed, they often appear as low-frequency tails to other well characterized bands, and it is possible that mechanisms 3 and 4 could proceed through vibronic interactions with nearby intense absorption. A recent detailed analysis of the photolysis of $[Fe(CO)_4]^{2-}$ was unable to discriminate between the possible primary absorption mechanisms (*41*).

Discrimination between the mechanisms just discussed can be achieved if detailed knowledge of the energetics of each process can be obtained. Of particular importance to mechanisms 3 and 4 is the energy predicted for the postulated primary absorptions of the $Fe^{2+}(H_2O)_6$ complex in solution. The internal iron 3d → 4s transition involves no substantial charge rearrangement, and hence the solvent shift obtained in taking this complex from the gas phase and dissolving it in water is expected (*42*) to be quite small, perhaps of the order of a few hundred wavenumbers. If the MLCT transition is delocalized over all of the ligands, then a similar picture would apply; alternatively, if the MLCT transition is localized, it will involve significant charge transfer and hence could be expected to show an appreciable solvent shift. Also, as strong hydrogen bonds to the solvent are involved, the solvent shift may be poorly described using dielectric continuum models for the solvent (*40, 42*). Although much is known about MLCT bands when the acceptor orbital is a low-lying ligand π orbital, little is known about such bands when the transfer is to a σ orbital, as the band center of such a transition lies in the vacuum ultraviolet region of the spectrum. Simple energetic considerations indicate that mechanisms 1 and 2 are plausible (*11, 12*). From the standard reduction potentials (*16, 43*) of Fe^{2+}–Fe^{3+} and e_g^-–e_{aq}^-, the free energy, ΔG, for the process of transferring an electron from equilibrated Fe^{2+} to produce equilibrated Fe^{3+} and an isolated solvated electron is 3.54 eV; corrections for entropy changes (*43, 44*) for this process give the estimated energy of the absorption origin at ΔH = 4.04 eV, at the foot of the observed (*1, 4, 6*) absorption band.

Our interest in the photooxidation of Fe^{2+} in aqueous solution derives from our more general interest in the effects of solvents on electronic transitions, particularly those in which strong specific interactions with solvent molecules are present (*42, 45, 46*). We proceed by performing electronic structure calculations, liquid structure simulations, and spectroscopic calculations for mechanisms 1, 3, and 4, investigating the nature of the photochemical processes of aqueous Fe^{2+}. In particular, we first require the gas-phase absorption frequencies and intensities of the $Fe^{2+}(H_2O)_6$ complex, using both ab initio and semi-empirical (INDO-MRSCI) techniques. Second, we need to determine the structure of water around the Fe^{2+} ion in solution. Third, we need to determine the solvent shifts of the absorption bands to evaluate transition energies in solution. This will lead to an estimation of relative importance of all but the charge transfer to solvent process (mechanism 2), calculation of which is beyond the capacity of our present computational facilities. The potential surfaces employed as well as other particulars of the methods used during the simulation of the liquid structure, and the algorithms used to determine the MLCT (mechanism 3) and photodissociation (mechanism 1) solvent shifts from the liquid structure are described in detail elsewhere (*47*). In the next section, the computational methods are outlined, and results for liquid structure are summarized. In the section after that, each of the proposed primary steps is discussed. The absorption spectrum for the direct photodetachment, mechanism 1, is calculated and compared to the experimental spectrum. Both INDO/S and ab initio calculations are performed for the energy of the MLCT and 3d → 4s bands in the gas phase, allowing the absolute solution absorption band center to be predicted, and results are given for the MLCT solvent shift, relating it closely to the solvent structure and specific issues that arise for this type of calculation.

Computational Methods and Liquid Structure

These are described in detail elsewhere (*42, 45–47*). The first requirement is generation of potential surfaces to be used in Monte Carlo simulations. We describe the aqueous ferrous ion as a Fe^{2+} ion interacting with nearest-neighbor water molecules through a potential derived from a force field of the type described by Tomasi and co-workers (*48*). The Monte Carlo simulation (*49, 50*) of a single Fe^{2+} ion in a solution of 190 water molecules was performed at constant pressure, density, and temperature (NPT ensemble), at a temperature of 298 K with periodic boundary conditions. Simulating charged systems is not straightforward because of the long-range nature of the Coulomb potential, and the impracticality of using a sample size whose extent exceeds its range. Here, we use a switch function to dampen the Coulomb interaction in the region of the box boundary. The use of other potentials and boundary conditions is considered elsewhere (*47*).

The radial distribution function $g_{Fe-O}(r)$ is shown in Figure 1 and is very similar to that obtained by Tomasi and co-workers (*48*). It indicates that there exists a well-defined first coordination shell in the region 1.9 Å $< r <$ 2.4 Å. The first maximum is at 2.12 Å and the coordination number is 6, both in agreement with experimental results (*51*). A second peak in the region 3.5 Å $< r <$ 5.0 Å indicates that a second coordination shell can be isolated, and integration of g_{Fe-O} shows that there are ca. 12 water molecules in this shell. This is in agreement with experimental measurements (*52*) and theoretical analysis (*48, 53, 54*). Beyond 5 Å, the structure is sensitive to the boundary conditions. Further details of structure, including the orientation of water molecules in the first shell, together with a detailed comparison of the results obtained with the different choices of potentials and boundary conditions, are given in reference 47. In what follows we shall simply refer to the results insofar as they are relevant to discrimination between the mechanisms listed above for the Fe^{2+} photooxidation primary step.

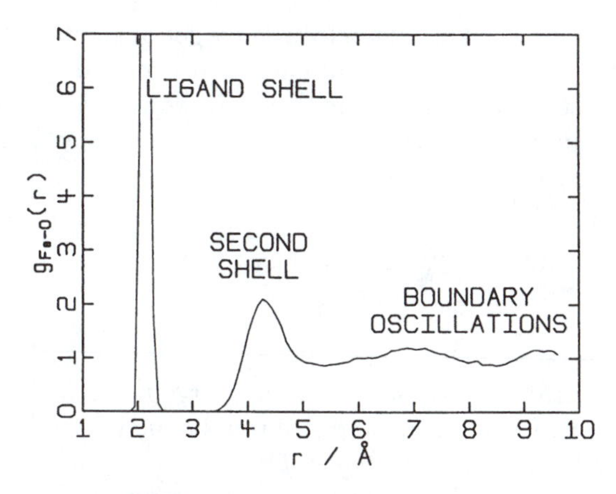

Figure 1. The radial distribution function, g_{Fe-O}(r), for iron with oxygen for a ferrous ion in aqueous solution as a function of the interatomic separation, r, *in angstroms, obtained using Tomasi's potential* (48).

The Possible Primary Steps

We discuss in turn the possibilities listed earlier in this chapter for the primary step in the photooxidation of aqueous Fe^{2+}.

Direct Photodetachment. According to postulated mechanism 1, the absorption spectrum arises from the direct photodetachment of an electron from the metal to a pre-existing cavity in the solvent lying beyond the ligand coordination shell. In the gas phase for a bare Fe^{2+} ion, this process is observed and the ionization energy is known (*55*) to be 30.64 eV. In solution, a process of this type undergoes an enormous solvent shift (later, we calculate ca. 30 eV or 240,000 cm^{-1}), principally because of the differences in the solvation energy of the di- and trivalent ion, and our method for evaluating solvent shifts may be applied for this process. We proceed by examining a selection of liquid configurations, determining the location of all solvent cavities, sequentially transferring an electron from the ion to each of these cavities, and determining the change in the electrostatic energy. Cavities are isolated by searching for locations that maximize the distance to the closest oxygen atom, and a cavity radius, r_c, is defined to be the average of all the distances from the center to the oxygen atoms of the water molecules in its first coordination shell; other definitions are possible (*19*).

Account must also be taken of the kinetic energy of the electron when it is confined to lie in a solvent cavity. This quantity has been calculated by quantum simulations of the equilibrated solvated electron (*56, 57*) and found to be 2.2 eV at 298 K. The equilibrated structure has sixfold coordination of water around the electron, and it has an average electron oxygen separation of 3.3 Å. Both diffuse (electron in a dielectric continuum) and confined (particle in a box) models for a solvated electron indicate that its kinetic energy increases proportionally to the square of the reciprocal cavity size, and hence we assume that the kinetic energy, E_K, is given by

$$E_K = \left(\frac{3.3}{r_c}\right)^2 \times 2.2 \text{ eV} \tag{5}$$

where r_c is in angstrom units.

Another contribution to the transition energy is the exchange repulsion between the solvated electron and the electrons bound within the solvent molecules. In general, the effects of these interactions are believed to be small provided that the electron does not approach the atoms too closely (*19, 56–58*), and the solvated electron has been likened to a F^- ion. We assume that the exchange repulsion can be represented using a hard-sphere potential, the consequence of which is that cavities are only considered if they lie no closer than 2 Å to any oxygen atom. Note that this criterion effectively eliminates any cavity that lies within the first coordination shell of the ion. The transition energy producing an electron located in a pre-existing solvent cavity is thus given by

$$h\nu = 30.64 \text{ eV} + E_K + h\Delta\nu \tag{6}$$

where $h\Delta\nu$ is the solvent shift.

In order to estimate the intensity and width of the absorption band, we postulate that intensity arises from only direct through-space electron transfer, and estimate the transition moment for this process. The initial state $|\Psi_i\rangle$ of the electron is a totally symmetric linear combination of the three metal t_{2g} orbitals, and we represent this simply as an iron 3s Slater orbital whose exponent is taken to be that of the iron 3d orbitals as adopted by Zerner (*59, 60*), 2.6 au. Using the analogy between the solvated electron and a F^- ion (*19*), we represent the final state of the electron $\langle\Psi_f|$ using a fluorine 2s Slater orbital, whose exponent (*60, 61*) is also 2.6 au. The one-electron matrix element coupling these two states is given in semi-empirical theories (*61*) as

$$V = \frac{-q_f S}{r_{\text{Fe-el}}} \times 14.4 \text{ eV} \tag{7}$$

where $r_{\text{Fe-el}}$ is the cavity center to ion distance in angstroms, $S = \langle\Psi_i|\Psi_f\rangle$ is the overlap of the two Slater orbitals, and $q_f = 3$ e is the charge of the final state of the ion. Perturbation theory then gives the transition moment, M, by

$$M = \frac{V}{h\nu} r_{\text{Fe-el}} \tag{8}$$

which is evaluated, leading to the calculation of the absorption spectrum. Note that this method ignores contributions to the width of the absorption band arising from Franck–Condon displacements of the water molecules around the ion between the initial and final electronic states.

An analysis of all the solvent cavities found in the liquid structure shows that the cavity to water radial distribution functions reproduce in detail those found in pure water (*19*). In a sample as large as ours, most cavities lie outside the second coordination shell. Our transition moment profile, however, contains an orbital overlap term that decreases exponentially with the distance of the cavity from the ion, effectively precluding transitions to cavities located outside the second shell. The cavities between the first and second shells have different structures from those in the bulk liquid (i.e., those outside the second shell), with four water molecules spanning each cavity. The average kinetic energies of the electron and solvent shift are calculated to be 5.2 eV and 29.7 eV, respectively.

The calculated spectrum, smoothed using Gaussian convolution at a resolution of 0.1 eV, is shown along with the observed spectrum (*4*) in Figure 2. The band is calculated to be centered at 6.0 eV (48,000 cm^{-1}) and has an oscillator strength of 0.0017; its foot is at 4.5 eV, slightly above the energy of 4.0 eV

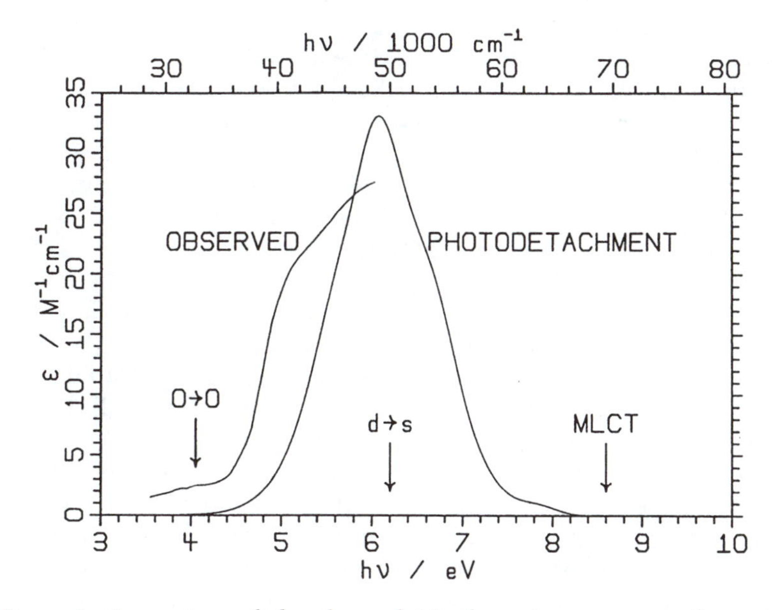

Figure 2. Comparison of the observed (4) absorption spectrum of aqueous $Fe^{2+}(H_2O)_6$ with the spectrum calculated for photodetachment of an electron into a pre-existing solvent cavity (at 0.1 eV resolution). The 0 → 0 transition energy, calculated from experimental data, and the band centers calculated for 3d → 4s and MLCT absorptions are indicated with arrows.

calculated from experimental data for the enthalpy difference between the ground state and the fully equilibrated excited state. These results do indeed suggest that direct photodetachment is the primary absorption process. We should note that the quality of the agreement seen between theory and experiment is beyond that which could reasonably be expected given the approximations used in generating the spectra. Nevertheless, the observed part of the absorption spectrum is qualitatively well represented, and the relative placing of the calculated band center to the experimental value for the energy of the $0 \rightarrow 0$ transition is very reasonable, and any large absorption at higher frequency than that observed could be attributable to a different mechanism.

Direct MLCT Absorption. Electronic structure calculations for the $Fe^{2+}(H_2O)_6$ complex have been performed (*47*) in order to determine the gas-phase (T_h symmetry) transition energy and oscillator strength for the MLCT absorption band using both INDO/S-CI and ab initio techniques. The INDO/S-CI calculations are performed using the rotationally invariant (*59*), spin-restricted, open-shell (*62*) (ROHF) formalism with configuration interaction developed by Zerner and co-workers (*59, 62–65*) using our own software. All possible excitations amongst an active space of the metal 3d and 4s orbitals are included in the configuration-interaction calculation (*65*), as are a large number of single excitations from each of those configurations. The result is a predicted weak gas-phase delocalized MLCT transition at ν = 71,000 cm^{-1} (8.8 eV) with an oscillator strength of 0.012. Ab initio calculations have also been performed using two different methods; both coincidently gave the same result, ν = 71,000 cm^{-1}, as obtained using INDO. One calculation, performed by the MOLCAS program (*66*), was a complete active space, self-consistent field (CASSCF) calculation using a STO–3G basis set for water and a triple-zeta basis set for iron (*67*). The other calculation was a multiconfigurational SCF (MCSCF) calculation performed using HONDO (*68*), with the active space being all singles and doubles excitations from the iron 3d orbitals into the lowest 15 virtual orbitals. For this, an effective-core potential was used for iron (*69*) and oxygen (*70*) atoms and a double-zeta basis for hydrogen (*71*).

The magnitude of the solvent shift when the complex is placed in water depends on localization or otherwise of the MLCT transition. If it remains delocalized over all six ligands, then little solvent shift is expected. If the transferring electron is localized to one of the ligands, an energy lowering and consequent appreciable solvent shift will be expected. Results actually obtained for a localized state are shown in Figure 3. These calculations use spherical boundary conditions, including explicitly all water molecules within a sphere of radius r_{Fe-O}, followed by an exclusion zone (*72, 73*) of radius 1 Å. Beyond this outer cavity radius $a = r_{Fe-O} + 1$ Å, a dielectric continuum is included. It is possible to expand a from a value so small that no explicit water molecules are included inside to the maximum radius permitted by the 190-molecule simulation, and the solvent shift is plotted in Figure 3 as a function of this radius. Below a =

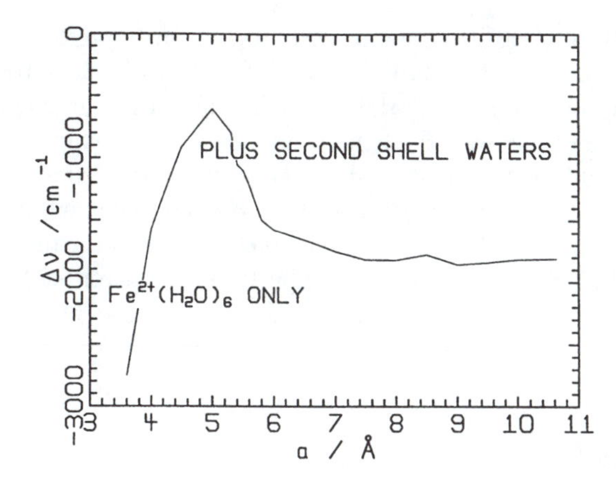

Figure 3. The MLCT solvent shift, $\Delta\nu$, as a function of the cavity radius, a, *obtained using Tomasi's potential* (48).

5 Å (r_{Fe-O} = 4 Å), no explicit water molecules outside the inner coordination shell are included, and the solvent shift shows the $1/a^3$ dependence expected from a classical continuum model (*74, 75*); its magnitude is significantly enhanced, however, because of the inclusion of polarizability centers within the solute complex (*74*). Between 5 Å < a < 6 Å (or 4 Å < r_{Fe-O} < 5 Å), the water molecules in the second coordination shell are explicitly included, and the magnitude of the solvent shift increases considerably. Beyond this radius, little change to the solvent shift occurs, indicating that it is insensitive to the long-range solvent structure. The solvent shift is thus dominated by the second coordination shell, and, within this shell, contributions from both specific hydrogen bonds and from dielectric solvation occur. The relative contributions of each to the overall solvent shift were calculated to be a blue shift of about 5900 cm^{-1} arising from specific hydrogen-bonding effects, offset by a larger red shift of 7700 cm^{-1} arising from dielectric solvation, resulting in a net solvent shift of about -1800 cm^{-1}.

Thus, if the MLCT transition is localized, the gas-phase transition energy will be lowered by about 2000 cm^{-1} (0.25 eV); in Figure 2, the lowest possible energy for the band center, 69,000 cm^{-1} (8.6 eV), is indicated by an arrow. It is thus unlikely that this transition could have a significant influence on the absorption spectrum in the region 40,000–50,000 cm^{-1}; this would require the observed intensity in the region (f = 0.0011) to comprise about 10% of that predicted (f = 0.012) for the MLCT band.

3d → 4s Absorption. INDO/S-CI calculations predict that the ^5S transition observed (*55*) in the free ion at 41,000 cm^{-1} is blue shifted to 50,000 cm^{-1} (6.2 eV) in the hexaquo complex. Similarly, the ab initio multiconfigurational SCF (MCSCF) calculations described previously predict an absorption frequency of 51,000 cm^{-1} (6.3 eV). It is difficult to estimate error bounds for these numbers, but general experience would suggest that 4000 cm^{-1} (0.5 eV) would be an upper limit, with the true frequency possibly lying to lower energy. This places the 3d → 4s transition well within the observed absorption band, as shown by an arrow in Figure 2, and it is possible that it either significantly inhibits (e.g., by not leading to electron release, thus reducing the observed quantum yield), aids (i.e., leads eventually to electron release), or facilitates (i.e., is the sole mechanism that leads to electron release) the photochemical reaction.

Conclusions

We have outlined a development of our methods, Parts I–III (*42, 45, 46*), for estimating solvent shifts of electronic spectra for species with strong specific solvent–solute interactions (e.g., hydrogen bonds) to treat inorganic charge-transfer spectra. The first application has been to the ultraviolet absorption spectrum of aqueous $Fe^{2+}(H_2O)_6$, an interesting problem which has remained unexplained for over 60 years. Our calculations predict that direct photodetachment (mechanism 1) will give rise to an absorption band of essentially the same frequency, intensity, and bandwidth as is experimentally observed. They also indicate that processes involving direct MLCT absorption (mechanism 3) do not contribute significantly to the observed photochemical process for this complex, and that the frequency of the formally forbidden 3d → 4s transition (mechanism 4) is close to that observed, but estimation of the magnitude of any contribution that this may make requires a knowledge of the relevant intensity-gaining mechanism, which is as yet unknown. The question of any possible contribution of the polaronic CTTS mechanism (mechanism 2) to the photochemical process must await developments in computer technology in order for a reliable calculation to be made. Such a calculation would need to treat fully quantum mechanically at least the metal 3d, 4s, and 4p orbitals, MLCT states, and direct photodetachment states, as well as the solvent-polarization-borne CTTS states.

Our study indicates that, for metal-containing systems, no one single mechanism is likely to be generally applicable to describe photochemical water decomposition. Indeed, all four mechanisms considered could in principle dominate in any given specific situation. This qualitatively explains the widely varying oscillator strengths observed for photochemically active bands for metallic complexes. Experimentally, absorption spectra (e.g., *see* reference 6) need to be determined to high enough energy to properly characterize the band shape (into the vacuum-ultraviolet region if necessary, e.g., for $Fe^{2+}(H_2O)_6$). Also, femtosecond dynamical studies (*28, 33*) of the motion of the excitation or electroabsorption spectroscopic studies (*76, 77*) (which allow excited-state dipole moments and polarizabilities to be determined (*78–80*), thus characterizing the excited state) should be made.

Our method for evaluating solvent shifts in strongly interacting systems is seen to be highly robust and capable of describing very different physical situations with quantitative accuracy. Previously, the method was tested by calculation of hydrogen-bonding blue shifts in azines like pyridine (*81*) and pyrimidine (*42*), where the solvent shift is ca. 0.25 eV. There, contributions from specific solvation and dielectric effects are of equal importance, and of the same sign. Here, for an MLCT process in which the donor and acceptor lie in the same solvent hole and the ground state has no dipole moment, a similarly low solvent shift is predicted, but in this case the specific solvation and dielectric effects oppose each other. We have recently investigated (*82*) the MLCT solvent shift in $Ru^{2+}(NH_3)_5$pyridine, in which the ground state has a large charge asymmetry; and found a sizable red shift of ca. 1 eV, which is quantitatively in agreement with experiment. Finally, in this work in what is an extreme limit, we investigate a photodetachment process in which the electron goes into a solvent cavity and the donor and acceptor states of the charge-transfer process must be thought of as residing in different solvent cavities.

Regardless of whether or not the intensity calculation presented is accurate (i.e., whether or not mechanism 1 is the major mechanism involved), the calculation of a solvent shift of ca. 30 eV must be quite accurate, as the gas-phase transition energy and the origin of any possible absorption band are both known precisely, and only the minor contribution from the electron kinetic energy is uncertain. Our general method is thus expected to be generally applicable to a wide range of problems. For the specific case of the Fe^{2+} photoprocess, as noted earlier in this section, a more definitive solution awaits further developments in computing capacity, but the wait, one hopes, will be less than another 60 years!

Acknowledgment

J. S. Craw and J. R. Reimers gratefully acknowledge support provided by the Australian Research Council for this project.

CAS Registry Numbers

$Fe^{2+}(H_2O)_6$ 15365-81-8; $Fe^{3+}(H_2O)_6$ 15377-81-8; $V^{2+}(H_2O)_6$ 15696-18-1; $Cr^{2+}(H_2O)_6$ 20574-26-9; $Ru^{2+}(NH_3)_6$ 19652-44-9; $[Fe(CN)_6]^{4-}$ 13408-63-8.

References

1. Potteril, R. H.; Walker, O. J.; Weiss, J. *Proc. R. Soc. London, Ser. A* **1936,** *156,* 561.

2. Weiss, J. *Trans. Faraday Soc.* **1941,** *37,* 463.
3. Farkas, A.; Farkas, L. *Trans. Faraday. Soc.* **1938,** *34,* 1113.
4. Rigg, T.; Weiss, J. *J. Chem. Phys.* **1952,** *20,* 1194.
5. Hayon, E.; Weiss, J. *J. Chem. Soc.* **1960,** 3866.
6. Dainton, F. S.; James, D. G. L. *Trans. Faraday Soc.* **1958,** *54,* 649.
7. Orgel, L. E. *J. Chem. Phys.* **1955,** *23,* 1004.
8. Jorgensen, C. K. *Acta Chem. Scand.* **1955,** 9, 919.
9. Orgel, L. E. *Discuss. Inst. Solvay* **1956,** *10,* 289.
10. Jortner, J.; Stein, G. *J. Phys. Chem.* **1962,** *66,* 1264.
11. Dainton, F. S.; Jones, F. T. *Trans. Faraday Soc.* **1965,** *61,* 1681.
12. Airey, P. L.; Dainton, F. S. *Proc. R. Soc. London Ser. A* **1966,** *291,* 340.
13. Sloper, R. W.; Braterman, P. S.; Cairns-Smith, A. G.; Truscott, T. G.; Craw, M. *J. Chem. Soc., Chem. Commun.* **1983,** 488.
14. Weiss, J. J. *Nature (London)* **1935,** *136,* 794.
15. Jortner, J.; Stein, G. *J. Phys. Chem.* **1962,** *66,* 1258.
16. Hart, E. J.; Aubar, M. *The Hydrated Electron;* Wiley-Interscience: New York, 1970; p 62.
17. Weiss, J. J. *Ber. Bunsen. Ges. Elektrochem.* **1969,** *73,* 131.
18. Fox, M. In *Concepts of Inorganic Photochemistry;* Adamson, A. W.; Fleischauer, P. D., Eds.; Wiley-Interscience: New York, 1975; p 355.
19. Schnitker, J.; Rossky, P. J.; Kenney-Wallace, G. A. *J. Chem. Phys.* **1986,** *85,* 2986.
20. Alfano, J. C.; Walhout, P. K.; Kimura, Y.; Barbara, P. F. *J. Chem. Phys.* **1993,** *98,* 5996.
21. Matsubara, T.; Efrima, S.; Metiu, H. I.; Ford, P. C. *J. Chem. Soc. Faraday Trans. II* **1979,** *75,* 390.
22. Blandamer, M. J.; Fox, M. F. *Chem. Rev.* **1970,** *70,* 59.
23. Lu, H.; Long, F. H.; Bowman, R. B.; Eisenthal, K. B. *J. Phys. Chem.* **1989,** *93,* 27.
24. Balzani, V.; Carassiti, V. *Photochemistry of Coordination Compounds;* Academic: London, 1970.
25. Nikogosyan, D. N.; Oraevsky, A. A.; Rupasov, V. I. *Chem. Phys.* **1983,** *77,* 131.
26. Migus, A.; Gaudel, Y.; Martin, J. L.; Antonetti, A. *Phys. Rev. Lett.* **1987,** *58,* 1559.
27. Rossky, P. J.; Schnitker, J. *J. Phys. Chem.* **1988,** *92,* 4277.
28. Long, F. H.; Lu, H.; Eisenthal, K. B. *Phys. Rev. Lett.* **1990,** *64,* 1469.
29. Murphrey, T. H.; Rossky, P. J. *J. Chem. Phys.* **1993,** *99,* 515.
30. Bowman, R. M.; Lu, H.; Eisenthal, K. B. *J. Chem. Phys.* **1988,** *89,* 606.
31. Stein, G.; Treinin, A. *Trans. Faraday Soc.* **1959,** *55,* 1086.
32. Jortner, J.; Ottolenghi, M.; Stein, G. *J. Phys. Chem.* **1964,** *68,* 247.
33. Long, F. H.; Lu, H.; Shi, X.; Eisenthal, K. B. *Chem. Phys. Lett.* **1990,** *169,* 165.
34. Sheu, W-S.; Rossky, P. *J. Am. Chem. Soc.* **1993,** *115,* 7729.
35. Sheu, W-S.; Rossky, P. *Chem. Phys. Lett.* **1993,** *202,* 186.
36. Sheu, W-S.; Rossky, P. *Chem. Phys. Lett.* **1993,** *213,* 233.
37. Truong, T. B. *J. Chem. Phys.* **1981,** *74,* 3561.
38. Gray, H. B.; Beach, N. A. *J. Am. Chem. Soc.* **1963,** *85,* 2922.
39. Alexander, J. J.; Gray, H. B. *Coord. Chem. Rev.* **1967,** *2,* 29.
40. Horvàth, A.; Uzonyi, Z. S. *Inorg. Chim. Acta* **1990,** *170,* 1.
41. Kunkely, H.; Vogler, A. *J. Organomet. Chem.* **1992,** *431,* C42.
42. Zeng, J.; Hush, N. S.; Reimers, J. R. *J. Chem. Phys.* **1993,** *99,* 1508.
43. *Standard Potentials in Aqueous Solution;* Bard, A. J.; Parsons, R.; Jordan, J., Eds; Marcel Dekker: New York, 1985.
44. Jortner, J.; Noyes, R. M. *J. Phys. Chem.* **1966,** *70,* 770.
45. Zeng, J.; Craw, J. S.; Hush, N. S.; Reimers, J. R. *J. Chem. Phys.* **1993,** *99,* 1482.
46. Zeng, J.; Hush, N. S.; Reimers, J. R. *J. Chem. Phys.* **1993,** *99,* 1495.
47. Zeng, J.; Craw, J. S.; Hush, N. S.; Reimers, J. R. *J. Phys. Chem.* **1994,** *98,* 11075.

48. Floris, F.; Persico, M.; Tani, A.; Tomasi, J. *Chem. Phys. Lett.* **1992,** *199,* 518.
49. Metropolis, N. A.; Rosenbluth, A. W.; Rosenbluth, M. N.; Teller, A. H.; Teller, E. *J. Chem. Phys.* **1953,** *21,* 1087.
50. Wood, W. W.; Parker, F. R. *J. Chem. Phys.* **1957,** *27,* 720.
51. Brunschwig, B. S.; Creutz, C.; Macartney, D. H.; Sham, T.-K.; Sutin, N. *Faraday Discuss. Chem. Soc.* **1982,** *74,* 113.
52. Ohtaki, H.; Radnai, T. *Chem. Rev.* **1993,** *93,* 1157.
53. Guàrdia, E.; Padró, J. A. *Chem. Phys.* **1990,** *144,* 353.
54. Curtiss, L. A.; Halley, J. W.; Hautman, J.; Rahman, A. *J. Chem. Phys.* **1987,** *86,* 2319.
55. Moore, C. E. *At. Energy Levels. Natl. Bur. Stand. (U.S.) Circ.* 1958, 467.
56. Schnitker, J.; Rossky, P. J. *J. Chem. Phys.* **1987,** *86,* 3471.
57. Wallqvist, A.; Martyna, G.; Berne, B. J. *J. Phys. Chem.* **1988,** *92,* 1721.
58. Schnitker, J.; Rossky, P. *J. Chem. Phys.* **1987,** *86,* 3462.
59. Bacon, A. D.; Zerner, M. C. *Theor. Chim. Acta* **1979,** *53,* 21.
60. Zerner, M. C. ZINDO Quantum Chemistry Package, University of Florida, Gainesville, FL.
61. Pople, J. A.; Beveridge; D. L. *Approximate Molecular Orbital Theory;* McGraw-Hill: New York, 1970.
62. Edwards, W. D.; Zerner, M. C. *Theor. Chim. Acta* **1987,** *72,* 347.
63. Ridley, J E.; Zerner, M. C. *Theor. Chim. Acta* **1973,** *32,* 111.
64. Zerner, M. C.; Loew, G. H.; Kirchner, R. F.; Mueller-Westerhoff, U. T. *J. Am. Chem. Soc.* **1980,** *102,* 589.
65. Anderson, W. P.; Edwards, W. D.; Zerner, M. C. *Inorg. Chem.* **1986,** *25,* 2728.
66. Andersson, K.; Füscher, M. P.; Lindh, R.; Malmqvist, P-A.; Olsen, J.; Roos, B. O.; Sadlej, A. J.; Widmark, P. O. University of Lund, Lund, Sweden.
67. Wachters, A. J. H. *J. Chem. Phys.* **1970,** *52,* 1033.
68. Dupuis, M.; Rys, J.; King, H. F. *J. Chem. Phys.* **1976,** *65,* 111.
69. Dolg, M.; Wedig, U.; Stoll, H.; Preuβ, H. *J. Chem. Phys.* **1987,** *86,* 866.
70. Preuβ, H.; Stoll, H.; Wedig, U.; Kruger, Th. *Int. J. Quant. Chem.* **1981,** *19,* 113.
71. Dunning, T. H., Jr.; Hay, P. J. In *Modern Theoretical Chemistry;* Schaefer, H. F., III, Ed.; Plenum: New York, 1976; Vol. 3.
72. Friedman, H. L. *Mol. Phys.* **1975,** *29,* 1533.
73. Rullmann, J. A. C.; van Duijnen, P. Th. *Mol. Phys.* **1988,** *63,* 451.
74. Liptay, W. In *Modern Quantum Chemistry;* Sinanoglu, O., Ed.; Academic: Orlando, FL, 1965; Vol. III, p 45.
75. Rettig, W. *J. Mol. Struct.* **1982,** *84,* 303.
76. Lockhart, D. J.; Goldstein, R. F.; Boxer, S. G. *J. Phys. Chem.* **1988,** *89,* 1408.
77. Oh, D. H.; Sano, M.; Boxer, S. G. *J. Am. Chem. Soc.* **1991,** *113,* 6880.
78. Liptay, W. Z. *Naturforsch. A: Phys. Phys. Chem. Kosmophys.* **1965,** *20,* 272.
79. Reimers, J. R.; Hush, N. S. In *Mixed Valence Systems: Applications in Chemistry. Physics, and Biology;* Prassides, K., Ed.; Kluwer Academic: Dordrecht, Netherlands, 1991; p 29.
80. Reimers, J. R.; Hush, N. S. *J. Phys. Chem.* **1991,** *95,* 9773.
81. Zeng, J.; Craw, J. S.; Hush, N. S.; Reimers, J. R. *Chem. Phys. Lett.* **1993,** *206,* 323.
82. Zeng, J.; Hush, N. S.; Reimers, J. R. *J. Phys. Chem.* **1995,** *99,* 10459.

17

Toward the Photoreduction of CO_2 with Ni(2,2′-bipyridine)$_n^{2+}$ Complexes

Yukie Mori[1], David J. Szalda[2], Bruce S. Brunschwig, Harold A. Schwarz, and Etsuko Fujita[2]

Chemistry Department, Brookhaven National Laboratory, Upton, NY 11973–5000

When an acetonitrile solution containing $Ni(bpy)_3^{2+}$ (bpy = 2,2′-bipyridine) triethylamine, and CO_2 is irradiated at 313 nm, CO is produced with a quantum yield of ~0.1% (defined as CO produced/photons absorbed). Flash photolysis, electrochemistry, and pulse radiolysis experiments provide evidence for the formation of $Ni^I(bpy)_2^+$, as an intermediate, in the photochemical $Ni(bpy)_3^{2+}$–triethylamine–CO_2 system. Although $Ni^0(bpy)_2$ does react with CO_2, $Ni^I(bpy)_2^+$ seems unreactive toward CO_2 addition. The X-ray structure of $(Ni_3(bpy)_6)(ClO_4)$, which crystallizes as blue–violet needles, reveals the existence of a dimer in the solid. UV–vis spectra also indicate that reduced $Ni(bpy)_3^{2+}$ solutions contain $Ni^I(bpy)_2^+$, $Ni^0(bpy)_2$, and $[Ni(bpy)_2]_2$ complexes in equilibrium.

The efficient reduction of CO_2 to fuels and organic chemicals is a fundamental chemical challenge. The activation of CO_2 by transition metal complexes continues to be the subject of considerable interest (*1*). Nickel(0) complexes have been previously used as catalysts for the C–C coupling reaction between alkenes and CO_2, and for CO_2 reduction to CO. Inoue et al. found that $Ni(COD)_2$ (COD = 1,5-cyclooctadiene) catalyzes the reaction of 1-hexyne and CO_2 into 4,6-dibutyl-2-pyrone along with 1-hexyne oligomers (*2*). A similar reaction, studied by Hoberg and Schaefer (*3*), indicated that an oxanickela five-membered ring complex is formed by condensation of CO_2 and alkyne with the Ni(0) complex. Addition of another alkyne yields a complex with the seven-membered ring structure suggested by Inoue. The 2-pyrone and the starting

---

[1] Current address: Molecular Photochemistry Laboratory, The Institute of Physical and Chemical Research (Riken), Wako, Saitoma 351–01, Japan

[2] Current address: Baruch College, City University of New York, New York, NY 10010

Ni(0) complex are formed upon heating this complex. Hoberg and co-workers further studied the C–C coupling reactions of CO_2 with alkynes, alkenes (including cycloalkenes), and 1,2- or 1,3-dienes (*4–10*). Unfortunately, most of these reactions produce stable five-membered metallacycle complexes and the catalytic reactions, involving insertion of activated alkynes (or other reagents) into the five-membered metallacycle followed by reductive elimination, have not been realized.

CO_2 copolymerization is another attractive approach to chemical utilization of CO_2. Recently, Tsuda and co-workers reported (*11–15*) the efficient copolymerization of CO_2 with diynes to produce poly(2-pyrone)s using $Ni(COD)_2$ as a catalyst.

Electrochemical methods offer an alternative for bringing about nickel-catalyzed CO_2 insertion into acetylenic derivatives under mild conditions (i.e., 1 atm CO_2 at 25 °C compared to 50 atm CO_2 at 90–120 °C in Tsuda's and Inoue's experiments). Duñach and co-workers successfully showed that the incorporation of carbon dioxide into alkynes catalyzed by electrogenerated nickel–bipyridine complexes gives α,β-unsaturated acids in moderate to good yields (*16–21*). The electrocatalytic carboxylation reaction was undertaken on a preparative scale in the presence of a sacrificial magnesium anode; the cleavage of the 5-membered nickelacycle by magnesium ions is thought to be the important step in this catalytic system.

The electrochemistry of $Ni(bpy)_3^{2+}$ (bpy = 2,2′-bipyridine) in acetonitrile (MeCN) or dimethylformamide (DMF) has been studied by several researchers (*16, 22–27*). However, there is no agreement on the identity of the redox active species. The first reduction wave is assigned to a variety of reactions involving $Ni^0(bpy)_2$, $Ni^0(bpy)_3$, $Ni^I(bpy)_2^+$, and $Ni^I(bpy)_3^+$. The majority of the studies indicate that the first reduction at −1.25 V vs. SCE (standard calomel electrode) is a two-electron reduction followed by loss of a bpy ligand. The reasons are (1) the current is twice that expected for a one-electron reduction, and (2) the difference between the anodic and cathodic peak potentials is ~40 mV, which is close to the theoretical value of 27 mV for a two-electron reduction process. Tanaka and co-workers (*22, 23*) have suggested that the first reduction is the result of two one-electron processes: $Ni^{II}(bpy)_3^{2+}$ to $Ni^I(bpy)_3^+$ followed by $Ni^I(bpy)_3^+$ to $Ni^0(bpy)_3$, based on the observation of an Ni(I) electron paramagnetic resonance (EPR) signal that they assign to $Ni^I(bpy)_3^+$. Prasad and Scaife (*24*) have isolated a blue solid, which they identify as $[Ni^I(bpy)_2]ClO_4$, from bulk electiolysis of $Ni^{II}(bpy)_3^{2+}$ (UV–vis wavelength of the solution: 400 nm (ϵ = 9000), 570 nm (ϵ = 6900)). They concluded that $Ni^{II}(bpy)_3^{2+}$ is first reduced to $Ni^I(bpy)_3^+$ followed by loss of a bpy ligand. However, the elemental analysis of their solid contains large errors for C, H, and Ni. Misono et al. reported (*28*) the spectrum of dark green $Ni^0(bpy)_2$ (with an absorption at 680 nm), prepared by de-ethylation from $NiEt_2$ (bpy) in the presence of bpy (*29*), in HMPT (hexamethylphosphoric triamide). This absorption maximum does not agree with that found by Prasad and Scaife. Dark violet crystals of $Ni^0(bpy)_2$ have been prepared by metal-vapor synthesis and characterized by IR and NMR spectroscopies; however, UV–vis data were not reported (*25*).

Although electrochemical CO_2 incorporation into unsaturated hydrocarbons is a significant advancement, it is not economical to fix CO_2 in this manner. We are interested in using $Ni(bpy)_3^{2+}$ to photochemically reduce CO_2. We have found that when an MeCN solution containing $Ni(bpy)_3^{2+}$, triethylamine (TEA), and CO_2 is irradiated at 313 nm, it produces CO with a quantum yield of ~0.1% (defined as CO produced/photons absorbed). Here we present results on photochemical CO_2 reduction using $Ni(bpy)_3^{2+}$ and discuss the nature of the various intermediates studied by electrochemistry, flash photolysis, and pulse radiolysis.

Experimental Details

Na–Hg Reduction. A solution of the reduced species in MeCN was prepared by successively reducing portions of $[Ni(bpy)_3](ClO_4)_2$ (0.1 mM–2.5 mM) with 0.5% Na–Hg under vacuum in sealed glassware. UV-vis spectra were monitored during the reduction, and when the absorption of the bands in the visible region reached a maximum, the reduction was stopped.

Photoreactions. A sample solution containing 0.33 mM $[Ni(bpy)_3](ClO_4)_2$ and 0.5 M triethylamine in MeCN (3.0 mL) was bubbled with CO_2 for 20 min and then irradiated at 313 nm (100 W Hg–Xe arc lamp with 1/4 m monochromator) in a 1-cm quartz cuvette under stirring. After photolysis, 0.1 mL of air and 0.1 mL of water were added to decompose the CO adduct. The CO evolved was analyzed using a Varian gas chromatograph [Model 3700, He carrier gas, 5-Å molecular sieve column (4-m length, 1/8-in. diameter)] equipped with a thermal conductivity detector. Each run was carried out two or three times.

Laser Flash Photolysis. A sample solution was prepared by vacuum-transfer techniques just before the measurements. Transient-absorption spectra and lifetimes of various intermediates were measured using the previously described apparatus (*30*). Excitation was with the fourth harmonic of a Nd:YAG laser with a pulse width of ca. 30 ps. The solutin was vigorously stirred for 10 s between laser shots.

Pulse Radiolysis. Pulse radiolysis was performed by using electrons from a 2-MeV Van de Graaff accelerator (*31*). The samples were thermostated, and the optical path length was generally 6.1 cm. About 1×10^{-6} M Ni(I) per pulse was produced in most studies.

Results

Photochemical Reduction of CO_2. When a solution containing $[Ni(bpy)_3](ClO_4)_2$ in TEA–MeCN was irradiated with 313-nm light under a

Table I. Photochemical CO Formation with $[Ni(bpy)_3](ClO_4)_2$

$[Ni(bpy)_3^{2+}]$ (mM)	[TEA] (M)	Added [bpy] (M)	Solvent	Atmosphere	CO
0.33	0.5	—	MeCN	CO_2	+
0.33	0.5	—	MeCN	CO_2(dark)	−
0.33	—	—	MeCN	CO_2	−
—	0.5	—	MeCN	CO_2	−
0.33	0.5	—	MeCN	Ar	−
0.33	0.5	—	water	CO_2	−
—	0.5	1.0	MeCN	CO_2	−

NOTE: Each solution (3.0 mL) was irradiated with 313-nm light under stirring, and kept in the dark for 2 h; then air and water (0.1 mL each) were added just before GC analysis. H_2 was not detected in any run.

CO_2 atmosphere, the color of the solution changed from colorless to orange. After photolysis for 50 min, no CO was detected in the gas phase. On addition of 0.1 mL of air and 0.1 mL of water to the solution, however, the orange color of the CO adduct disappeared and CO was observed. As shown in Table I, no CO was detected without TEA or $[Ni(bpy)_3](ClO_4)_2$, and the reaction did not take place in the dark. These results indicate that the nickel complex, TEA, and light are necessary for the reduction of CO_2 to CO.

Figure 1 shows the variation of the optical spectrum of the $[Ni(bpy)_3](ClO_4)_2$ solution observed during irradiation at 313 nm. A broad absorption band appeared around 450 nm and increased with irradiation for 20 min. Continued photolysis for 50 min yielded a decrease in the 450-nm band, an increased absorption to the blue with a shoulder at ~380 nm, and an isosbestic point at 430 nm. When the photolyzed solution was kept in the dark, the optical spectrum changed slowly and a broad absorption band was observed at around 480 nm. If the photolysis was continued longer than 50 min, the isosbestic point was lost, although the absorbance at 380 nm continued to increase.

The yield of CO is plotted versus irradiation time in Figure 2. The CO yield was not linearly correlated with irradiation time but showed an induction period. After irradiation for 20 min, when the absorbance at 450 nm reached a maximum, only trace amounts of CO were detected. It is noteworthy that after the 50-min irradiation, the CO yield significantly increased when the solution was stored in the dark before the addition of air. The CO yield reached 50% of the amount of $[Ni(bpy)_3](ClO_4)_2$ used at 100 min. These results suggest that a two-step reaction takes place in the presence of CO_2.

In order to obtain information on the mechanism of this photochemical reaction, the effects of some additives were investigated. The results are summarized in Table II. When the progress of the reaction was monitored in the visible region, addition of free bipyridine accelerated the reaction rate, whereas the presence of excess Ni(II) ion retarded it. However, no significant difference

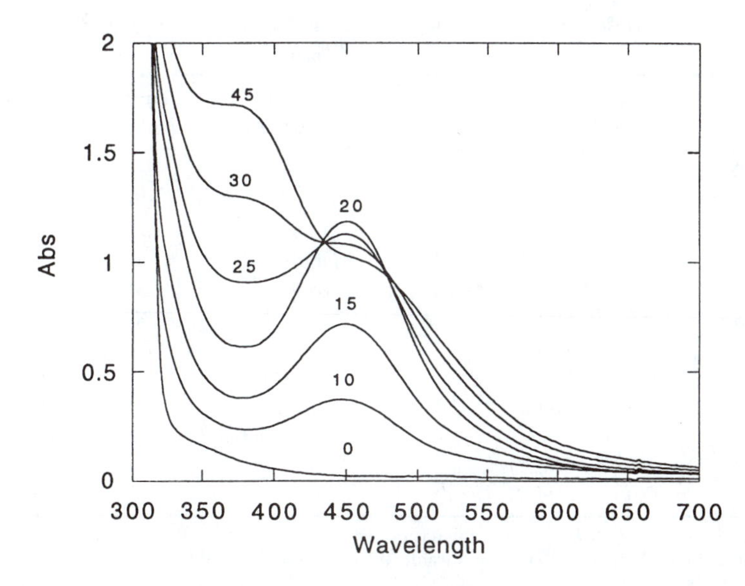

Figure 1. Variation of optical spectrum during photolysis of a solution containing 3.3×10^{-4} M $[Ni(bpy)_3](ClO_4)_2$ and 0.5 M TEA in CO_2-saturated MeCN at 313 nm. Numbers indicate irradiation time (min).

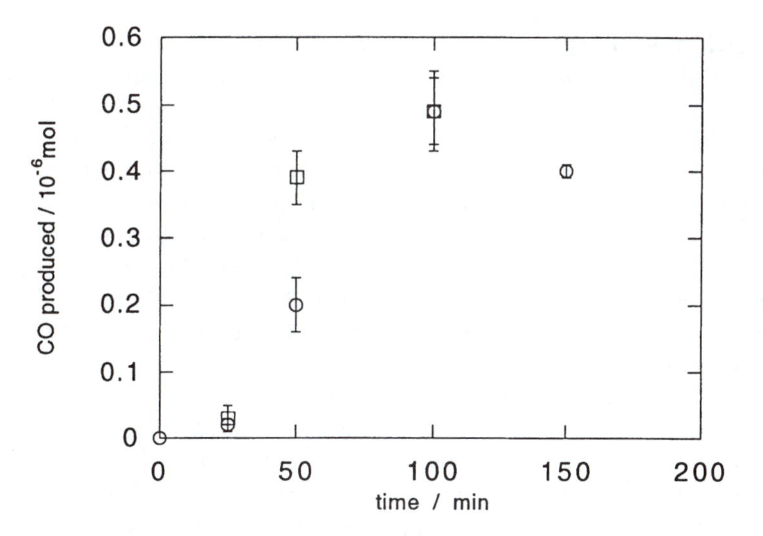

Figure 2. Relationship between CO yield and irradiation time on photolysis of a solution containing 3.3×10^{-4} M $[Ni(bpy)_3](ClO_4)_2$ and 0.5 M TEA in CO_2-saturated MeCN at 313 nm. ○: Analyzed immediately after photolysis. □: Analyzed after the photolyzed solution was kept in the dark for 2 h.

Table II. CO Yield from Photochemical Reaction of $[Ni(bpy)_3](ClO_4)_2$ Under Various Conditions

$[Ni(bpy)_3]$ (mM)	*[TEA] (M)*	*Solvent*	*Additive*	*Time (min)*	*CO produced (μmol)*
0.33	0.5	MeCN	None	50	0.38
0.33	0.5	MeCN	None	100	0.49
1.0	0.5	MeCN	None	100	0.51
0.33	0.5	MeCN–EtOH (1:1)	None	40	0.34
0.33	0.5	MeCN–H_2O (1:1)	None	50	0
0.33	0.5	MeCN	1 mM bpy	100	0.46
0.33	0.5	MeCN	None	120	0.30
0.33	0.5	MeCN	3.3 mM bpy	120	0.30
0.33	0.5	MeCN	0.33 mM $Ni(ClO_4)_2$	220	0.28
0.33	5.0	MeCN	None	120	0.19
0.33	0.05	MeCN	None	240	0

NOTE: Each solution (3.0 mL) was irradiated under 1 atm CO_2 with 313-nm light under stirring, and kept in the dark for 2 h; then air and water (0.1 mL each) were added just before GC analysis.

in the CO yield was observed. When water or water–MeCN mixture was used as a solvent instead of pure MeCN, neither a spectral change nor CO formation was detected.

Sodium–Amalgam Reduction of $[Ni(bpy)_3]^{2+}$. When a 0.1–2.5 mM $[Ni(bpy)_3](ClO_4)_2$ acetonitrile solution was treated with 0.5% Na–Hg under vacuum, the solution exhibited an intense olive–green color with absorption maxima at 422, 592, and 910 nm. As the reduction proceeded the color changed to blue–green, the absorption intensity increased throughout the visible region, with the peak at 592 nm red-shifting to 610 nm (Figure 3). The intensity increase of the band at 422 nm depended on the concentration of $[Ni(bpy)_3]^{2+}$. At low nickel concentration (<0.2 mM) the increase is almost negligible, with a final absorbance ratio of the bands at 422 and 610 nm of about 1:1. A new band at 1300 nm, whose intensity was also dependent on the concentration of $[Ni(bpy)_3]^{2+}$, appeared as the band at 910 nm disappeared. The molar absorptivity of the 1300-nm band is ~9×10^3 M^{-1} cm^{-1}, with 2.5 mM Ni. At the end of the experiment, the two-electron reduction of $[Ni(bpy)_3]^{2+}$ was confirmed by adding one equivalent mole of $Co^{III}dimBr_2ClO_4$ (dim = 2,3-dimethyl-1,4,8,11-tetraazacyclotetradeca-1,3-diene) to the reduced solution. The intense blue–green color disappeared and a band at 428 nm appeared because of the absorption (*32*) of $Co^{I}dim^+$. Therefore, the species with absorption bands at 422, 592, and 910 nm may be a monoreduced Ni(I) species, and the species with absorption bands at 422, 610, and 1300 nm may be a direduced Ni(0) species which partially dimerizes allowing π–π interaction of the bpy

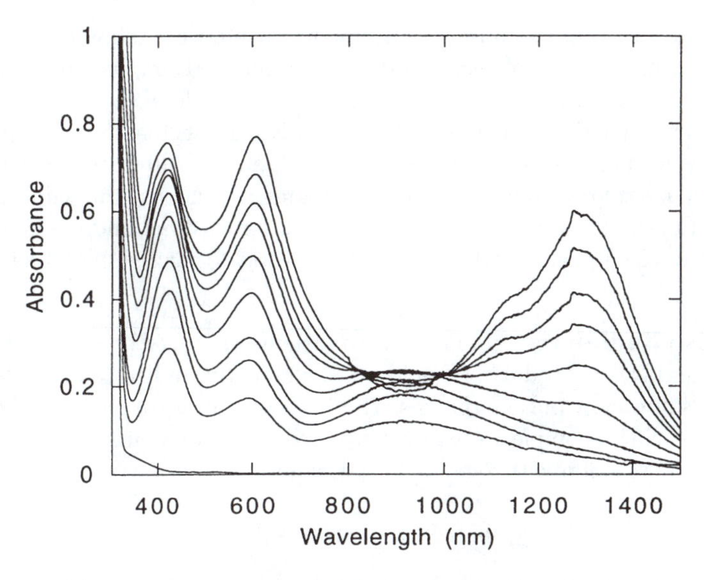

Figure 3. Absorption spectra recorded during the Na–Hg reduction of 1.24 mM $Ni(bpy)_3(ClO_4)_2$ in MeCN with a 0.1-cm cell.

ligands. (*See* the results of the X-ray structure determination.) Further reduction by Na–Hg leads to the production of the relatively unstable bpy radical anion (*33*) together with the loss of the bands at 422, 610, and 1300 nm, and eventually the solution loses its intense color almost completely.

When Na–Hg reduction of 2.5 mM $[Ni(bpy)_3](ClO_4)_2$ was performed in the presence of 25 mM bipyridine, the same spectrum was obtained. Nor did addition of TEA to the solution of the reduced species cause any significant differences to the absorption spectrum.

The reactions with CO and CO_2 were examined by adding each gas to the mono- and direduced species. The addition of CO to the monoreduced solution produced an unstable CO adduct (sh 350 nm and 470 nm), which decomposed in one hour. The addition of CO to the direduced solution produced a stable CO adduct (peaks at 382 nm and 466 nm). The addition of CO_2 to the monoreduced solution caused the slow decay of the reduced nickel species in one hour without any indication of intermediates. The addition of CO_2 to the direduced solution resulted in peaks at 350 nm and 470 nm, which indicates the formation of the "CO adduct." A study to identify the CO adduct by means of X-ray structure and IR is in progress.

Laser Flash Photolysis. A sample solution containing 1×10^{-4} M $[Ni(bpy)_3](ClO_4)_2$ and 0.5 M TEA in MeCN under vacuum was excited with the fourth harmonic of a Nd:YAG laser pulse (266 nm), and the transient absorption spectrum was observed. Immediately after excitation (~15 ns), two absorption bands were observed around 420 and 590 nm, which are similar to those obtained for Na–Hg reduction of $[Ni(bpy)_3](ClO_4)_2$. The observed spectrum remained unchanged for 100 μs. This suggests that the photoproduct is rapidly formed and has a relatively long lifetime. Although the transient spectrum measured under a CO_2 atmosphere was almost the same as that observed under vacuum, the transient spectrum measured under a CO atmosphere showed a peak at 470 nm, which indicates formation of CO adduct.

Pulse Radiolysis. Some of the transients produced by photolysis, electrolysis, and Na–Hg reduction could be conveniently studied in more detail in aqueous solution by pulse radiolysis. The Ni(I) species were produced by reaction of the Ni(II) complexes with the hydrated electron (with the $H^\cdot$ and $OH^\cdot$ removed by reaction with 2-methyl-2-propanol).

$$e^-_{aq} + Ni^{II}(bpy)_n{}^{2+} \rightarrow Ni^{I}(bpy)_n{}^{+} \quad (1)$$

These reactions were found to be so rapid ($k = 4.0 \times 10^{10}$ M^{-1} s^{-1} for $Ni^{II}(bpy)^{2+}$ and $Ni^{II}(bpy)_2{}^{2+}$, and 5.4×10^{10} M^{-1} s^{-1} for $Ni^{II}(bpy)_3{}^{2+}$) that there is no chance of further reduction on the time scale of the experiment. $Ni^{I}(bpy)^{+}$ has absorption maxima at 390 nm ($\varepsilon = 3100$ M^{-1} cm^{-1}) and 590 nm ($\varepsilon = 1900$). Reduction of either $Ni^{II}(bpy)_2{}^{2+}$ or $Ni^{II}(bpy)_3{}^{2+}$ produced $Ni^{I}(bpy)_2{}^{+}$ within the time scale of the experiment (maxima at 415 nm, $\varepsilon =$ 5300; and 570 nm, $\varepsilon = 3300$). These spectra are similar to those obtained in MeCN by Na–Hg reduction and by flash photolysis. Both Ni(I) species were found to react with CO with nearly the same rate constant (2.4×10^9 M^{-1} s^{-1}) and resulted in spectra similar to those observed in flash photolysis under a CO atmosphere and in Na–Hg reduction followed by the addition of CO.

The $CO_2{}^-$ radical, produced in solutions of $Ni(bpy)_n{}^{2+}$ with formate and either N_2O or CO_2 present, also reduces the Ni(II) to Ni(I), but much more slowly than e^-_{aq}, with rate constants of about 2×10^6 M^{-1} s^{-1}. In this case it is difficult to avoid some further reduction of the Ni(I) to Ni(0), and there is strong indication of an interaction of one of these species with CO_2.

Discussion

Nature of the Reduced Species. Although the electrochemistry of $Ni(bpy)_3{}^{2+}$ species is not well established, we believe that the pulse radiolysis results give a clear indication that $Ni(bpy)_3{}^{2+}$ can be reduced in a single-electron step and that bpy is rapidly lost from the monoreduced species.

In our pulse radiolysis study, we avoid the second step of the reduction by using a very small dose to the solution. The spectra of $Ni^{I}(bpy)_2{}^{+}$ and the product of the reaction in equation 1 ($n = 3$) are identical, with bands at 415 nm and 570 nm indicating loss of a bpy ligand from $Ni^I(bpy)_3^+$ in H_2O. Therefore, the following reactions need to be considered to explain the electrochemistry of $Ni(bpy)_3^{2+}$ species in MeCN.

$$Ni^{II}(bpy)_3^{2+} + e^- \rightleftharpoons Ni^I(bpy)_2^+ + bpy \quad (2)$$

$$Ni^I(bpy)_2^+ + e^- \rightleftharpoons Ni^0(bpy)_2 \quad (3)$$

$$2Ni^I(bpy)_2^+ + bpy \rightleftharpoons Ni^{II}(bpy)_3^{2+} + Ni^0(bpy)_2 \quad (4)$$

When a Na–Hg reduction of $Ni^{II}(bpy)_3^{2+}$ was performed in MeCN, bands at 422 and 592 nm increased in intensity. The spectrum is similar to that of $Ni^I(bpy)_2^+$ in H_2O except for a small red shift. When the Ni concentration is low (<0.15 mM), a clear change to a second reduction step is observed. While the absorption at 422 nm remains constant, the absorption at 592 nm shifts to 610 nm and a new absorption at 1300 nm appears. When the Ni concentration is higher, the change from the first to the second reduction step is not as clear, probably because of the increased rate of Ni(I) disproportionation shown in equation 4. The intensity of the absorption at 1300 nm shows a dependence on concentration. Certain reduced nickel complexes have a tendency to dimerize (*34*). The dimers have a near IR absorption due to the stacking interaction between ligands (*35*).

$$2Ni^0(bpy)_2 \rightleftharpoons [Ni^0(bpy)_2]_2 \quad (5)$$

The equilibrium constants of equations 4 and 5 are currently being investigated.

The above results indicate that the first reduction peak in the electrochemistry is the result of two single-electron reduction steps (equations 2 and 3) involving the loss of a bpy ligand from $Ni^I(bpy)_3^+$. The two steps appear as one peak in the voltammetry because E_1 ($Ni^{II}(bpy)_3^{2+}/Ni^I(bpy)_3^+$) is very close to E_2 ($Ni^I(bpy)_2^+/Ni^0(bpy)_2$). The net equation is shown as:

$$Ni^{II}(bpy)_3^{2+} + 2e^- \rightleftharpoons Ni^0(bpy)_2 + bpy \quad (6)$$

Blue–violet needles were grown at room temperature from the blue–green acetonitrile solution obtained by Na–Hg reduction of $Ni(bpy)_3(ClO_4)_2$. X-ray structure analysis revealed three $Ni(bpy)_2$ units, one perchlorate anion, and one acetonitrile of crystallization in the asymmetric unit. This yields a formula of $[Ni(bpy)_2]_3(ClO_4)\cdot NCCH_3$ that requires one Ni(I) and two Ni(0). Of the three $Ni(bpy)_2$ units, one is a monomer while the other two form a dimer with a Ni–Ni distance of 3.440 ± 0.004 Å. In the dimer, the bpy's of one unit are parallel to the bpy's of the other unit with a bpy–bpy distance of 3.50 Å, as shown in Figure 4. The existence of the dimer in the reduced solution is suggested by the similarity in color between the solution and crystals. The coordination geometries of the three $Ni(bpy)_2$ units are almost identical: each nickel atom is coordinated to the four nitrogen atoms of two bpy ligands (Ni–N 1.911 Å in the monomer, 1.931 and 1.997 Å in the dimer), and the dihedral angle between the two bpy's is about 40° in each unit.

It is attractive to assign the monomer as Ni(I) because: (1) the monomer has a shorter Ni–N distance than the dimer, and (2) the UV–vis spectrum of the dimer in solution only appears after the one-electron reduction is complete. However, it is difficult to distinguish Ni(I) from Ni(0) species using the X-ray structures. There are some reported nickel structures containing bpy: CpNi′-

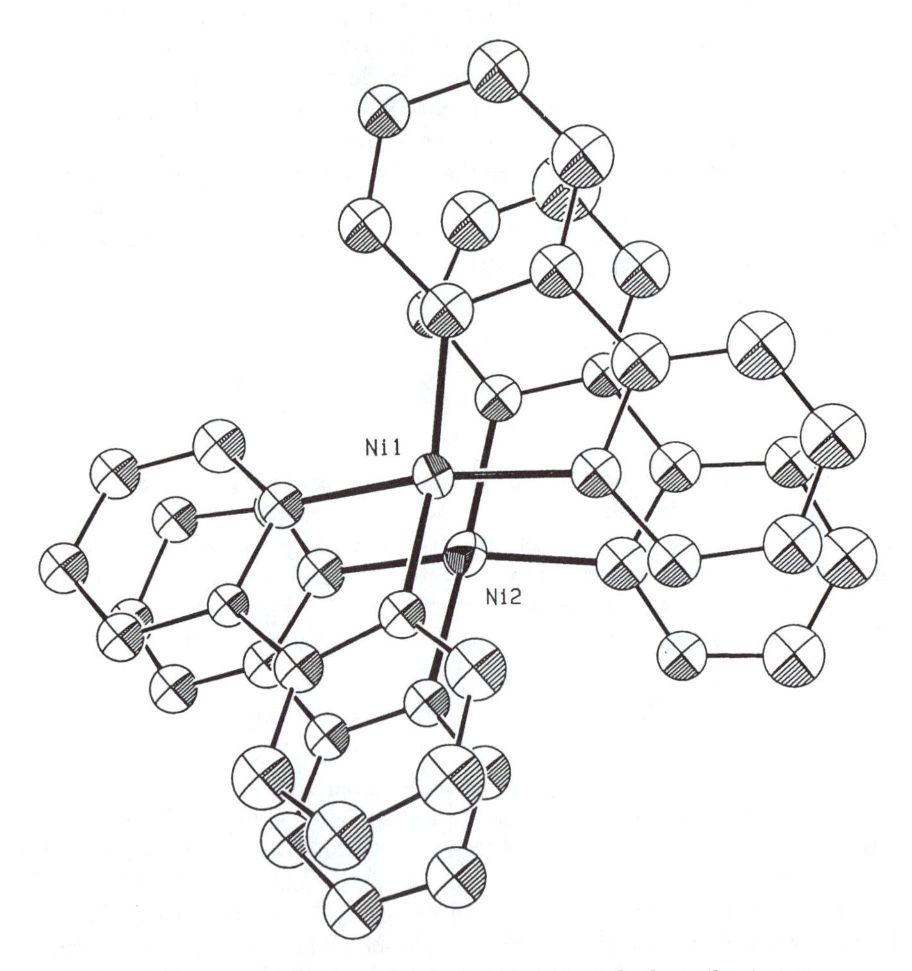

Figure 4. Structure of $[Ni(bpy)_2]_3(ClO)_4(CH_3CN)$*. Only the* $Ni(bpy)_2$ *units are shown.*

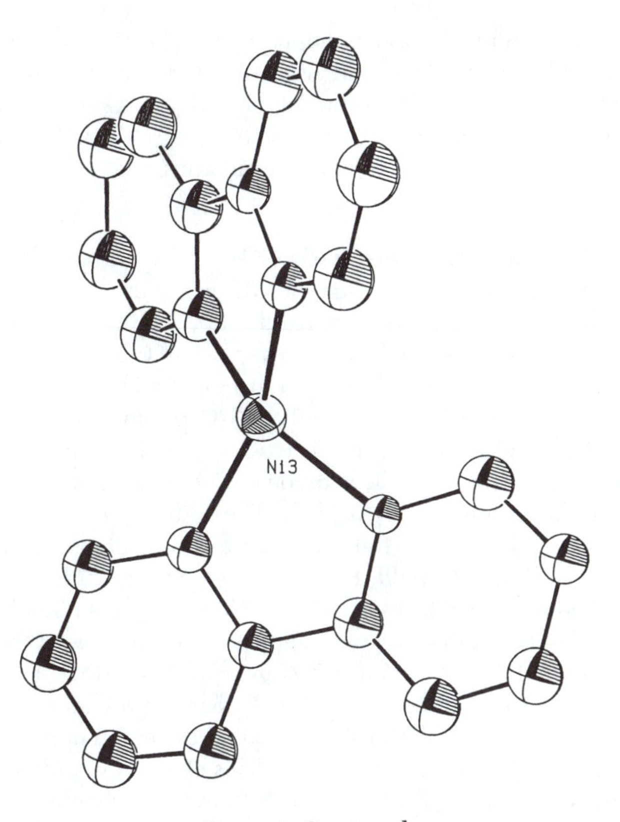

Figure 4. Continued.

(bpy) (Cp = cyclopentadienyl) (*36*), Ni–N 1.957 Å; Ni^0(COD)(bpy) (*37*), Ni–N 1.940 Å; Ni^0(phosphaalkene)(bpy) (*38*) Ni–N = 1.946 Å. The Ni–N distance in these compounds may differ from those of $Ni(bpy)_2^+$ and $Ni(bpy)_2$ because of the different coordination environment. The bridging C–C bond distance in most known tris(bipyridine) complexes is close to the 1.490 (3) Å found in free bpy (*39*). By contrast, the structure of $Ni(bpy)_2^+$ monomer reveals an extremely short C–C bond distance (avg. 1.42 Å), indicating a substantial transfer of electron density from nickel to the π^* orbital of bpy, as found in the structures of $Co^I(bpy)_3^+$ (1.42 (2) Å) (*39*), $Mo^{II}(O\text{-}i\text{-}Pr)_2(bpy)_2$ (1.425 (4) Å) (*40*), and $Fe^0(\eta^6\text{-tol})(bpy)$ (tol = $C_6H_5CH_3$) (1.417 (3) Å) (*41*). The distances in $[Ni(bpy)_2]_2$ (avg. 1.45 Å) are longer than that of $Ni(bpy)_2^+$ (probably because of the stacking interaction of bpy), but is as short as those found in other low-valent nickel complexes: 1.455 Å in $CpNi^I$(bpy), 1.459 (6) Å in Ni^0(COD)(bpy), and 1.480 Å in Ni^0(phosphaalkene)(bpy).

Flash Photolysis. The flash photolysis results show that the $Ni(bpy)_2^+$ is rapidly formed and stable for >100 μs. This indicates that the rate of reaction

4 is slow under flash photolysis conditions, where the $Ni(bpy)_2^+$ concentration is low. The addition of CO to the flash photolysis solution resulted in the immediate formation of the CO adduct of the $Ni(bpy)_2^+$. However, the addition of CO_2 did not affect the formation or stability of $Ni(bpy)_2^+$, indicating that the reaction does not occur under these conditions.

Photochemical Reaction with CO_2. The $Ni(bpy)_3^{2+}$–TEA system produces CO from CO_2 by irradiation at 313 nm with quantum yield ~0.1%. Because $Ni(bpy)_3^{2+}$ has an absorption band at 309 nm (ε = 41,700 M^{-1} cm^{-1}), over 95% of light was absorbed by $Ni(bpy)_3^{2+}$. The CO produced reacts with the reduced $Ni^I(bpy)_2^+$ and $Ni^0(bpy)_2$ to form CO adducts; therefore, photochemical reaction is stoichiometric and the CO production is 0.5 mole from 1.0 mole of $Ni^{II}(bpy)_3^{2+}$. The final spectrum of continuous photolysis (Figure 1) is similar to that observed in the addition of CO to the reduced nickel species, indicating the formation of a CO adduct. The addition of excess bpy (3 times that of $Ni(bpy)_3^{2+}$) accelerated the reaction rate; however, no significant difference was observed for CO yield. Emission from $Ni(bpy)_3^{2+}$ in MeCN was not observed at room temperature or at 77 K. However flash photolysis, electrochemistry, and pulse radiolysis experiments provide evidence of the intermediate, $Ni^I(bpy)_2^+$, in the photochemical $Ni(bpy)_3^{2+}$–TEA system. The mechanism of the photochemical formation of $Ni^I(bpy)_2^+$ has not yet been identified. The formation of $Ni^I(bpy)_2^+$ could involve the direct excitation of an electron from a donor (TEA) to the solvent (*30, 42, 43*). This electron would be expected to react rapidly with $Ni(bpy)_3^{2+}$ to produce $Ni^I(bpy)_2^+$. It should also be pointed out that $Ni^I(bpy)_2^+$ seems unreactive toward CO_2 addition. However, $Ni^0(bpy)_2$ does react with CO_2. The reduced $Ni(bpy)_3^{2+}$ solution contains various species such as $Ni^I(bpy)_2^+$, $Ni^0(bpy)_2$, and $[Ni(bpy)_2]_2$. Studies to determine the equilibrium constants between these species are in progress.

Acknowledgments

We thank Norman Sutin and Carol Creutz for their helpful comments. This research was carried out at Brookhaven National Laboratory under contract DE-AC02-76CH00016 with the U.S. Department of Energy and was supported by its Division of Chemical Sciences, Office of Basic Energy Sciences.

References

1. For example: (a) *Proceedings of the International Symposium on Chemical Fixation of Carbon Dioxide;* The Chemical Society of Japan, the Research Group on Fixation of Carbon Dioxide: Tokyo, Japan, 1991. (b) *Proceedings of the International Conference on Carbon Dioxide Utilization;* University of Bari: Bari, Italy, 1993.

2. Inoue, Y.; Itoh, Y.; Hashimoto, H. *Chem. Lett.* **1977,** 855.
3. Hoberg, H.; Schaefer, D. *J. Organomet. Chem.* **1982,** *238,* 383.
4. Hoberg, H.; Schaefer, D.; Burkhart, G.; Kruger, C.; Romao, M. J. *J. Organomet. Chem.* **1984,** *266,* 203.
5. Hoberg, H.; Oster, B. W. *J. Organomet. Chem.* **1984,** *266,* 321.
6. Hoberg, H.; Peres, Y.; Michereit, A. *J. Organomet. Chem.* **1986,** *307,* C41.
7. Hoberg, H.; Gross, S.; Milchereit, A. *Angew. Chem. Int. Ed. Engl.* **1987,** *26,* 571.
8. Hoberg, H.; Jenni, K.; Angermund, K.; Krüger, C. *Angew. Chem. Int. Ed. Engl.* **1987,** *26,* 153.
9. Hoberg, H.; Peres, Y.; Krüger, C.; Tsay, Y.-H. *Angew. Chem. Int. Ed. Engl.* **1987,** *26,* 771.
10. Hoberg, H.; Bärhausen, D. *J. Organomet. Chem.* **1989,** *379,* C7.
11. Tsuda, T.; Maruta, K.; Kitaike, Y. *J. Am. Chem. Soc.* **1992,** *114,* 1498.
12. Tsuda, T.; Morikawa, S.; Sumiya, R.; Saegusa, T. *J. Org. Chem.* **1988,** *53,* 3140.
13. Tsuda, T.; Morikawa, S.; Saegusa, T. *J. Chem. Soc., Chem. Commun.* **1989,** 9.
14. Tsuda, T.; Morikawa, S.; Hasegawa, S.; Saegusa, T. *J. Org. Chem.* **1990,** *55,* 2978.
15. Tsuda, T.; Hokazone, H. *Macromolecules* **1994,** *27,* 1289.
16. Derien, S.; Duñach, E.; Perichon, J. *J. Am. Chem. Soc.* **1991,** *113,* 8447.
17. Duñach, E.; Perichon, J. *J. Organomet. Chem.* **1988,** *352,* 239.
18. Derien, S.; Duñach, E.; Perichon, J. *J. Organomet. Chem.* **1990,** *385,* C43.
19. Derien, S.; Clinet, J.-C.; Duñach, E.; Perichon, J. *J. Chem. Soc., Chem. Commun.* **1991,** 549.
20. Derien, S.; Clinet, J.-C.; Duñach, E.; Perichon, J. *J. Organomet. Chem.* **1992,** *424,* 213.
21. Derien, S.; Clinet, J.-C.; Duñach, E.; Perichon, J. *J. Org. Chem.* **1993,** *58,* 2578.
22. Tanaka, N.; Sato, Y. *Inorg. Nucl. Chem. Lett.* **1968,** *4,* 487–490.
23. Tanaka, N.; Ogata, T.; Niizuma, S. *Inorg. Nucl. Chem. Lett.* **1972,** *8,* 965–968.
24. Prasad, R.; Scaife, D. B. *J. Electroanal. Chem.* **1977,** *84,* 373.
25. Henne, B. J.; Bartak, D. E. *Inorg. Chem.* **1984,** *23,* 369.
26. Daniele, S.; Ugo, P. *J. Electroanal. Chem.* **1987,** *219,* 259.
27. Bartlett, P. N.; Eastwick-Field, V. *Electrochim. Acta* **1993,** *38,* 2515.
28. Misono, A.; Uchida, Y.; Yamaguchi, T.; Kageyama, H. *Bull. Chem. Soc. Jpn.* **1972,** *45,* 1438.
29. Saito, T.; Uchida, Y.; Misono, A.; Yamamoto, A.; Morifuji, K.; Ikeda, S. *J. Am. Chem. Soc.* **1966,** *88,* 5198.
30. (a) Ogata, T.; Yanagida, S.; Brunschwig, B. S.; Fujita, E. *J. Am. Chem. Soc.* **1995,** *117,* 6708. (b) Milder, S. J.; Brunschwig, B. S. *J. Phys. Chem.* **1992,** *96,* 2189.
31. Schwarz, H. A.; Creutz, C. *Inorg. Chem.* **1983,** *22,* 707.
32. Fujita, E.; Creutz, C.; Sutin, N.; Szalda, D. J. *J. Am. Chem. Soc.* **1991,** *113,* 343–353.
33. Shida, T. *Electronic Absorption Spectra of Radical Ions; Physical Science Data;* Elsevier: Amsterdam, 1988.
34. (a) Peng, S.; Ibers, J. A.; Millar, M.; Holm, R. H. *J. Am. Chem. Soc.* **1976,** *98,* 8037. (b) Peng, S.; Goedkin, V. L. *J. Am. Chem. Soc.* **1976,** *98,* 8500.
35. Furenlid, L. R.; Renner, M. W.; Szalda, D. J.; Fujita, E. *J. Am. Chem. Soc.* **1991,** *113,* 883.
36. Barefield, E. K.; Krost, D. A.; Edwards, D. S.; Van Derveer, D. G.; Trytko, R. L.; O'Rear, S. P.; Williamson, A. N. *J. Am. Chem. Soc.* **1981,** *103,* 6219.
37. Dinjus, E.; Walther, D.; Kaiser, J.; Sieler, J.; Thanh, N. N. *J. Organometal. Chem.* **1982,** *236,* 123.
38. Spek, A. L.; Duisenberg, A. J. M. *Acta Cryst.* **1987,** *C43,* 1216.
39. Szalda, D. J.; Creutz, C.; Mahajan, D.; Sutin, N. *Inorg. Chem.* **1983,** *22,* 2372.

40. Chisholm, M. H.; Huffman, J. C.; Rothwell, I. P. *J. Am. Chem. Soc.* **1981,** *103,* 4945.
41. Radonovich, L. J.; Eyring, M. W.; Groshens, T. J.; Klabunde, K. J. *J. Am. Chem. Soc.* **1982,** *104,* 2816.
42. Hall, G. E.; Kenney-Wallace, G. A. *Chem. Phys.* **1978,** *28,* 205.
43. Hall, G. E.; Kenney-Wallace, G. A. *Chem. Phys.* **1978,** *32,* 313.

18

Kinetics of Electron Transfer in Solution Catalyzed by Metal Clusters

J. Khatouri, M. Mostafavi, and J. Belloni

Laboratoire de Physico-Chimie des Rayonnements associé au Centre National de la Recherche Scientifique, Bât. 350, Université Paris-Sud, Centre d'Orsay, 91405 Orsay Cedex, France

Cluster properties, mostly those that control electron transfer processes such as the redox potential in solution, are markedly dependent on their nuclearity. Therefore, clusters of the same metal may behave as electron donor or as electron acceptor, depending on their size. Pulse radiolysis associated with time-resolved optical absorption spectroscopy is used to generate isolated metal atoms and to observe transitorily the subsequent clusters of progressive nuclearity yielded by coalescence. Applied to silver clusters, the kinetic study of the competition of coalescence with reactions in the presence of added reactants of variable redox potential allows us to describe the autocatalytic processes of growth or corrosion of the clusters by electron transfer. The results provide the size dependence of the redox potential of some metal clusters. The influence of the environment (surfactant, ligand, or support) and the role of electron relay of metal clusters in electron transfer catalysis are discussed.

The increase of the redox potential of a metal cluster in a solvent with its nuclearity is now well established (*1–4*). The difference between the single atom and the bulk metal potentials is large (more than 2 V, for example, in the case of silver (*3*)). The size dependence of the redox potential for metal clusters of intermediate nuclearity plays an important role in numerous processes, particularly electron transfer catalysis. Although some values are available for silver clusters (*5, 6*), the transition of the properties from clusters (mesoscopic phase) to bulk metal (macroscopic phase) is unknown except for the gas phase (*7–9*).

The redox potential of short-lived metal clusters may be evaluated by the study of the electron transfer kinetics involving a donor–acceptor couple of known redox potential and used as a monitor (*5, 6*). The metal atoms and the electron donor are generated in the aqueous solution through a short electron pulse. During the coalescence of the clusters, their redox potential increases, as does their nuclearity, n, so that in early steps the potential of the smallest clusters is far below that of the donor and the transfer does not occur (Figure 1). Beyond a certain critical time, t_c, large enough to enable the growth of clusters and the increase of their potential above the threshold imposed by the donor, the electron transfer from the monitor to the supercritical clusters is allowed and detected by the absorbance decay of the monitor.

The observation of an effective transfer therefore implies that the potential of the critical cluster is slightly more positive than that of the reference system. The values of the nuclearity of this critical cluster allowing the transfer from the monitor and of the transfer rate constant are derived from the fit between the experimental results and the corresponding data calculated by numerical simulation obtained through adjusted parameters (*10, 11*). By changing the reference potential in a series of redox monitors, the dependence of the cluster potential on the nuclearity was obtained (*5, 6*). It was also shown that once formed, a critical cluster of silver, for example, behaves as a growth nucleus: alternate reactions of electron transfer and adsorption of surrounding metal ions make the redox potential more and more favorable to the transfer, so that autocatalytic growth is observed (Figure 1). These data enabled us to suggest a new explanation of the photographic development as resulting from (1) the size-dependence of the cluster potential (increasing with n in aqueous solution), and (2) the existence of a potential threshold and therefore a critical size imposed by the developer to the electron transfer (*5, 12*).

The aim of this work is to extend the kinetics study of electron transfer to monitor donors of more positive redox potential than previously studied, toward silver clusters, Ag_n^+, as acceptors and thus to approach the domain where clusters get metal-like properties (*13*). The selected donor is the naphtazarin hydroquinone, with properties similar to those of the hydroquinone used as a developer in photography. Its redox potential depends on pH, so that different monitor potentials are available through control of pH. Moreover, the reactivity of the donor may be followed by variation of absorbance when naphtazarin hydroquinone, almost transparent in the visible, is replaced by oxidized quinone with an intense absorption band.

Rate constants of the process and the nuclearity–redox potential correlation will be compared with corresponding data obtained in another environment, particularly when a surfactant or an associated ligand is present. The complete analysis of the autocatalytic transfer mechanism will also be compared with the photographic process of electron transfer from hydroquinone developer to clusters supported on silver bromide.

Experimental Details

All reagents were pure: silver salt (Ag_2SO_4) and 2-propanol were from Fluka, and naphtazarin (5,8-dihydroxy-1,4-naphthoquinone) (Q) was from Sigma Chemical Co. (*see* Figure 2) (*14, 15*). Electron pulses (3-ns duration) were

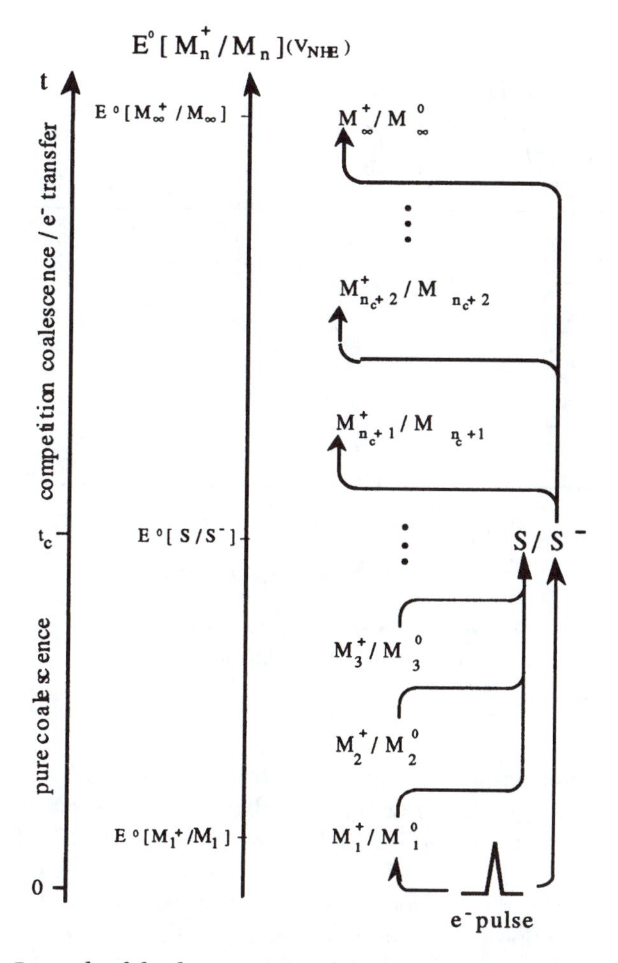

Figure 1. Principle of the determination of short-lived cluster redox potential by kinetics methods. The reference electron donor, S^-, of a given potential and the metal atoms are generated by a single pulse. During cluster coalescence, the redox potential of the couple $E°(M_n{}^+–M_n)$ progressively increases, so that an effective transfer is observed after a critical time when the cluster potential becomes higher than that of the reference, constituting a threshold. Repeatedly, a new adsorption of excess cations, M^+, onto the reduced cluster, M_n ($n \geq n_c$), allows another electron transfer from S^- with incrementation of nuclearity. The subcritical clusters M_n ($n < n_c$) may be oxidized by S, but the reference is selected so that this process is negligible.

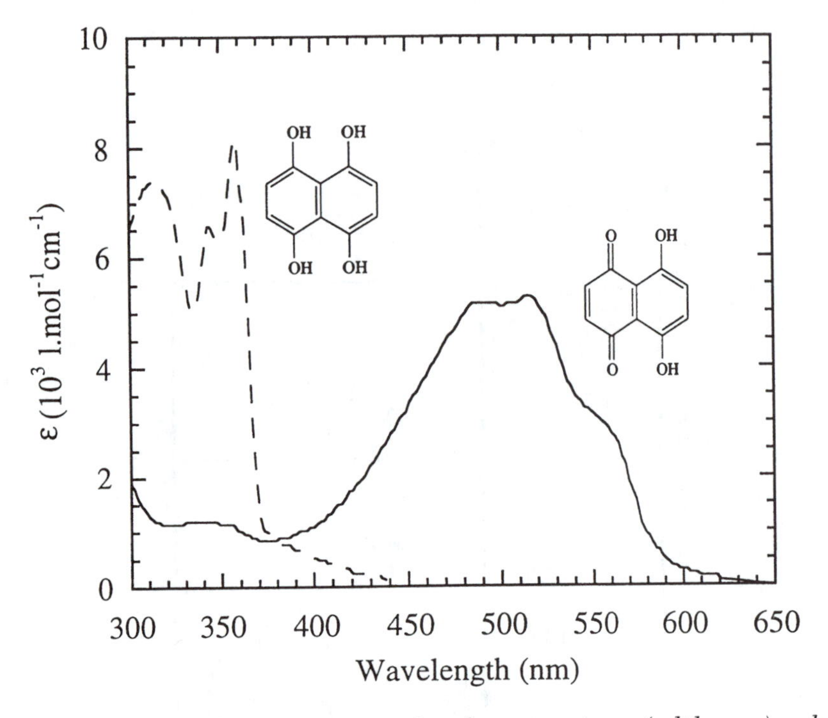

Figure 2. Optical absorption spectra of naphtazarin quinone (solid curve) and hydroquinone (dashed curve).

delivered by an electron accelerator (Febetron 706; 600-keV electron energy) to samples contained in a quartz Suprasil cell with a thin entrance window (0.2 mm) for the beam and an optical path length of 1 cm (*16, 17*). The dose per pulse was (0.05 to 4) $\times$ 10^{18} eV mL^{-1} or 9 to 650 Gy, which corresponds to (0.05 to 4) $\times$ 10^{-4} mol L^{-1} of reducing species per pulse. The cell was deaerated by using nitrogen flow. The solution was changed after each pulse. Absorption of transient species was analyzed by means of a classical xenon lamp monochromator and photomultiplier setup connected with a transient digitizer. Splitting of the beam allowed the signals to be recorded simultaneously at two different wavelengths and thus the spectra to be normalized.

The concentration of Ag_2SO_4 was typically 3 $\times$ 10^{-5} mol L^{-1} ($[Ag^+]$ = 6 $\times$ 10^{-5} mol L^{-1}), and that of naphtazarin was 10^{-4} mol L^{-1}. 2-Propanol was added at a concentration of 0.2 mol L^{-1} in order to scavenge the primary radicals $OH^{\cdot}$ and $H^{\cdot}$ and to replace them with strongly reducing radicals $(CH_3)_2C^{\cdot}OH$. The redox potential $E°(Q^{\cdot-}–QH_2)$ of the electron donor is pH-dependent. The kinetics were studied at pH 4.8 ($E°$ = $+0.22V_{NHE}$) and pH 3.9 ($E°$ = $+0.33$ V_{NHE}). The pK_a values of the naphtazarin are 7.85 and 10.7, so the neutral form predominates (*14, 15*). The redox and spectral properties of short-lived species of the bielectronic quinone systems have been studied themselves by kinetics methods through pulse radiolysis.

The evolution of the system was followed by time-resolved observation of the absorbance of the electron donor system. Because naphtazarin hydroquinone, QH_2, is transparent in the visible (Figure 2), the electron transfer was monitored by the absorbance of the oxidized forms, semiquinone, $Q^{\cdot-}$, and quinone, Q, which exhibit intense absorption bands with maximum wavelengths, λ_{max}, at 360 nm ($\varepsilon_{360}(Q^{\cdot-})$ = 10,500 L mol^{-1} cm^{-1}) and at 512 nm ($\varepsilon_{512}(Q)$ = 5000 L mol^{-1} cm^{-1}), respectively (Figure 2) (*14, 15*). The difference between extinction coefficients at 512 nm ($\Delta\varepsilon = \varepsilon_Q - \varepsilon_{QH_2}$) is at the maximum. We also observe the silver aggregates at λ = 380 nm, where the species Q and QH_2 absorb weakly and $\Delta\varepsilon$ = 0.

Results

The species involved in the electron transfer to be studied are both generated by the electron pulse interacting with the appropriate aqueous solution of the precursors. The electron pulse creates the following primary species in the solvent:

$$H_2O \rightsquigarrow \rightarrow e^-_{aq}, H_3O^+, H^{\cdot}, H_2, OH^{\cdot}, H_2O_2 (1)$$

In the presence of 2-propanol, the radicals $H^{\cdot}$ and $OH^{\cdot}$ are scavenged readily and replaced by the strong reducing radical $(CH_3)_2C^{\cdot}OH$ ($E°$ $[(CH_3)_2CO–(CH_3)_2C^{\cdot}OH] = -1.8\ V_{NHE}$) (*18*).

$$(CH_3)_2CHOH + H^{\cdot}\ (OH^{\cdot}) \rightarrow (CH_3)_2C^{\cdot}OH + H_2\ (H_2O) \quad (2)$$

Kinetics. ***Naphtazarin Solutions.*** In the absence of silver cation, all reducing species (e^-_{aq} and $(CH_3)_2C^{\cdot}OH$) are scavenged by naphtazarin. The solute naphtazarin used as the precursor of the electron donor reacts rapidly with the solvated electron (*14, 15*):

$$Q + e^-_{aq} \rightarrow Q^{\cdot-},\ k_3 = (3.1 \pm 0.2) \times 10^{10}\ L\ mol^{-1}\ s^{-1} \quad (3)$$

and with the alcohol radical, giving rise to the basic form of naphtazarin semiquinone (pK_a = 2.7):

$$Q + (CH_3)_2C^{\cdot}OH \rightarrow Q^{\cdot-} + (CH_3)_2CO + H^+,$$
$$k_4 = (5.1 \pm 0.3) \times 10^9\ L\ mol^{-1}\ s^{-1} \quad (4)$$

Then the semiquinone disproportionates into the previous quinone and the hydroquinone:

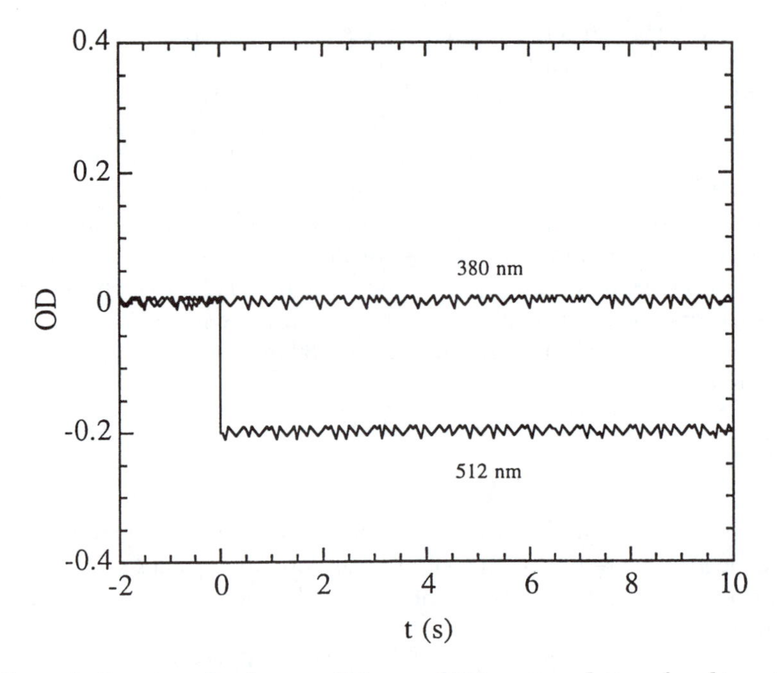

Figure 3. Transient absorbance at 512 nm and 380 nm in a solution of naphtazarin after a single 3-ns pulse. $[Q]_{t=0} = 10^{-4}$ mol L^{-1}, $[(CH_3)_2CHOH] = 0.2$ mol L^{-1}, and pH = 4.8. The total initial concentration of reducing species (e^-_{aq} and alcohol radicals) is 8×10^{-5} mol L^{-1}.

$$2Q^{\cdot -} \xrightarrow{H^+} Q + QH_2 \tag{5}$$

We observed the second-order decay of $Q^{\cdot -}$ at $\lambda = 360$ nm according to the published bimolecular rate constant (*14, 15*). The value of k_5 is pH dependent, but in the range $3.5 < pH < 6$, the change is negligible: $k_5 = (1.1 \pm 0.1) \times 10^9$ L mol^{-1} s^{-1}. Reactions 1 to 5 are achieved within a few microseconds. At that moment, the medium contains the naphtazarin quinone and hydroquinone. At 380 nm, the absorbance becomes zero.

As a result, the end-of-pulse bleaching at 512 nm shown in Figure 3 is caused by the overall reduction of the quinone into the transparent hydroquinone (Figure 2) via reactions 3–5. Then the absorbance, $OD_{\lambda=512}$, is constant because hydroquinone cannot react with any species: QH_2 is stable.

Silver and Naphtazarin Solutions. *Times up to t_c.* When the solution contains both naphtazarin and silver cations, the solvated electrons also react partly with silver ions shortly after the pulse (*19*):

$$Ag^+ + e^-_{aq} \rightarrow Ag^0, \qquad k_6 = 3.6 \times 10^{10} \text{ L mol}^{-1} \text{ s}^{-1} \tag{6}$$

The amounts of electrons reacting in the competitive processes 3 and 6 are proportional to the respective probabilities of reaction occurrence: $k_3 \times [Q]$ and $k_6 \times [Ag^+]$. The reduction of Ag^+ by $(CH_3)_2C^{\cdot}OH$ is negligible because $E°[(CH_3)_2CO–(CH_3)_2C^{\cdot}OH]$ is higher than $E°(Ag^+–Ag^0)$ (*3*), so that only about 20% of reducing species are scavenged by Ag^+ in mixed solutions. The total concentration of reducing species depends on the dose per pulse and was between 0.3 and 1.5×10^{-4} L mol^{-1} s^{-1}. The pH of a neutral solution may decrease from its initial value during the reduction of Ag^+ just after the pulse. The pH shifts from an initial value of pH = 5.6 to 4.8. In contrast, an initial pH of 3.9 is not affected.

Kinetics at 380 nm. The absorbance at 380 nm is only due to Ag_n, because an isobestic point exists at this wavelength for the spectra of QH_2 and Q: $\Delta\varepsilon = \varepsilon_{380}(QH_2)–\varepsilon_{380}(Q) = 0$ (Figure 2).

Generally, the total scavenger concentration is higher than that of reducing radicals. The reduction (reaction 6) therefore concerns only a portion of the silver cations: atoms or clusters are generated in the presence of an excess of Ag^+ ions. The silver atoms formed through reaction 6 associate with excess silver cations at a diffusion-controlled rate (*20*).

$$Ag^+ + Ag^0 \rightarrow Ag_2^+, \qquad k_7 = 5 \times 10^9 \text{ L mol}^{-1} \text{ s}^{-1} \tag{7}$$

These complexed species then coalesce progressively and give rise to clusters of increasing size according to the following mechanism (*21*):

$$Ag^+ + Ag_2^+ \rightarrow Ag_3^{2+} \tag{8}$$

$$Ag_2^+ + Ag_2^+ \rightarrow Ag_4^{2+} \tag{9}$$

$$Ag_{m+x}^{x+} + Ag_{n+y}^{y+} \rightarrow Ag_{p+z}^{z+} \tag{10}$$

Although clusters contain associated cations, their nuclearity will be defined by the number of zero valency atoms, *n,* exclusively. The total concentration of metal atoms does not change during the coalescence, but the average cluster, nuclearity increases and their concentration decreases (*10*). Their absorbance increases at 380 nm (Figure 4) because of an increasing extinction coefficient per atom to a certain size and then is constant (*22*). It corresponds to an initial concentration in silver atoms of $\approx 10^{-5}$ mol L^{-1}. Under the conditions of Figure 4, a plateau is observed up to 40 ms, which will be considered as the critical time, t_c. The kinetics of silver coalescence are similar before t_c to the coalescence in the absence of naphtazarin. Assuming that the coalescence reactions between atoms and clusters of any size occur with the same rate constant as for reaction 9, $k_{10} = k_9 = 2 \times 10^8$ L mol^{-1} s^{-1} (*22*), the kinetics of the abundance of all clusters in the case of pure coalescence may be derived (*10*). In fact, the cluster

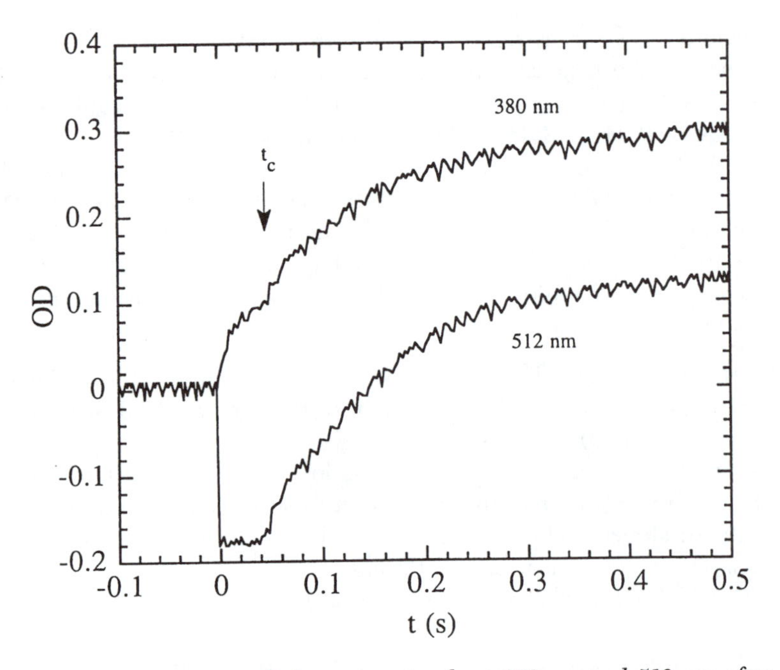

Figure 4. Transient optical absorption signals at 380 nm and 512 nm after a single pulse in a mixed solution of silver ions and naphtazarin at pH = 4.8. $[Ag^+]_{t=0} = 2 \times [Ag_2SO_4]_{t=0} = 6 \times 10^{-5}$ mol L^{-1}, $[Q]_{t=0} = 10^{-4}$ mol L^{-1}, and $[(CH_3)_2CHOH] = 0.2$ mol L^{-1}. The total initial concentration of reducing species (e_{aq}^- and alcohol radicals) is 8×10^{-5} mol L^{-1}.

nuclearity at 40 ms is on the order of a few tens, that is, beyond the range of the nuclearity dependence of the extinction coefficient.

Kinetics at 512 nm. In the meantime, naphtazarin hydroquinone is produced by reactions 3–5 as it was in the absence of silver cations, except that the amount is smaller because of the competition with Ag^+. The sudden bleaching at 512 nm after the pulse caused by naphtazarin reduction into hydroquinone corresponds in Figure 3 to an initial concentration of electron donor of 4×10^{-5} mol L^{-1} (8×10^{-5} mol L^{-1} in reduction equivalent).

The concentration is stable during almost 40 ms, which corresponds to the end of the first 380-nm plateau. During this time, the solution contains growing silver clusters, excess Ag^+, hydroquinone, and naphtazarin. However, no crossed reaction occurs between the silver and the naphtazarin species before $t_c = 40$ ms.

$$QH_2 + Ag_{n+1}^+ \rightarrow \text{No electron transfer } (0 \leq n < n_c) \quad (11)$$

Actually, hydroquinone is a bielectronic electron donor. The first step would be a monoelectronic transfer involving the couple $Q^{\cdot-}$–QH_2, with a potential value controlled by the pH of the medium: $E°(Q^{\cdot-}–QH_2) = 0.22$ V_{NHE} and 0.33 V_{NHE} at pH 4.8 and 3.9, respectively. The second monoelectronic donation is given by the radical semiquinone formed during the first step, with a standard potential almost constant between pH 5 and 3.9, $E°(Q–Q^{\cdot-}) = -0.105$ V_{NHE}. Thus, the semiquinone is a stronger electron donor than the parent molecule hydroquinone, but it cannot be involved in the mechanism unless hydroquinone has already been oxidized itself into the semiquinone.

A charged cluster may constitute an electron acceptor, but that depends on its own redox potential value, $E°(Ag_n{}^+–Ag_n)$ relative to the threshold imposed by the monitor potential, $E°(Q^{\cdot-}–QH_2)$. As the redox potential increases with cluster nuclearity (*5, 6*), a certain time after the pulse is required to allow the first supercritical clusters to be formed and their potential to reach the threshold value imposed by the hydroquinone. When time, t, is less than t_c, where $n < n_c$, the transfer is not allowed. During this induction period, the kinetics at 380 nm correspond to pure coalescence of clusters (Figure 4), and hydroquinone is stable (the bleaching OD_{512} is constant). That means, obviously, that none of the silver species present at that time can react with hydroquinone, especially free Ag^+ ions and Ag^+ ions associated with the smallest clusters.

Time Beyond t_c. *Kinetics at 512 nm.* As soon as n is higher than n_c, the competition starts between the cluster coalescence and the electron transfer from hydroquinone to the clusters. Now, charged clusters of potential higher than that of the threshold allow the electron transfer from hydroquinone, which is finally oxidized into naphtazarin. After the induction time, t_c, the bleaching at 512 nm decays (Figure 4):

$$QH_2 + Ag_{n+1}^+ \rightarrow Ag_{n+1} + Q^{\cdot-} + 2H^+ \qquad (n \geq n_c) \qquad (12)$$

The hydroquinone disappears by reaction with silver ions adsorbed on supercritical clusters and is oxidized into the semiquinone. The semiquinone is also readily oxidized into naphtazarin, which strongly absorbs at 512 nm:

$$Q^{\cdot-} + Ag_n{}^+ \rightarrow Q + Ag_n \qquad (13)$$

The concentration of silver clusters is quite low, so that the transfer (reactions 12 and 13) occurs within 0.5 s. The hydroquinone transitorily formed after the pulse is therefore acting as an electron relay (Figure 1), and the absorbance at 512 nm due to naphtazarin is partially restored. The silver cations react up to their complete disappearance.

Kinetics at 380 nm. Correlatively, the absorbance at 380 nm increases again at t_c in a second step because of the formation of supplementary silver atoms via the electron transfer process (reactions 12–13) up to a second plateau corresponding to the complete reduction of Ag^+ (Figure 4). Actually, the absorption spectrum of the large silver clusters at the end is somewhat different from that at short time (5): it is less intense at 380 nm and now also extends around 500 nm. The 512-nm absorbance at $t > 0.2$ s is partly due to Ag_n. For this reason, the total absorbance at 512 nm is eventually higher than the initial quinone absorbance of the solution [$2 \times \varepsilon_{512}(Ag_n) > \varepsilon_{512}(QH_2) - \varepsilon_{512}(Q)$]. Finally, the total reduced equivalent of excess hydroquinone and of silver atoms, generated directly after the pulse or indirectly by the hydroquinone, corresponds to the initial radiolytic yield of reduction.

Electron Transfer Mechanism. Thus, the first oxidation step of the hydroquinone occurs, provided the nuclearity is supercritical, $n > n_c$, which means when the potential $E°(Ag^+_{n_c+1}-Ag_{n_c+1})$ becomes higher than the threshold $E°(Q^{\cdot -}-QH_2)$. Then the supercritical cluster acts as a nucleus for its own growth through an autocatalytic electron transfer (reactions 5, 6) according to the mechanism that was summarized by reaction 12:

$$Ag_{n_c} + Ag^+ \rightarrow Ag^+_{n_c+1} \quad (14)$$

$$Ag^+_{n_c+1} + QH_2 \rightarrow Ag_{n_c+1} + Q^{\cdot -} + 2H^+ \quad (15)$$

$$Ag_{n_c+1} + Ag^+ \rightarrow Ag^+_{n_c+2} \quad (16)$$

$$Ag^+_{n_c+2} + QH_2 \rightarrow Ag_{n_c+2} + Q^{\cdot -} + 2H^+ \quad (17)$$

$$\vdots$$

As the adsorption reactions (14, 16, . . .) are fast, the turnover rate constant, k_t, of the overall process is almost the same as the rate constant of electron transfer reactions (15, 17, . . .). Again, the H_3O^+ concentration increases during the reduction (reaction 12), with the consequence of a positive shift for $E°(Q^{\cdot -}-QH_2)$. However, $E°(Ag^+_{n_c+1}-Ag_{n_c+1})$ also increases with n, and the autocalytic transfer continues up to the Ag^+ exhaustion.

The semiquinone is also an electron donor. The second monoelectronic potential, $E°(Q-Q^{\cdot -}) = -0.105\ V_{NHE}$, is even more negative than that of the first monoelectronic couple with hydroquinone. Once formed through reaction 12, the semiquinone is thus able, a fortiori, to also transfer electrons to $Ag^+_{n>n_c}$ by reaction 13.

However, the semiquinone does not act as a redox probe, because its concentration is zero as long as hydroquinone itself has not started the transfer. Then it is always produced in the close vicinity of the supercritical cluster already selected by the hydroquinone, and it reacts readily with the *same* cluster before diffusion. It is worth noting that the semiquinone essentially amplifies the catalytic transfer. The overall hydroquinone oxidation into the quinone produces twice as many silver atoms as the initial QH_2 concentration (reactions

12 and 13). Nonetheless, the transfer is controlled by the potential difference between $E°(Ag_n^+–Ag_n)$ and the first monoelectronic transfer potential $E°(Q^{\cdot-}/QH_2)$. This conclusion is confirmed by the observed effect of the pH (see the next section), which precisely controls $E°(Q^{\cdot-}–QH_2)$ but not $E°(Q–Q^{\cdot-})$ in the pH range of the study. The second fast monoelectronic transfer from $Q^{\cdot-}$ does not affect the overall turnover mechanism (reactions 12–17), except in yielding twice as many silver atoms.

It is also important to note *a posteriori* that the semiquinone generated just after the end of the pulse (reactions 3 and 4) did not react with any of the silver species and that it disappears totally through disproportionation. The situation concerning the size distribution of surrounding clusters just after the pulse is indeed quite different from that after t_c. At microsecond range, the atoms Ag_1, complexes Ag_2^+, and dimers Ag_2 are predominant. But their redox potentials are all more negative than that of $Q–Q^{\cdot-}$. The semiquinone was unable to transfer any electron, and the only possible reaction was disproportionation (5).

Influence of pH. ***Plateau up to 140 ms.*** Figure 5 presents the absorbance increase due to silver aggregates at 380 nm in a solution of pH 3.9

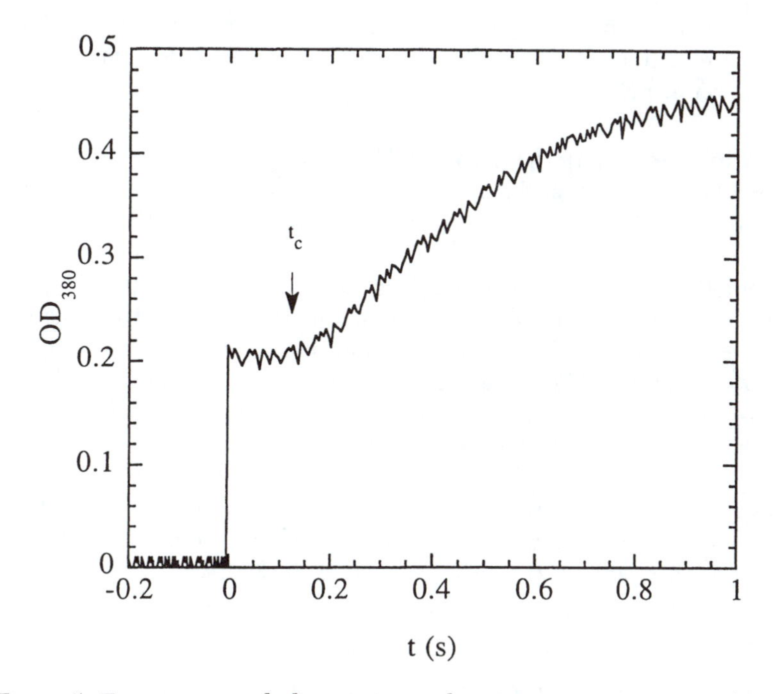

Figure 5. Transient optical absorption signal at 380 nm at pH = 3.9. (Same conditions as in Figure 4.)

with the same initial concentrations, $[Ag^+]_0$ and $[Q]_0$, as above (pH 4.8). At short time, the kinetics are similar to those at pH 4.8, and the early increase due to $\varepsilon(n)$ of Ag_n is achieved within 20 ms. Correlatively, hydroquinone is rapidly formed.

The redox potential of the monitor hydroquinone, $E°(Q^{\cdot-}–QH_2)$, is now $+0.33\ V_{NHE}$ at pH 3.9, whereas the potential $E°(Q–Q^{\cdot-})$ is the same as at pH 4.8. The absorbance of silver clusters at 380 nm is remarkably constant up to 140 ms (Figure 5), as is the hydroquinone absorbance. The change in kinetics of Ag_n growth occurs markedly later than at pH 4.8. The difference in t_c ($t_c \approx$ 40 ms at pH 4.8, and $t_c \approx$ 140 ms at pH 3.9) corresponds to different critical sizes, n_c, so that $E°(Ag^+_{n_c+1}–Ag_{n_c+1}) > E°(Q^{\cdot-}–QH_2)$, which is the required condition for the transfer from the hydroquinone to be allowed.

Because, at pH 3.9, $E°(Q^{\cdot-}–QH_2)$ is higher than at pH 4.8 and t_c is longer, it is clear that the critical nuclearity will be higher (and the redox potential of the critical cluster will also be more positive). This observation confirms previous results obtained with a stronger redox system ($E° = -0.41\ V_{NHE}$) (*5*): for similar initial conditions, the critical time was much shorter ($t_c \approx 1$ ms), implying a smaller critical nuclearity.

***Time* > t$_c$.** Not only is t_c longer at pH 3.9 than at pH 4.8, but the observed rate after t_c is smaller (Figures 4 and 5). This rate would depend linearly on the supercritical cluster concentration. This confirms that, under conditions yielding identical initial amounts of silver atoms, larger clusters are required to accept electrons from a weaker donor and that the nuclei concentration is much lower at pH 3.9.

Influence of $[Ag^0]_{t=0}$ and $[QH_2]_{t=0}$. The initial concentrations of Ag^0 and QH_2 depend on the concentration of the primary radiolytic species, and thus on the dose per pulse, and on the relative concentrations of precursors. For given concentrations of the precursors $[Ag^+]_{t=0}$ and $[Q]_{t=0}$, the initial concentrations x_0, which equals $[Ag^0]_{t=0}$, and s_0, which equals $[QH_2]_{t=0}$, increase with the dose but their ratio is constant. It is found that according to the model calculated by numerical simulation (*11*), the observed rate of absorbance decay at 512 nm and the reciprocal value of t_c are linear functions of $x_0 = [Ag^0]_{t=0}$ in the range $0.6–2.4 \times 10^{-5}$ mol L^{-1} for both pHs.

On the other hand, if x_0 is unchanged and s_0 different, as in Figure 6 compared with Figure 4, the critical time is the same (~40 ms) because it corresponds to the time required to reach the critical nuclearity fixed by the donor potential, and the growth kinetics depend only on the initial monomer concentration x_0 and on the coalescence rate constant k_d.

Discussion

Determination of Critical Size and k_t. It has been shown (*11*) that the features of the kinetics of cluster coalescence in competition with a

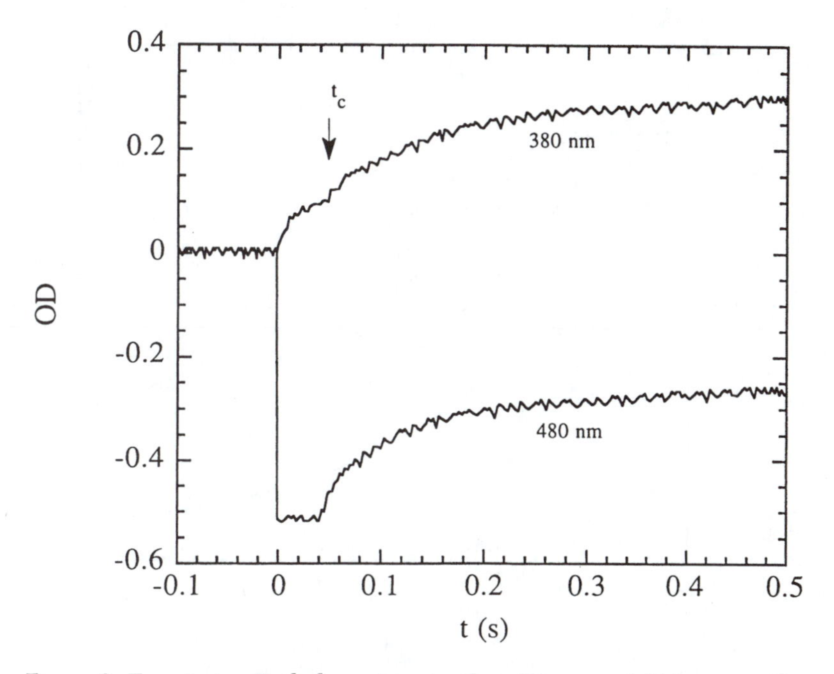

Figure 6. Transient optical absorption signal at 480 nm and 380 nm at pH = 4.8. $[Ag^+]_{t=0} = 2 \times [Ag_2SO_4]_{t=0} = 5 \times 10^{-5}$ mol L^{-1}, $[Q]_{t=0} = 2 \times 10^{-4}$ mol L^{-1}, and $[(CH_3)_2\,CHOH] = 0.2$ mol L^{-1}. The absorbance at 480 nm is the same as at 512 nm. The total initial concentration of reducing species (e^-_{aq} and alcohol radicals) is 2.2×10^{-4} mol L^{-1}.

growth process via an autocatalytic electron transfer may be reproduced by numerical simulation. The time-dependence of the fraction of donor molecules consumed or of the distribution of cluster abundances versus nuclearity is controlled by the initial conditions of concentration ($x_0 = [Ag^0]_{t=0}, s_0 = [QH_2]_{t=0}$) and by the parameters n_c, k_d, and k_t. The series of coalescence and electron transfer reactions (reactions 18 and 19, respectively) in competition at variable n are written for simplicity without the excess ions adsorbed on clusters (the subscripts only refer to reduced atoms):

$$Ag_i + Ag_j \rightarrow Ag_{i+j}\ (k_d) \tag{18}$$

$$Ag_i + 2Ag^+ + QH_2 \rightarrow Ag_{i+2} + Q + 2H^+\ (k_t), \qquad i \geq n_c \tag{19}$$

The general problem of this competition has been solved by numerical simulation for variable values of x_0, s_0, n_c, k_d, and k_t (*11*). The value of $k_d = 2 \times 10^8$ L mol^{-1} s^{-1} for silver clusters is taken as the same as for pure coalescence (reaction 9). The fixation of Ag^+ onto Ag_i is fast and does not interfere with

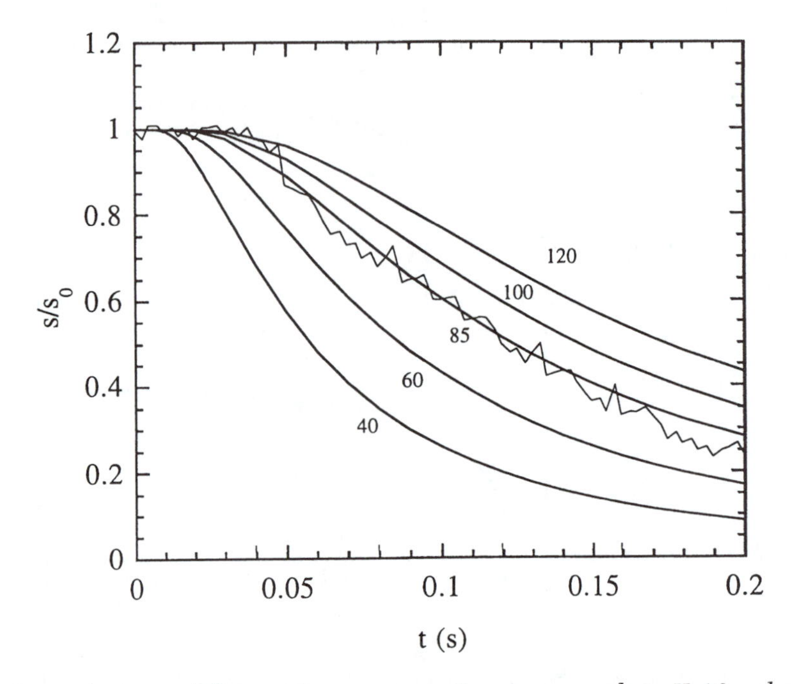

Figure 7. Decay of electron donor concentration as measured at pH 4.8 and as calculated by numerical simulation with dependence on n_c. *Experimental signal: same conditions as in Figure 4. Because the concentration of silver ions after the pulse is smaller than that of hydroquinone, the ordinate of the experimental plot is* $(OD-OD_{\infty})/OD_{t=0}$. *Numbers next to simulation curves correspond to* n_c. *The value of* k_t *is* 2.25×10^8 *mol* $L^{-1}s^{-1}$. *The best adjustment with numerical simulation is for* n_c *(pH = 4.8) = 85 ± 5.*

the mechanism. For $i < n_c$, the coalescence is identical to that in the absence of a donor (*11*). We also assume that the turnover rate constant k_t (reaction 19) will be the same for both pH values and that at each value the critical nuclearity must depend only on the donor potential and be independent of the initial concentrations of donor s_0 and acceptor x_0.

The best fit between experimental and calculated kinetic data is given by the adjusted values $k_t = (2.25 \pm 0.25) \times 10^8$ L mol^{-1} s^{-1}, n_c (pH = 4.8) = 85 ± 5, and n_c (pH = 3.9) = 500 ± 30 Figures 7–9). The different features of QH_2 decay, such as critical time, shape of the curves, and $[Ag^0]_{t=0}$ concentration dependence, are well reproduced by the numerical simulation for both pH values. It is interesting to observe the sensitivity of the simulation to the adjusted parameters. In Figures 7 and 9, the influence of n_c is shown for a given value of k_t, 2.25×10^8 L mol^{-1} s^{-1}, on the calculated decay at pH 3.9 and 4.8, respectively. The uncertainty is limited because the same k_t value must account for the decays at both pH values. Note that the increase of n_c, all other parameters being fixed, has the effect of delaying the decay, $s(t)/s_0$, toward higher values of t (Figure 7). The same $s(t)$ value is reached at times proportional to n_c. The influence of k_t is shown in Figure 8.

The study of the resolution of kinetics by numerical simulation (*11*) has shown that the regime of the competition (reactions 18 and 19) must be controlled by the ratio $k_t \times 2 \times [QH_2]_{t=0}/k_d \times [Ag^0]_{t=0}$. In the present case, the ratio is much higher than unity. That means that reaction 19 is much faster than reaction 18, so that the total concentration of supercritical clusters remains almost unchanged during the very rapid autocatalytic transfer. From the model, the results of Figures 7 and 9 correspond to a concentration of nuclei of $\approx 7 \times 10^{-8}$ mol L^{-1} at pH 4.8 and $\approx 1.2 \times 10^{-8}$ mol L^{-1} at pH 3.9. Note that the reservoir of excess silver ions after the pulse is $\approx 5 \times 10^{-5}$ mol L^{-1}, so that each nucleus receives about 700 and 4000 supplementary atoms, respectively, through electron transfer. Just after t_c, the supercritical clusters are still under formation, and the decay rate increases from zero to the turnover value. At long time, the coalescence is no longer negligible relative to the transfer, and the concentration of clusters decreases, so that the decay is slower (Figures 7 and 9).

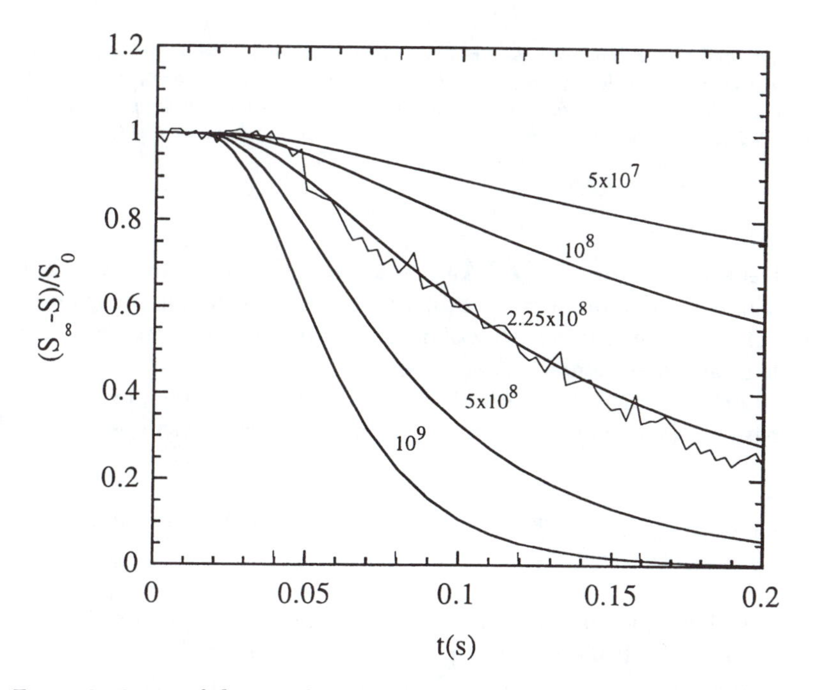

Figure 8. Decay of electron donor concentration as measured at pH 4.8 and as calculated by numerical simulation of dependence on k_t. *(Same conditions as in Figure 7.) Numbers next to simulation curves correspond to* k_t. *The value of* n_c *is 85. The best adjustment is for* $k_t = (2.25 \pm 0.25) \times 10^8$ *mol* L^{-1} s^{-1}.

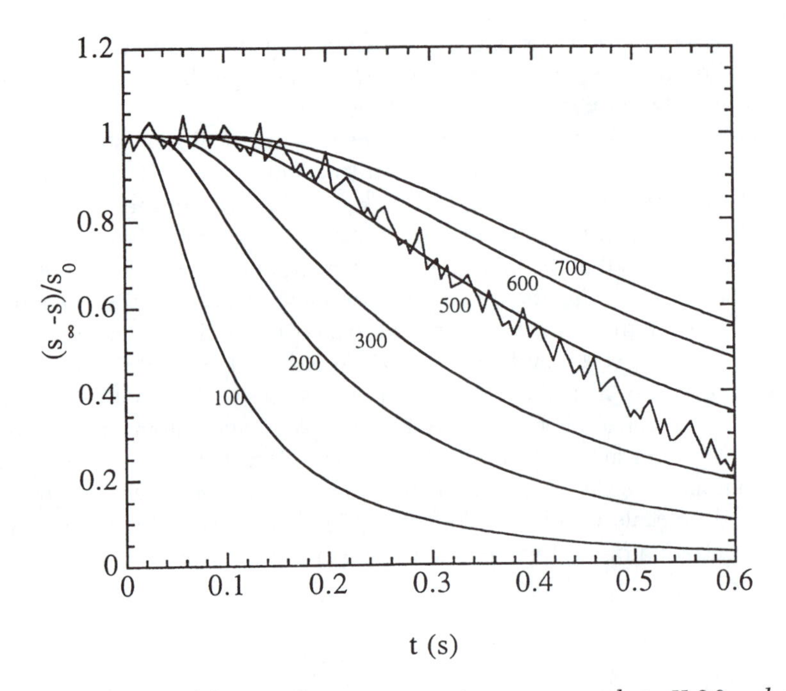

Figure 9. Decay of electron donor concentration as measured at pH 3.9 and as calculated by numerical simulation with dependence on n_c*. Experimental signal. same conditions as in Figure 5. Simulation curves: the same value of* k_t*, 2.25 ×* 10^8 *mol* $L^{-1}s^{-1}$*, was selected as at pH = 4.8 The best adjustment is for* n_c *(pH = 3.9) = 500 ± 30.*

Size Dependence of $E°(Ag_n^+ - Ag_n)$. The quantum–size effect on metal clusters redox properties in solution is the most important feature for cluster chemistry in solution. Most of data have been obtained on silver clusters suitable as an experimental model (*1–4, 22*).

Assuming in the results just given that the redox potential of the critical cluster is slightly higher than that of the monitor system used, we conclude that:

$$E°(Ag_{85}^+ - Ag_{85}) = +\ 0.22\ V_{NHE},\ \text{and}\quad E°(Ag_{500}^+ - Ag_{500}) = +0.33\ V_{NHE} \tag{20}$$

Figure 10 shows the nuclearity dependence of silver cluster redox potential in water: together with the data just presented, the previously published values are reported for n = 1 (*3*), 2 (*23*), 5 (*5*), 10 (*24*), and 11 (*25*). The $E°$ values for nuclearities n = 1 and n = 2 resulted from thermodynamic calculations. The value for n = 10 was obtained from electron transfer studies where the clusters were the donor and were corroded by H_3O^+. As a function of the nuclearity (Figure 10), the redox potentials of hydrated silver clusters are seen to increase with n, thus confirming the trend previously observed (*5, 6*). The density of values available so far is not sufficient to prove the existence of odd–even oscillations like those observed for ionization potentials, IP_g, of bare silver clusters in the gas phase (*26, 27*). In fact, $E°$ is correlated to the ionization potential of solvated clusters, IP_{solv}, by $IP[Ag_n]_{solv} = e \times E°[Ag_n{}^+-Ag_n] + 4.5$ (*5, 28*). However, it is obvious that the variations of $E°$ or IP_{solv} and IP_g do exhibit opposite trends versus n for the solution and gas phase, respectively. The difference between ionization potentials of bare and solvated clusters decreases with increasing n and corresponds fairly well to the solvation free energy deduced from the Born model (*29*).

The redox potential of copper clusters in aqueous solution also increases with n, and this trend is seemingly general for all metals (*13*).

The redox potential difference between the silver clusters $n = 85$ and $n = 500$ is not very large (0.11 V). This suggests that we approach an asymptotic value of $E°$ for the bulk metal $E°(Ag_\infty{}^+-Ag_\infty)$ close to $+0.40\ V_{NHE}$. Note that the redox potential of the bulk metal $E°(Ag_\infty{}^+-Ag_\infty)$ differs from the well known electrochemical potential $E'°(Ag^+-Ag_\infty) = +0.796\ V_{NHE}$ by the adsorption energy of the Ag^+ ion on the bulk metal:

$$E'°(Ag^+-Ag_\infty) = E°(Ag_\infty{}^+-Ag_\infty) - \Delta G_{ads}(Ag^+) \quad (21)$$

The free energy, $\Delta G_{ads}(Ag^+)$, is therefore equal to –0.4 V, which is a reasonable value. The results of Fig. 10 indicate that, at least concerning the redox proper-

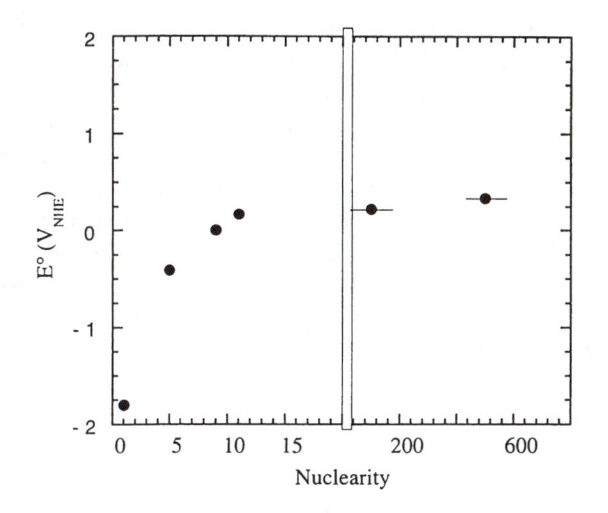

Figure 10. Size dependence of the redox potential of silver clusters in water (●). Data previously published are reported: n = *1 (3), 2 (23), 5 (5), 10 (24), and 11 (25);* n = *85 and 500 (this work).*

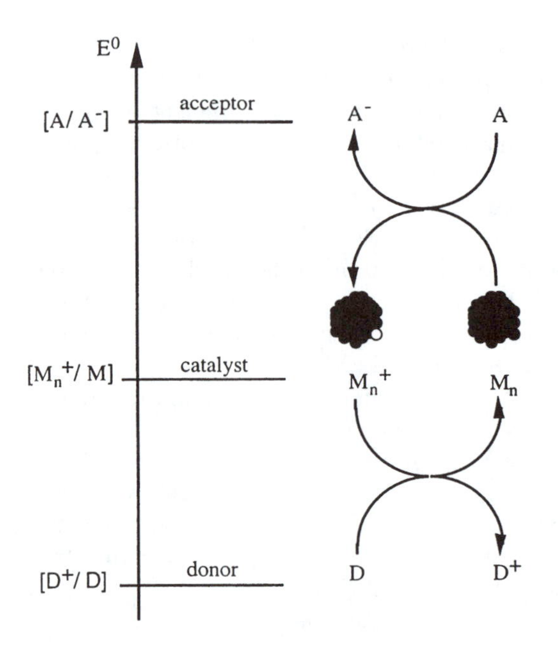

Figure 11. Mechanism of catalytic electron transfer involving metal clusters as relay. The thermodynamic conditions to be fulfilled are that the cluster redox potential be higher than the donor D and lower than the acceptor A potential, which implies that the cluster itself is in a size range that offers the efficient redox potential.

ties of silver clusters, the transition between the mesoscopic and the macroscopic phase occurs around the nuclearity $n = 500$ (diameter ≈ 0.8 nm).

The size-dependence of redox properties of metal clusters is crucial for their catalytic efficiency in electron transfer processes (*30, 31*). Actually, a metal cluster acting as a catalytic relay behaves alternatively as an acceptor and a donor of electrons (Figure 11). Therefore, the thermodynamics imply that the value of the redox potential of the couple $E°(M_n{}^+–M_n)$ would be intermediate between that of the donor system as the lower threshold and that of the acceptor system as the upper threshold. The reaction between these systems, negligible in the absence of the catalyst, becomes efficient because of the double electron transfer through the metal cluster (*32*). The strong efficiency of ultradivided metals is thus due not only to their high specific area but essentially to their appropriate thermodynamic properties. Note that the local roughness on even large clusters also creates variable potentials, which when selected by the double threshold are favorable to the transfer.

Influence of the Environment. In contrast with the ionization potential of clusters in the gas phase, which depends only on the nuclearity and the metal considered, the redox or the ionization potential of clusters in solution depends not only on the nuclearity but also strongly on all interactions with surrounding molecules: solvent, surfactant, ions, ligand, and support. The data are still more scarce than for just hydrated clusters. However, some optical absorption spectra of single silver atoms in different solvents suggest the influence of the environment (*30*).

Surfactant. It has been found that silver clusters generated in the presence of the surfactant polyacrylate (PA) (*33*) or of some other polyanions such as polyphosphate (PP) (*34*) coalesce slowly [k $(2Ag_4^{2+})(PA)$ = 10^5 L mol s^{-1}] and are stabilized at very small nuclearity, of only a few units. Scanning tunneling microscopy (STM) techniques have confirmed the stability of oligomeric species in the presence of PA ($n \leq 7$) (*35*). The kinetics study of electron transfer from the donor sulfonatopropylviologen anion to Ag_n(PA) showed that the transfer starts effectively at the critical size, n_c, of 4, which is similar to clusters without surfactant. However, the rate constant for the electron transfer is 2×10^8 and 5×10^7 L mol s^{-1} in neutral and acidic media, respectively, instead of 7×10^8 L mol^{-1} without surfactant (*36*). Moreover, the clusters Ag_4(PA) are not oxidized by molecular oxygen, so that their redox potential would be higher than -0.33 V_{NHE}. These differences in behavior are assigned to the complexing properties of polyacrylate, which stabilizes and protects small clusters from coalescence and corrosion (*37*).

Ligand. The redox potential of the single silver atom solvated in water was calculated with the aid of a thermodynamic cycle including the electrochemical potential of the bulk metal in aqueous solution and the sublimation energy of the metal (*3*). The hydration energy of the neutral species is considered negligible relative to that of the cation.

In the presence of a strong complexing agent such as the ligand CN^-, the metal–ligand binding energy must be taken into account. A recent study (*38*) has calculated the redox potential of the couple $Ag^I{}_1(CN)_2{}^-$–$Ag^0{}_1(CN)_2{}^{2-}$ using the self-consistent field method for the determination of the electronic structure of the gaseous species and the cavity model for the solvation energy. The results showed that the redox potential of the complexed atom is very negative, $E°$ $(Ag^I{}_1(CN)_2{}^-$–$Ag^0{}_1(CN)_2{}^{2-})$ = -2.6 V_{NHE}, and is lower by 0.8 V than the uncomplexed hydrated atom (*3*). The equilibrium constant of complexation of Ag^0 by $2CN^-$ has been evaluated to 10^7 at 298 K, which implies that the cyano complex is stable relative to dissociation. However, it is highly reactive as an electron donor. The results for the ligand CN^- illustrate again the influence of the local environment on cluster properties (*39, 40*). In the case of cyano complexation, the redox potential of the complexed atom at least decreases markedly relative to that of free solvated atoms. Recent results on the ligand NH_3 also lead to the conclusion of a more negative redox potential, $E°(Ag^I{}_1(NH_3)_2{}^+$–$Ag^0{}_1(NH_3)_2)$, than $E°(Ag^+$–$Ag^0{}_1)$ (*41, 42*).

Support. It is noteworthy to compare the potential of electron donors able to transfer electrons to unsupported clusters compared with the potential of a photographic developer able to develop AgBr-supported clusters of the same critical nuclearity. For instance, free clusters solvated in water of nuclearity $n_c = 4$ are developed by an electron donor of $E° = -0.41\ V_{NHE}$ (5), whereas a weaker reducing agent, such as hydroquinone (*p*-dihydroxybenzene) is able to develop clusters supported on an AgBr emulsion from the same critical nuclearity $n_c = 4$ (*43*). As discussed above in the case of naphtazarin hydroquinone, the dielectronic donor hydroquinone ($Q'H_2$) induces a two-step reduction via the semiquinone ($Q'^{\cdot-}$–$Q^{\cdot}$) with two standard redox potential values. Under the basic pH conditions of photographic development, successive potential values involved in the usual hydroquinone developer are $E°(Q'^{-}\text{–}Q'H_2) = +0.024\ V_{NHE}$ (or $-0.20\ V_{AgCl}$) and $E°(Q'\text{–}Q'^{\cdot-}) = +0.078\ V_{NHE}$ (or $-0.144\ V_{AgCl}$). Note that in this system the second step from the semiquinone corresponds to a potential higher than from $Q'H_2$. The first electron transfer from the hydroquinone determines the threshold potential for developability, that is, the potential of the critical cluster, here of nuclearity $n_c = 4$. Thus, we may conclude that the reduction of the free cluster Ag_5^+ requires a potential of the donor more negative by $[0.024\text{–}(-0.41)] = 0.43$ V than the same AgBr-supported cluster, the difference being induced by the stabilizing effect of the support. The difference is likely dependent on the nuclearity. However, it is expected that the redox potential of AgBr-supported clusters is systematically more positive than the potential of free clusters measured in water, as shown for this first evaluation at size $n = 4$.

Stabilizing effects have been directly observed by pulse radiolysis studies of silver cluster coalescence supported on 4-nm silica colloidal particles (*44*). Very small oligomeric clusters absorbing at 290 nm and 330 nm are stable in the presence of oxygen and even Cu^{2+} or $Ru(NH_3)_6Cl_3$ ($E° = 0.2\ V_{NHE}$). These supported the exhibition of higher redox potentials than for free oligomeric clusters.

Conclusions

The redox potentials of short-lived silver clusters have been determined through kinetics methods using reference systems. Depending on their nuclearity, the clusters change behavior from electron donor to electron acceptor, the threshold being controlled by the reference system potential. Bielectronic systems are often used as electron donors in chemistry. When the process is controlled by critical conditions as for clusters, the successive steps of monoelectronic transfer (and not the overall potential), of which only one determines the threshold of autocatalytical electron transfer (or of development) must be separately considered. The present results provide the nuclearity dependence of the silver cluster redox potential in solution close to the transition between the mesoscopic phase and the bulk metal-like phase. A comparison with other literature data allows emphasis on the influence of strong interaction of the environment (surfactant, ligand, or support) on the cluster redox potential and kinetics. Relative to free solvated clusters, these interactions may either lower (CN^-, NH_3) or increase (surfactants PA and PP; support) the reactivity and therefore control the cluster efficiency in the electron transfer catalysis.

Acknowledgment

We are indebted to A. J. Swallow for fruitful suggestions and discussions during this work.

References

1. Delcourt, M. O.; Belloni, J. *Radiochem. Radioanal. Lett.* **1973,** *13,* 329.
2. Basco, N.; Vidyarthi, S. K.; Walker, D. C. *Can. J. Chem.* **1973,** *51,* 2497.
3. Henglein, A. *Ber. Bunsen Ges. Phys. Chem.* **1977,** *81,* 556.
4. Henglein, A. *Chem. Rev.* **1989,** *89,* 1861.
5. Mostafavi, M.; Marignier, J. L.; Amblard, J.; Belloni, J. *Radiat. Phys. Chem.* **1989,** *34,* 605.
6. Mostafavi, M.; Marignier, J. L.; Amblard, J.; Belloni, J. *Z. Phys. D: At. Mol. Clusters* **1989,** *12,* 31.
7. Morse, M. D. *Chem. Rev.* **1986,** *86,* 4049.
8. Schumacher, E. *Chimia* **1988,** *42,* 357.
9. Haberland, H. *Clusters of Atoms and Molecules;* Springer-Verlag, Berlin, Germany, 1994; Vols. 1, 2.
10. Khatouri, J.; Mostafavi, M.; Ridard, J.; Amblard, J.; Belloni, J. *Z. Phys. D: At. Mol. Clusters* **1995,** *34,* 47.
11. Khatouri, J.; Ridard, J.; Mostafavi, M.; Amblard, J.; Belloni, J. *Z. Phys. D: At. Mol. Clusters* **1995,** *34,* 57.
12. Belloni, J.; Mostafavi, M.; Marignier, J. L.; Amblard, J. *J. Imaging Sci.* **1991,** *35,* 68.
13. Khatouri, J.; Mostafavi, M.; Amblard, J.; Belloni, J. *Proc. 6th Int. Symp. Small Part. Inorg. Clusters, Z. Phys. D: At. Mol. Clusters* **1993,** *26,* S82.
14. Land, E.; Mukherjee, T.; Swallow, A. J.; Bruce, J. M. *J. Chem. Soc. Faraday Trans. I* **1983,** *79,* 391.
15. Land, E.; Mukherjee, T.; Swallow, A. J.; Bruce, J. M. *J. Chem. Soc. Faraday Trans.* **1989,** *79,* 405.
16. Belloni, J.; Billiau, F.; Cordier, P.; Delaire, J. A.; Delcourt, M. O. *J. Phys. Chem.* **1978,** *82,* 532.
17. Saito, E.; Belloni, J. *Rev. Soc. Inst.* **1976,** *47,* 629.
18. Schwarz, H. A.; Dodson, R. W. *J. Phys. Chem.* **1989,** *93,* 409.
19. Buxton, G. V. *J. Phys. Chem. Ref. Data* **1988,** *17,* 513.
20. Von Pukies, J.; Roebke, W.; Henglein, A. *Ber. Bunsen Ges. Phys. Chem.* **1968,** *72,* 842.
21. Janata, E.; Henglein, A.; Ershov, B. G. *J. Phys. Chem.* **1994,** *98,* 10888.
22. Henglein, A. *Ber. Bunsen Ges. Phys. Chem.* **1995,** *99,* 903.
23. Tausch-Treml, R.; Henglein, A.; Lilie J. *Ber. Bunsen Ges. Phys. Chem.* **1978,** *82,* 1335.
24. Platzer, O.; Amblard, J.; Marignier, J. L.; Belloni, J. *J. Phys. Chem.* **1992,** *96,* 2334.
25. Henglein, A.; Tausch-Treml, R. *J. Colloid Interface Sci.* **1981,** *80,* 84.
26. Jackschath, C.; Rabin, I.; Schulze, W. *Z. Phys. D: At. Mol. Clusters* **1992,** *22,* 517.

27. Alameddin, G.; Hunter, J.; Cameron, D.; Kappes, M. M. *Chem. Phys. Lett.* **1992,** *192,* 122.
28. Reiss, H. *J. Phys. Chem.* **1985,** *89,* 3783.
29. Belloni, J.; Khatouri, J.; Mostafavi, M.; Amblard, J. In *Ultrafast Reaction Dynamics and Solvent Effects;* Rossky, P. J.; Gauduel, Y., Eds.; American Institute of Physics: College Park, MD, 1994; p 527.
30. Belloni. J.; Delcourt, M. O.; Marignier, J. L.; Amblard, J. In *Radiation Chemistry;* Hedwig, P.; Nyikos, L.; Schiller, R., Eds.; Akadémiai Kiädo: Budapest, Hungary, 1987; p 89.
31. Delcourt, M. O.; Kegouche, N.; Belloni, J. *Nouv. J. Chim.* **1983,** *21,* 177.
32. Belloni, J.; Lecheheb, A. *Radiat. Phys. Chem.* **1987,** *29,* 89.
33. Mostafavi, M.; Kegouche, N.; Delcourt, M. O. *Radiat. Phys. Chem.* **1992,** *40,* 445.
34. Henglein, A. *J. Phys. Chem.* **1993,** *97,* 5457.
35. Remita, S.; Orts, J. M.; Feliu, J. M.; Mostafavi, M.; Delcourt, M. O. *Chem. Phys. Lett.* **1994,** *218,* 115.
36. Mostafavi, M.; Delcourt, M. O.; Belloni, J. *J. Imaging Sci. Technol.* **1994,** *38,* 54.
37. Mostafavi, M.; Kegouche, N.; Delcourt, M. O.; Belloni, J. *Chem. Phys. Lett.* **1990,** *167,* 193.
38. Remita, S.; Archirel, P.; Mostafavi, M. *J. Phys. Chem.* **1995,** *99,* 13198.
39. Texier, I.; Mostafavi, M. *Radiat. Phys. Chem.* **1997,** *49,* 459.
40. Remita, S.; Mostafavi, M.; Delcourt, M. O. *J. Phys. Chem.* **1996,** *100,* 10187.
41. Texier, I.; Remita, S.; Archirel, P.; Mostafavi, M. *J. Phys. Chem.* **1996,** *100,* 12472.
42. Mostafavi, M.; Remita, S.; Delcourt, M. O.; Belloni, J. *J. Chim. Phys.* **1996,** *93,* 1828.
43. Rosche, Ch.; Wolf, S.; Leisner, T.; Granzer, F.; Wöste, L. *Proc. IS&T's 47th Annu. Conf./ICPS* **1994,** *1,* 54.
44. Lawless, D.; Kapoor, S.; Kermepohl, P.; Meisel, D.; Serpone, N. *J. Phys. Chem.* **1994,** *98,* 9619.

19

Electron Transfer Reactions Under High Pressure: Application of Spectroscopic and Electrochemical Techniques

Rudi van Eldik

Institute for Inorganic Chemistry, University of Erlangen–Nürnberg, Egerlandstrasse 1, Erlangen 91058, Germany

The effect of pressure (up to 200 MPa) on the kinetics and thermodynamics of electron transfer reactions involving metal complexes and cytochrome c has been studied in detail for a number of systems. The observed activation and reaction volume data enable the construction of a reaction volume profile for each investigated system, and allow a detailed analysis of the partial molar volume changes associated with the electron transfer process. The reported results demonstrate the excellent agreement between data obtained using UV–vis, stopped-flow, pulse-radiolysis, flash-photolysis, and electrochemical techniques. The ultimate goal of this work is to contribute toward a better understanding of long-distance electron transfer reactions.

We have developed an interest in the effect of pressure on chemical reactions in solution, and, in particular, in obtaining mechanistic information from such effects. Our work has involved the study of thermal, photoinduced, and radiation-induced reactions in inorganic, organometallic, and bioinorganic chemistry (*1–4*). In this chapter we focus on the effect of pressure on electron transfer reactions involving metal complexes and cytochrome *c*, that is, so-called "long-distance" electron transfer processes. The kinetics of the reactions are followed by stopped-flow, pulse-radiolysis, and flash-photolysis techniques at pressures of up to 200 MPa. The thermodynamics of the reactions are studied using spectrophotometric and electrochemical techniques at high pressure (*5*). A combination of the kinetic and thermodynamic data leads to the construction of a volume profile for the investigated electron transfer reactions, which forms the basis of the interpretation of pressure effects in terms of partial molar volume changes along the reaction coordinate. In this way further information regarding the intimate nature of the reaction mechanism is obtained (*2–4*).

This chapter demonstrates the good agreement between kinetic and thermodynamic data obtained using the various experimental techniques just mentioned. Furthermore, the kinetic and themodynamic data are consistent and underline the validity of the constructed volume profiles. The latter is used as a basis for the mechanistic discussion.

Effect of Pressure on Chemical Reactions

In kinetic and thermodynamic analyses of chemical processes in solution, there are basically two physical parameters that can be varied experimentally, viz, temperature and pressure. The temperature dependence is used to obtain reaction and activation free energy, enthalpy, and entropy, which are used to construct an energy profile for the reaction under study. The thermodynamic activation parameters reveal information regarding the energetics of the reaction and the nature of the transition state, especially in terms of structural order as obtained from the activation entropy. The pressure dependence is used to obtain reaction and activation volumes that are used to construct a volume profile for the reaction under study (*2, 3*).

The location of the transition state along the reaction coordinate can then be discussed in terms of partial molar volume changes associated with the chemical reaction. This concept has the advantage that the interpretation of volume changes in terms of intrinsic and solvational components is more reliable and more straightforward than in the case of the thermal activation parameters. This is partly due to the fact that we are dealing with an absolute partial molar volume scale for the reactant and product species in solution, on which basis molecular and solvent reorganization in the transition state can be visualized. In the case of electron transfer reactions, it will mainly be solvent reorganization due to changes in electrostriction associated with the electron transfer process that will determine the associated volume changes. Several reviews dealing with these aspects, as well as the instrumentation required to perform such measurements, have appeared in the literature (*5–8*). Some experimental details will be referred to where appropriate.

In the following sections the effect of pressure on different types of electron transfer processes will be discussed systematically. The main emphasis will fall on the most recent work that we have been involved in, dealing with "long-distance" electron transfer processes on cytochrome *c*. However, by way of introduction, we present a short discussion on the effect of pressure on self-exchange (symmetrical) and nonsymmetrical electron transfer reactions between transition metal complexes that have been reported in the literature.

Self-Exchange Electron-Transfer Reactions

Self-exchange reactions are the most simple electron transfer reactions. They are symmetrical processes for which both the reaction free energy and reaction volume are zero, and they are ideal for theoretical modeling. Swaddle and co-workers have made a significant contribution in this area (*9–18*). They have studied the effect of pressure on the self-exchange reactions (volumes of activation are quoted in parentheses in $cm^3\ mol^{-1}$) $Fe(H_2O)_6^{3+/2+}$ (–11.1) (*9*), $Fe(phen)_3^{3+/2+}$ (–2.2) (*10*), $Fe(CN)_6^{3-/4-}$ (+22) (*11*), $MnO_4^{2-/-}$ (–23) (*12*), $Co(sep)^{3+/2+}$ (–6.4) (*13*), $Co([9]aneS_3)^{3+/2+}$ (–4.8) (*13*), $Co(diamsarH_2)^{5+/4+}$ (–9.6) (*14*), $Co(diamsar)^{3+/2+}$ (–10.4) (*14*), $Co(en)_3^{3+/2+}$ (–15.5) (*15*), and $Co(phen)_3^{3+/2+}$ (–17.6) (*16*), where phen = 1,10-phenanthroline, sep = sepulchrate, diamsar = diaminosar-cophagine and en = ethylenediamine. In most cases the self-exchange reaction is significantly accelerated by pressure, with the exception of the $Fe(CN)_6^{3-/4-}$ system, which goes in exactly the opposite way, and the observed volume of activation is in agreement with the significantly negative entropy of activation reported for such systems. In addition, Swaddle and co-workers have also gone through impressive efforts to calculate the volumes of activation theoretically based on the Marcus–Hush–Stranks treatment and their own modifications and additions (*17, 18*). For a large number of systems, good agreement between the experimental and theoretically calculated volumes of activation was found. In most cases, solvent reorganization accounts for the largest contribution toward the observed volume of activation. Large deviations were found only for the $Co(en)_3^{3+/2+}$ and $Co(phen)_3^{3+/2+}$ systems, where the theoretical volume of activation is between 10 and 15 $cm^3\ mol^{-1}$ more positive than the experimental value (*15, 16*). This deviation is most probably related to the participation of a high-spin to low-spin changeover associated with the electron transfer process, which can account for an additional volume collapse of ca. 10 $cm^3\ mol^{-1}$ (*14*).

The reactions just mentioned all proceed via an outer-sphere electron transfer mechanism. By way of comparison, the volume of activation for the exchange reaction in $Fe(H_2O)_5OH^{2+}/Fe(H_2O)_6^{2+}$ was reported to be +0.8 $cm^3\ mol^{-1}$, that is, significantly more positive than that found for $Fe(H_2O)_6^{3+/2+}$ (*9*). This difference was ascribed to the release of a solvent molecule associated with the formation of a hydroxo-bridged intermediate in terms of an inner-sphere mechanism.

From Self-Exchange to Cross Reactions

The Marcus cross relation can in general be applied to correlate the rate constants of self-exchange reactions, K_{11} and K_{12}, with the rate constant, K_{12}, and equilibrium constant, K_{12}, of a cross reaction between two redox systems. For the self-exchange reactions

$$A^+ + A \rightarrow A + A^+, k_{11} \quad (1)$$

$$B^+ + B \rightarrow B + B^+, k_{22} \quad (2)$$

and the cross reaction

$$A^+ + B \rightarrow A + B^+, k_{12} \quad (3)$$

the simplified cross relation for reactions with low driving forces is given by

$$k_{12} = (k_{11}k_{22}K_{12})^{1/2} \quad (4)$$

where K_{12} is the equilibrium constant for the cross reaction. It follows that the pressure derivative of ln k_{12}, which can be related to the corresponding volume of activation, $\Delta V^{\#}_{12}$, leads to the following expression:

$$\Delta V^{\#}_{12} = 1/2(\Delta V^{\#}_{11} + \Delta V^{\#}_{22} + \Delta V_{12}) \quad (5)$$

The latter volume cross relation for self-exchange and cross reactions was recently tested for a number of reactions involving the reduction of $Fe(H_2O)_6{}^{3+}$ by $Co([9]aneS_3)_2{}^{2+}$ ($[9]aneS_3$ = 1,4,7-trithiacyclo-nonane) and $Co(sepulchrate)^{2+}$ (*18*). The results suggest the usefulness of the relation as a mechanistic criterion for reactions with a moderate driving force. Furthermore, the volume cross relation affords a means of obtaining experimentally inaccessible volumes of activation for adiabatic outer-sphere redox reactions. In a similar way, the volume of activation for the self-exchange reaction of cytochrome *c* could recently be extrapolated from self-exchange and reaction volume data using the volume cross relation (*19*). Thus, the applicability of the cross relation to reaction and activation volumes for electron transfer reactions could have fruitful applications in future mechanistic studies.

Nonsymmetrical Electron-Transfer Reactions

It has generally been the objective of many mechanistic studies dealing with inorganic electron transfer reactions to distinguish between outer-sphere and inner-sphere mechanisms (*6–8*). Along these lines, high-pressure kinetic methods and the construction of reaction volume profiles have also been employed to contribute toward a better understanding of the intimate mechanisms involved in such processes. The differentiation between outer-sphere and inner sphere mechanisms depends on the nature of the precursor species, Ox//Red, in the following scheme, which can either be an ion pair or encounter complex, or a bridged intermediate, respectively.

$$Ox + Red \rightleftharpoons Ox//Red, K \quad (6)$$

$$\text{Ox//Red} \rightarrow \text{Ox}^-\text{//Red}^+,\ k_{ET} \quad (7)$$

$$\text{Ox}^-\text{//Red}^+ \rightleftharpoons \text{Ox}^- + \text{Red}^+ \quad (8)$$

This means that the coordination sphere of the reactants remains intact in the former case and is modified by ligand substitution in the latter which will naturally affect the associated volume changes. In our earlier work we mainly concentrated on the analysis of nonsymmetrical electron transfer reactions (*8*), that is, reactions in which redox products are formed and for which the overall driving force and reaction volume will not be zero. Some typical examples of such studies will be discussed in more detail to form a basis for the next section, which deals with "long distance" electron transfer processes.

A general difficulty encountered in kinetic studies of outer-sphere electron transfer processes concerns the separation of the precursor formation constant (K) and the electron transfer rate constant (k_{ET}) in the scheme just outlined. In the majority of cases, precursor formation is a diffusion-controlled step, and it is followed by rate-determining electron transfer. In the presence of an excess of Red, the rate expression is given by

$$k_{obs} = k_{ET}K[\text{Red}]/(1 + K[\text{Red}]) \quad (9)$$

where k_{obs} is the observed rate constant. In many cases K is small, such that this equation simplifies to $k_{obs} = k_{ET}K[\text{Red}]$, which means that the observed second-order rate constant and the associated activation parameters are composite quantities, viz, $\Delta V^{\#} = \Delta V^{\#}(k_{ET}) + \Delta V(K)$. When K is large enough such that $1 + K[\text{Red}] > 1$, it is possible to separate k_{ET} and K kinetically and also the associated activation parameters, viz, $\Delta V^{\#}(k_{ET})$ and $\Delta V(K)$ (*8*).

One of the first systems investigated concerned the redox reaction between $Co(terpy)_2^{2+}$ and $Co(bpy)_3^{3+}$ in different solvents, where terpy = 2,2′; 6′,2″-terpyridine, and bpy = 2,2′-bipyridine (*20*). Because of the similar charge on these species, K for ion-pair formation in terms of an outer-sphere mechanism is very small, and the observed $\Delta V^{\#}$ is a composite quantity. The reported activation volumes for the investigated solvents are –9.4 (H_2O), –13.8 ($HCONH_2$), and –5.1 (CH_3CN) cm^3 mol^{-1}. Theoretical calculations based on the Marcus–Hush relationships resulted in a $\Delta V^{\#}$ value of –7.3 cm^3 mol^{-1} for the reaction in water, which is indeed close to the experimental value (*20*). A series of reactions were studied where it was possible to resolve K and k_{ET}, that is, $\Delta V(K)$ and $\Delta V^{\#}(k_{ET})$. In this case, oppositely charged reaction partners were selected, as indicated in the following scheme (*21–23*):

$$Co(NH_3)_5X^{(3-n)+} + Fe(CN)_6^{4-} \rightleftharpoons [Co(NH_3)_5X^{(3-n)+}\cdot Fe(CN)_6^{4-}],\ K \quad (10)$$

$$[Co(NH_3)_5X^{(3-n)+}\cdot Fe(CN)_6^{4-}] \rightarrow Co^{2+} + 5NH_3 + X^{n-} + Fe(CN)_6^{3-},\ k_{ET}$$

Table I. Summary of the Equilibrium Constants and Associated Thermodynamic Parameters for Ion-Pair Formation According to $Co(NH_3)_4(NH_2R)X^{(3-n)+} + Fe(CN)_6^{4-} \rightleftharpoons \{Co(NH_3)_4(NH_2R)X^{(3-n)+}, Fe(CN)_6^{4-}\}$

R	X^{n-}	*K* (25 °C) (M^{-1})	$\Delta H^{\circ a}$ (kJ mol^{-1})	$\Delta S^{\circ b}$ (J K^{-1} mol^{-1})	ΔV^c (cm^3 mol^{-1})
H	H_2O	480 ± 110			−15 ± 8
H	C_5H_5N	168 ± 7			
H	$(CH_3)_2SO$	34 ± 4	−8 ± 21	−240 ± 67	−11 ± 3
H	N_3^-	49 ± 1	+2 ± 14	−204 ± 46	−16 ± 2
H	Cl^-	37.9 ± 0.3	+28 ± 8	−120 ± 27	−3 ± 8
CH_3	Cl^-	37.2 ± 0.3	−20 ± 9	−280 ± 29	+3 ± 2
i-C_4H_9	Cl^-	13 ± 1	−4 ± 3	−240 ± 9	−6 ± 1

[a] ΔH° is reaction enthalphy.
[b] ΔS° is reaction entrophy.
[c] ΔV is reaction volume.
SOURCE: Data taken from reference 23.

$$X^{n-} = H_2O, Me_2SO, py, Cl^-, N_3^- \quad (11)$$

where py = pyridine. The thermodynamic parameters for ion-pair formation and the kinetic parameters for electron transfer are summarized in Tables I and II, respectively. The data in Table I indicate that ion-pair formation does not only depend on the charge of the participating species, and other types of interactions must be considered to account for the trend in *K*. Throughout the

Table II. Summary of Rate and Activation Parameters for Electron Transfer According to the Reaction $\{Co(NH_3)_4(NH_2R)X^{(3-n)+}, Fe(CN)_6^{4-}\} \rightarrow Co^{2+} + 4NH_3 + NH_2R + X^{n-} + Fe(CN)_6^{3-}$

R	X^{n-}	$10^2 k_{ET}$(25 °C) (s^{-1})	ΔH^a (kJ mol^{-1})	ΔS^b (J K^{-1} mol^{-1})	ΔV^c (cm^3 mol^{-1})
H	H_2O	12.7 ± 1.0	102 ± 5	+79 ± 15	+26.5 ± 2.4
H	C_5H_5N	0.89 ± 0.03	118 ± 8	+113 ± 29	+29.8 ± 1.4
H	$(CH_3)_2SO$	20 ± 1	84 ± 2	+25 ± 8	+34.4 ± 1.1
H	N_3^-	0.062 ± 0.004	104 ± 6	+44 ± 20	+18.8 ± 1.1
H	Cl^-	2.7 ± 02	85 ± 3	+11 ± 8	+25.9 ± 3.1
CH_3	Cl^-	5.0 ± 0.2	114 ± 4	+112 ± 14	+25.1 ± 1.5
i-C_4H_9	Cl^-	20 ± 2	103 ± 2	+87 ± 8	+31.3 ± 0.9

[a] ΔH is activation enthalphy.
[b] ΔS is activation entrophy.
[c] ΔV is activation volume.
SOURCE: Data taken from reference 23.

series, ion-pair formation is accompanied by significantly negative $\Delta S°$ (reaction entrophy) values and close-to-zero ΔV (reaction volume) values. The latter is rather surprising because it is generally accepted that ion-pair formation should involve considerable charge neutralization accompanied by strong desolvation due to a decrease in electrostriction. The values of ΔV therefore indicate that the reaction partners most probably exist as solvent-separated ion pairs, that is, with no significant charge neutralization accompanied by desolvation.

The activation parameters in Table II clearly demonstrate that the electron transfer steps exhibit a strong pressure deceleration; most systems have a $\Delta V^{\#}$ value of between +25 and +34 cm^3 mol^{-1}. These values indicate that electron transfer is accompanied by extensive desolvation, most probably related to charge neutralization associated with the electron transfer process (*23*). A simplified model based on partial molar volume data, in which electron transfer occurs from the precursor ion pair $[Co(NH_3)_5X^{(3-n)+} \cdot Fe(CN)_6{}^{4-}]$ to the successor ion pair $[Co(NH_3)_5X^{(2-n)+} \cdot Fe(CN)_6{}^{3-}]$, predicts an overall volume increase of ca. 65 cm^3 mol^{-1}. This means that according to the reported $\Delta V^{\#}$ values, the transition state for the electron transfer process lies approximately halfway between the reactant and product states on a volume basis for the precursor and successor ion pairs. The largest volume contribution arises from the oxidation of $Fe(CN)_6{}^{4-}$ to $Fe(CN)_6{}^{3-}$, which is accompanied by a large decrease in electrostriction and an increase in partial molar volume. Theoretical calculations also confirm that the transition state for these reactions lies approximately halfway along the reaction coordinate on a volume basis (*23*). This first information on the nature of the volume profile for an outer-sphere electron transfer reaction proved to be in good agreement with subsequently reported results for systems with low driving forces, in which it was possible to construct a complete volume profile by studying the effect of pressure on both the forward and reverse reactions, as well as on the overall equilibrium constant (*see* the following discussion).

The effect of pressure on a number of inner-sphere electron transfer reactions has also been investigated. By way of example, the reaction of $Co(NH_3)_5X^{2+}$ with $Fe(H_2O)_6{}^{2+}$ exhibits $\Delta V^{\#}$ values of +10.7 (X = F), +8.7 (X = Cl), +6.4 (X = Br), and +13.0 (X = $N_3{}^-$), which are mainly ascribed to the release of a solvent molecule during the formation of the bridged inner-sphere species, $[(NH_3)_5Co\text{–}X\text{–}Fe(H_2O)_5]^{4+}$ (*24*). Other examples of pressure effects on inner-sphere electron transfer reactions, also including some intramolecular reactions induced by pulse radiolysis, have been reported in the literature (*1, 25, 26*).

"Long Distance" Electron-Transfer Reactions

A challenging question concerns the feasibility of the application of high-pressure kinetic and thermodynamic techniques in the study of "long distance" electron transfer reactions. Do such processes exhibit a characteristic pressure dependence, and to what extent can a volume profile analysis reveal information on the intimate mechanism of the electron transfer process? Thus our hope is to be able to contribute to resolving (1) the discrepancy concerning the actual route taken by the electron, and (2) the extent to which the protein matrix plays a dominating role in such reactions.

The systems that we investigated in collaboration with the groups of J. F. Wishart (Brookhaven National Laboratory, Upton, NY), H. B. Gray (California Institute of Technology, Pasadena, CA), and T. W. Swaddle (University of Calgary, Alberta, Canada) involved intermolecular and intramolecular electron transfer reactions between ruthenium complexes and cytochrome *c*. In our own group we also studied a series of intermolecular reactions between chelated cobalt complexes and cytochrome *c*. A variety of high-pressure experimental techniques, including stopped flow, flash photolysis, pulse radiolysis, and voltammetry, were employed in these investigations. As the following presentation will show, remarkably good agreement was found between the volume data obtained with the aid of these different techniques, which clearly demonstrates the complementarity of these methods for the study of electron transfer processes.

Application of pulse-radiolysis techniques revealed that the following intramolecular and intermolecular electron transfer reactions all exhibit a significant acceleration with increasing pressure. The reported volumes of activation are -17.7 ± 0.9, -18.3 ± 0.7, and -15.6 ± 0.6 cm^3 mol^{-1}, respectively, and clearly demonstrate a significant volume collapse on going from the reactant to the transition state (*27*).

$$(NH_3)_5Ru^{II}\text{–}(His33)cyt\ c^{III} \rightarrow (NH_3)_5Ru^{III}\text{–}(His33)cyt\ c^{II} \quad (12)$$

$$(NH_3)_5Ru^{II}\text{–}(His39)cyt\ c^{III} \rightarrow (NH_3)_5Ru^{III}\text{–}(his39)cyt\ c^{II} \quad (13)$$

$$Ru^{II}(NH_3)_6^{2+} + cyt\ c^{III} \rightarrow Ru^{III}(NH_3)_6^{3+} + cyt\ c^{II} \quad (14)$$

At this stage it was uncertain what the negative volumes of activation really meant because overall reaction volumes were not available. There was, however, data in the literature (*28*) suggesting that the oxidation of $Ru(NH_3)_6^{2+}$ to $Ru(NH_3)_6^{3+}$ is accompanied by a volume increase of ca. 30 cm^3 mol^{-1}, which would mean that the activation volumes quoted above could mainly arise from volume changes associated with the oxidation of the ruthenium redox partner. In order to obtain further information on the magnitude of the overall reaction volume and the location of the transition state along the reaction coordinate, a series of intermolecular electron transfer reactions of cytochrome *c* with pentammine–ruthenium complexes were studied, where the sixth ligand on the ruthenium complex was selected in such a way that the overall driving force was low enough for the reaction kinetics to be studied in both directions (*29, 30*). The selected substituents were isonicotinamide (isn), 4-ethylpyridine

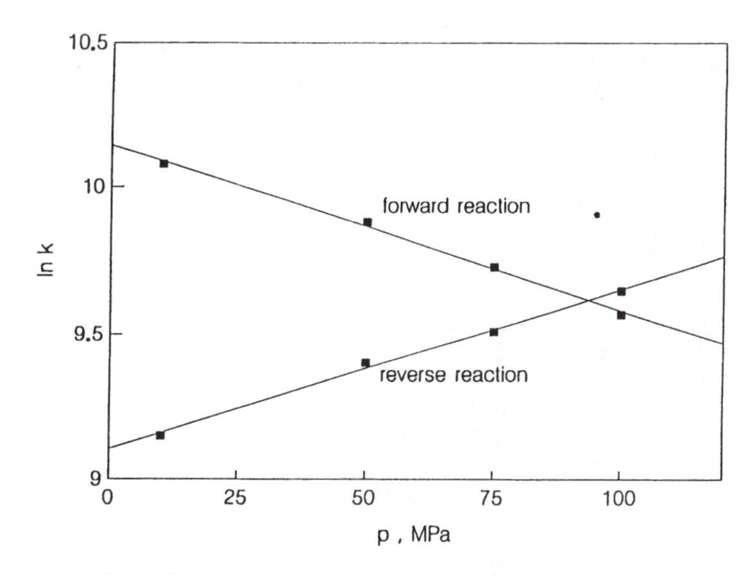

Figure 1. Plots of ln k *vs. pressure for the forward and reverse reactions in* $Ru^{III}(NH_3)_5etpy^{3+} + cyt\ c^{II} \rightleftarrows Ru^{II}(NH_3)_5etpy^{2+} + cyt\ c^{III}$ (30).

(etpy), pyridine (py), and 3,5-lutidine (lut). The overall reaction can be formulated as:

$$Ru^{III}(NH_3)_5L^{3+} + \text{cyt } c^{II} \underset{k_b}{\overset{k_f}{\rightleftharpoons}} Ru^{II}(NH_3)_5L^{2+} + \text{cyt } c^{III}, \quad (15)$$

where k_f and k_b are the forward and backward rate constants. For all the systems investigated, the forward reaction was significantly decelerated by pressure, whereas the reverse reaction was significantly accelerated by pressure. A typical example is shown in Figure 1. The absolute values of the volumes of activation for the forward and reverse processes were indeed very similar, demonstrating that a similar, rearrangement occurs in order to reach the transition state. In addition, the overall reaction volume for these systems could be determined spectrophotometrically by recording the spectrum of an equilibrium mixture as a function of pressure, and electrochemically by recording cyclic and differential pulse voltammograms as a function of pressure (*31*). These results are summarized along with the kinetic data and activation parameters in Table III. A comparison of the ΔV data demonstrates the generally good agreement between the values obtained from the difference in the volumes of activation for the forward and reverse reactions, and those obtained thermodynamically. Furthermore, the values also clearly demonstrate that $|\Delta V^{\#}| \approx 0.5|\Delta V|$, that is, the transition state lies approximately halfway between the reactant and product

Table III. Summary of Rate and Activation Parameters for the Electron Transfer Reaction Between Cytochrome c and Several Pentammine–Ruthenium Complexes: $Ru^{III}a_5L^{3+}$ + cyt c^{II} $\rightleftharpoons$ $Ru^{II}a_5L^{2+}$ + cyt c^{III}

Reaction	k $(M^{-1}\ s^{-1})^a$	$\Delta H^{\neq a}$ $(kJ\ mol^{-1})$	$\Delta S^{\neq b}$ $(J\ K^{-1}\ mol^{-1})$	$\Delta V^{\neq c}$ $(cm^3\ mol^{-1})$	ΔV^d $(cm^3\ mol^{-1})$	$-\Delta G$ (eV)	K_E [e]	K_{TD} [f]	K_{KIN} [f]
$Ru^{III}a_5$lutidine + cyt c^{II}	27,144 ± 1271	35.4 ± 0.3	−41 ± 1	+16.9 ± 1.4	33.6 ± 1.7[g]	0.012	1.6	2.6 ± 0.1	2.9 ± 0.4
$Ru^{II}a_5$lutidine + cyt c^{III}	9,448 ± 516	21 ± 1	−99 ± 5	−17.8 ± 1.6	34.7 ± 1.6[h]				
$Ru^{III}a_5$etpy + cyt c^{II}	26,823 ± 477	29 ± 2	−61 ± 7	+14.7 ± 0.9	26.9 ± 1.8[g]	0.011	1.5	2.2 ± 0.4	2.9 ± 0.1
$Ru^{II}a_5$etpy + cyt c^{III}	9,182 ± 124	25 ± 2	−86 ± 6	−14.9 ± 1.1	29.6 ± 1.0[h]				
$Ru^{III}a_5$py + cyt c^{II}	48,620 ± 1161	28 ± 1	−64 ± 5	+17.4 ± 1.5	33.4 ± 1.9[g]	0.045	5.8	6.4 ± 2.1	4.6 ± 0.4
$Ru^{II}a_5$py + cyt c^{III}	10,517 ± 494	33 ± 4	−59 ± 13	−17.7 ± 0.8	35.1 ± 1.0[h]				
$Ru^{III}a_5$isn + cyt c^{II}	1.15×10^5	22 ± 1	−75 ± 3	+16.0 ± 0.9	31 ± 1[g]	0.112	78	71 ± 7	75.7
					26.4 ± 0.9[g]				
$Ru^{II}a_5$isn + cyt c^{III}	1,520 ± 130	28 ± 4	−87 ± 12	−17.2 ± 1.5	33.2 ± 1.3[h]				

[a] ΔH is activation enthalpy.
[b] ΔS is activation entrophy.
[c] ΔV is activation volume.
[d] ΔV is reaction volume.
[e] Reaction conditions: T = 25°C, μ = 0.1 M, [cyt c] = 1×10^{-5} M, [Tris] = 0.05 M, [$LiClO_4$] = 0.05 M, pH = 7.1, and λ = 550 nm.
[f] Equilibrium constant for the oxidation of cytochrome c. K_E is calculated from the redox potential, K_{TD} from spectroscopic measurements, and K_{KIN} from kinetic measurements.
[g] Reaction volume is determined spectrophotometrically for the oxidation of cytochrome c.
[h] Reaction volume is determined kinetically for the oxidation of cytochrome c.
[e] Reference 31.
SOURCE: Data taken from reference 30.

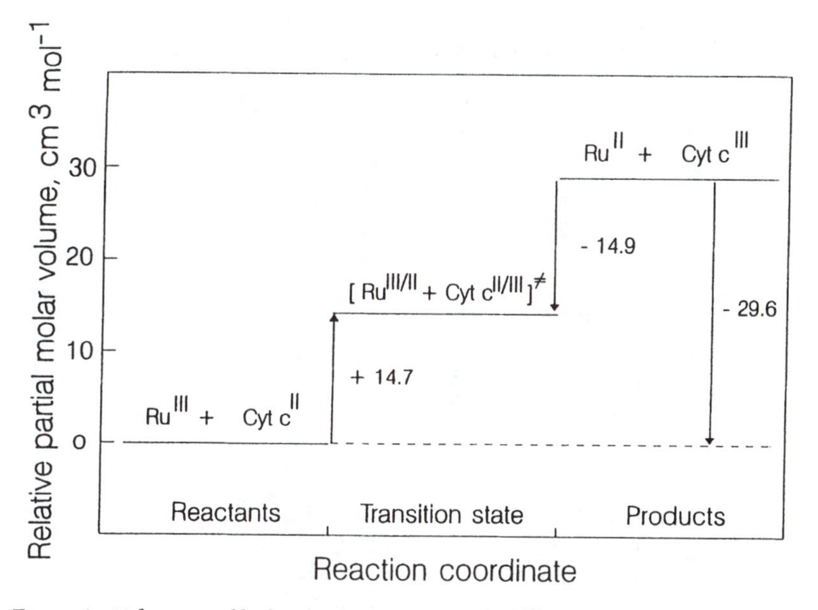

Figure 2. Volume profile for the overall reaction $Ru^{III}(NH_3)_5etpy^{3+}$ + *cyt* $c^{II} \rightleftarrows$ $Ru^{II}(NH_3)_5etpy^{2+}$ + *cyt* c^{III} (30).

states on a volume basis independent of the direction of electron transfer. The typical volume profile in Figure 2 presents the overall picture, from which the location of the transition state can clearly be seen.

Similar results were recently obtained for the redox reactions of a series of cobalt amine complexes with cytochrome *c*, for which the kinetic and thermodynamic parameters are summarized in Table IV (*32, 33*). In general, good agreement exists between the kinetically and thermodynamically determined parameters, and the typical volume profile in Figure 3 once again demonstrates the location of the transition state with respect to the reactant and product states. At this point it is important to ask the question, where do these volume changes really come from? We have always argued that the major volume change arises from changes on the redox partner and not on cytochrome *c* itself. This was suggested by the fact that the change in partial molar volume associated with the oxidation of the investigated Ru(II) and Co(II) complexes, as obtained from electrochemical and density measurements, almost fully accounted for the observed overall reaction volume. Thus the reduction of cytochrome *c* can only make a minor contribution toward the overall volume change.

These arguments were apparently in contradiction with electrochemical results reported by Cruanes et al. (*34*), according to which the reduction of cytochrome *c* is accompanied by a volume collapse of 24 cm^3 mol^{-1}. This value is so large that it almost represents all of the reaction volume found for the investigated reactions discussed above. A reinvestigation of the electrochemistry

Table IV. Summary of Rate and Activation Parameters for the Electron Transfer Reaction Between Cytochrome c and Several Cobalt Complexes: cyt c^{II} + $Co(L)^{3+}$ ⇌ cyt c^{III} + $Co(L)^{2+}$

Reaction	k $(M^{-1}\ s^{-1})$ [a]	$\Delta H^{\neq}$ $(kJ\ mol^{-1})$	$\Delta S^{\neq}$ $(J\ K^{-1}\ mol^{-1})$	$\Delta V^{\neq}$ $(cm^3\ mol^{-1})$	ΔV $(cm^3\ mol^{-1})$	$-\Delta G$ (eV)	K_E [b]	K_{TD} [b]	K_{KIN} [b]
$Co(bpy)_3^{3+}$ + cyt c^{II}	582 ± 13	49.9 ± 0.7	−28 ± 2	+12.5 ± 0.9	21.8 ± 0.7 [c]	0.028	3.0	3.3 ± 0.4	3.4 ± 0.3
$Co(bpy)_3^{2+}$ + cyt c^{III}	169 ± 5	28 ± 1	−107 ± 5	−12.6 ± 1.5	25.1 ± 1.7 [d]				
					27.5 ± 1.4 [e]				
$Co(phen)_3^{3+}$ + cyt c^{II}	3,753 ± 39	44 ± 3	−28 ± 9	+17.0 ± 0.9	37.9 ± 2.0 [c]	0.085	32	20 ± 3	17.3 ± 0.6
$Co(phen)_3^{2+}$ + cyt c^{III}	217 ± 5	14 ± 1	−136 ± 4	−16.2 ± 1.0	34.2 ± 1.7 [d]				
					35.4 ± 2.0 [e]				
$Co(terpy)_2^{3+}$ + cyt c^{II}	1,427 ± 36	40 ± 1	−47 ± 4	+18.4 ± 1.2	33 ± 3 [c]	−0.003	0.9	0.7 ± 0.2	0.9 ± 0.1
$Co(terpy)_2^{2+}$ + cyt c^{III}	1,704 ± 46	14 ± 1	−136 ± 4	−18.0 ± 1.4	36 ± 2 [d]				
					36.3 ± 2.0 [e]				

[a] Reaction conditions: T = 25°C, μ = 0.1 M, [cyt c] = 1×10^{-5} M, [Tris] = 0.05 M, $[LiClO_4]/[LiNO_3]$ = 0.05, pH = 7.1, λ = 550 nm, and A = NH_3.
[b] Equilibrium constant for the oxidation of cytochrome. K_E is calculated from the redox potential, K_{TD} from spectroscopic measurements, and K_{KIN} from kinetic measurements.
[c] Reaction volume is determined spectrophotometrically for the oxidation of cytochrome c.
[d] Reaction volume is determined kinetically for the oxidation of cytochrome c.
[e] Reaction volume is determined electrochemically for the oxidation of cytochrome c.
SOURCE: Data taken from reference 33.

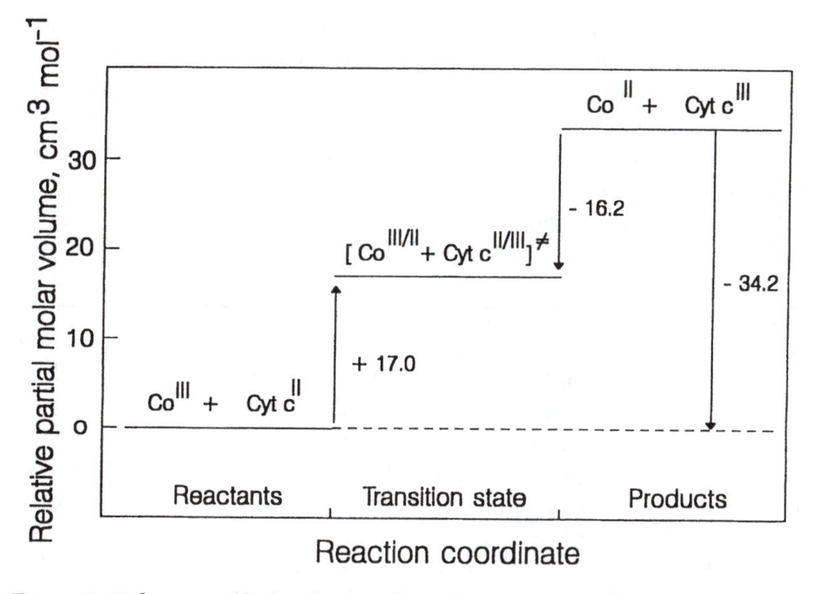

Figure 3. Volume profile for the overall reaction $Co(phen)_3^{3+}$ + *cyt* $c^{II} \rightleftarrows$ $Co(phen)_3^{2+}$ + *cyt* c^{III} (33).

of cytochrome *c* as a function of pressure, using cyclic and differential pulse voltammetric techniques (*31*), revealed a reaction volume of -14.0 ± 0.5 cm^3 mol^{-1} for the reaction

$$\text{cyt } c^{III} + Ag(s) + Cl^- \rightarrow \text{cyt } c^{II} + AgCl(s) \quad (16)$$

A correction for the contribution from the reference electrode can be made on the basis of the data published by Tregloan and co-workers (*35*) and a series of measurements of the potential of the Ag–AgCl(KCl saturated) electrode relative to the $Ag–Ag^+$ electrode as a function of pressure. The contribution of the reference electrode turned out to be -9.0 ± 0.6 cm^3 mol^{-1}, from which it followed that the reduction of cytochrome c^{III} is accompanied by a volume decrease of 5.0 ± 0.8 cm^3 mol^{-1}. This contribution is significantly smaller than concluded by Cruanes et al. (*34*) and is also in line with the other arguments referred to above. Thus, we conclude that the observed activation and reaction volumes arise mainly from volume changes on the Ru and Co complexes, which in turn will largely be associated with changes in electrostriction in the case of the ammine complexes. The oxidation of the Ru(II) ammine complexes will be accompanied by a large increase in electrostriction and almost no change in the metal–ligand bond length, whereas in the case of the Co complexes a significant contribution from intrinsic volume changes associated with the oxidation of Co(II) will partially account for the observed effects.

The results in Tables III and IV nicely demonstrate the complementarity of the kinetic and thermodynamic data obtained from stopped-flow, UV–vis, electrochemical and density measurements. The resulting picture is consistent and allows a further detailed analysis of the data. The overall reaction volumes determined in four different ways are surprisingly similar and underline the validity of the different methods employed. The volume profiles in Figures 2 and 3 demonstrate the symmetric nature of the intrinsic and solvational reorganization in order to reach the transition state of the electron transfer process. In these systems, the volume profile is controlled by effects on the redox partner of cytochrome *c*, but this does not necessarily always have to be the case. The location of the transition state on a volume basis will reveal information concerning the "early" or "late" nature of the transition state and reveal details of the actual electron transfer route followed.

One system was investigated recently in which the effect of pressure on the electron transfer rate constant revealed information on the actual electron transfer route. We investigated the effect of pressure on distant electronic coupling in $Ru(bpy)_2(im)$-modified His33 and His72 cytochrome *c* derivatives, for which the electron transfer from Fe(II) to Ru(III) is activationless (*36*) (im is imidazole). In the case of the His33-modified system, the electron transfer rate constant exhibited no dependence on pressure within experimental error limits. However, the rate constant for the His72-modified protein increased significantly with increasing pressure, corresponding to a $\Delta V^{\#}$ value of -6 ± 2 cm^3 mol^{-1}. Because this value is exactly opposite to that expected for the reduction of Ru(III), the result was interpreted as an increase in electronic coupling at elevated pressure. The application of moderate pressures will cause a slight compression of the protein that in turn shrinks the through-space gaps that are key units in the electron-tunneling pathway between the heme and His72. A decrease of 0.46 Å in the tunneling path length at a pressure of 150 MPa can account for the observed increase in the rate constant. This in turn means that there is an average decrease in the space-gap of 0.1 Å. The absence of an effect for the His33-modified species is understandable because electronic coupling through covalent and hydrogen bonds will be less pressure sensitive than coupling via van der Waals gaps (*36*).

Conclusions

The examples presented in this chapter clearly demonstrate that electron transfer reactions exhibit a characteristic pressure dependence that can be employed to gain further insight into the mechanism of the electron transfer process. The pressure dependence of self-exchange reactions can be used to develop the theoretical interpretation of the observed volumes of activation because the overall reactions involve no net volume change. In the case of nonsymmetrical reactions, the volume profile treatment can reveal information regarding the reorganization involved in going from the reactant to the transition and product state. A systematic variation of the nature of the redox partners can enable a detailed analysis of the factors that control the driving force, the rate and activation parameters, and so assist the interpretation of the volume profile. The quality and mutual consistency of the volume data are such that they offer, in principle, the possibility of a detailed mechanistic treatment on the basis of partial molar volume changes associated with the redox processes. In this respect, it is presently the dependence of such profiles on the actual electron-transfer route that offers especially interesting possibilities. Along these lines, future investigations will focus on reactions that occur over long distances and involve redox partners that will exhibit minor solvational changes during the redox process. In this way, the focus will be on the redox pathway itself once the other contributions can be minimized.

The experiments referred to in this chapter have also assisted the theoretical analysis of the role of temperature and pressure on biological electron transfer reactions. In the case of cytochrome *c*, an empirical approach could show that the heme iron is screened more efficiently from surface charges in the oxidized state (*37*). In general, solute–solvent interactions are directly influenced by temperature and pressure, and these interactions will affect the electrostatic interaction energies, which can be accounted for in terms of changes in the dielectric constant of both the solute and the solvent. Such interactions are of major importance in the understanding of electron transfer processes.

Acknowledgments

I gratefully acknowledge financial support from the Deutsche Forschungsgemeinschaft and the Volkswagen–Stiftung. Stimulating collaborations with various groups, as mentioned in this chapter, are highly appreciated.

References

1. van Eldik, R.; Asano, T.; le Noble, W. J. *Chem. Rev.* **1989,** *89,* 549.
2. van Eldik, R.; Merbach, A. E. *Comments Inorg. Chem.* **1992,** *12,* 341.
3. van Eldik, R. In *Perspectives in Coordination Chemistry;* Williams, A. F.; Floriani, C.; Merbach, A. E., Eds.; VHCA: Basel, Switzerland, 1992; p 55.
4. van Eldik, R. *Pure Appl. Chem.* **1993,** *65,* 2603.
5. Hubbard, C. D.; van Eldik, R. *Instrum. Sci. Technol.* **1995,** *23,* 1.
6. Stranks, D. R. *Pure Appl. Chem.* **1974,** *38,* 303.
7. Swaddle, T. W. In *Inorganic High Pressure Chemistry: Kinetics and Mechanisms;* van Eldik, R.; Ed.; Elsevier: Amsterdam, Netherlands, 1986; Chapter 5.
8. van Eldik, R. *High Pressure Res.* **1991,** *6,* 251.
9. Jolley, W. H.; Stranks, D. R.; Swaddle, T. W. *Inorg. Chem.* **1990,** *29,* 1948.
10. Dione, H.; Swaddle, T. W. *Can. J. Chem.* **1988,** *66,* 2763.
11. Takagi, H.; Swaddle, T. W. *Inorg. Chem.* **1992,** *31,* 4669.
12. Spiccia, L.; Swaddle, T. W. *Inorg. Chem.* **1987,** *26,* 2265.
13. Doine, H.; Swaddle, T. W. *Inorg. Chem.* **1991,** *30,* 1858.
14. Shalders, R. D.; Swaddle, T. W. *Inorg. Chem.* **1995,** *34,* 4815.

15. Jolley, W. H.; Stranks, D. R.; Swaddle, T. W. *Inorg. Chem.* **1990,** *29,* 385.
16. Grace, M. R.; Swaddle, T. W. *Inorg. Chem.* **1993,** *32,* 5597.
17. Swaddle, T. W. *Inorg. Chem.* **1990,** *29,* 5017.
18. Grace, M. R.; Takagi, H.; Swaddle, T. W. *Inorg. Chem.* **1994,** *33,* 1915.
19. Meier, M.; van Eldik, R. *Inorg. Chim. Acta* **1996,** *242,* 185.
20. Braun, P.; van Eldik, R. *J. Chem. Soc., Chem. Commun.* **1985,** 1349.
21. Krack, I.; van Eldik, R. *Inorg. Chem.* **1986,** *25,* 1743.
22. Krack, I.; van Eldik, R. *Inorg. Chem.* **1989,** *28,* 851.
23. Krack, I.; van Eldik, R. *Inorg. Chem.* **1990,** *29,* 1700.
24. van Eldik, R. *Inorg. Chem.* **1982,** *21,* 2501.
25. Meyerstein, D.; Zilbermann, I.; Cohen, H.; van Eldik, R. *High Pressure Res.* **1991,** *6,* 287.
26. Goldstein, S.; Czapski, G.; van Eldik, R.; Cohen, H.; Meyerstein, D. *J. Phys. Chem.* **1991,** *95,* 1282.
27. Wishart, J. F.; van Eldik, R.; Sun, J.; Su, C.; Isied, S. S. *Inorg. Chem.* **1992,** *31,* 3986.
28. Sachinidis, J. I.; Shalders, R. D.; Tregloan, P. A. *Inorg. Chem.* **1996,** *35,* 2497.
29. Bänsch, B.; Meier, M.; Martinez, M.; van Eldik, R.; Su, C.; Sun, J.; Isied, S. S.; Wishart, J. F. *Inorg. Chem.* **1994,** *23,* 4744.
30. Meier, M.; Sun, J.; Wishart, J. F.; van Eldik, R. *Inorg. Chem.* **1996,** *35,* 1564.
31. Sun, J.; Wishart, J. F.; van Eldik, R.; Shalders, R. D.; Swaddle, T. W. *J. Am. Chem. Soc.* **1995,** *117,* 2600.
32. Meier, M.; van Eldik, R. *Inorg. Chim. Acta* **1994,** *225,* 95.
33. Meier, M.; van Eldik, R. *Chem. Eur. J.* **1997,** *3,* 33.
34. Cruanes, M. T.; Rodgers, K. K.; Sligar, S. G. *J. Am. Chem. Soc.* **1992,** *114,* 9660.
35. Sachinidis, J. I.; Shalders, R. D.; Tregloan, P. A. *Inorg. Chem.* **1994,** *33,* 6180.
36. Meier, M.; van Eldik, R.; Chang, I-J.; Mines, G. A.; Wuttke, D. S.; Winkler, J. R.; Gray, H. B. *J. Am. Chem. Soc.* **1994,** *116,* 1577.
37. Smith, E. T. *J. Am. Chem. Soc.* **1995,** *117,* 6717.

20

Ultrafast Electron Transfer and Short-Lived Prereactive Steps in Solutions

Y. Gauduel and H. Gelabert

Laboratoire d'Optique Appliquée, Centre National de la Recherche Scientifique Unité de Recherche Associée 1406, Institut National de la Santé et de la Recherche Médicale U451, Ecole Polytechnique–ENS Techniques Avancées, 91120 Palaiseau, France

With the intensive development of ultrafast spectroscopic methods, reaction dynamics can be investigated at the subpicosecond time scale. Femtosecond spectroscopy of liquids and solutions allows the study of solvent-cage effects on elementary charge-transfer processes. Recent work on ultrafast electron-transfer channels in aqueous ionic solutions is presented (electron–atom or electron–ion radical pairs, early geminate recombination, and concerted electron–proton transfer) and discussed in the framework of quantum theories on nonequilibrium electronic states. These advances permit us to understand how the statistical density fluctuations of a molecular solvent can assist or impede elementary electron-transfer processes in liquids and solutions.

The investigation of elementary chemical processes in solution can be performed by using the interaction of ionizing radiation (electron or photon beams) with ions or molecules. During the last two decades, significant experimental advances have been made with the help of picosecond pulse radiolysis and femtosecond ultraviolet–infrared (UV–IR) spectroscopy (*1–8*). Pulse radioloysis experiments have been mainly devoted to the study of electron-transfer and radical reactions in molecular liquids (*2, 9–12*). In radiation chemistry, the pulse duration represents a limiting factor for the investigation of subnanosecond events (*2, 13*) or the identification of ultra-short-lived states (proton transfer, electron–ion pair deactivation). The temporal sequence presented in Figure 1 shows that the elementary chemical steps in solution would occur in less than 2 or 3 ps. Ultrafast photophysical investigations are more appropriate for the study of some fundamental aspects of radiation chemistry and photochemistry: formation of a hydration cage around excess electrons, encounter pair formation, ion–molecule reactions, electron attachment to solvent or mo-

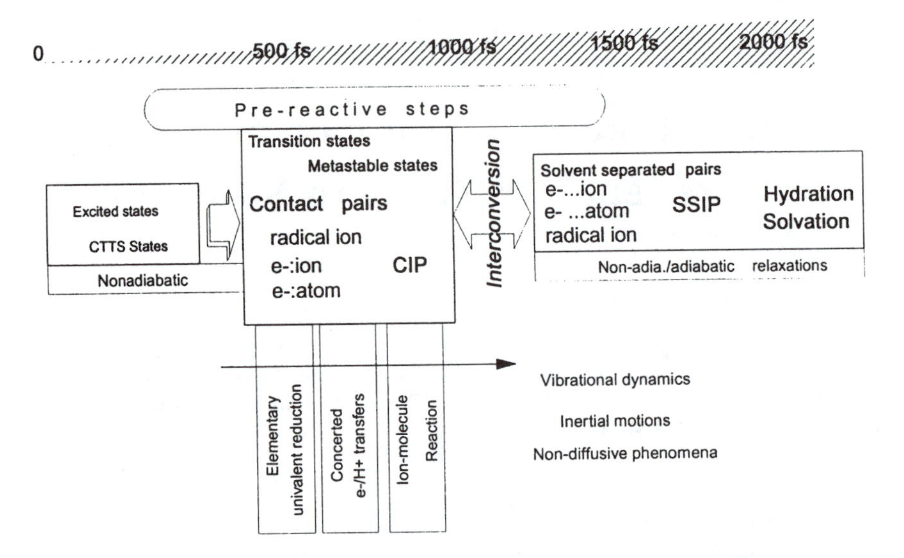

Figure 1. Time dependence of elementary chemical processes in solutions. In polar solutions, most of these primary events (electron detachment, ion–ion pair interconversion, concerted electron–proton transfer, and electron solvation) occur in less than 2×10^{-12} s and are controlled by vibrational or electronic responses of the reaction medium.

lecular acceptor, early electron–proton recombination, and effects of librational, vibrational, or rotational motions or dielectric relaxation during electron transfer in large clusters, solutions, and organized assemblies (*7, 14–18*).

Time-resolved spectroscopic methods combining ultrashort laser pulses of typically less than 100-fs duration (1 fs $= 10^{-15}$ s) and different pump–probe configurations are very efficient for the investigation of ultrafast elementary steps in solution chemistry (photoexcitation of molecular probes, photoejection of subexcitation electrons). Femtosecond investigations can use different nonlinear spectroscopic configurations such as fluorescence, hole burning, four-wave mixing, pump–probe absorption dichroism, photon echoes, and Raman spectroscopy (*6, 19–25*). In this way, femtosecond spectroscopy of nonequilibrium electronic states in liquids permits us to obtain unique information on transient solute–solvent interactions or short-lived solvent-cage effects during nonadiabatic–adiabatic electronic transitions.

At the same time, intensive theoretical developments have been devoted to the microscopic description of a solvent cage around ground or excited states (*25–31*). An understanding of solvent motions during electron-transfer reactions requires the investigation of elementary steps in conjunction with the time dependence of structural and energetic parameters of solute solvation shells (*32, 33*).

Electrolyte solutions with halide ions represent a particular class of chemical systems for the investigation of transient couplings between different solute electronic states and surrounding solvent molecules (*34–36*). In aqueous solution, the ground state of a halide is localized in polar-solvent cavities and exhibits a characteristic absorption band in the ultraviolet. This electronic absorption spectrum contains complex subbands and represents the signature of a charge transfer to solvent (CTTS) (*36, 37*). The stabilization of this electronic state by surrounding solvent molecules represents a typical solvent effect. In CTTS states, one of the electrons is partially delocalized between the atomic core and the first hydration shells. During the photoexcitation of an aqueous halide, short-lived couplings can take place between a newly created electronic state of the solute (excited CTTS states) and solvent molecules. These transient couplings are controlled by librational motions, short-range polarization effects, and molecular reorganization of solvent in the vicinity of newly created electronic configurations of the solute.

The behavior of CTTS states is dependent on energy levels of the ion–solvent molecular couplings. These levels can lead to internal relaxation and/or complete electron detachment via adiabatic or nonadiabatic electron transfer. The ultrafast spectroscopic investigations of electronic dynamics in ionic solutions would permit us to learn more about the primary steps of an electron-transfer reaction within a cationic atmosphere. The influence of counterions on early electron photodetachment trajectories from a halide ion can be considered as prereactive steps of an electron transfer.

This chapter is organized as follows. First, we present a background survey of electron-transfer theories in solution. Then, we describe femtosecond spectroscopic investigations of electron-transfer processes and prereactive steps in pure and ionic solutions.

Background on Electron-Transfer Theory in Solutions

The description of an oxidoreduction reaction at the microscopic level represents a fundamental challenge in physical chemistry because the elementary steps of an electron transfer between reactants and products involve very short-lived states ($t \leq 10^{-12}$ s) and angström or subangström displacements. At the molecular level, vibrational and rotational motions can influence chemical bond formation or breaking. One of the fundamental questions on chemical reactions in solution is the role of microscopic solvent dynamics during charge-transfer reactions. An understanding of solvent frictions in the definition of reaction frequency factors encompasses the synergy between high-time-resolution photochemical studies and advances in computational solution chemistry (*25, 38–41*).

The importance of dynamical solvent effects on the rate of charge-transfer reactions is particularly evident for reactions with small activation barriers: activationless, solvent-controlled, fast intramolecular electron transfer for which the free energy (ΔG) of the reaction is small compared to thermal agitation energy (kT) (*26, 42*). The solvent part is dominant in the contribution of the free activation energy. When the coupling zone between reactants and products is weak, the energy profiles for the electronic wave functions of the initial and final states (Ψ_i, Ψ_f) cross in single point. Under this condition, equalization of energies for the reactants and products remains occasional and largely governed by solvent fluctuations. For a nonadiabatic process, there is no involvment of solvent dynamics in the rate-determining step. The other situation corresponds to the adiabatic process for which the crossing zone of potential energy surfaces is large owing to strong coupling between reactants and products. This reaction zone defines a single potential energy surface for which the rate constant, k_{et}, is proportional to the inverse of the longitudinal time (T_L^{-1}) or the experimental solvation time (T_{obs}^{-1}) (*31–33*).

Electron-transfer kinetics in solutions have often been analyzed and interpreted in the framework of the general adiabatic theory of Marcus (*43*). Although electron-transfer dynamics are not always characterized by a classical rate constant (*44*), a general formulation of the chemical reaction concerns the rate constant k, which can be expressed as:

$$k = \nu_{eff} \cdot K_{el} \cdot \Gamma_n \exp(-\Delta G/k_B T) \quad (1)$$

where ν_{eff} is the effective frequency for motion along the reaction coordinates, K_{el} is the electronic transmission factor, Γ_n is the nuclear tunneling factor, ΔG is the free energy of activation, k_B is the Boltzmann constant, and T is the temperature. The dynamical solvent effects are expressed by friction parameters (of collisional or dielectric origin), which appear in the three frequency prefactors of equation 1. The electronic transmission coefficient, K_{el}, can be expressed by using the probability for a transition from the initial state to the final state adiabatic energy surface through the crossing region:

$$K_{el} = 2P_0/(1 + P_0) \quad (2)$$

where $P_0 = 1 - \exp(-2\pi\gamma)$, the surface hopping probability.

The nonadiabatic limit is characterized by the following expression of K_{el} (*42, 43*):

$$k_{el} = 2\backslash H_{if}\backslash^2 \pi^{3/2} \backslash h \nu_{eff} \cdot k_B \cdot T \cdot E_\lambda{}^{1/2} \quad (3)$$

where H_{if} is the transfer integral, h is Planck's constant, and E_λ is the reorganization energy of the reaction system.

The analysis of reaction dynamics at the microscopic level requires the real-time discrimination of short-lived transition states by high-time spectroscopic methods. Ultrashort optical pulses offer the opportunity of direct discrimination of elementary steps during an intermolecular charge-transfer process (*44–46*). The investigation of solvent effects during chemical reactions has been mainly devoted to the study of the time-dependent solvent response triggered by a sudden change of solute electronic state and the investigation of resonant electron reactions in Debye solvents (*25, 31*). A fundamental point concerns the excess electron because this elementary charge can exhibit several delocalized or localized states, and there are fascinating issues on transient couplings with neutral or ionic acceptors and solvent molecules.

Ultrafast Spectroscopy of the Hydrated Electron

The hydrated electron represents a ubiquitous entity in irradiated aqueous solutions, and its experimental discovery by pulse radiolysis has raised considerable interest in investigations of electron-transfer reactions in chemistry and radiobiology (*1, 47–49*). The dynamical component of an excess electron in a polar solution is directly dependent on transient electron–solvent couplings. The excess electron is equivalent to a microprobe that can test the inhomogeneous structures of a reaction area. This elementary charge exhibits several delocalized or localized states in the condensed phase (*14, 16, 18, 29*).

The complex couplings occurring between water molecules and an excess electron can be influenced by the short-lived hydrogen-bond network, the protonated configurations (protic character of water), or the fluctuations of the short-range polarization energy. During the last decade, numerous aspects of the dynamical properties of liquid water (*50–54*) and solvation dynamics (*55, 56*) have been explored. Water molecules can influence the quantum aspects of a delocalized (unbound state) or localized (bound state) electron. Figure 2 shows some experimental spectra of a localized electron in a solvent cavity (hydrated electron ground state). The ejection of excess electrons can be performed by the ionization of an aqueous medium with a pulse electron beam (radiolysis) or laser pulses (photolysis) (*1, 34, 57, 58*). The different relaxation processes lead to a complete electron stabilization in the bulk. In pure water at room temperature, this fully relaxed radical (e^-_{hyd}) is completely developed in less than 2 ps and exhibits an asymmetric broadband peaking around 1.72 eV at 294 K (*59–61*).

The direct photoexcitation of water molecules by ultrashort laser pulses is used for the investigation of primary events occurring from 10^{-14} s (thermal orientation of water molecules and ultrafast proton transfer) to 10^{-10} s (primary reactions of a solvated electron with protic species) (*57, 58, 61–65*). The nonlinear interaction of ultrashort UV pulses (typically less than 100 fs in duration and having a power of ~10^{12} W cm^{-2}) with water molecules triggers multiple electron photodetachment channels within a hydrogen bond network (see equations 4–7). An initial energy deposition via a two-photon absorption process (2 × 4 eV) leads to the formation of nonequilibrium states of an excess electron

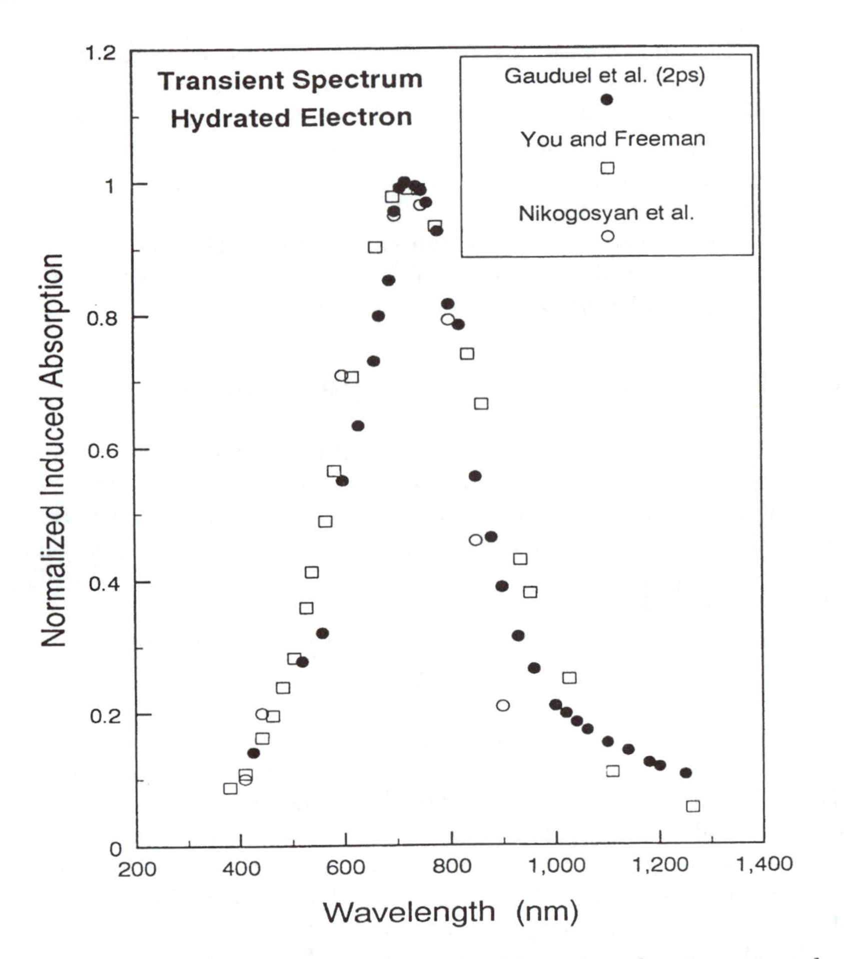

Figure 2. Transient absorption spectrum of visible aqueous electrons generated by different excitation methods of pure liquid water: pulse radiolysis (60), picosecond pulse photolysis (58), and femtosecond UV photolysis (6). The long-lived spectra obtained by three different pulsed methods correspond to a broad absorption band of relaxed hydrated electrons (in an s-like ground state) centered around 720 nm (1.72 eV).

(quasi-free or dry electron $\{e^-_{qf}\}$, thermalized electron $\{e^-_{th}\}$, or localized electron $\{e^-_{loc}$ or $e^-_{trap}\}$), a fully relaxed hydrated electron $\{e^-_{hyd}\}$, and prototropic species $\{H_3O^+, OH\}$ (29). Computer analysis of femtosecond UV–IR absorption spectroscopy in pure liquid water (56) allows us to obtain the time dependence of different electron–water couplings (Figure 3).

$$H_2O + (h\nu, 310\text{ nm}) \xrightarrow{\tau < 50\text{ fs}} [H_2O]^* \rightarrow H_2O^+ + e^-_{qf} \qquad (4)$$

$$H_2O^+ + H_2O \xrightarrow{100\ fs} H_3O^+ + OH \tag{5}$$

$$e^-_{qf} + nH_2O \xrightarrow{110\ fs} e^-_{IR}(e^-_{prehyd}) \xrightarrow{240\ fs} e^-_{vis}(e_{hyd}) \tag{6}$$

Ten years ago, femtosecond IR spectroscopy of an excess electron in pure water showed the existence of an ultrashort-lived prehydrated state (*61*). This IR nonequilibrium electronic configuration is built up in less than 120 fs in H_2O and represents a direct precursor of the hydrated electron ground state (equation 6). In the infrared (0.99 eV), the monoexponential relaxation of the signal toward an s-like ground state of the hydrated electron (240 ± 20 fs) has been analyzed in the framework of a two-state model (*61, 65*). With a similar model, an indirect estimate of the infrared electron relaxation in the red spectral region gives a deactivation rate of $2 \times 10^{12}\ s^{-1}$ (*62, 66*). The very fast appearance of the infrared electron (e^-_{IR}) is comparable to any nuclear motion, solvent dipole orientation, or thermal motion of water molecules. The relaxation of

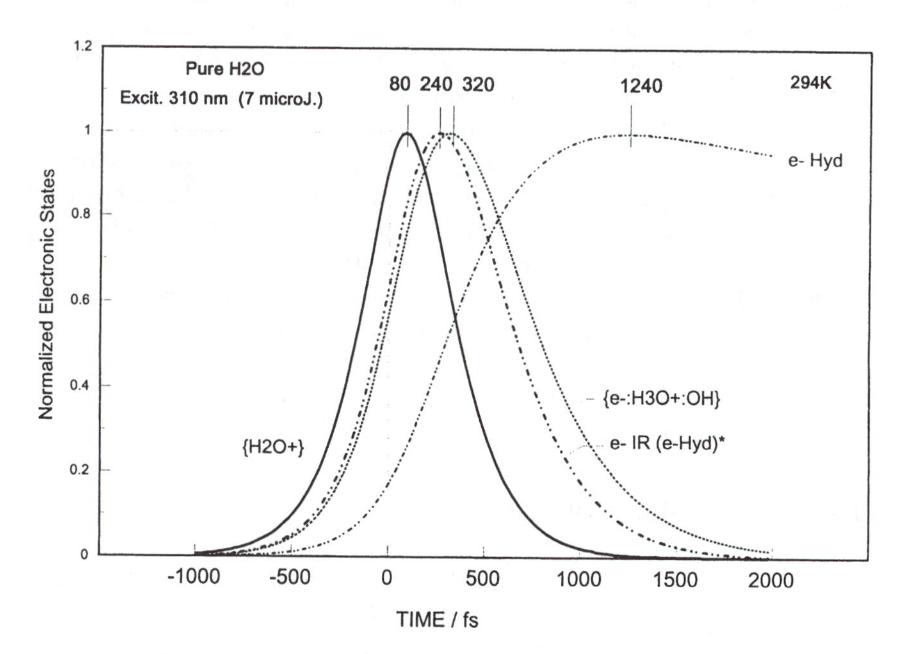

*Figure 3. Time dependence of different electron-transfer trajectories in molecules of pure liquid water at room temperature. The femtosecond UV excitation of water molecules (2 × 4 eV) triggers either an ultrafast electron photodetachment with the formation of hydronium ions and a nonadiabatic relaxation of excited p-like hydrated electrons (high photochemical channel), or concerted electron–proton transfer (low photochemical channel) (*56, 72*). The characteristic time of each trajectory is reported on the curve.*

e^-_{IR} (e^-_{prehyd}) is not accompanied by a significant continuous shift from the infrared to the visible (*61, 65*).

In agreement with pulse radiolysis and photolysis experiments (see Figure 2), the computed long-lived ground state exhibits a maximum optical transition in the red and a 0.8-eV bandwidth (*38*). Investigators have performed intensive semiquantum molecular dynamics simulations of electron presolvation and solvation in water (*28, 29, 67, 68*). In agreement with femtosecond IR spectroscopy (*61*), quantum simulations of IR signal relaxation would correspond to an internal transition of a p-like hydrated electron (excited state) toward an s-like ground state of the hydrated electron. The relaxation dynamics of nonadiabatic stabilization of the hydrated electron occur in the subpicosecond regime (*29, 67, 70*). Coupling between the excess electron and the solvent molecules is estimated via the quantum expectation value of the electron–water interaction potential. An ultrafast response of the solvent (~20–40 fs) is attributed to molecular rotations of water in the first solvation shell, and two slower components (~240 fs, 1100 fs) are on the order of the dielectric relaxation time.

More recently, hole-burning experiments on the ground state of the hydrated electron have shown that the internal conversion from an excited state (p → s de-excitation) occurs with a time constant of 310 fs and is followed by a second component of 1.1 ps. This slow component is assigned to a cooling phenomenon (*63*). Even if the nonmonotonic behavior of time-resolved spectra agrees with this cooling effect, a recent statistical theory on the femtosecond pump-probe spectroscopy of electron hydration argues for the contribution of bleaching and absorption dynamics (*69*). Moreover, experimental and theoretical developments in electron solvation dynamics emphasize the role of anisotropic fluctuations of surrounding water molecules (*70, 71*).

Within the time scale of the electron hydration process, the primary water molecular cation (H_2O^+) reacts with surrounding water molecules (ion–molecule reaction, equation 5). This ultrafast proton transfer is faster than a second electron stabilization channel for which a concerted electron–proton transfer is under consideration (equation 7) (*56, 72, 73*).

$$n(H_2O) + h\nu \rightarrow (H_3O^+\cdots OH)_{hyd} + e^-_{qf} \xrightarrow{130\ fs} (H_3O\cdots OH)^*hyd \leftrightarrow (H_3O^+{:}e^-{:}OH)_{hyd} \quad (7)$$

$$\begin{aligned}(H_3O\cdots OH)^*hyd \leftrightarrow [(H_3O^+{:}e^-{:}OH)hyd &\xrightarrow{340\ fs} H + H_2O + OH \\ &\rightarrow H_3O\cdots OH? \\ &\rightarrow e^-_{hyd} + (H_3O^+\cdots OH)? \\ &\rightarrow (H_2O)_2?\end{aligned}$$

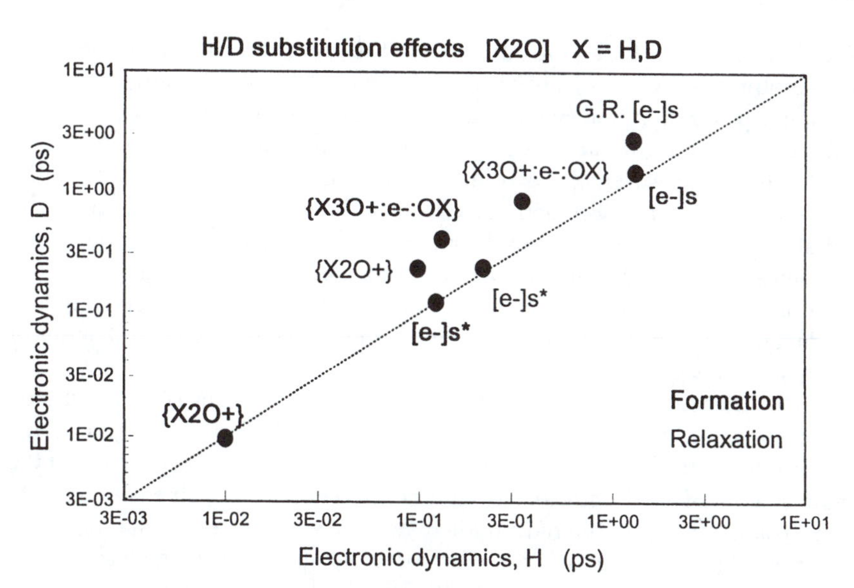

Figure 4. Comparative analysis of H–D isotope effects on elementary charge transfer, including electron photodetachment and localization, and electron-protonated radical couplings in pure water at 294 K. The dotted line represents the characteristic limit for which the electronic dynamics are independent of H–D isotope substitution.

Figure 4 shows significant hydrogen–deuterium (H–D) substitution effects on UV, visible, or infrared spectroscopy of pure water. H–D isotope substitution represents a useful tool for testing the influence of the energy vibrational modes (OH, OD) during the relaxation of nonequilibrium electronic states. Significant differences in the microscopic structure of H_2O and D_2O have been reported: deuterated water exhibits stronger hydrogen bonds than light water, the energetic vibrational mode (OD versus OH) is $2^{1/2}$ times lower in D_2O than in H_2O (*59*), and the lifetime of protropic species is about two times longer in D_2O (*74*). Figure 4 shows that multiple electronic relaxation channels can be tested by very short UV–IR pulses in liquid water, and underscores the absence of significant isotope effects on nonadiabatic $\{e_s^-\}^* \rightarrow \{e_s^-\}$ deactivation in the infrared (*61*). This experimental observation is not captured by the quantum simulation of relaxation of an excess electron within a bath of classical rigid water molecules (*70, 75*). More realistic treatment of water molecules, including quantum modeling of flexible molecules, and ab initio molecular dynamic modeling of electronic structure, intramolecular energy transfer, and short-range solvent polarization effects in water, are required to enhance the computer simulations of electron hydration via nonadiabatic transitions (*53, 76–78*).

An interesting point raised by Figure 4 concerns the significant effects played by H–D isotopic substitution on the dynamics of concerted electron–proton transfer involving transient couplings between nonrelaxed electrons and prototropic radicals or early geminate recombination between the hydrated electron ground state and the hydrated proton (hydronium ion or H_3O^+). In liquid water, these primary reactions are likely governed by the electron energy level, vibronic interactions, and multiple configurations of the hydrated proton (*73, 79*). An understanding of transient couplings occurring between electron–hydronium ion pairs and surrounding water molecules needs to be precise about whether the dynamics of these electron–proton couplings are dependent on intracomplex structural changes (geometric perturbations of the hydration cage).

It is interesting to note that the cleavage rate constant of the hydrated electron–proton pairs occurs on the same time scale as the H-bond or hydronium-ion mean lifetimes in liquid water (*56, 74*). We have suggested that the limiting factor of the deactivation dynamics of the encounter pair corresponds to the activation energy of the radical-ion bond cleavage reaction, including either a proton migration from a hydronium ion to neighbor water molecules or a local polarization effect on H bonds (*72*). The elementary reactivity of an excess electron with a hydrated hydronium ion $(H_3O^+ + e^-)_{hyd}$ would depend on (1) the local structure of water molecules in the vicinity of this cation, (2) the initial electron-hole pair distributions, and (3) the H-bond dynamics between H_3O^+ and water molecules or proton migration from hydronium to neighboring water molecules. In this last case, the relaxation of the encounter pair can be compatible with the vibrational modes of water molecules in the femtosecond range, namely, vibrational OH bonds, and librational and translational modes of the hydrated proton (hydronium ion). Complex spectroscopic investigations of nonequilibrium electronic states produced by nonlinear photoexcitation (two-photon process: 2×4 eV) of water molecules have demonstrated that only one electronic deactivation mode ($p \rightarrow s$ transition) leads to the complete hydration process of an excess electron in pure water (equation 6). The second electronic channel, which we assigned to a concerted electron–proton transfer (equation 7), represents a competitive deactivation phenomenon, involving transient anisotropic solvent-cage effects within a protic molecular liquid (*72, 80*). Such short-lived primary events cannot be directly observed in pulse radiolysis of water but would influence early chemical steps in tracks (*81*).

Early Electron Photodetachment Steps from an Aqueous Ionic Solute

The interactions of halide ions (X^-, X = F, Br, I, Cl) with polar solvent molecules correspond to specific solvent-cage effects. In aqueous solutions, the ground state of halide ions is localized in solvent cavities and exhibits a strong absorption band in the ultraviolet. This electronic absorption spectrum is the signature of a charge transfer to solvent (CTTS), for which an electron interacts with both polar solvent molecules and the atomic parent (*36*). Within the Franck–Condon approximation, the early steps of a photoinduced electron detachment from an aqueous halide can be governed by the fluctuations of solvent density states in the vicinity of excited CTTS states or near the bottom of the conduction band of the solvent. The behavior of these excited electronic states is dependent on the energy of the ion–solvent couplings and ion–ion interactions. Three decades ago, the photochemistry of inorganic anions (halide ions) in solutions was explored using single charge-transfer reactions at the macroscopic levels, and strong evidence was found for the formation of solvated electrons through thermally activated processes (*34, 35, 82*). Microsecond and nanosecond photophysical work has demonstrated that an electron photodetachment channel would involve an early deactivation of dissociative and nondissociative excited states through ultrafast recombination reactions (*35*).

More recently, femtosecond photophysical and photochemical investigations of electron-transfer processes in ionic aqueous solutions have been performed in diluted and concentrated aqueous ionic solutions (*83–86*). These experimental advances provide guidance for quantum molecular (MD) simulations of short-lived couplings between newly created solute electronic states and solvent molecules (*87–89*). The femtosecond spectroscopy of charge-transfer processes in ionic solutions represents a good tool for the investigation of elementary chemical steps at the microscopic level.

The next paragraphs focus on the most recent advances in electron photodetachment processes in aqueous ionic solutions. Interesting results on ultrafast UV–IR spectroscopy of photoexcited aqueous chloride ions are presented in Figure 5–8. A complex photokinetic model of time-resolved data has been considered and explained in detail in recent publications (*85, 86*). The primary photophysical and photochemical events triggered by one- or two-photon processes can be summarized with the following equations:

One- or two-photon excitation Cl^- and electron photodetachment:

$$(Cl^-)_{hyd} + h\nu \xrightarrow{<<50\ fs} CTTS^*\ 190 \xrightarrow{190\ fs} CTTS^* \qquad (8)$$

$$(Cl^-)_{hyd} + 2h\nu \xrightarrow{<<50\ fs} CTTS^{**} \xrightarrow{50\ fs} CTTS^* \text{ or charge transfer} \qquad (9)$$

Electronic relaxation and electron hydration:

$$(Cl^-)_{hyd} + 2h\nu \xrightarrow{<<50\ fs} (Cl^-)^*_{hyd} \xrightarrow{130\ fs} (Cl) + e^-_{prehyd} \xrightarrow{300\ fs} (e^-) \qquad (10)$$

Solvent-cage effects and electron–atom pair formation:

$$(Cl^-)_{hyd} + 2h\nu \xrightarrow{270\ fs} (Cl{:}e^-)\text{pair} \qquad (11)$$

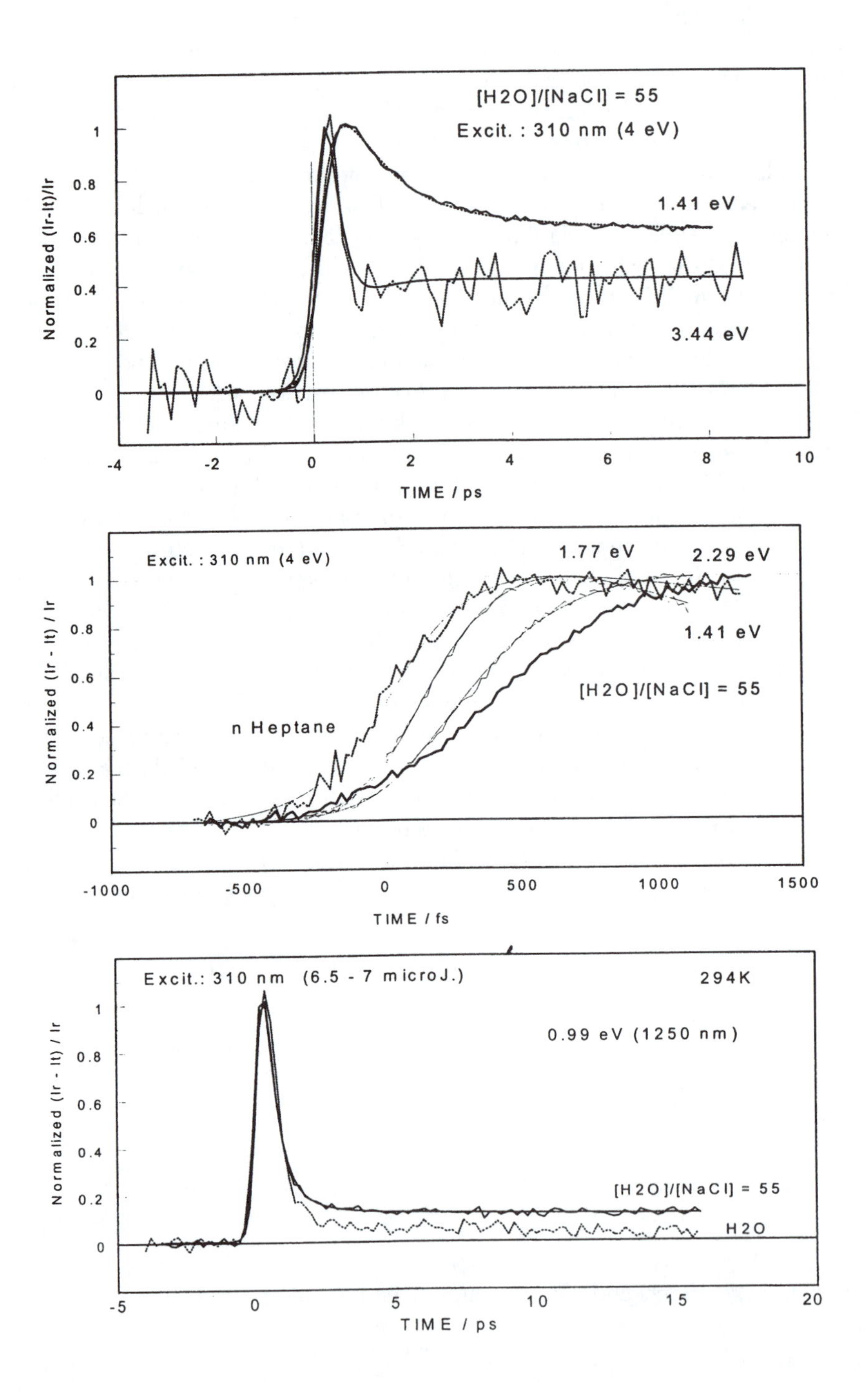
[H2O]/[NaCl] = 55
Excit. : 310 nm (4 eV)
1.41 eV
3.44 eV
Normalized (Ir-It)/Ir
TIME / ps
Excit. : 310 nm (4 eV)
1.77 eV
2.29 eV
1.41 eV
[H2O]/[NaCl] = 55
n Heptane
Normalized (Ir - It) / Ir
TIME / fs
Excit.: 310 nm (6.5 - 7 microJ.)
294K
0.99 eV (1250 nm)
[H2O]/[NaCl] = 55
H2O
Normalized (Ir - It) / Ir
TIME / ps

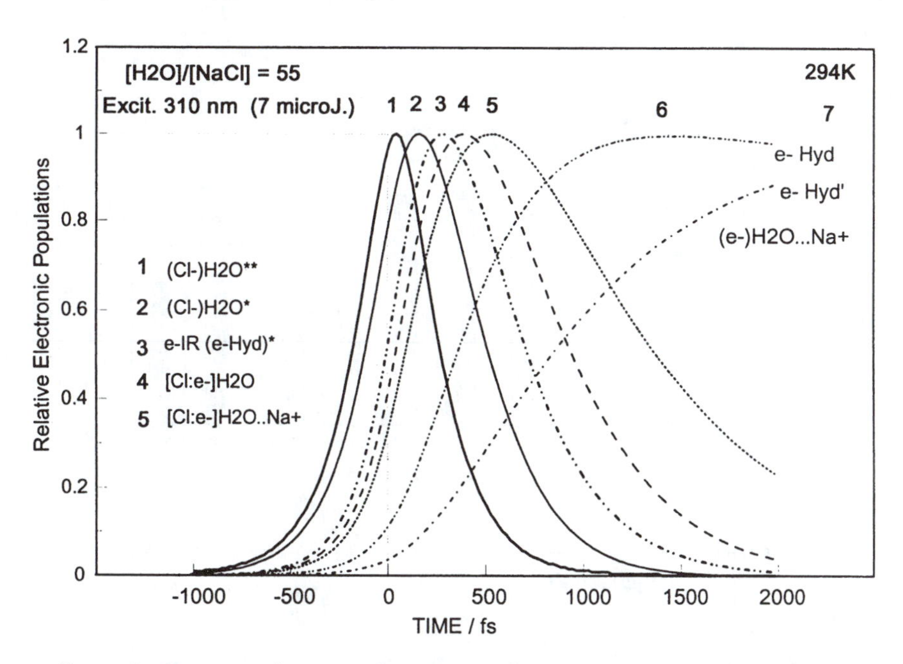

Figure 6. Dynamics of primary electron-transfer processes triggered by the femtosecond UV excitation of an aqueous sodium chloride solution ($[H_2O]/[NaCl] = 55$). The different steps of an electron photodetachment from the halide ion (Cl^-) involve charge transfer to the solvent state (1, 2), *transient electron–atom couplings* (4, 5), *and the nonequilibrium state of excess electrons* (3). *The final steps of the multiple electron photodetachment trajectories* (6, 7) are also reported. These data are obtained from time-resolved UV–IR femtosecond spectroscopic data published in references 85 and 86.

Early geminate recombination reaction:

$$(Cl) + (e^-)_{hyd} \xrightarrow{1.3 \text{ ps}} (Cl^-)_{hyd} \tag{12}$$

Ultrafast electron–atom pair deactivations:

←

Figure 5. Set of time-resolved UV–near-IR spectroscopic data (3.44–0.99 eV) following the femtosecond UV excitation of an aqueous sodium chloride solution ($[H_2O]/[NaCl] = 55$). An instrumental response of the pump-probe configuration at 1.77 eV (n-heptane) *is also shown in the middle part of the figure. The ultrashort-lived components discriminated by UV and IR spectroscopy correspond to low or high excited CTTS states (CTTS*, CTTS**), electron–atom pairs ($Cl{:}e^-$ pairs), and excited hydrated electrons (e^-_{hyd}*). The spectral signature of relaxed electronic states (ground state of a hydrated electron, (e^-_{hyd}), and electron–cation pairs, $\{Na^+{:}e^-\}_{hyd}$) are observed in the red spectral region.*

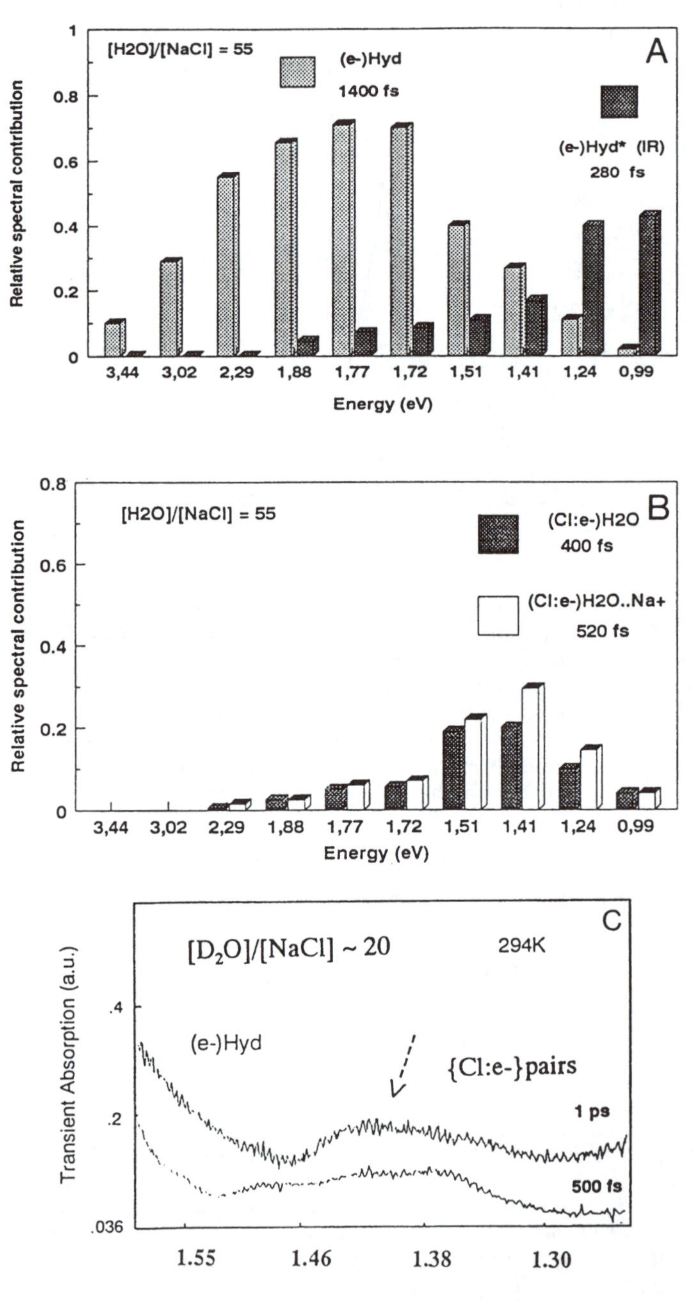

[H2O]/[NaCl] = 55
(e-)Hyd
1400 fs
A
(e-)Hyd* (IR)
280 fs
Relative spectral contribution
Energy (eV)
3,44 3,02 2,29 1,88 1,77 1,72 1,51 1,41 1,24 0,99
[H2O]/[NaCl] = 55
(Cl:e-)H2O
400 fs
B
(Cl:e-)H2O..Na+
520 fs
Relative spectral contribution
Energy (eV)
3,44 3,02 2,29 1,88 1,77 1,72 1,51 1,41 1,24 0,99
C
[D2O]/[NaCl] ~ 20
294K
(e-)Hyd
{Cl:e-}pairs
1 ps
500 fs
Transient Absorption (a.u.)
.4
.2
.036
1.55 1.46 1.38 1.30
Energy (eV)

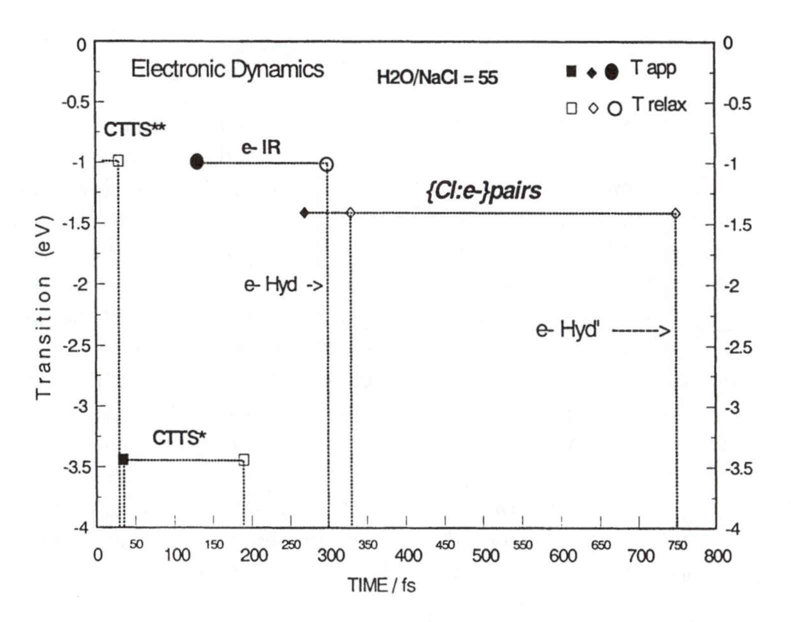

*Figure 8. Energy-level diagram of ultrafast electron-transfer processes in aqueous sodium chloride solution. Transitions (eV) correspond to experimental spectroscopic data obtained for different test wavelengths. The abscissa represents the appearance and relaxation dynamics of nonequilibrium electronic populations (CTTS**, CTTS*, $(e^-_{hyd})^*$\{Cl:e$^-$\} pairs). The two channels involved in the formation of an s-like ground hydrated electron state (e^-_{hyd}, $e^-_{hyd'}$) are also reported in the figure. From these data, it is clear that the high excited CTTS state (CTTS**) corresponds to an ultrashort-lived excited state of aqueous chloride ions preceding an electron photodetachment process.*

Figure 7. Spectral contributions of transient electronic configurations triggered by the femtosecond UV excitation of aqueous chloride ions. The relative spectral contributions are obtained from the computed analysis of time-resolved UV–IR femtosecond spectroscopic data. A: first photophysical channel, including a nonadiabatic transition from a p-like excited hydrated electron state (e^-_{hyd}) to an s-like ground hydrated electron state. B: spectral contributions of two well-defined transient \{e^-:Cl\} pairs. The presence of counterions (Na^+) influences the dual behavior of these transient electronic configurations. C: Direct identification of the spectral band assigned to near-infrared \{e^-:Cl\} pairs, made by using a cooled Optical Multichannel Analyzer (OMA 4) equipped with CCD detectors (1024 × 256 pixels).*

$$(Cl^-)_{hyd} \xleftarrow{330\ fs} (Cl{:}e^-)\ pairs \xrightarrow{750\ fs} (e^-)'_{hyd} \qquad (13)$$

Two ultrashort-lived transitions corresponding to excited CTTS states (CTTS* and CTTS**) have been identified by UV and IR spectroscopy within the first picosecond (see equations 8 and 9). Figures 5 and 6 show the time dependence of these transient states. A low-lying excited CTTS state of the chloride ion (UV CTTS*) is obtained by (1) a direct pumping in the low-energy tail of the CTTS ground state via a monophotonic process ($3p \rightarrow 4s$ transition), (2) and/or an ultrafast relaxation of infrared high excited states of the aqueous halide ion. The second transient electronic configuration of an aqueous chloride ion would correspond to a high-lying excited CTTS state (IR CTTS**). This electronic configuration is due to a two-photon process and its deactivation via an internal conversion (CTTS** $\rightarrow$ CTTS* transition), or a direct electron photodetachment occurs in 50 ± 20 fs (*85*). In agreement with recent quantum molecular dynamic simulations of two-photon excitation of an aqueous halide (I^-) (*87*), femtosecond UV and IR spectroscopy demonstrate that the transient absorption spectra of the lower and higher excited CTTS states of Cl^- are well separated by about 2 eV (*85*).

The time-dependent energy-level diagram of Figure 8 emphasizes that deactivation of the infrared high excited CTTS states (CTTS**) can compete with ultrafast electron photodetachment channels (*85, 86*). Experimental time-resolved IR spectroscopy of aqueous sodium chloride solutions underlines the existence of two-electron photodetachment channels (Figures 5 and 6). Regarding the first IR channel tested at 0.99 eV, the time scale for internal conversion of an excited hydrated electron (infrared e^-_{prehyd}) is 300 ± 20 fs (equation 10). As in pure liquid water, this electron hydration channel represents a nonadiabatic relaxation of (e^-_{prehyd}) toward the s-like ground state of the hydrated electron (e^-_{hyd}). The spectral contribution of this fully relaxed state is at a maximum for $\tau = 1400$ fs, and it exhibits a broadband centered at 1.77 eV (*90*). As previously shown by pulse radiolysis and photolysis experiments (*58, 60*), this band is asymmetric and contains a high-energy tail in the UV spectral region (Figure 2). The low-energy tail of this broadband overlaps the high-energy component of the infrared prehydrated electron (Figure 7). This short-lived nonequilibrium hydrated state presents a maximum at 280 fs, and its spectral band peaks above 1 eV. Similar femtosecond spectroscopic results have been obtained in pure liquid water by using three-dimensional spectral representation (*90*). This femtosecond IR photophysical channel characterized by the photodetachment of an excess electron and its subsequent $p \rightarrow s$ deactivation can represent a general electron solvation process in pure water and aqueous ionic solutions.

The second electron photodetachment channel occurring from excited electronic states of an ionic solute has been discriminated by femtosecond near-infrared spectroscopy (1.24–1.41 eV). This channel is characterized by the presence of two transient subbands peaking around 1.41 eV and wholly built in 500 fs (Figure 7). The estimates of these spectral bands are confirmed by using an Optical Multichannel Analyzer (OMA4) equipped with a cooled charge-control device (CCD) detector (1024 × 256 pixels). On the low-energy tail of a fully hydrated electron, a transient hump peaking in the near-infrared (850–950 nm) is clearly observed at two different times: $\tau \sim 500$ fs and 1 ps. These near-IR bands are assigned to transient inhomogeneous populations of electron–chlorine atom pairs ($\{Cl{:}e^-\}{:}Na^+$, $\{Cl{:}e^-\}\cdots Na^+$), which lead either to a complete electron photodetachment (formation of a polaron-like state, $\{Na^+{:}e^-\}_{hyd}$) or to an ultrafast geminate electron–atom recombination (equations 11 and 13). The two deactivation processes of these transient electronic states have been assigned to different solvent-cage effects in the vicinity of electron–chlorine atom pairs (*85, 86*). The electron localization channel yielding the blue-shifted polaron-like state ($\{Na^+{:}e^-\}_{hyd}$) would be governed by short-range effects of the alkali ion (Na^+) and by dynamical responses of solvent molecules to a sudden change of charge repartition. The existence of selective H–D isotope effects on the subpicosecond $\{Cl{:}e^-\}{:}Na^+ \xrightarrow{\tau} \{Na^+{:}e^-\}_{hyd}$ transition ($\tau_D/\tau_H = 2.27$) argues for a direct effect of intramolecular vibrational modes (OH, OD) on early electron-transfer trajectories in ionic solutions (*86*). The dynamics of this delayed electron hydration channel are likely controlled by an adiabatic relaxation of near-infrared electron–chlorine atom pairs (Figures 6–8).

The existence of branching between short-lived electron photodetachment pathways (metastable electron–atom pairs) has also been investigated by quantum MD simulations of excited aqueous halides (*87–89*). Transient electron–atom pairs ($\{e^-{:}Cl\}_{n'H_2O}$) can be characterized by bound–bound transitions below the conduction band of the liquid and would implicate partial electronic distribution between the chlorine atom and water molecules within the first solvation shells (*89*). In the framework of recent developments in quantum MD simulations of excited iodide and chloride ions in water, electron–atom pairs are understood as metastable states whose lifetimes are estimated to be less than a few picoseconds (*87–89*). In aqueous sodium chloride solution, water motion or fluctuations in density states within the first solvation shells of the halide can promote either a confinement of the electron–atom pair or an interconversion mechanism leading to the ground state of a hydrated electron (*86, 89*). With the treatment of solvent electronic polarization around chloride ions, quantum MD simulations of an electron photodetachment from a 3p → 4s transition of Cl^- emphasize that electron–atom pairs are metastable states for which the eigenstates of the photodetached electron extend less than 8 Å from the atomic core (*87, 89*). This radius of gyration is compatible with a distance of 6–7 Å, for which the first solvation shells "remember" the presence of polarizable Cl^- (*91*). A branching from electron–Cl pairs would lead either to an electron hydration through an adiabatic electron detachment trajectory (*85, 86, 89*) or to a picosecond geminate recombination (*85–89*).

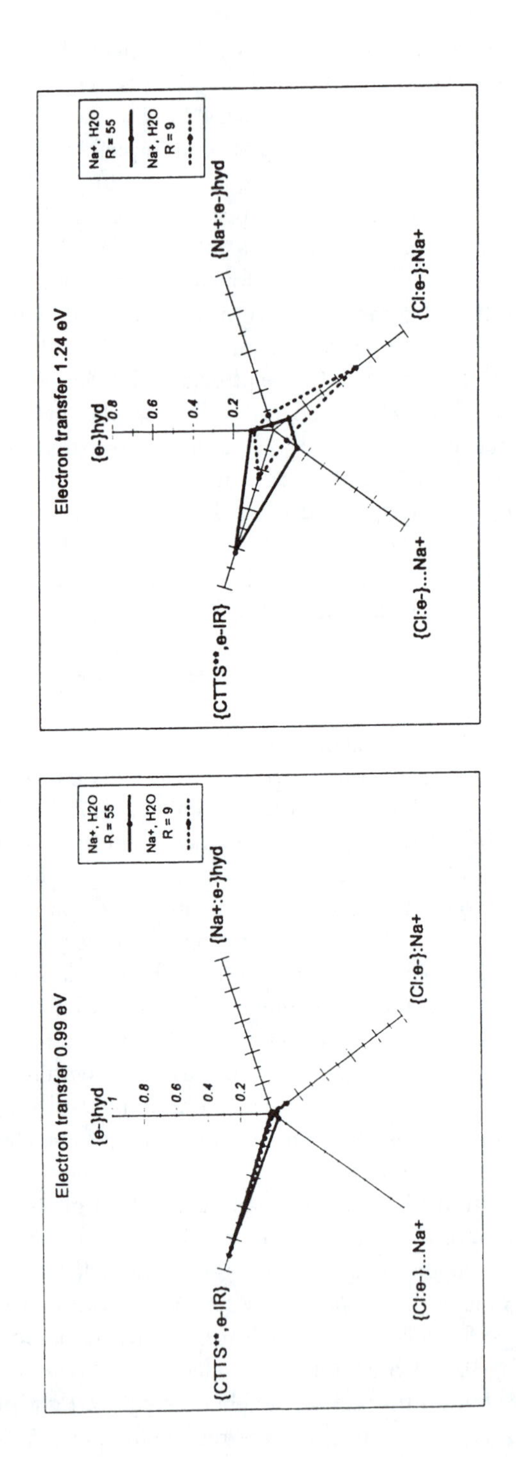
Electron transfer 1.24 eV
Na+, H2O R = 55
Na+, H2O R = 9
{e-}hyd
0.8
0.6
0.4
0.2
{Na+:e-}hyd
{Cl:e-}:Na+
{Cl:e-}...Na+
{CTTS**,e-IR}
Electron transfer 0.99 eV
Na+, H2O R = 55
Na+, H2O R = 9
{e-}hyd
1
0.8
0.6
0.4
0.2
{Na+:e-}hyd
{Cl:e-}:Na+
{Cl:e-}...Na+
{CTTS**,e-IR}

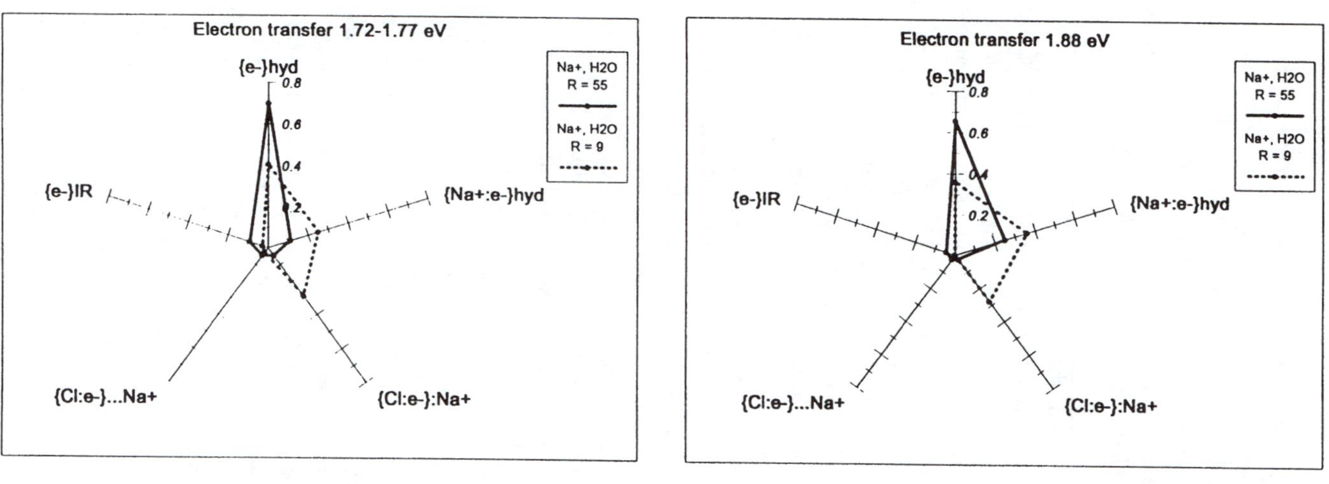

Figure 9. Influence of ionic strength, ⟨⟨R⟩⟩ *(molecular ratio), on the relative spectral contributions of femtosecond photoinduced electron-transfer processes in aqueous sodium chloride solutions. The ionic strength is defined by the molecular ratio,* R, *which equals* $[H_2O]/[XC1]$. *The different test wavelengths (0.99, 1.24, 1.72, 1.77, and 1.88 eV) permit the discrimination of transient electronic states (CTTS**,* e^-_{IR}, *electron–atom pairs* $\{Cl{:}e^-\}{:}Na^+$) *and two configurations of the hydrated electron ground state* ($\{e^-\}_{hyd}$). *In concentrated ionic solution* (R = 9), *an electron photodetachment channel favors the formation of polaron-like states* ($\{Na^+{:}e^-\}_{hyd}$) (see *reference 86*).

Counterion Effects and Ultrafast Electron Transfers

Aqueous ionic solutions represent a paradigm for the study of early branching between ultrafast nonadiabatic and adiabatic electron transfers. The very recent experimental observations of specific counterion effects on electronic dynamics provide direct evidence of complex influences of inhomogeneous ion–ion distributions on ultrafast electron-transfer processes. These microscopic effects are particularly evident in IR electronic trajectories in sodium chloride solution (*86, 92*).

An important aspect concerns the influence of solvation shells on the total solvation energy of anions, cations, and excess electrons. Monte Carlo calculations and MD simulations of ionic solutions have established that the microscopic structure of ion hydration shells and the solvent properties are connected to each other (*93–97*). In aqueous sodium chloride solutions, the rate constant for the transition between contact and solvent-separated ion pairs (CIP–SSIP states) has been estimated to be 50–200 ps (*94*). The lifetime of aqueous $\{Cl^-: Na^+\}$ pairs (CIP–SSIP states) is long enough for an electron photodetachment from transient electron–chlorine atom pairs to take place without a significant temporal change of the mean force potential profile of $\{Cl^-:Na^+\}$ pairs (W_r). Consequently, the counterion effects on ultrafast electron transfers in solution cannot be discussed only in the framework of ion-pair dynamics but more likely in connection with the ion–solvent correlation function, short-range ordering of water molecules, and anisotropic electric field effects (*98–100*). MD simulations of aqueous sodium chloride solutions have shown that the dynamics of water molecules are dependent on the angular distribution of solvent within the internal solvation shells entrapped between Na^+ and Cl^- (solvent bridge bonding) and the external solvation shells (*98*).

Femtosecond pump-probe UV–IR spectroscopy of electron transfer in ionic solutions allows us to establish the role of inhomogeneous microscopic structures on early branchings of short-lived trajectories. The short-range effects of the alkali ion (Na^+) on early electron photodetachment channels would be governed by the dynamic response of water molecules to a change of charge distribution in the vicinity of the anion. These cationic effects are particularly important during the relaxation of short-lived IR hydrated electrons (*72*) and near-IR electron–atom pairs (*86*). Figure 9 shows that the charge-switching process between the Cl atom (ultrafast electron–chlorine atom recombination) and the counterion (polaron-like state) is dramatically sensitive to the molecular ratio, *R*, of the ionic solution (equation 15). In concentrated aqueous NaCl solution ($R = 9$), the near-IR contribution of $\{Cl{:}e^-\}\cdots Na^+$ pairs is decreased contrary to the relative weight of the $\{Cl{:}e^-\}{:}\ Na^+ \xrightarrow{\tau} \{Na^+{:}e^-\}_{hyd}$ transition. Considering an interconversion mechanism between different electron–atom pairs, the attractive modes would be involved in ultrafast, barrierless, electron–atom reactions, and the repulsive modes in an activated electron–atom separation with subsequent adiabatic formation of the ground state of the hydrated electron. Femtosecond near-IR spectroscopy demonstrates that a change in the ion–ion distribution in concentrated aqueous NaCl solutions increases the electric field effect of Na^+ on the electron detachment from transient electron–atom pairs. This charge transfer corresponds to an adiabatic electron photodetachment and leads to a polaronlike entity, that is, a fully relaxed electron within the solvation shell of the sodium ion ($e^-_{hyd}\cdots Na^+$). This electronic ground state remains very similar to a solvent-separated electron–ion pair and exhibits a broad absorption band in the visible (*86, 92*). The ion–ion pair distribution in ionic solutions can be influenced by the molecular ratio or a change of counterion valence.

Figure 10 shows the significant spectroscopic differences that we have observed in aqueous NaCl and $MgCl_2$ solutions. For the same Cl^- concentration (1 equiv), the effects of counterions on the global signal rise time at 1.77 eV are fastest in aqueous $MgCl_2$ solution. This difference does not correspond to a direct dynamical effect but to a balance between the spectral contributions of the hydrated electron and the $\{Cl{:}e^-\}{:}Mg^{++} \xrightarrow{T} \{Mg^{++}{:}e^-\}_{hyd}$ transition. The complex effects of counterion valence on elementary one-electron transfer reactions will be discussed in forthcoming papers (*101*). In this way, counterion effects in polar solutions represent an interesting challenge for quantum simulations of reaction dynamics in dissipative condensed media.

Conclusions

The investigation, at the microscopic level, of solvent-cage effects on elementary chemical processes in polar liquids is fundamental for an understanding of ion–molecule reactions, electron–ion or ion–ion pair interconversions, adiabatic or nonadiabatic electron-transfer processes, and primary recombination mechanisms. Figure 11 illustrates some fundamental aspects of radical ion pairs in connection with SN_1 ionization reactions. In this way, multiple elementary steps such as transient CIP and interconversion between CIP and SSIP pairs would share some similarities with ultrafast electron-transfer processes discussed in this chapter. In aqueous sodium chloride solution, the formation of radical ion pairs $\{(Na^+)_{nH2O}\ (e^-)\}$ occurs in less than 4.5×10^{-12} s. Within this temporal window, transient electron-transfer states such as $\{e^-{:}Cl\}_{nH2O}$ pairs have been discriminated. The complete electron detachment from the chlorine atom to water molecules or a hydrated cationic atmosphere needs to cross a small free energy of activation barrier. The direct characterization of ultrafast electron–atom reactions in a polar liquid provides a further basis for (1) the investigation of ultrashort-lived solvent-cage effects at the microscopic level, (2) a better understanding of branching processes during ultrafast electron-transfer reactions, and (3) a knowledge of the role of the reorientational correlation function of solvent molecules around semi-ionized states and metastable electronic states in polar solutions.

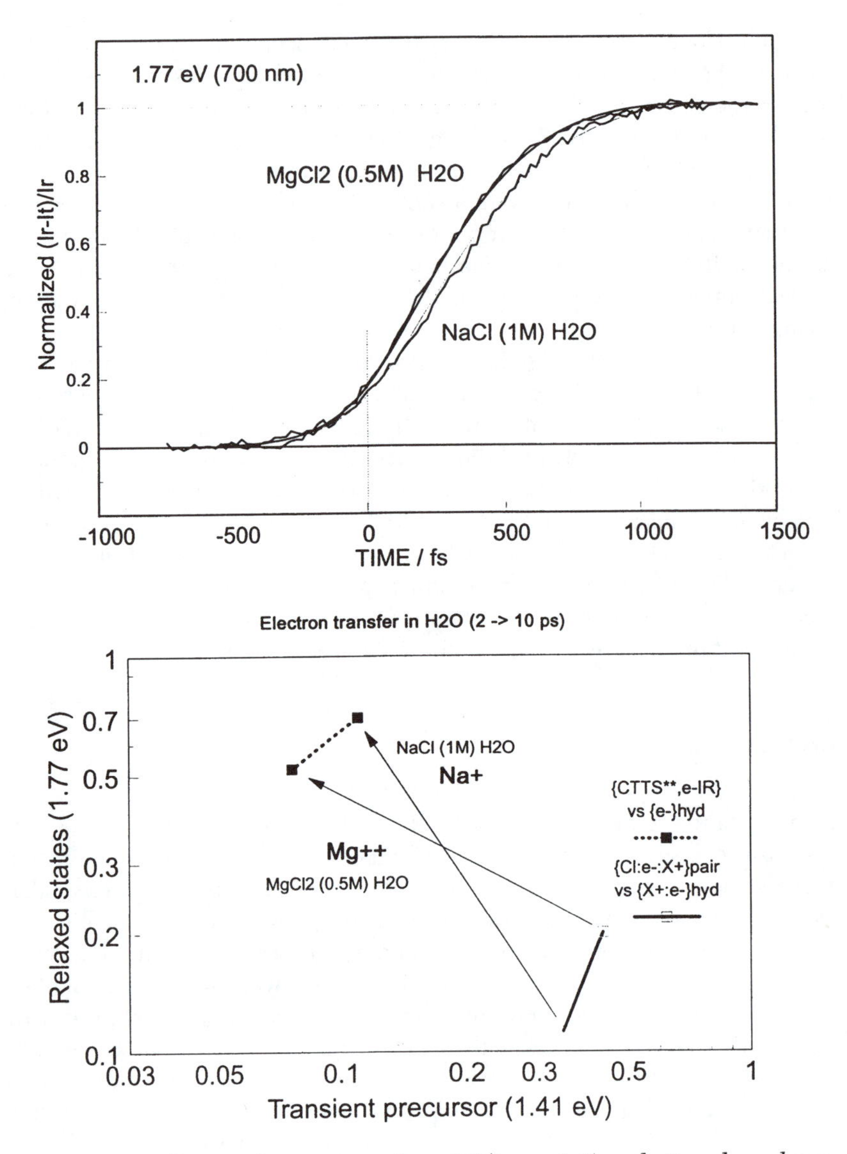

Figure 10. Influence of counterion valence (X^{n+}, n = 1, 2) on the time dependence of femtosecond, photoinduced, electron-transfer trajectories in aqueous ionic solutions (X^n, nCl^-, $X = Na^+$, Mg^{++}). The upper part represents the absorption signal rise time at 1.77 eV following the femtosecond UV excitation of aqueous Cl^- *(2 × 4 eV). The difference in the signal rise times of Na^+ and Mg^{++} is due to the balance between two electronic transitions: $e^-_{IR} \rightarrow \{e^-\}_{hyd}$ and $\{Cl{:}e^-\}{:}X^{n+} \rightarrow \{e^-{:}X^{n+}\}_{hyd}$. The second electronic channel is more efficient in the presence of a divalent cation (Mg^{++}). The relationships between the relative spectral contributions of near-IR transient states and equilibrium electronic states in the visible are represented in the lower part of the figure.*

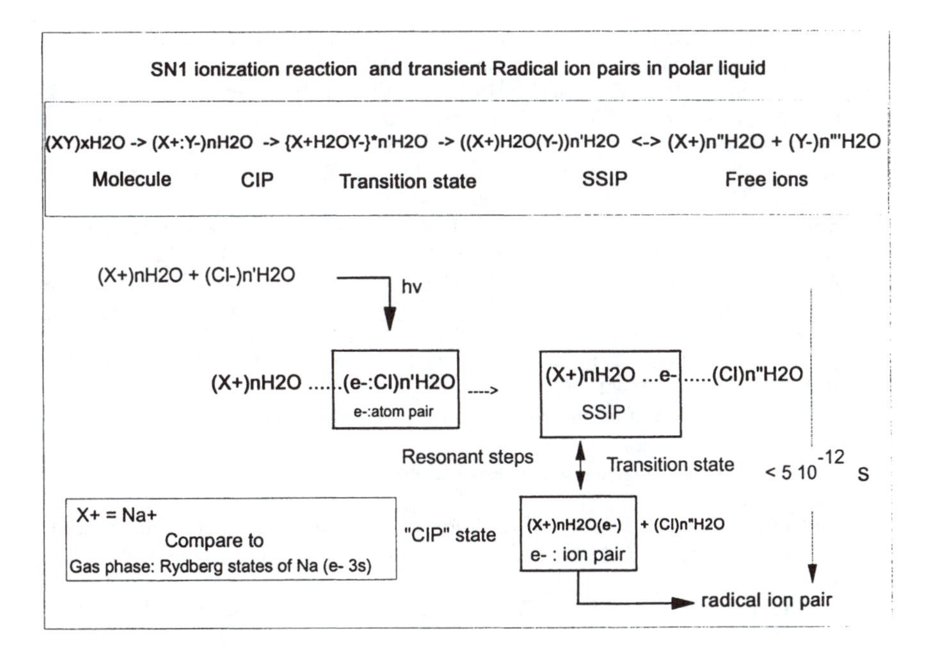

Figure 11. Schematic representation of sequential events of an SN_1 ionization reaction in a polar liquid. Elementary events involve contact ion pairs (CIP) and solvent-separated ion pairs (SSIP). In ionic aqueous solutions, the influence of different ion-pair configurations on early electron-transfer trajectories can be considered through the investigation of ultrafast electronic dynamics and radical ion-pair formation.

Acknowledgments

This work is supported by grants from the Chemical Department of the Centre National de la Recherche Scientifique, GDR No. 1017 (France), and the Commission of the European Communities. We thank S. Bratos, D. Borgis, and A. Staib for fruitful discussions before the preparation of this manuscript.

References

1. Hart, E. J.; Anbar, M. In *The Hydrated Electron;* Wiley Intersciences: New York, 1970.
2. *The Study of Fast Processes and Transient Species by Electron Pulse Radiolysis;* Baxendale, J. H.; Busi, F., Eds.; Nato Adv. Study Inst. Series., V. 86; D. Reidel; Dordrecht, Netherlands, 1982.
3. *Radiation Chemistry;* Farhataziz, Rodgers, M. A. J., Eds.; VCH: New York, 1987.
4. "The Chemical Physics of Solvation", Parts A, B, C. In *Studies in Physical and Theoretical Chemistry;* Dogonadze, R. R.; Kalman, E.; Kornyshev, A. A.; Ulstrup, J., Eds.; Elsevier: New York, 1988; Vol. 38.

5. Mataga, N.; Hirata, Y. In *Advances in Multiphoton Processes and Spectroscopy;* Lin, S. H., Ed.; World Scientific: London, 1989; Vol. 5.
6. *Applications of Time-Resolved Optical Spectroscopy;* Bruckner, V.; Feller, K. H.; Grummt, U. W., Eds.; Elsevier: New York, 1990; Su, S.; Simon, J. J. *J. Phys. Chem.* **1989,** *93,* 753.
7. Wynne, K.; Hochstrasser, R. M. *Chem. Phys.* **1995,** *193,* 211–236; Richert, R.; Wagner, H. *J. Phys. Chem.* **1995,** *99,* 10948.
8. Mukamel, S. *Ann. Rev. Phys. Chem.* **1990,** *41,* 647 and references therein; *see also* the special issue of *J. Opt. Soc. Am.* **1990,** *B7;* Potter, E. D.; Liu, Q.; Zewail, A. H. *Chem. Phys. Lett.,* **1992,** *200,* 605; Wang, N.; Chernyak, V.; Mukamel, S. *J. Chem. Phys.* **1994,** *100,* 2465; *Principles of Nonlinear Optical Spectroscopy;* Mukamel, S., Ed.; Oxford University Press: Oxford, England, 1995.
9. Turner, J. E.; Magee, J. L.; Weight, H. A.; Chatterjee, A.; Hamm, R. N.; Ritchie, R. H. *Radiat. Res.* **1983,** *96,* 437.
10. Klassen, N. V. In *Radiation Chemistry;* Farhataziz; Rodgers, M. A. J., Eds.; VCH: New York, 1987; p 29.
11. *Kinetics of Nonhomogeneous Processes;* Freeman G. R., Ed.; Wiley Interscience: New York, 1987.
12. *Radiation Research, A Twentieth-Century Perspective;* Dewey, W.; Edington, M.; Fry, R. J. M.; Hall, E. J.; Whitmore, G. F., Eds.; Academic: Orlando, FL, 1992; Vol. II.
13. Lin, Y.; Jonah, C. D. *J. Phys. Chem.* **1992,** *96,* 10119; *Chem. Phys. Lett.* **1992,** *191,* 357.
14. Guaduel, Y.; Migus, A.; Martin, J. L.; Antonetti, A. *Chem. Phys. Lett.* **1984,** *108,* 319; Gauduel, Y.; Migus, A.; Martin, J. L.; Lecarpentier, Y.; Antonetti, A. *Ber. Bunsen Ges. Phys. Chem.* **1985,** *89,* 218; Gauduel, Y.; Migus, A.; Antonetti, A. In *Chemical Reactivity in Liquids; Fundamental Aspects;* Moreau, M.; Turcq, P., Eds.; Plenum: New York, 1988; Vol 15; Gauduel, Y.; Berrod, S.; Migus, A.; Yamada, N.; Antonetti, A. *Biochemistry* **1988,** *27,* 2509; Gauduel, Y.; Pommeret, S.; Yamada, N.; Migus, A.; Antonetti, A. *J. Am. Chem. Soc.* **1989,** *111,* 4974.
15. Asahi, T.; Mataga, N. *J. Phys. Chem.* **1991,** *95,* 1956.
16. Barbara, P. F.; Jarzeba, W. In *Advances in Photochemistry;* Volman, D. H.; Hammond, G. S.; Gollnick, K., Eds.; Wiley Interscience: New York, 1989; Vol. 15, p 1; Kahlow, M. A.; Jarzeba, W.; Kang, T. J.; Barbara, P. F. *J. Chem. Phys.* **1989,** *90,* 151.
17. In *Structure and Dynamics of Solutions;* Ohtaki, H.; Yamatera, H., Eds.; Studies in Physical and Theoretical Chemistry, Vol. 79; Elsevier: New York, 1992; Part 4.
18. Long, F. H.; Lu, H.; Shi, X.; Eisenthal, K. B. *J. Chem. Phys.* **1989,** *91,* 4103; *Chem. Phys. Lett.* **1990,** *169,* 165.
19. Cho, M.; Scherer, N. F.; Fleming, G. R.; Mukamel, S. *J. Chem. Phys.* **1992,** *96,* 5618.
20. Duppen, K.; de Haan, F.; Nibbering, E. T. J.; Wiersma, D. A. *Phys. Rev. A* **1993,** *47,* 5120.
21. Diels, J. C.; Rudolph, W. *Ultrashort Laser Pulse Phenomena;* Academic: Orlando, FL, 1996.
22. Becker, P. C.; Fragnito, H. L.; Bigot, J. Y.; Brito Cruz, C. H.; Fork, R. L.; Shank, C. V. *Phys. Rev. Lett.* **1989,** *63,* 505.
23. Nibbering, E. J.; Wiersma, D. A.; Duppen, K. *Phys. Rev. Lett.* **1991,** *66,* 2462.
24. Maroncelli, M.; Fleming, G. R. *J. Chem. Phys.* **1988,** *89,* 875; Papazyan, A.; Maroncelli, M. *J. Chem. Phys.* **1993,** *98,* 6431; Horng, M. L.; Gardecki, J. A.; Papazyan, A.; Maroncelli, M. *J. Phys. Chem.* **1995,** *99,* 17311.
25. *Ultrafast Reaction Dynamics and Solvent Effects;* Gauduel, Y.; Rossky, P. J., Eds.;

AIP Conference Proceedings, Vol. 298, American Institute of Physics: New York, 1994.

26. Hynes, J. T. *Ann. Rev., Phys. Chem.* **1985,** *36,* 573; Kim, H. J.; Hynes, J. T. *J. Phys. Chem.* **1990,** *94,* 2736; Keirstead, W. P.; Wilson, K. R.; Hynes, J. T. *J. Chem. Phys.* **1991,** *95,* 5256; Van der Zwan, G.; Hynes, J. T. *Chem. Phys.* **1991,** *152,* 169.
27. Fonseca, T.; Ladanyi, B. *J. Phys. Chem.* **1991,** *95,* 2116; *J. Mol. Liq.* **1994,** *60,* 1; Ladanyi, B. M.; Stratt, R. M. *J. Phys. Chem.* **1995,** *99,* 2502; **1996,** *100,* 1266.
28. Barnett, R. N.; Landman, U.; Cleveland, C. L.; Jortner, J. *J. Chem. Phys.* **1988,** *88,* 4421; **1988,** *88,* 4429; Barnett, R. N.; Landman, U.; Nitzan, A. *J. Chem. Phys.* **1990,** *93,* 6535.
29. Rossky, P. J. *J. Opt. Soc. Am.* **1990,** *B7,* 1727; Webster, F.; Rossky, P. J.; Friesner, R. A. *Comp. Phys. Com.* **1991,** *63,* 494; Webster, F.; Schnitker, J.; Friedrich, M. S. M.; Friesner, R. A.; Rossky, P. J. *Phys. Rev. Lett.* **1991,** *66,* 3172.
30. Bagchi, B. *Ann. Rev. Phys. Chem.* **1989,** *40,* 115; Bagchi, B.; Chandra, A. *Adv. Chem. Phys.* **1991,** *80,* 1; Roy, S.; Bagchi, B. *J. Chem. Phys.* **1993,** *99,* 1310, 3139, 9938; Roy, S.; Bagchi, B. *Chem. Phys.* **1994,** *183,* 207; Nandi, N.; Roy, S.; Bagchi, B. *J. Chem. Phys.* **1995,** *102,* 1390; Ravichandran, S.; Roy, S.; Bagchi, B. *J. Phys. Chem.* **1995,** *99,* 2489; Biswas, R.; Bagchi, B. *J. Phys. Chem.* **1996,** *100,* 1238.
31. Rips, I.; Klafter, J.; Jortner, J. *J. Chem. Phys.* **1988,** *88,* 3246; Stell, G.; Zhou, Y. *J. Chem. Phys.* **1989,** *91,* 4869, 4879, 4885; Rips, I.; Pollack, E. *J. Chem. Phys.* **1995,** *103,* 7912; Rips, I. *Chem. Phys. Lett.* **1995,** *245,* 79; *J. Chim. Phys.* **1996,** *93,* 1591; Yoshihara, K.; Tominaga, K.; Nagasawa, Y. *Bull. Chem. Soc. Jpn.* **1995,** *68,* 696.
32. Castner, E. W.; Maroncelli, M.; Fleming, G. R. *J. Chem. Phys.* **1987,** *86,* 1090; Castner, E. W.; Fleming, G. R.; Bagchi, B.; Maroncelli, M. *J. Chem. Phys.* **1988,** *89,* 3519.
33. Cole, R. H. *Ann. Rev. Phys. Chem.* **1989,** *40,* 1 and references therein.
34. Grossweiner, L. I.; Matheson, M. S. *J. Phys. Chem.* **1957,** *61,* 1089; Matheson, M. S.; Mulac, W. A.; Rabani, J. *J. Phys. Chem.* **1963,** *67,* 2613.
35. Jortner, J.; Ottolenghi, M.; Stein, G. *J. Phys. Chem.* **1964,** *68,* 247; Ottolenghi, M. *Chem. Phys. Lett.* **1971,** *12,* 339.
36. Blandamer, M. J.; Fox, M. F. *Chem. Rev.* **1969,** *70,* 59.
37. Sheu, W. S.; Rossky, P. J. *J. Am. Chem. Soc.* **1993,** *115,* 7729.
38. *Chemical Reactivity in Liquids; Fundamental Aspects;* Moreau, M.; Turcq, P., Eds.; Plenum: New York, 1988.
39. *Ultrafast Dynamics of Chemical Systems;* Simon, J. D., Ed.; Kluwer: Dordrecht, Netherlands 1994.
40. *Femtosecond Chemistry;* Manz, J.; Woste, J. L., Eds.; VCH: Heidelberg, Germany, 1995.
41. *Computer Simulation of Liquids;* Allen, M. P.; Tildesley, D. J., Eds.; Oxford Science: Oxford, England, 1987.
42. Newton, M. D. *Chem. Rev.* **1991,** *91,* 767.
43. Marcus, R. A. *J. Chem. Phys.* **1956,** *24,* 966; Marcus, R. A. *Ann. Rev. Phys. Chem.* **1964,** *15,* 155; Sumi, H.; Marcus, R. A. *J. Chem. Phys.* **1986,** *84,* 4894; Marcus, R. A. *Rev. Mod. Phys.* **1993,** *65,* 599.
44. Joachim, C. *Chem. Phys.* **1987,** *116,* 339; Reimers, J. R.; Hush, N. S. *Chem. Phys.* **1990,** *146,* 89 and references therein; Kosloff, R.; Ratner, M. A.; *Isr. J. Chem.* **1990,** *30,* 45.
45. Southland, P. O.; Dyer, R. B.; Woodruff, W. H. *Science (Washington, D.C.)* **1992,** *257,* 1913.
46. *See* a recent Special Issue "Elementary Chemical Processes in Liquids and Solutions" of *J. Chim. Phys.* **1996,** *93,* 1577–1938.

47. Hart, E. J.; Boag, J. W. *J. Am. Chem. Soc.* **1962,** *84,* 4090.
48. Brodsky, A. M.; Tsarevsky, A. V. *Adv. Chem. Phys.* **1980,** *44,* 483.
49. Lam, K. Y.; Hunt, J. W. *Int. Radiat. Phys. Chem.* **1975,** *7,* 317; Okazaki, K.; Freeman, G. *Can. J. Chem.* **1978,** *56,* 2313; Kondo, Y.; Aikawa, M.; Surniyoshi, T.; Katayama, M.; Kroh, J. *J. Phys. Chem.* **1984,** *80,* 2544.
50. Walrafen, G. E.; Hokmabadi, M. S.; Yang, W. H. *J. Chem. Phys.* **1986,** *85,* 6964.
51. Sasai, M. *J. Chem. Phys.* **1990,** *93,* 7329.
52. Mizoguchi, K.; Tominaga, Y. *J. Chem. Phys.* **1992,** *97,* 1961.
53. Ohmine, I.; Tanaka, H. *Chem. Rev.* **1993,** *93,* 2545.
54. Cho, M.; Fleming, G. R.; Saito, S.; Ohmine, I. *J. Chem. Phys.* **1994,** *100,* 6672.
55. Chandler, D.; Leung, K. *Annu. Rev. Phys. Chem.* **1994,** *45,* 557.
56. Gauduel, Y. *Ultrafast Reaction Dynamics and Solvent Effects;* Gauduel, Y.; Rossky, P. J., Eds.; AIP Conference Proceedings, Vol. 298; American Institute of Physics: New York, 1994; pp 191–204.
57. Wiesenfeld, J. M.; Ippen, E. M. *Chem. Phys. Lett.* **1980,** *73,* 47.
58. Nikogosyan, D. N.; Oraevsky, A. O.; Rupasov, V. I. *Chem. Phys.* **1983,** *77,* 131.
59. Anbar, M.; Meyerstein, D. *Trans. Faraday Soc.* **1966,** *62,* 2121.
60. Jou, F. Y.; Freeman, G. R. *J. Phys. Chem.* **1979,** *83,* 2383.
61. Gauduel, Y.; Martin, J. L.; Migus, A.; Antonetti, A. In *Ultrafast Phenomena V;* Fleming, G. R.; Siegman, Eds.; Springer Verlag: New York, 1986; Vol. 308; Migus, A.; Gauduel, Y.; Martin, J. L.; Antonetti, A. *Phys. Rev. Lett.* **1987,** *108,* 318.
62. Long, F. H.; Lu, H.; Eisenthal, K. B. *Chem. Phys. Lett.* **1989,** *160,* 464; *Phys. Rev. Lett.* **1990,** *64,* 1469; Long, F. H.; Lu, H.; Shi, X.; Eisenthal, K. B. *Chem. Phys. Lett.* **1991,** *185,* 47.
63. Alfano, J. C.; Walhout, P. W.; Kimura, Y.; Barbara, P. F. *J. Chem. Phys.* **1993,** *98,* 5996; Kimura, Y. *J. Phys. Chem.* **1994,** *98,* 3450.
64. Sander, M. U.; Luther, K.; Troe, J. *J. Phys. Chem.* **1993,** *97,* 11489; McGowen, J. L.; Ajo, H. M.; Hang, J. Z.; Schwartz, B. I. *Chem. Phys. Lett.* **1994,** *231,* 505.
65. Gauduel, Y.; Pommeret, S.; Migus, A.; Yamada, N.; Antonetti, A. *J. Opt. Soc. Am., B* **1990,** *7,* 1528.
66. Reuther, A.; Laubereau, A.; Nikogosyan, D. N. *J. Phys. Chem.* **1996,** *100,* 16794.
67. Webster, F.; Wang, E. T.; Rossky, P. J.; Friesner, R. A. *J. Chem. Phys.* **1994,** *100,* 4835.
68. Romero, C.; Jonah, C. D. *J. Chem. Phys.* **1989,** *90,* 1877.
69. Reid, P. J.; Silva, C.; Walhout, P. K.; Barbara, P. F. *Chem. Phys. Lett.* **1994,** *228,* 658.
70. Schwartz, B. J.; Rossky, P. J. *Phys. Rev. Lett.* **1994,** *72,* 3282; *J. Chem. Phys.* **1994,** *101,* 6902, 6917; *J. Phys. Chem.* **1995,** *99,* 2953; Schwartz, B. J.; Rossky, P. J. *J. Chem. Phys.* **1996,** *105,* 6997.
71. Bratos, S.; Leicknam, J. C. *Chem. Phys. Lett.* **1996,** *261,* 117; *J. Chim. Phys.* **1996,** *93,* 1737.
72. Gauduel, Y.; Pommeret, S.; Migus, A.; Yamada, N.; Antonetti, A. *J. Am. Chem. Soc.* **1990,** *112,* 2925; Pommeret, S.; Antonetti, A.; Gauduel, Y. *J. Am. Chem. Soc.* **1991,** *113,* 9105; Gauduel, Y.; Pommeret, S.; Antonetti, A. *J. Phys. Chem.* **1993,** *97,* 134; Gauduel, Y. *J. Mol. Liq.* **1995,** *63,* 1.
73. Muguet, F. *J. Mol. Struct.* (*Theochem.*) **1996,** *368,* 173; Muguet, F.; Gelabert, H.; Gauduel, Y. *J. Chim. Phys.* **1996,** *93,* 1808.
74. Halle, B.; Karlstrom, G. *J. Chem. Soc. Faraday Trans.* **1983,** *70,* 1047.
75. Barnett, R. B.; Landman, U.; Nitzan, A. *J. Chem. Phys.* **1989,** *90,* 4413.
76. Zhu, S. B.; Zhu, J. B.; Robinson, G. W. *Phys. Rev. A* **1991,** *44,* 2602.
77. Sprik, M.; Klein, M. L. *J. Chem. Phys.* **1988,** *89,* 7556; Sprik, M. *J. Phys. Chem.* **1991,** *95,* 2283; Laasonen, K.; Sprik, M.; Parrinello, M.; Car, R. *J. Chem. Phys.*

1993, *99,* 9080; Tuckerman, M.; Laasonen, K.; Sprik, M.; Parrinello, M. *J. Phys. Chem.* **1995,** *99,* 5749; *J. Chem. Phys.* **1995,** *103,* 150.
78. Trokhymchuk, A. D.; Holovko, M. F.; Heinzinger, K. *J. Chem. Phys.* **1993,** *99,* 2964; Liu, K.; Loeser, J. G.; Elrod, M. J.; Host, B. C.; Rzepiela, J. A.; Pugliano, N.; Saykally, R. J. *J. Am. Chem. Soc.* **1994,** *116,* 3507; Sremaniak, L. S.; Perera, L.; Berkowitz, M. L. *J. Chem. Phys.* **1996,** *105,* 3715.
79. Edison, A. S.; Markley, J. L.; Weinhold, F. *J. Phys. Chem.* **1995,** *99,* 8013; Jensen, J. H.; Gordon, M. S. *J. Phys. Chem.* **1995,** *99,* 8091; Kiefer, P. M.; Leite, V. B. P.; Whitnell, R. M. *Chem. Phys.* **1995,** *194,* 33; Agmon, N. *Chem. Phys. Lett.* **1995,** *244,* 456; *J. Chim. Phys.* **1996,** *93,* 1714.
80. Bhattacharya, I.; Voth, G. A. *J. Phys. Chem.* **1993,** *97,* 11253; Straus, J. B.; Calhoun, A.; Voth, G. A. *J. Chem. Phys.* **1995,** *102,* 529; Lobaugh; Voth, G. A. *J. Chem. Phys.* **1996,** *104,* 2056 and references therein.
81. Schmidt, K. H.; Han, P.; Bartels, D. M. *J. Phys. Chem.* **1995,** *99,* 10530.
82. Stein, G. *Adv. Chem. Ser.* **1965,** *50,* 230.
83. Gauduel, Y.; Migus, A.; Chambaret, J. P.; Antonetti, A. *Rev. Phys. Appl.* **1987,** *22,* 1755; Gauduel, Y.; Pommeret, S.; Migus, A.; Yamada, N.; Antonetti, A. *J. Am. Chem. Soc.* **1990,** *112,* 2925.
84. Long, F. H.; Shi, X.; Lu, H.; Eisenthal, K. B. *J. Phys. Chem.* **1994,** *98,* 7252.
85. Gauduel, Y.; Gelabert, H.; Ashokkumar, M. *Chem. Phys.* **1995,** *197,* 167.
86. Gelabert, H.; Gauduel, Y. *J. Phys. Chem.* **1996,** *100,* 13993.
87. Sheu, W. S.; Rossky, P. J. *Chem. Phys. Lett.* **1993,** *213,* 233; Sheu, W. S.; Rossky, P. J. *Chem. Phys. Lett.* **1993,** *202,* 186; *J. Phys. Chem.* **1996,** *100,* 1295.
88. Borgis, D.; Staib, A. *Chem. Phys. Lett.* **1994,** *230,* 405; Staib, A.; Borgis, A. *J. Chem. Phys.* **1995,** *103,* 2642.
89. Borgis, D.; Staib, A. *J. Chem. Phys.* **1996,** *104,* 4776; **1996,** *104,* 9027.
90. Gauduel, Y.; Gelabert, H.; Ashokkumar, M. *J. Mol. Liq.* **1995,** *64,* 57.
91. Sprik, M.; Klein, M. L.; Watanabe, K. *J. Phys. Chem.* **1990,** *94,* 6483; Sprik, M. *J. Phys. Chem.* **1991,** *95,* 2283; *J. Phys.* **1991,** *IV,* C5–99.
92. Gelabert, H.; Sander, M.; Gauduel, Y. *J. Chim. Phys.* **1996,** *93,* 1608.
93. Guardia, E.; Rey, R.; Padro, J. A. *J. Chem. Phys.* **1991,** *95,* 2823; Guardia, E.; Rey, R.; Padro, J. A. *Chem. Phys.* **1991,** *155,* 187; Guardia, E.; Padro, J. A. *J. Phys. Chem.* **1990,** *94,* 6049; Rey, R.; Guardia, E. *J. Phys. Chem.* **1992,** *96,* 4712.
94. Karim, O. A.; McCammon, A. *Chem. Phys. Lett.* **1986,** *132,* 219.
95. Ciccotti, G.; Ferrario, M.; Hynes, J. T.; Kapral, R. *J. Chem. Phys.* **1990,** *93,* 7137.
96. Impey, R. W.; Madden, P. A.; McDonald, I. R. *J. Phys. Chem.* **1983,** *87,* 5071; Engström, S.; Jönsson, B.; Impey, R. W. *J. Chem. Phys.* **1984,** *80,* 5481.
97. Caillot, J. M.; Levesque, D.; Weiss, J. J. *J. Chem. Phys.* **1989,** *91,* 5544.
98. Belch, A.; Berkowitz, M.; McCammon, J. A. *J. Am. Chem. Soc.* **1986,** *108,* 1755.
99. Hartman, R. S.; Konitsky, W. M.; Waldeck, D. H. *J. Am. Chem. Soc.* **1993,** *115,* 9692; Ueno, M.; Tsuchihashi, N.; Yoshida, K.; Ibuki, K. *J. Chem. Phys.* **1996,** *105,* 3662.
100. Basilevsky, M. V.; Parsons, D. F. *J. Chem. Phys.* **1996,** *105,* 3734; Lu, D.; Singer, S. J. *J. Chem. Phys.* **1996,** *105,* 3700.
101. Gelabert, H.; Sander, M.; Gauduel, Y. *J. Mol. Fig.*, in press.

21

Photochemically Induced Charge Separation in Electrostatically Constructed Organic–Inorganic Multilayer Composites

Steven W. Keller[1], Stacy A. Johnson, Edward H. Yonemoto, Elaine S. Brigham, Geoffrey B. Saupe, and Thomas E. Mallouk*

Department of Chemistry, The Pennsylvania State University, University Park, PA 16802

*A multilayer film growth technique, in which single anionic sheets derived from inorganic solids are interleaved with cationic polyelectrolytes, has recently been developed. This method allows for the growth of concentric monolayers of redox-active polymers on high-surface-area silica supports, and for vectorial electron transfer reactions through the layers of the "onion". Transmission electron microscopy was used to probe the morphology of these lamellar heterostructures. Photoinduced charge separation has been observed in composites consisting of an inner polycationic layer of poly(styrene-*co*-N-vinylbenzyl-N′-methyl-4,4′-bipyridine) (PS–MV^{2+}), and an outer polycationic layer of poly[$Ru(bpy)_2$(vbpy)]$^{2+}$, vbpy = 4-vinyl-4′-methyl-2,2′-bipyridine, bpy = 2,2′-bipyridine, which are separated by a thin inorganic sheet of α-$Zr(PO_4)_2^{2-}$. The thickness of the individual polymer layers was determined by ellipsometry for equivalent structures on planar supports. Electron transfer quenching of the Ru(II) polymer luminescence occurs upon addition of a solution-phase, reversible electron donor, disodium methoxyaniline-*N,N′*-diethylsulfonate ($MDESA^{2-}$). In the absence of an inner viologen layer, this simple donor–acceptor charge-separated state decays in several microseconds to regenerate the ground state. In the triad system, which contains an inner viologen polymer layer, rapid electron transfer from Ru(I) to viologen creates a charge-separated state with a half-life*

[1] Current address: Department of Chemistry, University of Missouri–Columbia, Columbia, MO 65211

* Corresponding author.

of 21 μs. The simultaneous second-order decay ($k_{recomb} = 1 \times 10^9 M^{-1} s^{-1}$) of signals from both MDESA$^{\cdot -}$ and reduced MV$^{\cdot +}$ is consistent with escape of the former from the Ru(II) polymer and subsequent diffusion to MV$^{\cdot +}$ sites. Quantum yields for charge separation are ca. 30%.

The reaction centers found in natural photosynthetic systems are remarkable not only for their unit quantum yields, but also for the long lifetimes of their charge-separated states (*1–5*). This feat is accomplished by an intricate vectorial arrangement of redox-active molecules that absorb light and separate charge across a lipid bilayer membrane. The membrane proteins that confine the reaction center play a key role in maximizing the branching ratio between forward- and back-electron transfer rates for each successive step in the charge-separation process (*6–9*).

The determination of the crystal structures of bacterial photosynthetic reaction centers has given chemists a blueprint from which to design biomimetic, artificial photosynthetic systems. In these designs, the energetics of electron transfer reactions, and the strength of electronic coupling between redox-active subunits, are key parameters in controlling electron transfer rates. The dependence of electron transfer rates on energetics, which was first predicted theoretically by Marcus (*10–12*), has now been studied experimentally in a host of geometrically well defined donor–acceptor systems (*13–27*). Because electronic coupling strength is a strong function of intermolecular distance, there have also been many fundamental studies of the distance dependence of electron transfer rates in various media (*28–57*).

In the last decade, many of the fundamental questions regarding electron transfer in both biological and model donor–acceptor systems have been substantially answered. Thanks to these fundamental studies, it is now possible to make elegant supermolecules that rival natural reaction centers in terms of their electron transfer rates and quantum yields (*58–64*). The success of these biomimetic systems is extremely impressive. Still, the synthesis of these multicomponent molecules is demanding, and it is difficult to couple them to catalytic particles in order to produce energy-rich chemicals from transiently stored free energy. In order to achieve similar control over the distance (and therefore over electron transfer rates) between subunits, but with less synthetic effort, self-assembling and microheterogeneous photoredox systems have been investigated. These organizing media include porous glasses (*65–67*), micelles (*68–70*), zeolites (*71, 72*), and polyelectrolytes (*73*), to name a few. Catalytic activity, typically in the form of water reduction to hydrogen, has manifested in some of these materials (*65, 68, 74*). However, most microheterogeneous media do not offer much flexibility in designing complex electron transfer chains, because typically only two moieties can be juxtaposed in a controlled fashion.

A new technique for growing multilayer thin films composed of oppositely charged polyelectrolytes has been described by Decher and co-workers

(*75–80*). Recently, we (*81*) and others (*82–84*) have shown that similar heterostructures can be prepared by using two-dimensional inorganic sheets (made by exfoliation of various lamellar solids) in place of the organic polyanion. This technique offers a potentially powerful alternative to the construction of multicomponent electron transfer systems, because it can, in principle, be used to stack up an arbitrary number of redox-active polymers without interpenetration (*85*). This chapter describes the preparation and photochemistry of simple multilayer composites on high-surface-area silica. Specifically, the synthesis and electron transfer kinetics of systems containing a polycationic sensitizer, poly-[Ru(bpy)$_2$(vbpy)(Cl)$_2$] (**1**), (abbreviated [Ru(bpy)$_3^{2+}$]$_n$; bpy = 2,2′-bipyridine and vbpy = 4-vinyl-4′-methyl-2,2′-bipyridine), and an electron-acceptor polycation poly[(styrene-*co*-*N*-vinylbenzyl-*N*′-methyl-4,4′-bipyridine)(Cl)$_2$] (**2**), (PS–MV^{2+}) are presented. Using a solution-phase electron donor, **3,** as the third electroactive component, it was possible to prepare and study the photoinduced electron transfer reactions of several different diad and triad combinations.

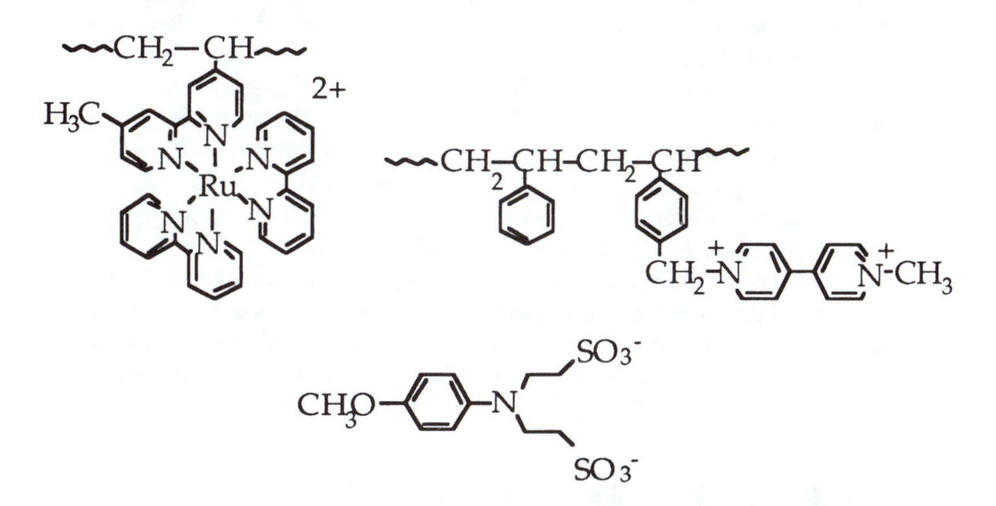

The single layers of inorganic phosphate in this system physically separate the oxidized and reduced species in the polymer layers, but they are thin enough to allow electron transfer between polymers on opposite sides. The growth of these composites on solid supports also, in principle, makes possible the incorporation of catalytic sites at which energy-storing chemical reactions could occur.

Experimental Details

General Materials. 4,4′-Bipyridine (Aldrich) was purified by dissolving it in dichloromethane and stirring with activated carbon for 30 min. The mixture was passed through a short silica–alumina column and eluted with additional dichloromethane, and the solvent removed under reduced pressure. *p*-Anisidine was purified by sublimation, and sodium 2-bromoethanesulfonate was recrystallized before use. All other chemicals and organic solvents used were Aldrich reagent or high-performance liquid chromatography (HPLC) grade and were used without further purification. Distilled water was passed through a Millipore deionizer to a resistance of 18.2 MΩ.

[Ru(2,2′-bipyridine)$_2$(4-vinyl-4′-methyl-2,2′-bipyridine)]-(PF_6^-)$_2$. Ru(2,2′-bipyridine)$_2Cl_2 \cdot 2H_2O$ (*86*) (0.5 g, 0.96 mmol) and 4-vinyl-4′-methyl-2,2′-bipyridine (*87*) (0.22 g, 1.12 mmol) were refluxed in 70 mL of a 4:6 mixture of H_2O/EtOH for 1h and allowed to cool to room temperature. The solvent mixture was removed under reduced pressure. The remaining solid was redissolved in water only and added to an aqueous solution of NH_4PF_6, from which a red–orange precipitate immediately formed. The product was collected by vacuum filtration, and washed with water and then ether to remove excess 4-vinyl-4′-methyl-2,2′-bipyridine that did not chelate. The PF_6^- salt was dissolved in a minimum amount of acetonitrile and chromatographed on silica gel, using 5:4:1 CH_3CN/H_2O/saturated aqueous KNO_3 as the eluant, which yielded one major fluorescent band. The eluted band was collected and the solvent removed to the point where the KNO_3 started to crystallize. At this point, acetone was added to precipitate the rest of the KNO_3. Some ether can be added to the cold solution, but excess will begin to precipitate the product and color the KNO_3 crystals. The KNO_3 was filtered off and washed with acetone. The bright red acetone solution (containing the nitrate salt) was evaporated and the solid dissolved in water, reprecipitated by adding a concentrated aqueous solution of NH_4PF_6, and vacuum filtered. The precipitate was then washed with water to remove excess NH_4PF_6 and KNO_3, then with ether, and then air dried to yield 0.61 g (84%).

Poly[Ru(2,2′-bipyridine)$_2$(4-vinyl-4′-methyl-2,2′-bipyridine) dichloride]. Compound **1** was prepared via free-radical polymerization of the monomer in acetone using azo-bis-isobutyronitrile (AIBN) as an initiator after Abruña and co-workers (*88*). Specifically, 0.100 g (0.11 mmol) of [Ru(2,2′-bipyridine)$_2$(4-vinyl-4′-methyl-2,2′-bipyridine)](PF_6^-)$_2$ and 5.0 mg of AIBN were dissolved in 5 mL of acetone, transferred to a Pyrex tube (13 mm o.d.), and subjected to three freeze-pump-thaw cycles to remove dissolved oxygen. The tube was sealed under reduced pressure and heated at 60 °C for 72 h. After the tube was cooled to room temperature it was broken open and the orange–red solution collected; no solid precipitate was evident. An ion-exchange column was prepared with Dowex anion exchange resin by loading the column with 1 M HCl and washing with water until the pH of the eluant was neutral. The acetone solution was introduced to the column, eluted with H_2O, and the small amount of residual acetone removed by rotary evaporation to form the water-soluble Cl^- salt (**1**). The ^1H-NMR peaks observed in d_6-dimethyl sulfoxide (d_6-DMSO) were broadened significantly from those of the monomer.

Poly(chloromethylstyrene-*co*-styrene). Polystyrene was chloromethlyated using the procedure of Merrifield (*89*). Polystyrene (25.0 g Aldrich, 45,000 MW) was dissolved in 150 mL of $CHCl_3$ in a 250-mL round bottom flask equipped with an addition funnel and placed in an ice bath at 0 °C. The entire volume was flushed with Ar for 30 min and continuously during the reaction. To the addition funnel was added 3.75 mL of $SnCl_4$ (0.032 mol) and 25 mL (0.329 mol) of chloromethyl methyl ether, also at 0 °C. The $SnCl_4$–chloromethyl methyl ether solution was added dropwise to the polystyrene solution with constant stirring over 10 min, resulting in a reddish-brown solution. After an additional 30 min, the reaction was quenched by dropwise addition of H_2O, the reddish coloration disappeared, and the solution was subsequently washed three times with H_2O. The cloudiness of the organic layer (due to trace amounts of SnO_2) was eliminated by vacuum filtration. The solvent was removed by rotary evaporation, which first left an oily substance and finally a glassy solid. After further drying under reduced pressure, the solid puffed up and was easily recoverable. ^1H-NMR spectroscopy in $CDCl_3$ showed two broad peaks in the aromatic region, a broad singlet in ~5 ppm (CH_2Cl), and several unresolved, overlapping peaks resembling profile views of snakes who had swallowed elephants in the aliphatic region. From ratios of aromatic to chloromethylene peaks, the amount of chloromethylation was estimated at 35%, which was confirmed by elemental analysis.

***N*-Methyl-4,4′-bipyridine.** We dissolved 1.5 g (10 mmol) of 4,4′-bipyridine in 150 mL of dichloromethane to which 5.0 g (35 mmol) of iodomethane (Fisher Scientific) was added. The solution was stirred for 2 h, after which a yellow precipitate was evident. Stirring for an additional 8 h at room temperature completed the precipitation, and the resulting yellow solid was collected, washed with CH_2Cl_2 and diethyl ether, and dried in air. ^1H-NMR in d_6-DMSO resulted in four sets of doublets in the aromatic region. *N*-Methyl-4,4′-bipyridine iodide was converted to the PF_6^-salt by dissolving it in water and adding an aqueous solution of NH_4PF_6. The white solid was filtered, washed with water and ether, and dried in vacuum.

Poly(styrene-*co*-N-vinylbenzyl-*N*′-methyl-4,4′-bipyridine) dichloride (2). *N*-Methyl-4,4′-bipyridine was attached to chloromethylated polystyrene via a quaternization reaction. Chloromethylated polystyrene (1.00 g) and 1.26 g of *N*-methyl-4,4′-bipyridine were dissolved in 100 mL of $CHCl_3$ and 100 mL of CH_3CN. The pale yellow solution was refluxed for 72 h, after which much of the solvent was removed. An ion-exchange column was prepared with Dowex anion exchange resin by loading the column with 1 M HCl and washing with water until the pH of the eluant was neutral. The solid residue was taken up in CH_3CN and CH_3OH, placed on the ion-exchange column, and eluted with water. The solvents were removed by rotary evaporation and the solid dissolved in 80:20 $H_2O:CH_3OH$ to make the stock solution.

Disodium Methoxyaniline-*N*-*N′*-diethylsulfonate (3). Sodium 2-bromoethanesulfonate (20.0 g, 0.189 mol) and *p*-anisidine (11.67 g, 0.095 mol) were placed in a 500-mL round-bottom flask and dissolved in 400 mL of dry ethanol. To this solution was added 13.13 g of K_2CO_3, and the mixture was heated at reflux for 12 h. The reaction was cooled and any solids remaining were filtered off, collected, and recrystallized first from boiling absolute ethanol, and finally from 20:80 water:ethanol mixtures, before being dried in air.

Layering Procedure. We have previously described briefly the procedure for growing the multilayer films on planar (*81*) as well as on small-particle supports (*90*). The silica-based composites were prepared by initially derivatizing high-surface-area silica (Cab-O-Sil, Kodak; 150 m^2/g) with 3-aminopropyltriethoxysilane. We have also used 4-aminobutyldimethylmethoxysilane (United Technologies) as a nonpolymerizable anchoring agent. No significant differences in the electron transfer kinetics have been observed between the two different composites. In a typical anchoring procedure, 5.0 g of Cab-O-Sil was suspended in 150 mL of anhydrous toluene in a poly(tetrafluoroethylene) bottle, to which was added 5.0 mL of silane. The solution was deaerated with flowing Ar, and the bottle sealed and heated with stirring at 60 °C for 72 h. The derivatized silica was isolated by centrifugation with intermittent washings with toluene, methanol, and finally water, and dried in air at 80 °C. Suspending the derivatized particles in water protonates the amine terminus and creates a cationic surface. Small-particle α-$Zr(HPO_4)_2$ (α-ZrP) was synthesized by using the method of Berman and Clearfield (*91*) and exfoliated with tetrabutylammonium hydroxide (TBAOH) to a constant $8 \leq pH \leq 8.5$. The faintly milky suspensions (~1–5 mmol/L) were centrifuged before use to remove any large particle agglomerates from the solution.

Silane-derivatized Cab-O-Sil was suspended in exfoliated α-ZrP solution (~1 g of silica per 200 mL of solution) and stirred for 12 h. The solid was isolated by centrifugation with three subsequent washings using 50 mL of water, before being dried in air at 75 °C. Deposition of the polycationic sensitizer was performed similarly, by dispersing typically 1 g of α-ZrP-anchored Cab-O-Sil in 250 mL of 10^{-5} M aqueous solutions of **1**. If smaller volumes of polyelectrolyte solution (50 mL per gram of composite) were used in this step, the resulting loading of polymer on the surface decreased by approximately 65%. Larger volumes aid the removal of the *tert*-butyl alcohol (TBA^+) into the solution because they bias the ion-exchange equilibrium in favor of surface-adsorbed polyelectrolyte. We note that this mass-action effect is important with high-surface-area substrates, where bulk amounts of counterions are released, but is probably unimportant with low-area planar substrates.

In order to deposit a thicker layer of the MV^{2+}-containing polymer (**2**), the deposition was carried out in 1 M NaCl. Higher ionic strength results in coiling of the polymer, as explained below. Under these conditions there was always excess polymer in the supernatant solution, and the solid was washed and centrifuged until the supernatant was clear. Loadings of the polymers were followed qualitatively by UV–visible diffuse reflectance spectroscopy of the dried powders, and quantified by measuring the change in polymer concentration in solution after each deposition step.

Instrumentation. Transmission electron micrographs were taken using a JEOL 1200EX II microscope operating at an accelerating voltage of 120 kV. Samples were prepared by dispersing the dried powders on carbon coated copper grids. Solution UV–visible spectra were collected on a Hewlett-Packard 8452A Diode Array Spectrometer, and diffuse reflectance spectra on a Varian DMS-300 equipped with an integrating sphere attachment. All ^{1}H-NMR spectra were recorded on a Bruker AM-300 spectrometer. Film thicknesses were measured with a Gaertner two-wavelength ellipsometer, using Si(001) wafers, cleaned as described elsewhere (*81*), as substrates. Typically, 5–8 spots per wafer were measured and averaged.

Transient flash photolysis and time-resolved luminescence experiments were performed using a system similar to one described previously (*92*), which is diagramed in Figure 1. Aqueous suspensions of ca. 100 mg of solid sample

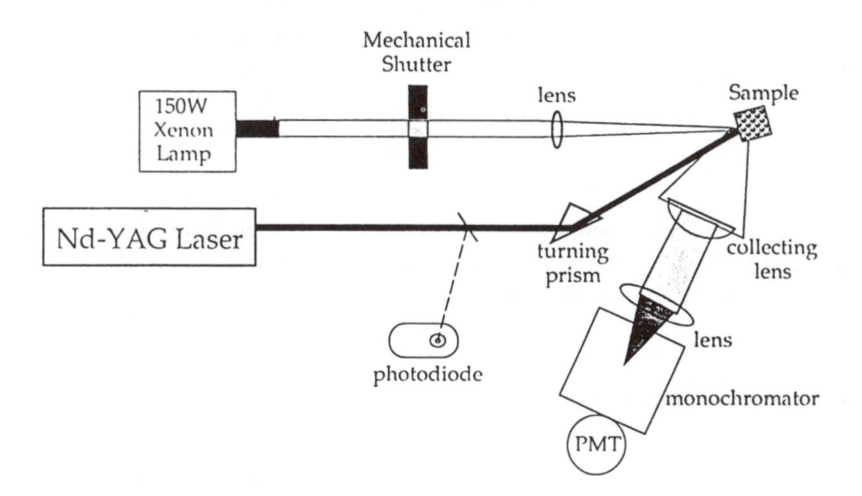

Figure 1. Schematic of the nanosecond transient diffuse reflectance and transient fluorescence instrumentation.

were placed in 1-cm quartz cuvettes, and continually stirred and purged with argon throughout the experiment. The second harmonic from a Spectra Physics Quanta Ray Nd:YAG laser [λ = 532 nm, ~20 ns full width of half-maximum (FWHM)] was used as the photoexcitation source in all experiments. An Oriel 150-W Hg–Xe lamp provided the analyzing light. To improve throughput to the detector, a high-voltage pulse generator (Kinetic Instruments, Austin, TX), which increased the brightness of the white light a factor of 50, was used. The diffusely reflected light from the sample was collected and collimated by two lenses and focused into an Oriel Model 77250 grating monochromater (with appropriate optical filters), to which was attached a Hammumatsu Model R928 photomultiplier tube (PMT) biased at 600–800 V with a Bertan Model 205B–03R high-voltage power supply. Signals from the PMT were recorded with a Tektronics TDS540–A digital oscilloscope. Timing of the laser flash lamps and Q-switch, mechanical shutter, and lamp pulser was controlled by in-house built electronics. Data collection was initiated by impinging ~5% of the incident laser beam on a photodiode trigger, and typically 20–30 shots were averaged by the oscilloscope before being sent to a personal computer (PC) for further analysis.

Because of the diffuse reflectance geometry of the experiment, raw voltage data from the PMT was first converted to Δ(absorbance), as in a transmission experiment, via:

$$\Delta A = \log(V/V_0)$$

where A is absorbance, V_0 is the averaged PMT voltage before the laser trigger, and V is the signal voltage. These data were then converted to relative absorbance, $-\Delta J/J_0$, which is the quantity that is proportional to the amount of excited-state species, by:

$$-\Delta J/J_0 = 1 - 10^{-\Delta A}$$

For spectra, an additional empirical correction was included (93) to account for the attenuation of the shorter-wavelength signals due to increased scattering of the exciting and analyzing light.

Results and Discussion

In order to probe the microstructure of "onion" structures containing polyelectrolytes **1** and **2**, model composites containing the same components were grown on planar Si supports, and the thickness of each successive layer was measured by ellipsometry. Shown in Figure 2a is the thickness evolution of a multilayer film of **1** separated by sheets of α-ZrP. The average thickness of both **1** and the α-ZrP is ca. 8 Å, and it is fairly consistent throughout the entire film. Although this value is reasonable for an α-ZrP sheet, it is smaller than that expected for the $Ru(bpy)_3^{2+}$ complex itself. Estimates of the surface coverage (assuming a cross-sectional area of 150 $Å^2$ for the $Ru(bpy)_3$ repeat unit, and 100 $Å^2$/g for the area of exposed Cab-O-Sil) indicate that this layer is incomplete, perhaps because of inefficient charge-compensation by the embedded dication, which must compete with TBA^+ for surface ion-exchange sites on the α-ZrP.

Figure 2b shows the dependence of the thickness of the PS–MV^{2+} layer on ionic strength; data for films prepared in both high ionic strength (1 M NaCl solution) and in zero ionic strength are presented. The average thicknesses for 2 are ca. 38 Å and 7 Å, respectively, in agreement with the findings of Decher and co-workers, who controlled the thicknesses of poly(allylamine) and poly(styrenesulfonate) layers by addition of NaCl to the polyelectrolyte solutions (*78*). The explanation for this effect is that polyelectrolytes adopt a more coiled conformation in high-ionic-strength solutions, because like charges can be screened effectively by the excess ions. The coiled conformation persists upon deposition onto the solid substrate, producing thicker layers. Polymers that are completely extended, in zero-ionic-strength media, are deposited as thinner layers onto planar substrates.

Figure 3a shows a transmission electron micrograph of the pristine silica, illustrating clearly that the nominally 200-Å-diameter spheres are agglomerated into irregularly shaped particles. Examination of many different particles indicates that some additional degree of agglomeration occurs during the anchoring process, likely due to partial polymerization of the trialkoxysilane anchoring agent. However, the surface area, measured by nitrogen adsorption (Brunauer–Emmett–Teller (BET) method) drops only slightly after this treatment, from 150 m^2/g to 146 m^2/g. Micrographs of the **1**/**2**/Cab-O-Sil composite (slash marks indicate a layer of α-ZrP) in Figure 3b show further agglomeration, as well as visible evidence of multilayer formation. Although we were unable to image the individual layers that make up the composite, there are significant textural differences. There appears to be a mottled coating around most of the silica particles in Figure 3b, the thickness of which is 50–80 Å. This thickness is reasonable for the **1**/**2**/film that is present. In addition, there are some regions on the particles that remained uncoated, possibly because of incomplete formation of the anchoring layer on the irregularly shaped substrate. To alleviate this complication, we have begun synthesizing SiO_2 particles within reverse micelles (*94*), and preliminary TEM images confirm a high degree of monodispersity and good separation of individual spheres in samples prepared in this fashion.

Deposition of the redox-active polymers onto high-surface-area silica was monitored using diffuse reflectance UV–visible spectroscopy. Shown in Figure 4a are the spectra for the PS-MV^{2+}/Cab-O-Sil composites, both coiled and

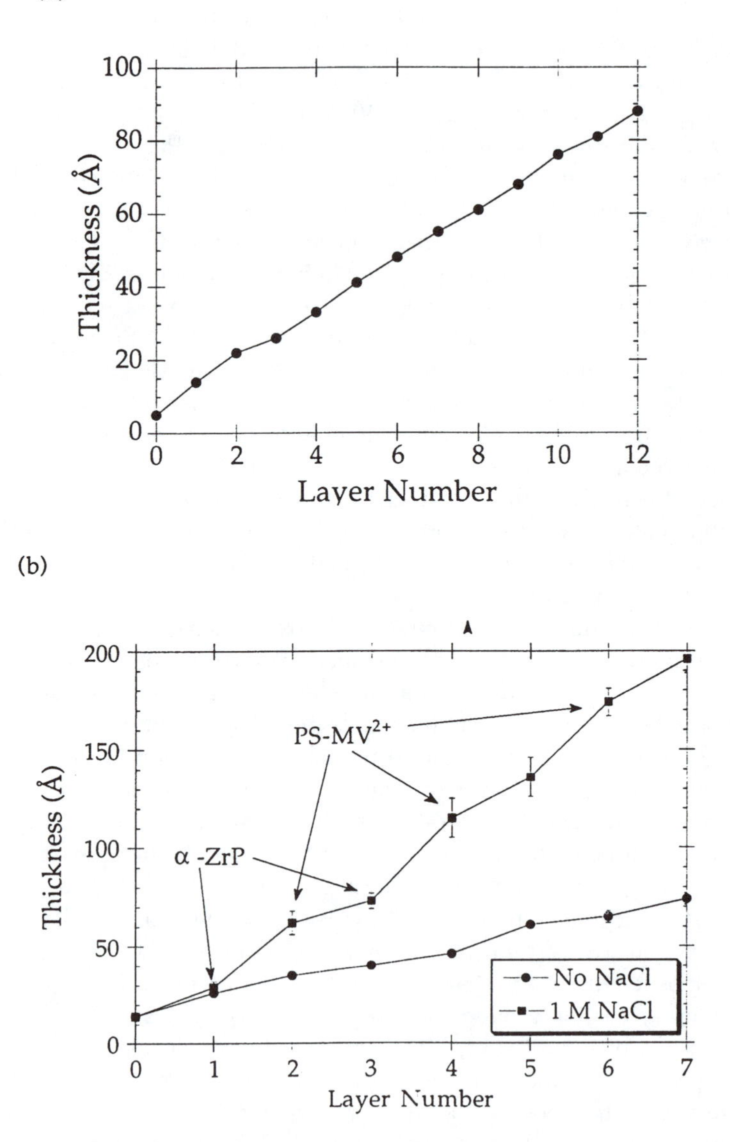

Figure 2. (a) Ellipsometric data for multilayer composites of ***1–1–1–*** *. . . grown on a planar Si wafer. (b) Ellipsometric data for ZrP–***2*** *multilayer films deposited with (circles) and without (squares) added electrolyte. The average thicknesses are 38 and 7 Å, respectively.*

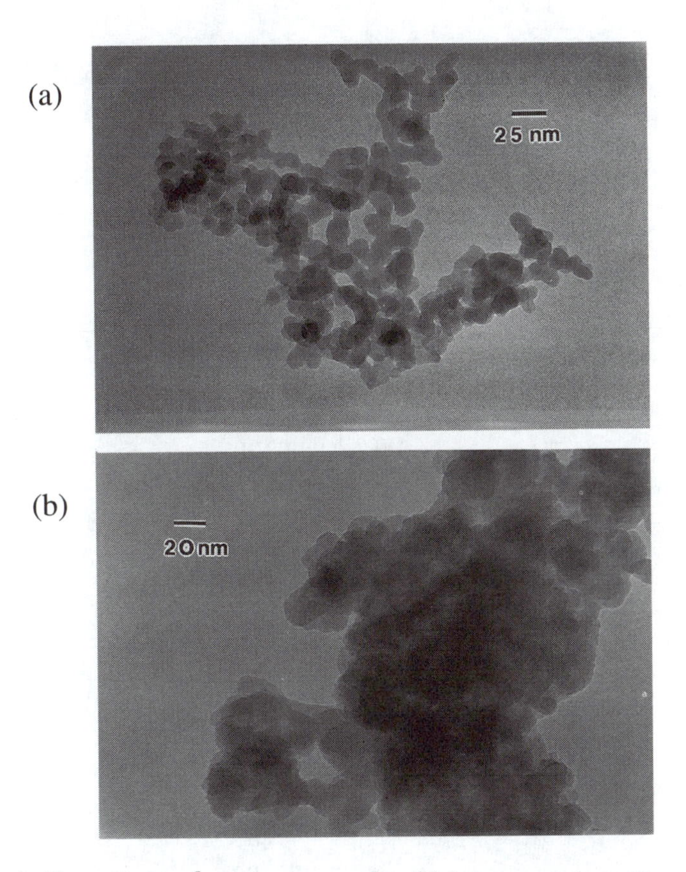

*Figure 3. Transmission electron micrographs of (a) reagent Cab-O-Sil, and (b) the **1/2**/Cab-O-Sil composite.*

uncoiled, before the addition of polymer **1** to the surface. It is clear that the coiling of **2** in solution results in increased loading on the high-surface-area particles, as was the case for the planar supports. The results of addition of a layer of α-ZrP and a layer of **1** to both composites are shown in Figure 4b. Roughly the same amount of $[Ru(bpy)_3{}^{2+}]_n$ is deposited regardless of the thickness of the first layer, although there is some loss of PS-MV^{2+} absorption (~15%) after the two subsequent steps. The loadings of the polymers were determined quantitatively from solution UV–visible data of the supernatant solutions from the deposition steps, and typically were $1\text{–}2 \times 10^{-5}$ mol/g composite.

The redox-potential diagram for the donor–$Ru(bpy)_3{}^{2+}$–MV^{2+} triad system in Figure 5 is helpful in understanding the sequence of electron transfer events in these composites. Although excitation of the sensitizer is always the initial step in the overall process, there are several possible pathways for subsequent reactions. Transient absorbance measurements on the donor–sensitizer and

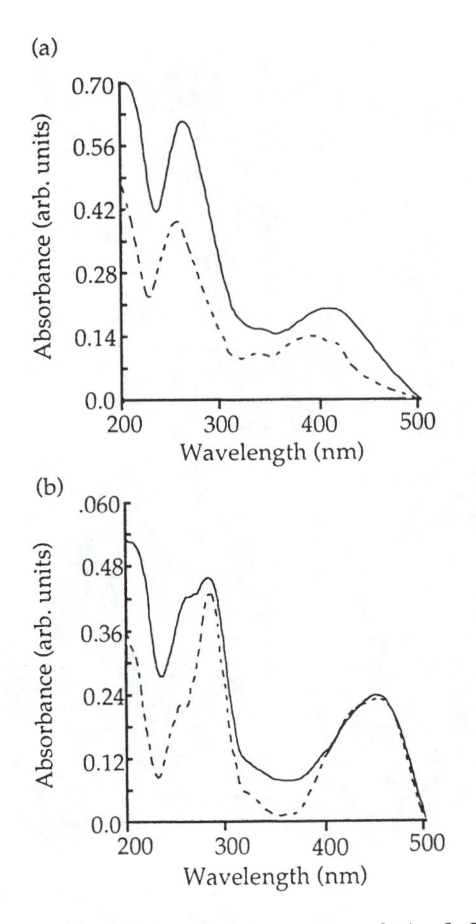

Figure 4. (a) UV–visible diffuse reflectance spectra of ***2(coiled)****/Cab-O-Sil and* ***2(uncoiled)****/Cab-O-Sil composites, and (b) the same samples after deposition of* ***1****.*

sensitizer–acceptor diad composites (**1**/Cab-O-Sil with MDESA and **1/2 (coiled)**–Cab-O-Sil, respectively) help to elucidate the operative electron transfer pathway in the triad system.

The transient absorption spectra of the **1/2(coiled)**/Cab-O-Sil diad shown in Figure 6 indicate that no oxidative quenching of the *Ru(II) occurs under these conditions. All of the spectral features can be attributed to metal-to-ligand charge transfer (MLCT) absorption and decay. The lifetimes of the 360-nm absorption and the 450-nm bleaching are identical and similar to **1**/Cab-O-Sil alone. More directly, no-peak associated with reduced viologen species is seen at ca. 390–400 nm (95). We postulate that although the reduction of MV^{2+} by *Ru(II) in reaction 1 is energetically favorable by ~ 300 mV and occurs

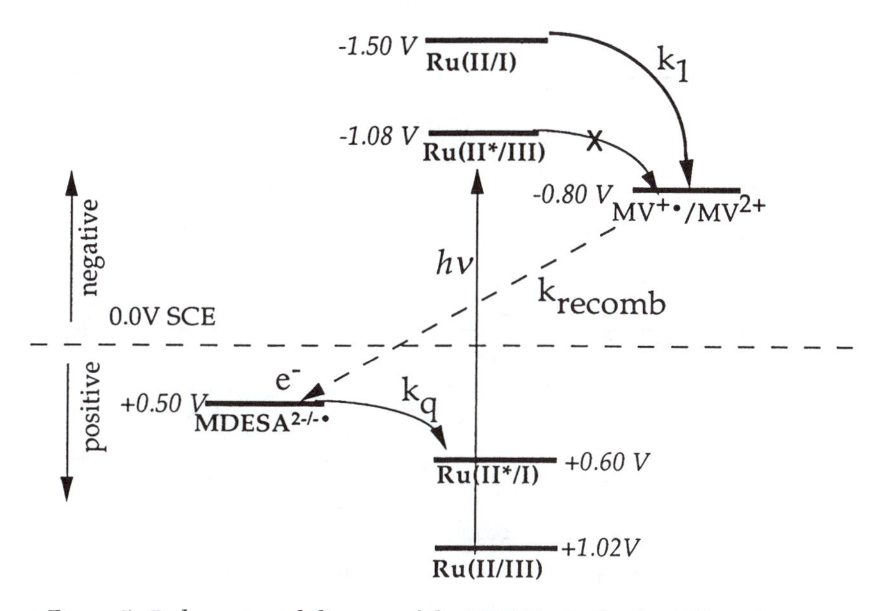

Figure 5. Redox potential diagram of the MDESA–Ru(bpy)$_3$–MV triad system, with the important electron transfer steps indicated. All potentials are referenced to SCE in aqueous solution.

rapidly in aqueous solution ($k = 5 \times 10^8\ M^{-1}\ s^{-1}$) (*96*), the intervening 10 Å of insulating α-ZrP slows the electron transfer reaction to the point where it cannot compete with MLCT decay of *Ru(II).

$$MV^{2+} + {}^*Ru(II) \rightarrow MV^{\cdot +} + Ru(III) \tag{1}$$

The other diad system, consisting of sensitizer **1** deposited on Cab-O-Sil and electron donor **3** in solution, shows more interesting photochemistry. Referring to Figure 5, the redox potential of **3** allows for the *reductive* quenching of excited-state *Ru(II) via reaction 2. The formal potential of **3** is +0.50 V versus SCE (standard calomel electrode) in aqueous solution (0.1 M KCl, Pt working electrode); by way of comparison, the formal potential of the *Ru-(II)–Ru(I) couple is ca. +0.6 V (*97, 98*).

$$MDESA^{2-} + Ru(II)^* \rightarrow MDESA^{\cdot -} + Ru(I) \tag{2}$$

Shown in Figure 7 are the Stern–Volmer plots of emission intensities and lifetimes, monitored at 630 nm, as a function of MDESA concentration. Both the static (intensity) and the dynamic (lifetime) components are nonlinear and indicate that the quenching mechanism is complicated. The extent of the static reaction (attributed to $MDESA^{2-}$ anions associated with $Ru(bpy)_3{}^{2+}$ cations)

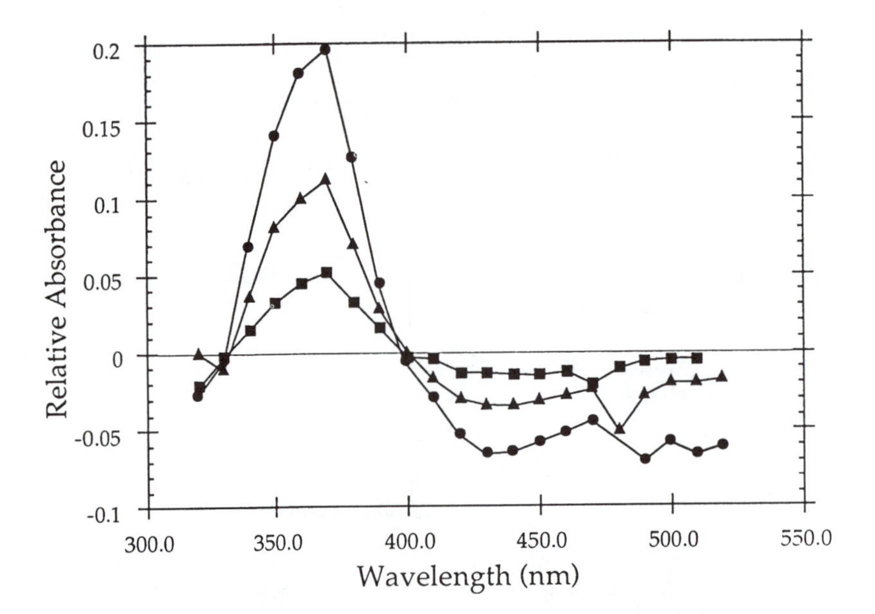

*Figure 6. Transient diffuse reflectance spectra for the $[Ru(bpy)_3^{2+}]_n$ $-(PS-MV^{2+}$, coiled)–Cab-O-Sil composite taken 200 ns (circles), 500 ns (triangles), and 1 μs (squares) after the laser flash. The peak at 360 nm corresponds to the absorption of the * Ru(II) excited state and decays with a lifetime of ca. 550 ns.*

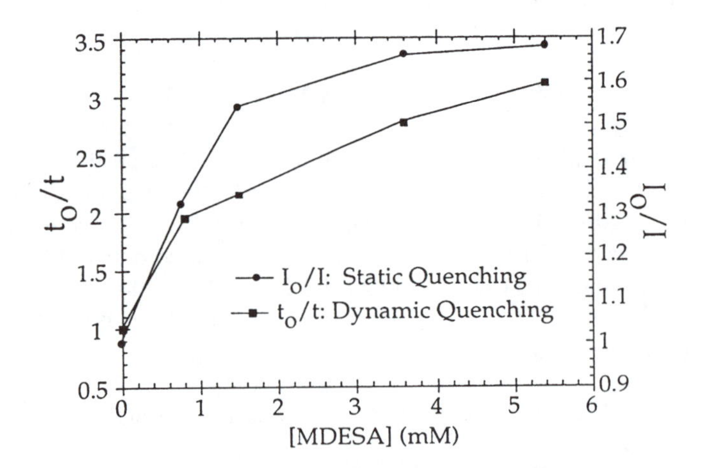

*Figure 7. Lifetime (t) and steady-state emission intensity (I) data for the $[Ru(bpy)_3^{2+}]_n$–Cab-O-Sil composite suspended in MDESA solution. Lifetime data were obtained by fitting the decay of the 630 nm *Ru(II) transient observed with 532-nm laser excitation.*

levels off at 3.5 mM donor, and quenches roughly 40% of the *Ru(II) centers. Some of the $*Ru(bpy)_3^{2+}$ ions may not be completely accessible to the solution-phase donors, accounting for the incomplete static quenching. Dynamic quenching yields were measured from emission lifetimes (also at 630 nm). The dynamic reaction quenches the luminescence of about 65% of the remaining *Ru(II), with a rate constant at low donor concentrations of $k_q = 3.7 \times 10^9$ M^{-1} s^{-1}. There is an additional complication in that upon addition of donor the emission decay is biexponential, with one component <100 ns and one ~200 ns. The kinetics in the absence of donor are described well by a single exponential decay.

By combining the two diads, it is possible to position Ru(I), which is a more powerful reducing agent by ca. 400 mV than *Ru(II), adjacent to the viologen layer. This arrangement accelerates electron transfer across the inorganic sheet as shown by the transient spectra of the **1/2(coiled)**/Cab-O-Sil composite with added aqueous MDESA in Figure 8b; an idealized cross-sectional view of the triad indicating the orientation of the components is given in Figure 8a. The time slices recorded 10 and 40 μs after the laser flash both have spectral features consisting of peaks at 390 nm and 550 nm (assigned to $MV^{\cdot +}$) and a shoulder at 510 nm, which is characteristic of the MDESA radical anion. It should be noted that *isolated* $MV^{\cdot +}$ radicals absorb at 400 nm and 600 nm. The blue-shifting of the long-wavelength peak to 550 nm is an indication that dimerization of the reduced viologen centers is occurring:

$$2MV^{\cdot +} \leftrightarrow (MV^{+})_2 \tag{3}$$

This reaction is observed in electrochemically reduced viologen polymers, in which the local concentration of $MV^{\cdot +}$ is high (*99, 100*). The situation here is similar because $MV^{\cdot +}$ is formed in relatively high quantum yield in an electron transfer sequence initiated by an intense laser flash, which converts most of the Ru(II) to *Ru(II). Although the fast formation of reduced **2** cannot be temporally resolved, the results from the diads give strong indirect evidence that the initial step is reduction of *Ru(II) to Ru(I), followed by rapid electron transfer from Ru(I) to **2.** The spectral features seen in Figure 8b are consistent with a charge-separated state formed by reaction 4, and consisting of oxidized solution-phase **3** and reduced **2,** which is "buried" within the inner layer of the composite and separated by the thin $[\alpha\text{-}Zr(OPO_3)_2^{2-}]_n$ sheet and the ground-state **1** layer.

$$Ru(I) + MV^{2+} \rightarrow Ru(II) + MV^{\cdot +} \tag{4}$$

The quantum yield for the formation of the charge-separated state is ~30%. Recombination of the oxidized **3** and reduced **2** occurs via equal-concentration second-order kinetics with a rate constant of 1×10^9 M^{-1} s^{-1}. The recombination reaction (5) has essentially optimal driving force (ca. 1.3 V; *see* Figure 5) and should therefore be rapid even over large separation distances.

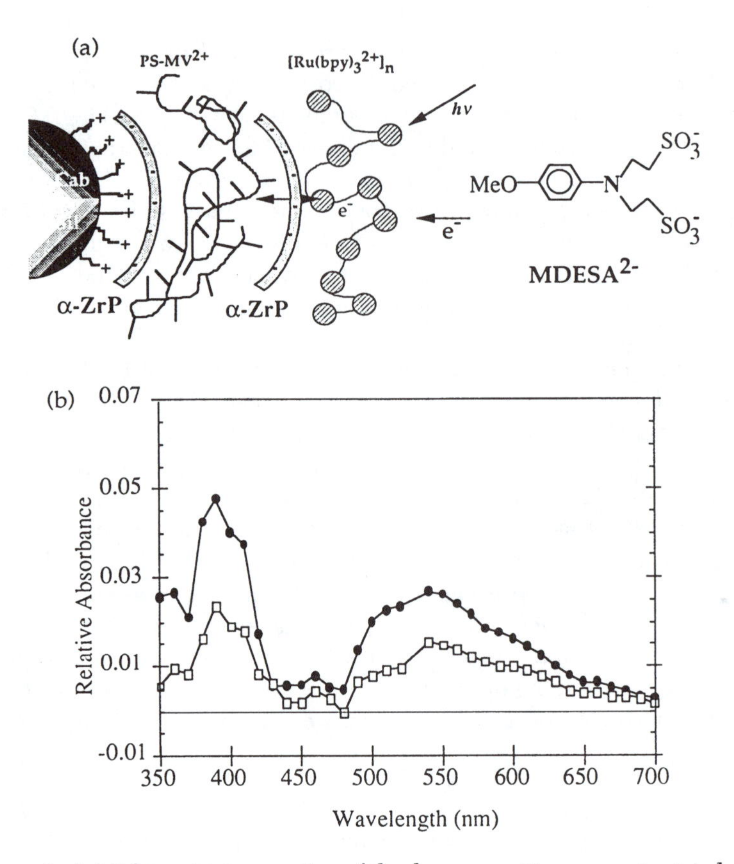

Figure 8. (a) Schematic cross section of the donor–sensitizer–acceptor triad. (b) Transient diffuse reflectance spectra for $[Ru(bpy)_3^{2+}]_n$–($PS–MV^{2+}$, coiled)–Cab-O-Sil composite with 3.5 mM MDESA in solution 10 μs (circles) and 40 μs (squares) after the 532-nm laser flash.

$$MDESA^{\cdot -} + MV^{\cdot +} \rightarrow MDESA^{2-} + MV^{2+} \quad (5)$$

Although we cannot rule out the possibility that oxidized **3** is able to diffuse into the inner $PS–MV^{2+}$ layer, a control experiment shows that this is probably not the case. An inverted composite (**2**/**1**/Cab-O-Sil) showed no quenching of the *Ru(II) excited state either with or without **3** in solution. This result argues that oxidized **3** probably does not have physical access to the inner layers of the composites.

If the $PS–MV^{2+}$ layer is deposited from a zero-ionic-strength solution, that is, in a more extended conformation, the quantum yield for viologen photoreduction decreases. Shown in Figure 9 are transient diffuse reflectance decays recorded at 400 and 510 nm for **1/2(coiled)**/Cab-O-Sil and **1/2(uncoiled)**/

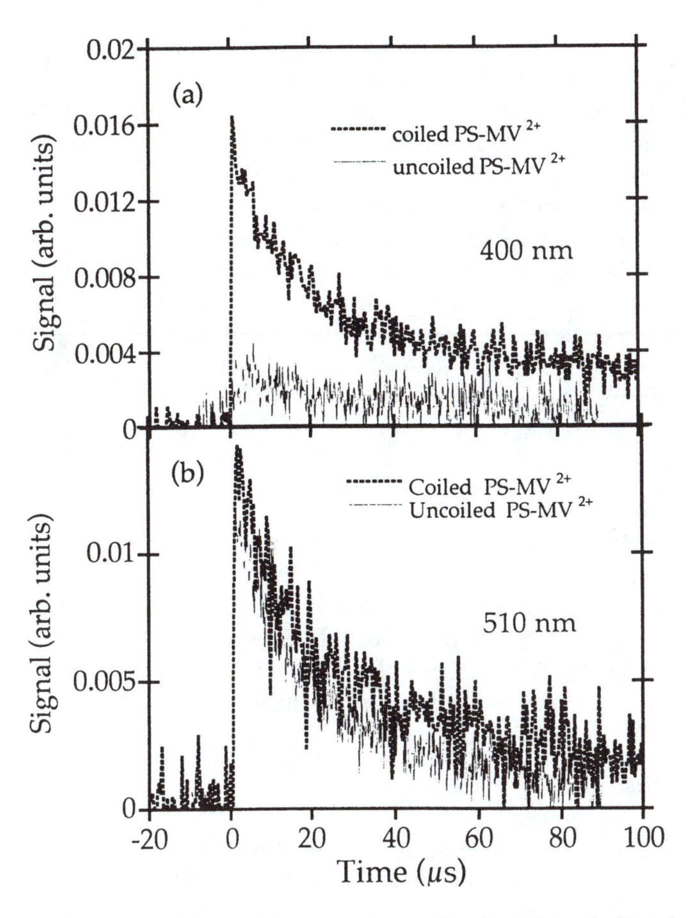

*Figure 9. Transient diffuse reflectance decays for the **1/2(coiled)**/Cab-O-Sil (dashed curves) and **1/2(uncoiled)**/Cab-O-Sil composites. In (a), data were recorded at 400 nm, where the predominant absorbing species is reduced viologen radical cation. In (b), the 510-nm transient attributed to oxidized **3** is shown.*

Cab-O-Sil, both containing similar amounts of **1.** The substantial difference in reduced viologen yield between the uncoiled and coiled structures (Figure 9a) is curious because the initial charge-transfer step does not involve the viologen polymer. Figure 9b shows that the 510-nm transient, attributed to oxidized 3, is not substantially affected by the conformation of the viologen polymer. It appears that in the case of the uncoiled polymer, the charge-separated state formed is Ru(I)–oxidized **3**. Apparently the viologen groups are physically inaccessible to Ru(I) in this case, and reaction (4) occurs to a very limited extent. Continued exploration of the photochemical properties of these polymers as a function of chain conformation, and a more detailed understanding of the microstructure of these composites, will be needed to optimize their efficiency.

Our current efforts are focused on preparing more complex photoredox chains, which incorporate additional donor and/or acceptor layers. By increasing the number of electron transfer steps in the overall sequence, the spatial separation of the ultimately oxidized and reduced layers, and therefore the lifetime of charge separation, should be enhanced. One can also imagine incorporating photon-harvesting polymers into these systems in order to create hybrid energy/electron transfer cascades. Finally, replacing the insulating α-ZrP layers with semiconducting sheets and appropriate catalytic particles may make possible the visible-light photodecomposition of water or hydrogen halides. Such extensions of this chemistry are currently in progress, as is fuller characterization of films grown on both planar and high-surface-area substrates. An understanding of the factors that control the details of these "onion-like" structures is a prerequisite for the rational design of any of the complex systems mentioned in this chapter.

Acknowledgments

This work was supported by the Division of Chemical Sciences, Office of Basic Energy Sciences, U.S. Department of Energy, contract number DE–FG02–93ER14374. T. E. Mallouk also thanks the Camille and Henry Dreyfus Foundation for support in the form of a Teacher–Scholar Award.

References

1. Boxer, S. G. *Biochim. Biophys. Acta* **1983,** *726,* 265.
2. Martin, J. L; Breton, J.; Hoff, A. J.; Migus, A.; Antonetti, A. *Proc. Natl. Acad. Sci. U.S.A.* **1986,** *83,* 957.
3. Breton, J.; Martin, J. L.; Migus, A.; Antonetti, A.; Orszag, A. *Proc. Natl. Acad. Sci. U.S.A.* **1986,** *83,* 5121.
4. Kaufman, K. J.; Dutton, P. L.; Netzel, T. L.; Leigh, J. S.; Rentzepis, P. M. *Science (Washington, D.C.)* **1975,** *188,* 1301.
5. Gunner, M. R.; Robertson, D. E.; Dutton, P. L. *J. Phys. Chem.* **1986,** *90,* 3783.
6. Deisenhofer, J.; Epp, O.; Miki, K.; Huber, R.; Michel, H. *J. Mol. Biol.* **1984,** *180,* 385.
7. Deisenhofer, J.; Epp, O.; Miki, K.; Huber, R.; Michel, H. *Nature (London)* **1985,** *318,* 618.
8. Allen, P.; Feher, G.; Yeates, T. O.; Rees, D. C.; Deisenhofer, J.; Michel, H.; Huber, R. *Proc. Natl. Acad. Sci. U.S.A.* **1986,** *83,* 8589.
9. Chang, C. H.; Tiede, D.; Tang, J.; Smith, U.; Norris, J.; Schiffer, M. *FEBS Lett.* **1986,** *205,* 82.
10. Marcus, R. A. *J. Chem. Phys.* **1956,** *24,* 966.
11. Marcus, R. A. *Disc. Faraday Soc.* **1960,** *29,* 21.
12. Marcus, R. A. *Annu. Rev. Phys. Chem.* **1964,** *15,* 155.
13. Calcaterra, L. T.; Closs, G. L.; Miller, J. R. *J. Am. Chem. Soc.* **1983,** *105,* 670.
14. Closs, G. L.; Miller, J. R. *Science (Washington, D.C.)* **1988,** *240,* 440, and references therein.
15. Liang, N.; Miller, J. R.; Closs, G. L. *J. Am. Chem. Soc.* **1990,** *112,* 5353.

16. Wasielewski, M. R.; Niemczyk, M. P.; Svec, W. A.; Pewitt, E. B. *J. Am. Chem. Soc.* **1985,** *107,* 1080.
17. Ohno, T.; Yoshimura, A.; Shioyama, H.; Mataga, N. J. *Phys. Chem.* **1987,** *91,* 4365.
18. Gould, I. R.; Ege, D.; Mattes, S. L.; Farid, S. *J. Am. Chem. Soc.* **1987,** *109,* 3794.
19. Gould, I. R.; Moody, R.; Farid, S. *J. Am. Chem. Soc.* **1988,** *110,* 7242.
20. Gould, I. R.; Farid, S. *J. Am. Chem. Soc.* **1988,** *110,* 7883.
21. Gould, I. R.; Moser, J. E.; Armitage, B.; Farid, S.; Goodman, J. L.; Herman, M. S. *J. Am. Chem. Soc.* **1989,** *111,* 1917.
22. Mataga, N.; Kanda, Y.; Asahi, T.; Miyasaka, H.; Okada, T.; Kakitani, T. *Chem. Phys.* **1988,** *127,* 239.
23. Zou, C.; Miers, J. B.; Ballew, R. M.; Dlott, D. D.; Schuster, G. B. *J. Am. Chem. Soc.* **1991,** *113,* 7823.
24. Chen, P.; Duesing, R.; Tapolsky, G.; Meyer, T. J. *J. Am. Chem. Soc.* **1989,** *111,* 1917.
25. MacQueen, D. B.; Schanze, K. S. *J. Am. Chem. Soc.* **1991,** *113,* 7470.
26. Fox, L. S.; Kozik, M.; Winkler, J. R.; Gray, H. B. *Science (Washington, D.C.)* **1990,** *247,* 1069.
27. Yonemoto, E. H.; Riley, R. L.; Kim, Y. I.; Atherton, S. J.; Schmehl, R. H.; Mallouk, T. E. *J. Am. Chem. Soc.* **1992,** *114,* 8081.
28. Miller, J. R.; Beitz, J. V.; Huddleston, R. K. *J. Am. Chem. Soc.* **1984,** *106,* 5057.
29. Beitz, J. V.; Miller, J. R. *J. Chem. Phys.* **1979,** *71,* 4579.
30. Miller, J. R. *Science (Washington, D.C.)* **1975,** *189,* 221.
31. Closs, G. L.; Johnson, M. D.; Miller, J. R.; Piotrowiak, P. *J. Am. Chem. Soc.* **1989,** *111,* 3751.
32. Hush, N. S.; Paddon-Row, M. N.; Cotsaris, E.; Oevering, H.; Verhoeven, J. W.; Heppener, M. *Chem. Phys. Lett.* **1985,** *117,* 8.
33. Verhoeven, J. W.; Paddon-Row, M. N.; Hush, N. S.; Oevering, H.; Heppener, M. *Pure Appl. Chem.* **1986,** *58,* 1285.
34. Oevering, H.; Paddon-Row, M. N.; Heppener, M.; Oliver, A. M.; Cotsaris, E.; Verhoeven, J. W.; Hush, N. S. *J. Am. Chem. Soc.* **1987,** *109,* 3258.
35. Oevering, H.; Verhoeven, J. W.; Paddon-Row, M. N.; Cotsaris, E.; Hush, N. S. *Chem. Phys. Lett.* **1988,** *150,* 179.
36. Paddon-Row, M. N.; Verhoeven, J. W. *New J. Chem.* **1991,** *15,* 107.
37. Jordan, K. D.; Paddon-Row, M. N. *Chem. Rev.* **1992,** *92,* 395.
38. Larrson, S. *J. Am. Chem. Soc.* **1981,** *103,* 4034.
39. McGuire, M.; McLendon, G. *J. Phys. Chem.* **1986,** *90,* 2549.
40. Heiler, D.; McLendon, G.; Rogalskyj, P. *J. Am. Chem. Soc.* **1987,** *109,* 7540.
41. Conklin, K. T.; McLendon, G. *J. Am. Chem. Soc.* **1988,** *110,* 3345.
42. McLendon, G. *Acc. Chem. Res.* **1988,** *21,* 160.
43. Helms, A.; Heiler, D.; McLendon, G. *J. Am. Chem. Soc.* **1991,** *113,* 4325.
44. Schmidt, J. A.; McIntosh, A. R.; Weedon, A. C.; Bolton, J. R.; Connolly, J. S.; Hurley, J. K.; Wasielewski, M. R. *J. Am. Chem. Soc.* **1988,** *110,* 1733.
45. Wasielewski, M. R.; Niemczyk, M. P. *ACS Symp. Ser.* **1986,** *321,* 154.
46. Hofstra, U.; Schaafsma, T. J.; Sanders, G. M.; Van Dijk, M.; Van der Plas, H. C.; Johnson, D. G.; Wasielewski, M. R. *Chem. Phys. Lett.* **1988,** *151,* 169.
47. Wasielewski, M. R.; Johnson, D. G.; Svec, W. A.; Kersey, K. M.; Minsek, D. W. *J. Am. Chem. Soc.* **1988,** *110,* 7219.
48. Wasielewski, M.; Niemczyk, M. P.; Johnson, D. G.; Svec, W. A.; Minsek, D. W. *Tetrahedron* **1989,** *45,* 4785.
49. Heitele, H.; Michel-Beyerle, M. E. *J. Am. Chem. Soc.* **1985,** *107,* 8286.
50. Finckh, P.; Heitele, H.; Volk, M.; Michel-Beyerle, M. E. *J. Phys. Chem.* **1988,** *92,* 6584.
51. Finckh, P.; Heitele, H.; Michel-Beyerle, M. E. *Chem. Phys.* **1989,** *138,* 1.

52. Heitele, H.; Poellinger, F.; Weeren, S.; Michel-Beyerle, M. E. *Chem. Phys. Lett.* **1990,** *168,* 598.
53. Fox, L. S.; Kozik, M.; Winkler, J. R.; Gray, H. B. *Science (Washington, D.C.)* **1990,** *247,* 1069.
54. Gray, H. B; Malmström, B. G. *Biochemistry,* **1989,** *28,* 7499.
55. Bowler, B. E.; Raphael, A. L.; Gray, H. B. *Prog. Inorg. Chem.* **1990,** *38,* 259.
56. Ryu, C. K.; Wang, R.; Schmehl, R. H.; Ferrere, S.; Ludwikow, M.; Merkert, J. W.; Headford, C. L.; Elliott, C. M. *J. Am. Chem. Soc.* **1992,** *114,* 430.
57. Yonemoto, E. H.; Saupe, G. B.; Schmehl, R. H.; Hubig, S. M.; Riley, R. L.; Iverson, B. L.; Mallouk, T. E. *J. Am. Chem. Soc.* **1994,** *116,* 4786.
58. Gust, D.; Moore, T. A.; Moore, A. L. *Acc. Chem. Res.* **1993,** *26,* 198, and references therein.
59. Gust, D.; Moore, T. A.; Moore, A. L.; Lee, S.-J.; Bittersman, E.; Juttrull, D. K.; Rehms, A. A.; DeGraziano, J. M.; Ma, X. C.; Gao, F.; Belford, R. E.; Trier, T. T. *Science (Washington D.C.)* **1990,** *248,* 199.
60. Gust, D.; Moore, T. A.; Moore, A. L.; MacPherson, Lopez, A.; DeGraziano, J. M.; Gouni, I.; Bittersman, E.; Seely, G. R.; Gao, F.; Nieman, R. A.; Ma, X. C.; Demanche, L.; Hung, S.-C.; Luttrull, D. K.; Lee, S.-J.; Kerrigan, P. K. *J. Am. Chem. Soc.* **1993,** *115,* 11141.
61. Wasielewski, M. R.; Johnson, D. G.; Svec, W. A.; Kersey, K. M.; Minsek, D. W. *J. Am. Chem. Soc.* **1988,** *110,* 7219.
62. Wasielewski, M. R.; Johnson, D. G.; Niemczyk, M. P.; Gaines, G. L., III; O'Neil, M. P.; Svec, W. A.; Niemczyk, M. P. *J. Am. Chem. Soc.* **1990,** *112,* 4559.
63. Johnson, D. G.; Niemczyk, M. P.; Minsek, D. W.; Wiederrecht, G. P.; Svec, W. A.; Gaines, G. L., III; Wasielewski, M. R. *J. Am. Chem. Soc.* **1993,** *115,* 5692.
64. Wasielewski, M. R.; Gaines, G. L., III; Wiederrecht, G. P.; Svec, W. A.; Niemczyk, M. P. *J. Am. Chem. Soc.* **1993,** *115,* 10442.
65. Slama-Schwok, A.; Avnir, D.; Ottolenghi, M. *J. Phys. Chem.* **1989,** *93,* 7544.
66. Slama-Schwok, A.; Avnir, D.; Ottolenghi, M. *J. Am. Chem. Soc.* **1991,** *113,* 3984.
67. Slama-Schwok, A.; Avnir, D.; Ottolenghi, M. *Nature (London)* **1992,** *355,* 240.
68. Brugger, P.-A.; Infelta, P. P.; Braun, A. M.; Grätzel, M. *J. Am. Chem. Soc.* **1981,** *103,* 320.
69. Brugger, P.-A.; Grätzel, M. *J. Am. Chem. Soc.* **1980,** *102,* 2461.
70. Kang, Y. S.; McManus, H. J. D.; Liang, K.; Kevan, L. *J. Phys. Chem.* **1994,** *98,* 1044.
71. Yonemoto, E. H.; Kim, Y.-I.; Schmel, R. H.; Wallin, J. O.; Shoulders, B. R.; Haw, J. F.; Mallouk, T. E. *J. Am. Chem. Soc.* **1994,** *116,* 10557.
72. Borja, M; Dutta, P. K. *Nature (London)* **1993,** *362,* 43.
73. Lee, P. C.; Matheson, M. S.; Meisel, D. *Isr. J. Chem.* **1982,** *22,* 133.
74. Brugger, P.-A.; Cuendet, P.; Grätzel, M. *J. Am. Chem. Soc.* **1981,** *102,* 2923.
75. Decher, G.; Hong, J.-D. *Makromol. Chem., Macromol. Symp.* **1991,** *46,* 321.
76. Decher, G.; Hong, J.-D.; Schmitt, J. *Thin Solid Films* **1992,** *210/211,* 504.
77. Decher, G.; Schmitt, J. *Prog. Colloid Polym. Sci.* **1992,** *89,* 160.
78. Lvov, Y.; Decher, G.; Möhwald, H. *Langmuir* **1993,** *9,* 481.
79. Lvov, Y.; Decher, G.; Sukhorukov, G. *Macromolecules* **1993,** *26,* 5396.
80. Lvov, Y.; Essler, F.; Decher, G. *J. Phys. Chem.* **1993,** *97,* 13773.
81. Keller, S. W.; Kim, H.-K.; Mallouk, T. E. *J. Am. Chem. Soc.* **1994,** *116,* 8817.
82. Kleinfeld, E. R.; Ferguson, G. S. *Science (Washington, D.C.)* **1994,** *265,* 370.
83. Kleinfeld, E. R.; Ferguson, G. S. *Adv. Mater.* **1995,** *7,* 414.
84. Kleinfeld, E. R.; Ferguson, G. S. *Solid State Ionics IV;* Nazri, G.-A.; Tarascon, J. M.; Schrieber, M., Eds.; MRS Symposium Proceedings 369; Materials Research Society: Pittsburgh, PA, 1995; pp 697–702.

85. Schmitt, J.; Grünewald, T.; Kjaer, K.; Pershan, P.; Decher, G.; Lösche, M. *Macromolecules* **1993,** *26,* 7058.
86. Lay, P. *Inorg. Synth.* **1986,** *24,* 292.
87. Abruna, H. *Inorg. Chem.* **1985,** *24,* 988.
88. Bommarito, S. L.; Lowery-Bretz, S. P.; Abruña, H. *Inorg. Chem.* **1992,** *31,* 495.
89. Merrifield, R. B. *J. Am. Chem. Soc.* **1963,** *85,* 2149–2154.
90. Keller, S. W.; Johnson, S. A.; Yonemoto, E. H.; Brigham, E. S.; Mallouk, T. E. *J. Am. Chem. Soc.* **1995,** *117,* 12879–12880.
91. Berman, R.; Clearfield, A. *J. Inorg. Nucl. Chem.* **1981,** *43,* 2141.
92. Kim, Y. I.; Mallouk, T. E. *J. Phys. Chem.* **1992,** *96,* 2879.
93. Kim, Y. I.; Atherton, S. J.; Brigham, E. S.; Mallouk, T. E. *J. Phys. Chem.* **1993,** *97,* 11802.
94. Osseo-Asare, K.; Arriagada, F. J. *Colloids Surf.* **1990,** *50,* 321.
95. Wantanabe, T.; Honda, K. *J. Phys. Chem.* **1982,** *86,* 2617.
96. Kalyanasundaram, K.; Kiwi, J.; Grätzel, M. *Helv. Chim. Acta* **1978,** *61,* 2720.
97. Bock, C. R.; Connor, J. A.; Gutierrez, A. R.; Meyer, T. J.; Whitten, D. G.; Sullivan, B. P.; Nagel, J. K. *J. Am. Chem. Soc.* **1979,** *101,* 4815.
98. Creutz, C.; Sutin, N. *Inorg. Chem.* **1976,** *15,* 496.
99. Bookbinder, D. C.; Wrighton, M. S. *J. Electrochem. Soc.* **1983,** *5,* 1080.
100. Kosower, E. M.; Cotler, J. L. *J. Am. Chem. Soc.* **1964,** *86,* 5524.

Author Index

Subject Index

D

G

H

M

N

O

P

Q

R

T

U

V

W

X

Bestsellers from ACS Books

The ACS Style Guide: A Manual for Authors and Editors (2nd Edition)
Edited by Janet S. Dodd
470 pp; clothbound ISBN 0–8412–3461–2; paperback ISBN 0–8412–3462–0

Writing the Laboratory Notebook
By Howard M. Kanare
145 pp; clothbound ISBN 0–8412–0906–5; paperback ISBN 0–8412–0933–2

Career Transitions for Chemists
By Dorothy P. Rodmann, Donald D. Bly, Frederick H. Owens, and Anne-Claire Anderson
240 pp; clothbound ISBN 0–8412–3052–8; paperback ISBN 0–8412–3038–2

Chemical Activities (student and teacher editions)
By Christie L. Borgford and Lee R. Summerlin
330 pp; spiralbound ISBN 0–8412–1417–4; teacher edition, ISBN 0–8412–1416–6

Chemical Demonstrations: A Sourcebook for Teachers, Volumes 1 and 2, Second Edition
Volume 1 by Lee R. Summerlin and James L. Ealy, Jr.
198 pp; spiralbound ISBN 0–8412–1481–6
Volume 2 by Lee R. Summerlin, Christie L. Borgford, and Julie B. Ealy
234 pp; spiralbound ISBN 0–8412–1535–9

The Internet: A Guide for Chemists
Edited by Steven M. Bachrach
360 pp; clothbound ISBN 0–8412–3223–7; paperback ISBN 0–8412–3224–5

Laboratory Waste Management: A Guidebook
ACS Task Force on Laboratory Waste Management
250 pp; clothbound ISBN 0–8412–2735–7; paperback ISBN 0–8412–2849–3

Reagent Chemicals, Eighth Edition
700 pp; clothbound ISBN 0–8412–2502–8

Good Laboratory Practice Standards: Applications for Field and Laboratory Studies
Edited by Willa Y. Garner, Maureen S. Barge, and James P. Ussary
571 pp; clothbound ISBN 0–8412–2192–8

For further information contact:
Order Department
Oxford University Press
2001 Evans Road
Cary, NC 27513
Phone: 1-800-445-9714 or 919-677-0977
Fax: 919-677-1303

Highlights from ACS Books

Desk Reference of Functional Polymers: Syntheses and Applications
Reza Arshady, Editor
832 pages, clothbound, ISBN 0–8412–3469–8

Chemical Engineering for Chemists
Richard G. Griskey
352 pages, clothbound, ISBN 0–8412–2215–0

Controlled Drug Delivery: Challenges and Strategies
Kinam Park, Editor
720 pages, clothbound, ISBN 0–8412–3470–1

Chemistry Today and Tomorrow: The Central, Useful, and Creative Science
Ronald Breslow
144 pages, paperbound, ISBN 0–8412–3460–4

Eilhard Mitscherlich: Prince of Prussian Chemistry
Hans-Werner Schutt
Co-published with the Chemical Heritage Foundation
256 pages, clothbound, ISBN 0–8412–3345–4

Chiral Separations: Applications and Technology
Satinder Ahuja, Editor
368 pages, clothbound, ISBN 0–8412–3407–8

Molecular Diversity and Combinatorial Chemistry: Libraries and Drug Discovery
Irwin M. Chaiken and Kim D. Janda, Editors
336 pages, clothbound, ISBN 0–8412–3450–7

A Lifetime of Synergy with Theory and Experiment
Andrew Streitwieser, Jr.
320 pages, clothbound, ISBN 0–8412–1836–6

Chemical Research Faculties, An International Directory
1,300 pages, clothbound, ISBN 0–8412–3301–2

For further information contact:
Order Department
Oxford University Press
2001 Evans Road
Cary, NC 27513
Phone: 1-800-445-9714 or 919-677-0977
Fax: 919-677-1303